STRENGTH AND RELATED
PROPERTIES OF CONCRETE

STRENGTH AND RELATED PROPERTIES OF CONCRETE

A Quantitative Approach

Sándor Popovics

Drexel University

JOHN WILEY & SONS, INC.

New York / Chichester / Weinheim / Brisbane / Singapore / Toronto

This book is printed on acid-free paper. ∞

This publication is designed to provide accurate and authoritative information in regard to the subject matter covered. It is sold with the understanding that the publisher is not engaged in rendering legal, accounting, or other professional services. If legal advice or other expert assistance is required, the services of a competent professional person should be sought.

Library of Congress Cataloging-in-Publication Data:

Popovics, Sándor, 1921–
    Strength and related properties of concrete : a quantitative
approach / Sándor Popovics.
        p.   cm.
    Includes index.
    ISBN 0-471-14903-9 (alk. paper)
    1. Concrete—Testing.   2. Concrete—Mechanical properties.
3. Cement.   I. Title.
TA440.P595 1998
620.1'36—dc21   97-31766

Printed in the United States of America.

10 9 8 7 6 5 4 3 2

Always use the method that works best. Whether it is scientific or empirical or in-between, that is secondary.

CONTENTS

PREFACE

It appears to this writer that the engineering kind of concrete technology that spans roughly a hundred years, from R. Feret through T.C. Powers to Bryant Mather, is finished. It has gradually been taken over by what may be called "concrete science." The activities in the "old" concrete technology have produced many important results. Thus, it seems timely and worthwhile to look back a century and produce an in-depth summary, a synthesis of the results of some of those activities that made concrete strong and the most widely used construction material.

From this view, does it follow necessarily that a new book should be written about concrete technology? After all, good books are available in English, among others by Neville (Longman/Wiley); Mehta and Monteiro (Prentice Hall); Mindess and Young (Prentice Hall); and Orchard (Wiley); in French by Duriez and Arambide (Dunod) and Venuat; in German by Wesche (Bauverlag) and Graf et al. (Springer). The similarities among these books are much larger than the differences in content and approach. Therefore, I agree that indeed there is no need, or market, for another, similar book. The point is, however, that this book is different from several aspects.

First, this book is not a substitution for but rather the continuation of the books mentioned above. Second, it concentrates on the *strength* of concrete. The other books deal with concrete technology in general, and strength is just one of the many topics discussed. For instance, only one chapter out of ten is devoted to strength in the latest edition of Neville's book, 35 pages in the Mehta–Monteiro book, and two of 22 chapters in the Mindess–Young book. Thus, these books cannot present in-depth details on strength or include recent developments. Such details, however, are needed because strength is the most frequently required property of concrete. So the writer's intention is to fill a gap in the literature by writing a continuation of the presently available books on concrete strength.

The third special feature is that no attempt was made in this book to cover all the strength-related topics evenly. For example, in certain areas, such as the concrete strength versus composition relationship, many more details are presented than in some other important areas such as the fracture mechanism of concrete. That is, the book concentrates on topics where the writer has had the most research or practical experience, where he has the most to say meaningfully, and where his heart is.

Fourth, and most important, the above-mentioned books on concrete are *not* numerically oriented. A quick look into any of them is enough to realize that most of the relationships are presented and discussed in descriptive, *qualitative* terms, tables, or graphs at best. This should not be the exclusive approach, regardless of how good these books may be otherwise. In this book the *quantitative* approach is emphasized, as in most books in other branches of engineering. This means that formulas are developed and used, as they are engineers' needed tool for their work in practically every branch of engineering. After all, one can do more with mathematics than without it in concrete technology, just as in other branches of engineering. Since this is the basic premise of the book, perhaps it is worthwhile to summarize here the author's opinion on quantification expressed earlier (S. Popovics, 1982b).

When properly used, *mathematics* in engineering is not a kind of elitism, nor is it an attempt to cover up the lack of new solutions with abstruse presentation. It is not mathematics for the sake of mathematics but rather, mathematics for the sake of engineering. It is the result of the recognition that practitioners—that is, concrete manufacturers, construction industry, and other consumers of concrete—need formulas to pave the way for improved application of concrete. Also, when the knowledge has reached a stage of clear numerical formulation, further deductions can be made by the process of mathematics. The start in this direction, however, is, of necessity, modest. In an ideal world, all engineering formulas would be based on accepted theories, that is on physics, chemistry, materials science, and so on. Unfortunately, in the real world only a few theoretically sound formulas are available in concrete technology at present. Thus, admittedly, the analytical methods applied in this book are usually phenomenological in nature. Concrete is such a complex material that the theory has not reached the level that could always produce formulas supported adequately by experiments. Also, the number of variables that significantly affect concrete behavior is so large, and some of these variables are so imponderable, that the exact mathematics gives way in many cases to approximate statistics. History tells us that further development gradually refines, justifies, or replaces empirical formulas, and I hope that this will happen to my formulas. However, the advancement of the theory would be much slower without the temporary existence of empiricism. In this spirit, it was felt important to present not only the formula but also to show, in many cases, the derivation process. The demonstration of how, or how not, to develop formulas should encourage researchers to derive improved numerical relationships. The derivations also assist in a better understanding of the principles, recognition of the weaknesses, and establishment of the limits of validity—thus, intelligent use of the formula.

An example for useful empiricism is the relationship between concrete strength and composition. Engineers need such a relationship. Yet as of today, fracture mechanics cannot give us good estimates for the strength of concrete from its composition as, say, the more-or-less empirical Abrams formula. This is not a happy situation. Perhaps concrete science will come up with a theoretically sound formula in the future for strength estimation, and most of us will be eager to embrace it. Until then, however, Abrams' formula or another suitable fitted curve is offered in this book without apology, because it gives the best estimates of concrete strength, rather than the theory. This is engineering!

To be more specific, the mathematics used is mostly in the form of mathematical modeling which results in formulas. Such a formula is developed typically from a working hypothesis or from a plausible assumption via the engineering way of thinking; if the formula is justified by experimental results, it will be accepted.

To put it in another way, in this book:

- Numerical relationships of concrete technology are presented that according to the judgment of this writer, are the *best* presently available, regardless of whether they are theoretical, empirical, or in-between.
- The validity of the formulas is examined experimentally.
- Their applications are illustrated either for practical purposes, or for development of additional formulas, or for providing a better insight into the nature of concrete.

This approach is closer to the methods used in studies on reinforced concrete than in theories of physics. By the way, this writer started using mathematical modeling long before he heard the term "mathematical modeling."

So it is not pretended that all or even most of the formulas in this book represent exact or final solutions to problems of concrete technology. Nevertheless, they are useful at present within certain limits for laboratory and/or practical use, for computers, for automation of the production, for quality control, and for raising concrete technology from the technician level of trial mixes to the engineering level equivalent to the level of reinforced concrete. Such formulas would also provide a basis for further improvement of the fundamentals of concrete technology and would improve the quality of teaching of concrete technology for engineering students.

Availability of formulas also helps computerization. It is amazing that no major effort has been made to computerize concrete technology, while in practically every aspect of the application of concrete, such as design of concrete structures, their construction, their economy, and so on, computers have been used, resulting in considerable benefits. As far as I know, this will be the first book on concrete technology that is computer oriented. *The usefulness of computerization is illustrated by the software, called Prop 21, provided on the accompanying disk.*

All this does not mean that an understanding of the *fundamental principles* is not important—on the contrary. Therefore, the fundamentals are also discussed in the book to produce a complete picture of concrete for engineers. Also, the fundamentals are illustrated by numerical relationships whenever it is possible.

The flip side of this is that this book is not for undergraduates. It is for people who teach concrete technology, graduate students, and concrete professionals who deal with concrete production, consumption, teaching, research, development, sale, specification, testing, and supervision.

The writer has invested considerable time to collect pertinent literature. So it is hoped that the listing of numerous references will save time for readers interested in additional details. This should also help in further literature search. The ultimate goal of this book is to help in the improved use of concrete.

STRENGTH AND RELATED PROPERTIES OF CONCRETE

1

COMPRESSIVE STRENGTH OF HARDENED CONCRETE

1.1 MEANING AND SIGNIFICANCE OF STRENGTH

Origin of Concrete Strength

The strength of concrete originates from the strength of the hardening cement paste, which, in turn, originates from the hydration products. The major portion of the hydration products is in the form of a rigid gel, the *cement gel*. Although there is no adequate theory yet as to the source of the strength of the cement gel itself, it is reasonable to assume that the bonds of the gel particles to each other, to the aggregate particles, and to other bodies in the concrete are responsible for the strength. In Chapters 3 and 4 we present details concerning the hydration of cement and the structure and strength of cement paste.

Meaning of Strength

The most sought-after property of a concrete is probably strength, despite the fact that in many cases other characteristics, such as durability, may be equally or even more important. This is understandable because concrete is, after all, a structural material, because the concrete strength appears to be a good index of a number of other technically important properties, and because routine strength tests in general are relatively simple to make (Kesler, 1966).

On the other hand, one must face the unfortunate fact that despite the apparent simplicity, concrete strength is an elusive property. Even if all the many factors that are known to affect the strength potential—the properties

1

and inherent variability in the concrete-making materials, proportions, air content, mixing, temperature, and others—are absolutely constant, there still can be a wide dispersion in the numerical values of the measured strength depending on how and how well the measurement is made. The number will depend on the size, shape, and method of fabrication of the specimen; its age; its treatment prior to testing; its physical condition, particularly moisture distribution and content at the time of test; and the extent of correctness of the testing procedure (Bloem, 1965). The engineer may consider this confusing situation as a big mess, or view it philosophically, as Bloem did: "What is the strength of concrete? That question is as unfathomable as Pilate's question for 'truth,' and those who seek the answer are often unaware as he was of the elusiveness of their quarry" (Bloem, 1968). Regardless of the point of view, however, the engineer must face the fact that the available strength tests provide, at best, results that are good enough only for comparative purposes. They can be used, with care, for the measure of quality and uniformity of the concrete, but they do not represent the "true" strength of the concrete with the same unambiguity that one may expect from many other fundamental measurements, such as mass or length measurements. Nevertheless, the strength of concrete specimens and the load-carrying capacity of concrete structures have been correlated by the tests made with full-scale structural members.

Significance of Various Concrete Strengths

The *compressive strength* of concrete is one of the most important technical properties. In most structural applications, concrete is employed primarily to resist compressive stresses. In those cases where other stresses (tensile, etc.) are of primary importance, the compressive strength is still frequently used as a measure of the resistance because this strength is the most convenient to measure. For the same reason, the compressive strength is generally used as a measure of the overall quality of the concrete, even when strength itself may be relatively unimportant. Finally, the concrete-making properties of the various ingredients of the mixture are usually measured in terms of the compressive strength.

In addition to its practical significance, the *tensile strength* of concrete has a fundamental role in the fracture mechanism of hardened concrete. It is an accepted view that the fracture in concrete occurs through cracking. This means that concrete fracture is essentially a tensile failure regardless of whether the fracture is caused by compression (or other loading), freezing, or by other factors. Therefore, the mechanical properties of a hardened concrete are controlled to a great extent by the fact that its tensile strength is about one-tenth of the compressive strength. In the practice, tensile and flexural strengths are most commonly utilized in beams and slabs. Here the tensile stresses are caused by tensile or flexural loads, temperature changes, shrinkage, and moisture changes. To avoid undue cracking in the concrete of such structures, tensile strength is of special importance, despite its low magnitude. The case of uniaxial tension is rarely encountered in practice and in laboratory tests can be obtained only with care. However, significant principal tensile stresses may be associated with multiaxial states of stress.

Perhaps the most important utilization of the *shearing strength* of concrete is in reinforced concrete beams to control the diagonal cracking of the beam under flexure. This is, however, not a pure shear situation. The case of pure shear acting on a plane has only theoretical interest. The pure shear strength of a concrete is about 20% of the compressive strength. The importance of the *torsion strength* of concrete is evident from the fact that the behavior of a reinforced concrete beam in pure torsion before cracking is similar to its corresponding plain concrete beam (Hsu, 1968a). The torsion strength of a concrete calculated by the plastic theory is usually somewhat higher than the tensile strength. The *impact strength* is important mainly in concrete piles and in certain military establishments. As a rough approximation, one may assume that the impact strength of a concrete is about half of its static compressive strength. An adequate *bond* between the hardened concrete and the embedded steel reinforcement is very important because it controls not only the cracking and deformation characteristics but also the load-carrying capabilities of structural elements (Bartos, 1982). In other words, this bond makes possible efficient use of a reinforced concrete member by transferring the stresses from the concrete to the steel.

The behavior of hardened concrete under *multiaxial stresses* is important because concrete in structures is practically always subjected to some multiaxial combination of compressive, tensile, shearing, and torsion stresses. The *fatigue strength* of a concrete has practical significance since the majority of structures is subjected to repeated loading. In certain structures, such as highway pavements and bridges, the loading cycles can reach several millions.

The two principal reasons for determining the strength of a concrete are (Bloem, 1968):

1. To provide, through testing standard laboratory specimens, a measure of the quality and/or uniformity of the concrete produced; this may be called the strength potential or *control strength.*
2. To provide a measure of load-carrying capacity in structures by, for example, testing specimens taken from the structure; this may be called the *design strength.*

The relation of the design strength to the control strength is extremely variable (Bloem, 1965; Petersons, 1964b; Campbell and Tobin, 1967; Bungey, 1982) depending mainly on the temperature situation and on the efficiency of curing in the field (Mather and Tynes, 1961) as well as on the differences in the degree of compaction, size and geometry of the specimens, and so on. As a rule, the strength in place is less than that of comparable laboratory specimens, sometimes not more than half of it. In other words, the standard-made and standard-cured strength specimens do not provide a quantitative measure of a concrete's *in-place* or *in-situ* load-bearing capacity.

Numerous test methods have been recommended for the determination of concrete strengths. Some of these methods measure fundamental properties of the concrete whereas others do not. Also, various simplifying assumptions are used in the procedures of the different test methods to convert a measured load to a calculated failure stress. Some of these procedures require more doubtful assumptions than others, which may considerably influence the relationships between various concrete strengths. For instance, the results of a standard compressive strength test are less sensitive to the fundamental assumptions (zero eccentricity, homogeneity, etc.) than the results of the direct tensile test. On the other hand, the direct tensile test provides more realistic values for the actual tensile strength of concrete than the flexural test because the usual manner of computation of flexural stresses assumes Hooke's law to hold. Thus, it is understandable that there exist empirical relations between the various types of concrete strength, but it is equally clear that the approximation of any of these relations is acceptable only for a limited range of concrete composition as well as for specified curing and testing conditions.

It is impossible to overemphasize the importance of proper preparation and test of the specimens, including sampling of the concrete. Improperly made specimens are worse than no specimen at all because they give false information that may have serious consequences. Therefore, the technician should strictly follow the pertinent points of the pertinent specification. The most important test methods for compressive, flexural, tensile, and splitting tensile strengths are summarized in the following sections. There are no generally accepted methods for determination of the torsion, shear, and other special strengths of concrete.

1.2 DIRECT TESTING OF COMPRESSIVE STRENGTH
(S. Popovics, 1971a)

General Procedure

The compressive strength of concrete or mortar is usually determined by submitting a specimen of constant cross section to a uniformly distributed increasing axial compression load in a suitable testing machine until failure occurs. This strength is expressed as the ultimate compression load per cross-sectional area, usually in psi, Pa, or kg/cm^2. The testing of the compressive strength of concrete started about 100 years ago.

Only the first step toward international standardization of the determination of the compressive strength of concrete has taken place in the form of a recommendation of RILEM (RILEM, 1975a). At present, various countries are using test methods that differ in details, including the shape and size of the specimen. For instance, the European countries use mainly 150- or 200-mm (6- or 8-in.) *cubes,* whereas in the United States and Canada *cylinders* are generally used for concrete, with a length equal to twice the diameter (Fig. 1.1). The standard cylinder size is 6 × 12 in. (150 × 300 mm) if the coarse aggregate does not exceed 2 in. (50 mm) nominal size, although the more economical 4 × 8-in. (100 × 200-mm) or 3 × 6-in. (75 × 150-mm) cylinders are also suitable for many purposes (Malhotra, 1973a; Forstie and Schnormeier, 1981; Nasser and Kenyon, 1984; Date and Schnormeier, 1984). Portions of beams remaining from flexural test can also be used for the determination of compressive strength. The standard form of this is the *modified cube method* (Fig. 1.1), used by Fuller and Thompson (1907). Another specimen used in special investigations is the *prism.* This has the advantage that not only the compressive strength but also a number of related properties can be determined from a single concrete prism (Best, 1964; Palotas, 1960; Palotas and Halasz, 1963). Such properties are flexural strength, splitting strength, direct shear strength, static and dynamic moduli of elasticity, and the stress–strain curve (Fig. 1.2). The compressive strength can also be de-

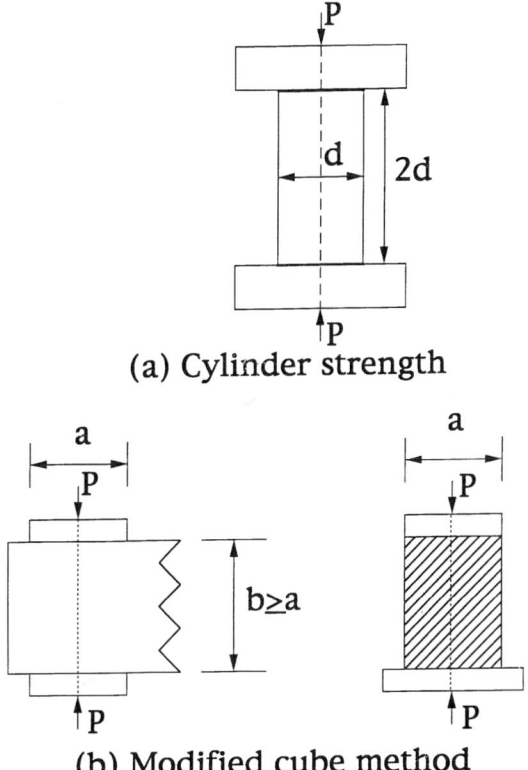

(a) Cylinder strength

(b) Modified cube method

Figure 1.1   Two standard methods for determination of the compressive strength of concrete.

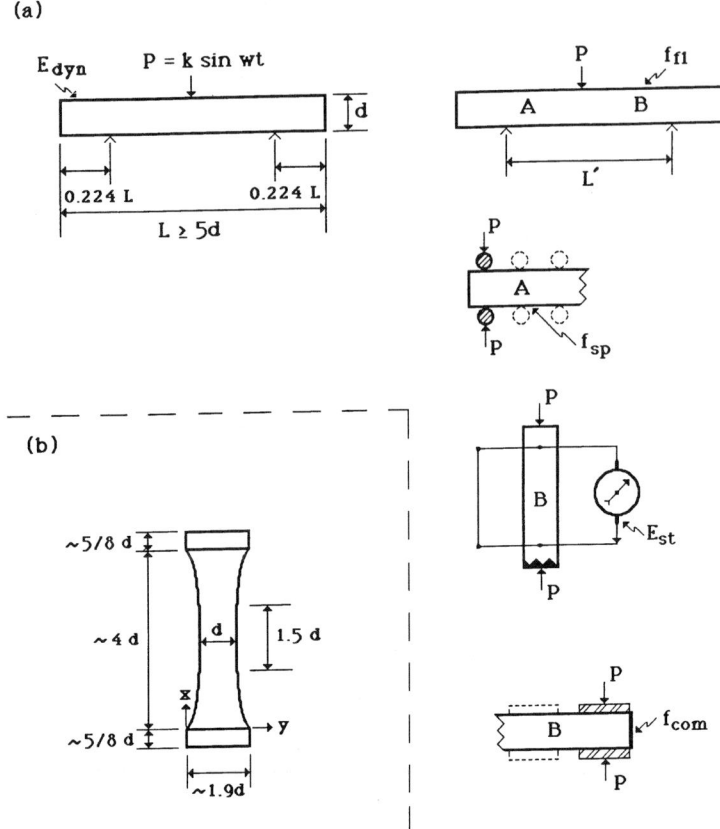

Figure 1.2 (a) Sequences of testing five different properties of concrete from the same prism specimen. (b) Jet-line specimen for the compression test of concrete.

termined by the *flexural test of over-reinforced beams* (Slater and Lyse, 1930). Despite the simplicity of the principle of the compressive strength test, proper performance requires great care (Concrete Society, 1975; Hester, 1980).

Control Strength

Specimens for the determination of the *control compressive strength,* that is, for verification of the potential strength and uniformity of the produced concrete mixtures, can be made in the laboratory or in the field. The standard method of making and curing concrete compression test cylinders in the *laboratory* is described in ASTM C192 and BS 1881, Part 110. Accordingly, having batched and mixed properly, the concrete is placed in serviceable molds in three equal layers. Each layer in a 6 × 12-in. (150 × 300-mm) cylinder is rodded with 15 strokes using a $\frac{5}{8}$-in. (15-mm) steel rod. An alternative possibility is to consolidate the specimen by appropriate vibration, in which case the concrete is placed in two layers in the mold. Concretes with slump greater than 3 in. (75 mm) should be rodded, while concretes with slump less than 1 in. (25 mm) should be consolidated by vibration; concretes with intermediate slumps may be either rodded or vibrated. Subsequently, the specimens should be stored at 73.4 ± 3°F (23 ± 1.7°C). The specimens should be removed from the molds about 24 hours and stored in moist conditions at the same temperature until the time of the test. The ends of all compressive specimens that are not plane within 0.002 in. (0.05 mm) should be properly capped or ground plane before testing. Compressive tests on the specimens should be made as soon as practicable after removal from the curing room and they should be tested in moist conditions in a testing machine in compliance with the specifications of ASTM C39.

An essentially similar procedure should be followed when the control compressive specimens are made and cured in the *field* (ASTM C31), with two exceptions: (1) it is desirable that the specimens be consolidated to the same air content that is intended in the concrete in the structure; (2) they should be stored in or on the structure as near the point of use as possible and should receive the same curing; that is, the same protection from the elements and all surfaces as given to the portion of the structure that they represent (ASTM C31).

The presently valid ASTM compressive strength test does not yet contain a general precision statement. However, for well-controlled laboratory conditions the following is specified by ASTM C192-90a: The single-operator standard deviation for 7-day compressive strength (C39) of trial batches has been found to be 203 psi (1.40 MPa). Therefore, the results of properly conducted tests on two trial batches made in the same laboratory should not differ by more than 574 psi (3.96 MPa). The multilaboratory standard deviation for such compressive strength of trial batches has been found to be 347 psi (2.39 MPa). Therefore, the results of properly conducted tests on single trial batches made in two different laboratories should not differ by more than 981 psi (6.77 MPa). Both these precision statements are considered applicable to laboratory trial batches proportioned to contain prescribed quantities of materials and to have a prescribed water-cement ratio. The values should be used with caution for air-entrained concrete, concrete with slump less than 2 in. (50 mm) or over 6 in. (150 mm), or concrete made with other than normal-weight aggregate, or aggregate larger than 1 in. (25 mm) nominal maximum particle size.

The diameter of a cylinder specimen should be at least three times the maximum nominal size of the coarse aggregate in the concrete. This requirement leads to quite large specimens in the case of mass concrete. The concomitant cumbersomeness frequently can be eliminated by testing only that part of the mass concrete that passes the $1\frac{1}{2}$-in. (38-mm) sieve. Here the following rule of thumb is helpful: The compressive strength of a concrete containing 6-in. (150-mm) aggregate and tested in large cylinders is approximately three-fourths of the compressive strength of similarly cured 6×12-in. (150×300-mm) cylinders of the same concrete from which the aggregate particles larger than $1\frac{1}{2}$ in. (38 mm) have been removed by wet screening (U.S. Department of the Interior, 1981).

For laboratory conditions, ASTM C192 requires that three or more compression specimens be made for each test age and test condition. Specimens involving any given variable should be made, if possible, from at least three separate batches mixed on different days. For field conditions, Waddell recommends (Waddell, 1962) that for each class of concrete one set of specimens be made daily from each 100 cu yd (76.5 m³) or fraction thereof. Each mix being used should be sampled every day it is used. The number of specimens in each set depends on the prevailing conditions (Cordon, 1966) (Section 1.6).

The preparation and test of concrete *cubes* for control strength are essentially similar to those of cylinders, except that there is no need for capping, and are described in numerous more-or-less-differing foreign specifications. Reference can be made to the British BS 1881, Part 108 and the German DIN 1048. The test method for breaking the cylinders for determination of the compressive strength is specified in ASTM C39 and that of cubes in BS 1881, Part 116.

The *modified cube method* utilizes portions of beams remaining from flexural test by applying the compression through square steel plates not smaller than the cross section of the beam (Fig. 1.1*b*). The plates are placed vertically above each other on the opposing sides of the beam portion (ASTM C116 and BS 1881, Part 119). This method is not an alternative for cylinder testing but can be useful in cases when procurement of extra cylinders is difficult or impossible.

Other kinds of compressive strength values can be obtained by loading the specimen on an area much smaller than its cross section (Lieberum and Reinhardt, 1989; Richardson, 1989). Another, recently developed method is the *diameter-compression test* for cylindrical core specimens (Ruijie, 1996). The cylinder is tested in horizontal position and the load is applied vertically, as in the splitting test (Section 2.2). In the new test, however, the load is not acting along lines but rather is distributed on the cylinder surface with the help of two fitted half-cylinder metal pieces. This arrangement simplifies the core testing because there is no need for cutting and capping the core specimen. Linear relationship was observed between the results of the diameter-compression tests and cube strengths.

Compressive strength values obtained by cylinders (f_c), prisms, cubes (f_{cu}), and modified cubes (f_{cm}), and compressive strengths determined by other methods are neither interchangeable nor necessarily comparable. As indicated by Tables 1.1 and 1.2, the f_c/f_{cu} and f_c/f_{cm} ratios of these compressive strengths vary within wide limits depending on the properties of the concrete, curing and testing conditions, and so on. The same can be said

TABLE 1.1 Comparison between compressive strength of 6 × 12-in. (150 × 300-mm) cylinders and that of cubes[a]

f_c/f_{cu}	Source[b]	Remarks
0.75	British Standard BS 1881:52	6-in. cubes
0.85	Hofsoy	10-cm cubes
0.86	Lyse and Johansen	10- or 20-cm cubes
0.88	Hummel	20-cm cubes
0.62 to 0.865	L'Hermite	15-cm cubes
0.64 to 0.994	Bignoli	8-in. cubes, overall
0.75 to 0.80	Bignoli	8-in. cubes, high-slump concrete
0.65 to 0.84	Gyengo	20-cm cubes
0.71 to 0.95	Gonnerman	6-in. cubes
0.77 to 1.04	Gonnerman	8-in. cubes
0.74 to 0.83	Symons	15-cm cubes
0.80 to 0.90	Graf	Cube side equals cylinder diameter
0.80 to 0.96	Walz et al.	20-cm cubes; lightweight concrete
0.81 to 0.89	Walz	20-cm cubes
0.82 to 0.88	Beton Kalender	20-cm cubes
$0.66 + 0.20f_{cu}/500$	Australian Standard ASA-78	15-cm cubes; f_{cu} in kg/cm^2
$0.68 + 0.22f_{cu}/500$	Australian Standard ASA-78	20-cm cubes; f_{cu} in kg/cm^2
$0.85 - 0.21f_{cu}/1000$	Petersons	15-cm cubes; f_{cu} in kg/cm^2
$0.82 - 21.1/f_{cu}$	Hernandez	Rounded siliceous aggregate; 15- or 20-cm cubes; f_{cu} in kg/cm^2
$0.87 - 17/f_{cu}$	Hernandez	Crushed limestone aggregate; 15- or 20-cm cubes; f_{cu} in kg/cm^2
$0.85 - 12/f_{cu}$	Poijarvi and Syrjala	20-cm cubes; f_{cu} in kg/cm^2
$0.85 \pm 50/f_{cu}$	Campus et al.	20-cm cubes; f_{cu} in kg/cm^2
$50/(77.5 - \sqrt{f_{cu}})$	Vourinen	20-cm cubes; f_{cu} in kg/cm^2
$0.76 + 0.20 \log{(f_{cu}/200)}$	L'Hermite	20-cm cubes; f_{cu} in kg/cm^2

Source: S. Popovics (1967a). Copyright TRB. Reprinted with permission.
[a] 1 kg/cm^2 = 14.2 psi = 0.098 MPa; 1 in. = 2.54 cm = 25.4 mm.
[b] References for the sources listed in this column are given in S. Popovics (1967a).

about the prism strength–cube strength and prism strength–cylinder strength ratios (S. Popovics, 1967a). For instance, the cylinder strength–cube strength ratio varies with specimens molded in laboratory within 0.62 and 0.96 (Table 1.1). Thus, for lack of a more accurate solution, one may simply assume that the cylinder strength of a concrete is slightly over 70% of the cube strength, provided that 6 × 12-in. (150 × 300-mm) cylinders and 8-in. (200-mm) cubes are used; they are made from the same concrete; prepared, cured, and tested similarly; and the composition and strength of the concrete is within the structural range. In this case the approximate value of the cylinder strength can be obtained in psi by multiplying the cube strength in kg/cm^2 by 10 or

TABLE 1.2. **Comparison between compressive strength of cylinders and that of modified cubes**[a]

f_c/f_{cm}	Source[b]	Remarks
0.88	Klieger	6-in. modified cubes
1.0	Koenitzer	6-in. modified cubes
1.0	Klieger	6-in. modified cubes
1.0	Shideler	6-in. modified cubes; sand and gravel concrete
1.2	Mather	$6\frac{1}{4}$-, 8-, 10-, and 12-in. modified cubes; specimens cast in laboratory
0.79 to 1.39	Sen and Bharara	6-in. modified cubes; ratios larger than 1 belong to strengths lower than 2000 psi (14 MPa)
0.85 to 0.97	Withey	6-in. modified cubes
1.0 to 1.16	Mather	f_c for cores; f_{cm} for specimens cut from slabs; ratio about 1 up to 5000 psi (35 MPa), above which it increases
$0.9 - 100/f_{cm}$	Kesler	6-in. modified cubes; f_{cm} in psi
$1.2 - 485/f_{cm}$	Shideler	6-in. modified cubes; structural lightweight concrete; f_{cm} in psi

Source: S. Popovics (1967a). Copyright TRB. Reprinted with permission.
[a]1 kg/cm^2 = 14.2 psi = 0.098 MPa; 1 in. = 2.54 cm = 25.4 mm.
[b]References for the sources listed in this column are given in S. Popovics (1967a).

the strength in MPa by 100. For instance, a cube compressive strength of 500 kg/cm^2 (49 MPa) is equivalent to a cylinder strength of 5000 psi (35 MPa).

Specimens for control compressive strength are tested at prescribed ages. Tests are recommended at the age of 28 days or if the specimens contain Type III cement, at the age of 7 days (ASTM C192). Despite the fact that additional tests are often made at earlier ages, there is a serious objection to this approach: that it takes too long to obtain information on the quality (i.e., on the potential compressive strength) of the concrete in question for it to be of real value for either concrete construction control or acceptance purposes. In other words, if low-strength concrete is not detected until 28 days, or even 7 days after it is placed, it is usually very expensive and difficult to correct the mistake. Furthermore, the 28- or 7-day strength data are usually too late to be useful to reveal the need for an adjustment in the composition of the substandard concrete. The main reasons for still using the 28-day strength in specifications are:

1. The expectation that by this age the strength of the concrete will be characteristic of its final strength potential regardless of the types of the concrete. Experience has shown indeed that this expectation is fulfilled in most cases.

2. This age has long been the decisive criterion in the legal system.

In any case, it would be very useful if the strength potential of a concrete would be determined substantially sooner than 28 or even 7 days. Several methods have been recommended for this purpose. These are discussed in Section 1.3.

Design Strength

Knowledge of the actual strength of concrete in structures, called the *design strength* or *in-situ strength* or *in-place strength*, is of great importance for the engineer, since this strength can be low even when the strength potential of the concrete used is satisfactory. Monitoring the strength development of a concrete in a structure at very early ages, that is, while the concrete is still in the mold or shortly after, is important because this controls, for instance, the form removal time, the release time of prestressed elements, determining when the structure may be put into service, and so on. Another important area is establishment of the load-carrying capacity of existing structures, especially if the structure is old and/or damaged. The success of any maintenance, repair, or rehabilitation measure depends on the accuracy with which the cause and extent of the deterioration have been established. *Once a specific conclusion as to the cause and extent of the damage has been reached, then and only then can a rational selection be made among alternative maintenance and repair strategies* (Stowe and Thornton, 1984).

The concrete strength in the structure can be estimated directly, measuring the strength of suitable specimens, or indirectly, measuring a concrete property other than the strength in question from which the strength can be estimated (Bungey, 1982). Indirect methods, such as nondestructive tests, are discussed in Sections 1.4 and 1.5. The British specification BS 6089:1981 provides guidelines for the assessment of concrete strength in existing structures, and so does RILEM (Szoke, 1976, 1979).

Cylinders for design strength determination should be molded in the same way as for control strength. Here, however, it is desirable that the specimens be consolidated to the same air content that is intended in the concrete in the structure. Also, they should be stored in or on the structure as near to the point of use as possible and should receive the same curing (i.e., the same protection from the elements and all surfaces) as is given to the portion of the structure which they represent (ASTM C31). Nevertheless, it has been recognized that the concrete strengths obtained with these cylinders can be significantly higher or lower than the strength of the corresponding concrete in the structure, for several reasons. For instance, the temperature of the concretes in the two places can be different because of differences in the ambient temperatures and wind factors, respectively; differences in the humidity; and differences in the temperature rise in the two concretes resulting from the heat of hydration. It is likely also that the degree of compaction is not identical for concretes in the cylinders and in the structure, respectively. Thus the results of the design cylinders should be interpreted with caution when they are used for the estimation of the strength of concrete in the structure.

An improved version that promises to eliminate most of the above-mentioned differences is when the design concrete cylinders are remotely cured at structure temperature and moisture conditions with the help of suitable equipment. This can be achieved by equipment which includes a temperature controller, a remote-sensing reference thermocouple in the structure, a remote thermocouple in a test cylinder, a chart recorder, assorted switches, and a cylinder curing chamber (Brander, 1980). The controller senses the temperature at the appropriate thermocouple in the structure, compares it to the test cylinder temperature, and if the cylinder temperature is lower than the structure temperature, the controller switches on a heater element. Not only does this equipment duplicate the structure temperature and moisture, thus bringing the cylinder strength closer to the strength of concrete in the structure, but the temperature record with cylinder breaks represents *maturity* data which can be used to predict the concrete strength in situ: for example, for determining when the next set of cylinders should yield the minimum form removal strength. The concept of maturity represents the combined effect of age and curing temperature and was defined originally as the age of the concrete multiplied by the average curing temperature above freezing that it maintained (Saul, 1951). Saul formulated the rule of *gain of strength with maturity* as follows (Section 3.7): "Concrete of the same mix at the same maturity (reckoned in temperature-time) has approximately the same strength whatever combination of temperature and time go make up that maturity." This one-to-one relationship makes it possible to develop formulas for the estimation of concrete strength from the value of maturity. Examples are Eqs. 1.9 and 1.10. Later investigators found that the validity of this rule is very limited. Therefore, several modifications have been recommended (Malhotra, 1971; Carino, 1991a), none completely successful. Nevertheless, the rule of maturity has practical applicability within established limits.

There is another method for estimating the in-situ strength of concrete from the correlation of the maturity to strength. The method includes the casting of cylinders in the usual manner and determination of the maturity of the concrete in the structure. The latter is normally done by using either thermocouples cast into the concrete pour and reading the temperature at intervals or maturity meters installed during placing of the concrete. In the latter case the thermocouples are connected to a computer that integrates the temperature with time and provides the maturity digitally directly on a display. The maturity meter consists of a bridge circuit incorporating a thermistor that varies with temperature in a predetermined way. The circuit is so arranged that for any temperature above the base temperature of $-10°C$ (14°F), the bridge is unbalanced by the action of the thermistor. The integrator circuits contain a capacitor which electronically creates a pulse current that drives the display counter. Hence, if the concrete is at a high temperature, a large number of pulses are fed to the counter in a given time and the maturity will be greater than if the temperature is low and the pulses are fewer. The standard-cured cylinder strength and maturity are then correlated, and the in-situ strength is estimated from the field maturity by a suitable formula. Once a correlation between maturity and corresponding strength has been established for a particular type of concrete, the concrete strength in the structure can be estimated from the maturity of the in-situ concrete alone (Mukherjee, 1975; Hulshizer et al., 1984). When the concrete composition changes, the correlation should be reexamined.

The limitations of this maturity method are as follows (Malhotra and Carette, 1980a): (1) The limits of validity for maturity are those represented by about 3 to 28 days at normal temperature; (2) the initial temperature of the concrete should be between 15 and 30°C; and (3) no loss of moisture by drying should occur during the curing period. Within these limits, however, experience with this relatively new method has generally been good (Bickley, 1975; Sadgrove, 1975; Lew and Reichard, 1978; Malhotra, 1973b).

Design strength determination of a concrete in an existing structure requires different techniques. Here the strength can be obtained with concrete specimens that are taken out of the finished structure by drilling or sawing. This is, however, an expensive and time-consuming procedure (Bungey, 1982; Concrete Society, 1976; Munday and Dhir, 1984; Swamy and Al-Hamed, 1984a). Also, the core extraction may reduce the load-bearing capacity of the concrete member even if the hole is refilled with concrete (Kirtschig, 1968). In certain cases, these difficulties can be reduced by using a push-out method.

The drilled compression specimens are cylinders, or *cores;* the sawed specimens are usually cubes or beams. The diameter of the core specimen should be at least three times the maximum nominal size of the coarse aggregate used in the concrete (ASTM C42). Note, however, that strength of cores of $2.5D$ diameter gave good correlation with strengths of larger cores from the same structure (Keiller, 1982). The length-diameter ratio of such cores should be kept between 1.0 and 1.2 for practical reasons. The beams are usually tested first for flexural strength and then for compression, using the modified cube method according to ASTM C116. The *push-out method* involves installing cylindrical plastic inserts in the forms which will be filled with concrete during the regular placement operation. The insert can be pushed out of the structure to provide a strength test specimen at any time after the concrete has hardened. In any case, if the ratio of length to diameter of the tested cylinder is appreciably less than 2, the measured compressive strength should be corrected (Concrete Society, 1976). According to ASTM C42, if this ratio is 1.75, 1.50, 1.25, or 1.00, the measured compressive strength should be multiplied by 0.98, 0.96, 0.93, or 0.87, respectively, but these factors may be too low (Bungey, 1979). BS 1881 Part 120 contains similar specifications.

For the relationship between the various kinds of design strengths, Bloem (1968) found that regardless of the curing conditions core strengths averaged about 7% lower than the corresponding push-out cylinders. He also found that compressive strengths of cores were less than corresponding molded field-cured cylinders by an average of about 10% for good curing and about 20% for poor curing. Although other investigators have reported similar tendencies (Keiller, 1982; Brink, 1970; Petersons, 1968) the applicability of these numbers is limited by the fact that all his cores were taken from slabs. There are other concrete structures, such as columns and mass concrete, where poor curing is less detrimental to the core strength. In other words, the looseness of the relationship between the results of these two types of tests should not be surprising. After all, the molded cylinders are prepared in a strictly standardized manner which eliminates many factors that exist during construction of the structure and influence the strength of the concrete cores. Since these effects usually cannot be expressed numerically, a dependable correlation between the cylinder design strength and core strength cannot exist. Because of this, Malhotra (1977) even suggested abandoning the core strength test. Nevertheless, there are cases where core strength is still the only practical value for the design strength. Note also that despite the variations in the core strength–cylinder strength relations reported by various investigators (Bloem, 1965; Mather and Tynes, 1961; Petersons, 1964a, 1968; Wagner, 1963; Hofsoy, 1964; Gaynor, 1974; Swamy and Ali, 1984b), there is evidence to indicate not only that cores taken from highways or reinforced concrete structures produce lower strengths than comparable molded cylinders or cubes but also that this difference increases with increasing concrete strength (Fig. 1.3). For these reasons, it has been suggested that the standard core strength can be considered as the lower limit of the actual strength of a traditional concrete in the structure, whereas the strength estimate obtained by the standard Windsor probe (Section 1.4) is considered the upper limit (Lee et al., 1992).

When cores are drilled from a bridge deck or from other structural concrete members, they often contain pieces of reinforcing steel. Gaynor (1965) demonstrated that when this steel is located in a plane perpendicular to the vertical axis of the specimen, it reduces the measured compressive strength by at least 10%. The selection of locations of cores is discussed in Section 1.4.

Experimental results also indicate that the variance of the compressive strengths determined by core tests is larger than that of molded cylinders (Bloem, 1968; Petersons, 1964b; Keiller, 1982; Bungey, 1979; Meininger, 1968; DiLeo et al., 1984), and this variance increases with a decrease of the size of the core. Consequently, if one wants to obtain as precise an estimate of strength with cores as that given by molded cylinders, then either the number or the size of the cores tested should be increased considerably (Mather and Tynes, 1961).

Failure of Compression Specimens Under Uniaxial Load

Apart from Szabo's dissenting opinion (Szabo, 1966, 1967), the general feeling is that the main reason for the difference between the cube compressive strength and the cylinder strength of a concrete is the differing magnitude of

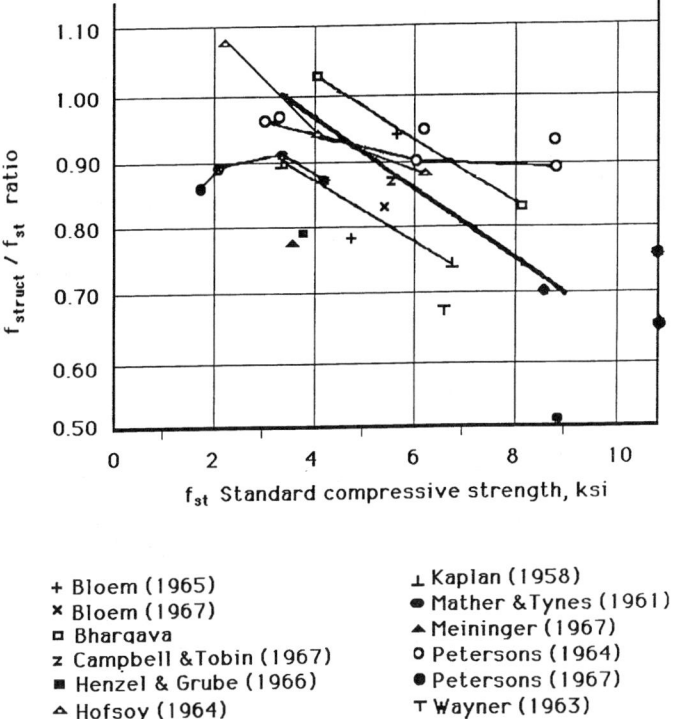

+ Bloem (1965)
× Bloem (1967)
□ Bhargava
z Campbell &Tobin (1967)
■ Henzel & Grube (1966)
⌃ Hofsoy (1964)

⊥ Kaplan (1958)
● Mather &Tynes (1961)
▲ Meininger (1967)
○ Petersons (1964)
● Petersons (1967)
⊤ Wayner (1963)

Figure 1.3 Relative strength of concrete in the structure f_{struct}, as measured on drilled cores, as a function of the corresponding compressive strength determined on standard specimens f_{st}. 1 ksi = 6.90 MPa. (From Petersons, 1968. Copyright RILEM. Reprinted with permission.)

the effect of the friction between the steel compression plates of the testing machine and the concrete specimen (Taylor, 1971; Taylor et al., 1972). As a result of the differences in the elastic constants of concrete and steel, the lateral strain in the plates is smaller during the test than that of the concrete specimen tested (if it were free to move); consequently, friction develops. The frictional forces counteract the lateral elongation of the specimen, creating lateral compressive stresses in the concrete adjacent to the compression plates in addition to the stresses produced by the uniaxial compressive load. This triaxial compressive stress condition increases artificially the *measured* compressive strength of the specimen.

This mechanism implies the following:

1. The typical failure of a specimen in a uniaxial compression test takes place in the middle of the height because the restricting effect of the frictional forces is minimum here.
2. The more slender the specimen, the lower the compressive strength measured because the effect of the frictional forces in the middle sec-

tions becomes smaller. However, when the specimen is slender enough, a further increase in the slenderness, up to the point where the stability starts controlling the failure, does not cause further significant reduction in the strength.

3. If the friction is reduced, for instance by applying appropriate interlayers between the loading surfaces (Hsu, 1969; Erdei, 1980), this not only reduces the measured value of the compressive strength but also reduces the variation of strength caused by the variation in slenderness of the specimen. This effect is especially spectacular when Teflon layers are applied as interlayer due to the low friction of this material (RILEM, 1997).

4. The magnitude of the cube strength–cylinder strength ratio is not a constant; it can depend on the the the type of concrete or aggregate.

All these implications are supported by many experimental data.

A theoretical distribution of the combined stresses of the uniaxial compression and friction was presented by Ros (1950), but there are additional discussions of this problem, including deformation measurements and photoelastic studies. Several pertinent papers are mentioned in an article by S. Popovics (1967a). These theoretical or semitheoretical discussions, as well as experimental data, clearly show that under usual circumstances the effect of a change in the height-width ratio on the measured compressive strength is strong when this ratio is less than 1.5, but the effect becomes slight when the height-width ratio is larger than 2.

Other factors that can influence the relation between cylinder strength and cube strength of a concrete are the age of the concrete as well as the composition (the type of mineral aggregate, water-cement ratio, air content, etc.), perhaps through their effects on the modulus of elasticity, Poisson's ratio, and propagation of the internal microcracking.

Another factor is that the compression load frequently is not uniformly distributed on the specimen during the test because of an unsatisfactory condition in the specimen ends or testing plates, insufficient rigidity of the testing machine, or load eccentricity. These affect the strength results to differing degrees, depending on the size and shape of the specimen. Fundamental aspects of the failure mechanism of concrete based on internal crack propagation are summarized in Section 2.4.

The question has frequently been raised whether cubes or cylinders are more appropriate for determination of the compressive strength of concrete. From the engineering point of view, the answer is the one that reflects better the load-carrying capacity of concrete in the structure. Unfortunately, we do not know which this is. Since the cylinder strength–cube strength ratio is affected by the type of concrete, only extensive testing of the load-carrying capacity of structural members made with various concretes could provide the answer through testing the same concretes in cylinders and cubes, and such data are not yet available. The technical literature contains pros and cons on both sides, although the cylinder appears slightly superior to the cube (Best, 1964; Rusch, 1956, 1957; Hansen et al., 1962; De Larrard et al., 1994). The main pragmatic objection to conventional concrete cylinders is that the ends must be capped to obtain satisfactory test surfaces; this takes time and may even introduce a distortion of the strength. Serviceable cube molds automatically provide two pairs of satisfactory test surfaces, with the possible exception of high-strength concretes (De Larrard et al., 1994). Also, a standard 8-in. (200-mm) cube is more suitable for testing concrete of larger maximum size than a 6 × 12-in. (150 × 300-mm) cylinder. On the other hand, cylinder capping can be eliminated by using the special mold and casting method recommended by Thaulow (1952). Also, the cylinder specimen seems superior to cubes in certain aspects. For instance, for conventional cylinder specimens the molds are lighter, the amount of concrete needed is less, a smaller and cheaper testing machine is satisfactory, the compression plates of the machine are less subject to wear, the stress distribution under uniaxial load is more uniform, the measured compressive strength is less influenced by friction on the test surfaces, modulus of elasticity and splitting strength determinations are applicable, and so on. Best also quotes Mather to the effect that a cube has more weak spots in the form of edges with 90° angles; furthermore, that the compression load is applied to a cylinder in the direction of placement, whereas to a cube it is applied normal to this direction, which may make the result somewhat ambiguous. Yet the advantage of using cylinders is most obvious when the strength of concrete should be checked in finished structures because such specimens can be bored out from the hardened concrete with relative ease.

On the basis of the discussion above, L'Hermite (1937, 1955) recommended another specimen for laboratory investigations related to compressive loads on concrete. This specimen, shown in Figure 1.2, may be called a *jet-line* specimen because its upper and lower parts are shaped like a jet, while the middle section is cylindrical. The advantage of this specimen shape is that the magnitude of the friction forces from compression is almost zero in the cylindrical part, and therefore the jet-line specimen is suitable for certain type of research. A somewhat similar specimen for compression strength is recommended by Nadai (1950). The effects of the shape and size of the specimen on the measured strength of concrete, along with other factors affecting the results of strength tests, are discussed more generally in Section 3.8.

Testing Concrete Under Triaxial Loads

ASTM C801-86 describes a recommended practice for the determination of mechanical properties of hardened concrete under triaxial loads. The device in which a right-circular-cylindrical specimen may be enclosed in an impermeable, flexible membrane is placed between two hardened bearing blocks and subjected to hydraulic pressure as well as deviator stress, as shown in Figure 1.4 (Bellamy, 1961; Krahl, 1965). Alternatively, devices may be used that combine the function of loading device and pressure chamber.

This procedure is for determination of the triaxial compressive strength of concrete, but it can also be used with suitable instrumentation for obtaining deformation characteristics under triaxial loads (Richart 1928; Hannant, 1969), shear strength at various lateral pressures, angle of shear resistance, strength in pure shear, deformation modulus, and creep behavior.

1.3   EARLY DETERMINATION OF THE CONTROL STRENGTH

As mentioned earlier, judgment on the adequacy of a concrete is needed on the construction site much sooner than 28 or even 7 days. There have been many attempts for the meaningful early determination of the strength potential of a concrete, including the control strength, several of which are discussed below.

Accelerated Curing

One approach is to accelerate the strength development in the concrete so as to obtain a high-enough strength early, not later than, say, 48 hours, which is

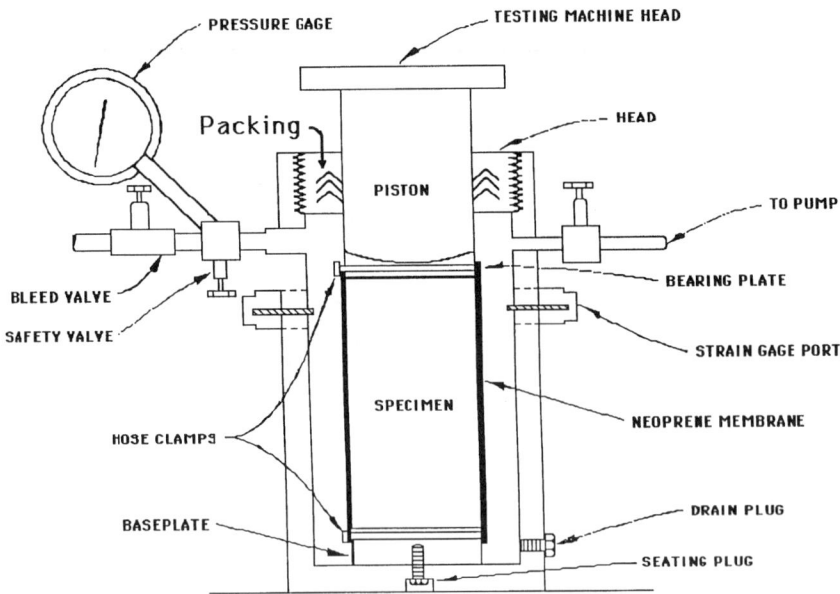

Figure 1.4 Triaxial testing apparatus as specified in ASTM C801-91. Copyright ASTM. Reprinted with permission.

characteristic of the strength potential of the concrete. This could be done, for instance, by using an accelerating admixture such as calcium chloride. However, this method has not proved suitable in general for checking the strength potential of concrete because different cements may react very differently with such chemical admixtures (S. Popovics, 1992). A better approach is to raise the temperature of the concrete early in the strength specimens, although it is true here too to a certain extent that strength development of concretes made with different materials and/or of different compositions may react differently to heat treatment. This heating is called the *accelerated curing* method and is quite popular, which is reflected by the large amount of pertinent literature available (RILEM, 1966; Transportation Research Board, 1975, ACI, 1978, Ramakrishnan, 1978; S. Popovics, 1982a).

There are numerous variants of the accelerated curing method using an external or internal source for the heat development; low, medium, or high temperature rises for curing; different durations for the heat treatment or delay periods; and so on (Berio, 1966). From the various possibilities four procedures have been standardized in ASTM C684-95 for accelerated curing and testing of compression test specimens: procedure A, in which warm water is used for curing; procedure B, in which boiling water is used for curing; procedure C, which is an autogenous method; and procedure D, which uses high temperature and pressure.

The primary function of the moderately heated water used in procedure A is to serve as insulation to conserve the heat generated by the hydration. The thermal acceleration is provided by the temperature level in procedure B. Procedure C involves storage of specimens in insulated curing containers in which the elevated curing temperature is obtained from heat of hydration of the cement. The sealed containers also prevent moisture loss. Procedure D involves simultaneous application of elevated temperature and pressure to the concrete in a closed container. Due to the special nature of procedure D, only the other three procedures are discussed here. Further characteristics of these procedures are presented in Table 1.3. Note that the strengths of a concrete obtained by procedures A, B, or C are fairly close to each other, as demonstrated in Figure 1.5 for various water-cement ratios.

The single-laboratory coefficient of variation for the ASTM methods of accelerated curing and testing of concrete compression test specimens has been determined as 3.6% for a pair of cylinders cast from the same batch. Therefore, results of two properly conducted strength tests by the same laboratory on two individual cylinders made with the same materials should not differ by more than 10% of their average. The single-laboratory multiday coefficient of variation has been determined as 8.7% for the average pairs of cylinders cast from single batches mixed on two days. Therefore, results of two properly conducted strength tests each consisting of the average of two cylinders from the same batch made in the same laboratory on the same materials should not differ more than 25% of their average.

TABLE 1.3 Brief description of accelerated curing procedures

Procedure	Molds	Accelerated Curing Medium	Accelerated Curing Temperature [°F(C°)]	Age Accelerated Curing Begins	Duration of Accelerated Curing	Age at Testing
A. Warm water	Reusable or single use	Water	95(35)	Immediately after casting	$23\frac{1}{2}$h ± 30 min	24 h ± 15 min
B. Boiling	Reusable or single use	Water	Boiling	23 h ± 15 min after casting	$3\frac{1}{2}$ h ± 5 min	$28\frac{1}{2}$ h ± 15 min
C. Autogenous	Single use	Heat of hydration	Initial concrete temperature augmented by heat of hydration	Immediately after casting	45 h ± 15 min	49 h ± 15 min
D. High temperature and pressure	Reusable	External heat and pressure	300 (149)	Immediately after casting	5 h ± 5 min	$5\frac{1}{4}$ h ± 5 min

Source: ASTM C684-95. Copyright ASTM. Reprinted with permission.

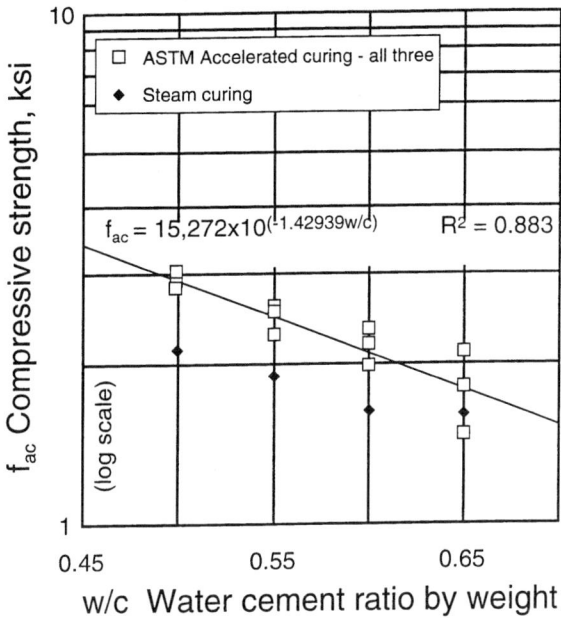

Figure 1.5 Compressive strength obtained with the three ASTM accelerated curing proce-dures f_{ac} as a function of the water–cement ratio w/c. Cement: Type I. Cement content: 500 lb/cu yd (300 kg/m³). 1 ksi = 6.90 MPa. (From S. Popovics and Pfeifer, 1992.)

Calibration Curves

The strengths obtained by accelerated curing are lower than the corresponding *normal* 28-day control strength, that is, the concrete strength obtained after 28 days of standard wet curing at 73°F (23°C) temperature. So can we use the results of the test with accelerated curing for judging the strength potential of the concrete considering that all our specifications and experience have been based on the 28-day normal strength? In other words, the critical ques-tion is whether there is a relationship between the strength obtained with accelerated curing and the normal 28-day strength of the concrete. The answer is a qualified yes: There is a relationship for each accelerated test method; however, it is not a strict mathematical relationship but rather, a statistical one. A typical example for the relationship between the 1-day accelerated strength obtained by the *boiling-water method* and the corresponding normal 28-day strength is shown in Figure 1.6.

The fitted line representing this relationship, called a *calibration curve*, can be used for estimating the strength after normal curing from the result of the accelerated test. Note that under other circumstances different calibration curves (i.e., relationships) may be valid, as shown in Figure 1.7.

Calibration curves are usually established by curve fitting to corresponding strength data obtained on testing concrete specimens after normal curing, and

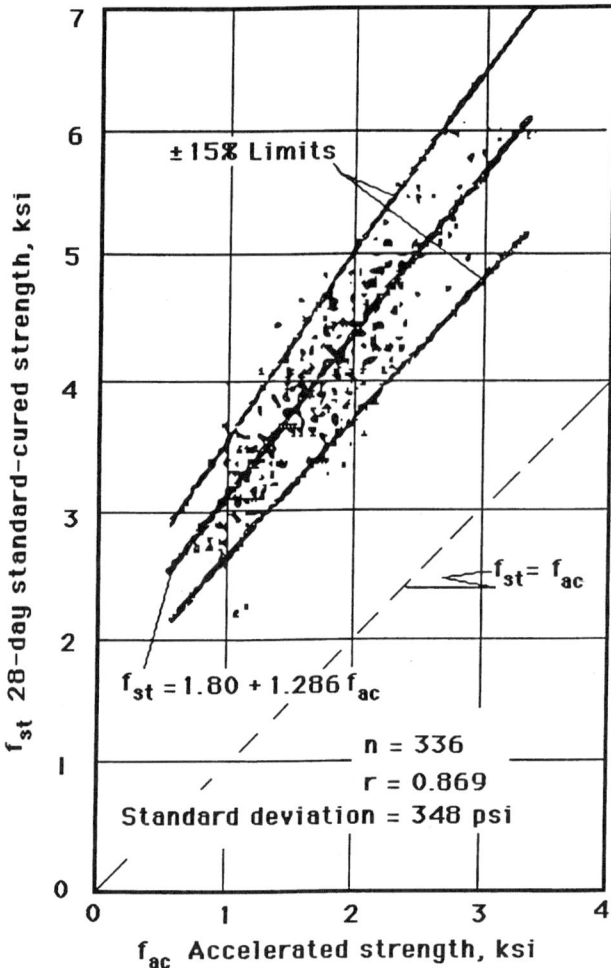

Figure 1.6 Field data of and relationship between 1-day accelerated compressive strengths obtained by the boiling-water method f_{ac} and the corresponding normal 28-day strengths f_{st} for a group of concretes using several brands of Type I cement. 1 ksi = 6.90 MPa. (From Malhotra, 1975a. Copyright TRB. Reprinted with permission.)

specimens from the same concrete after accelerated curing. Such statistical methods should be used for this curve fitting that can take into consideration the random variations in both strength types (ACI Committee 228, 1996). The form of the equation of the curve is usually selected in an arbitrary manner as a formula with one independent variable, the accelerated strength, f_{ac}, the dependent variable being the normal 28-day strength, f_{st}. When the strength range is narrow, say, a couple of ksi, a straight line provides as good a fit as any curve (Figs. 1.6 and 1.10). In the case of

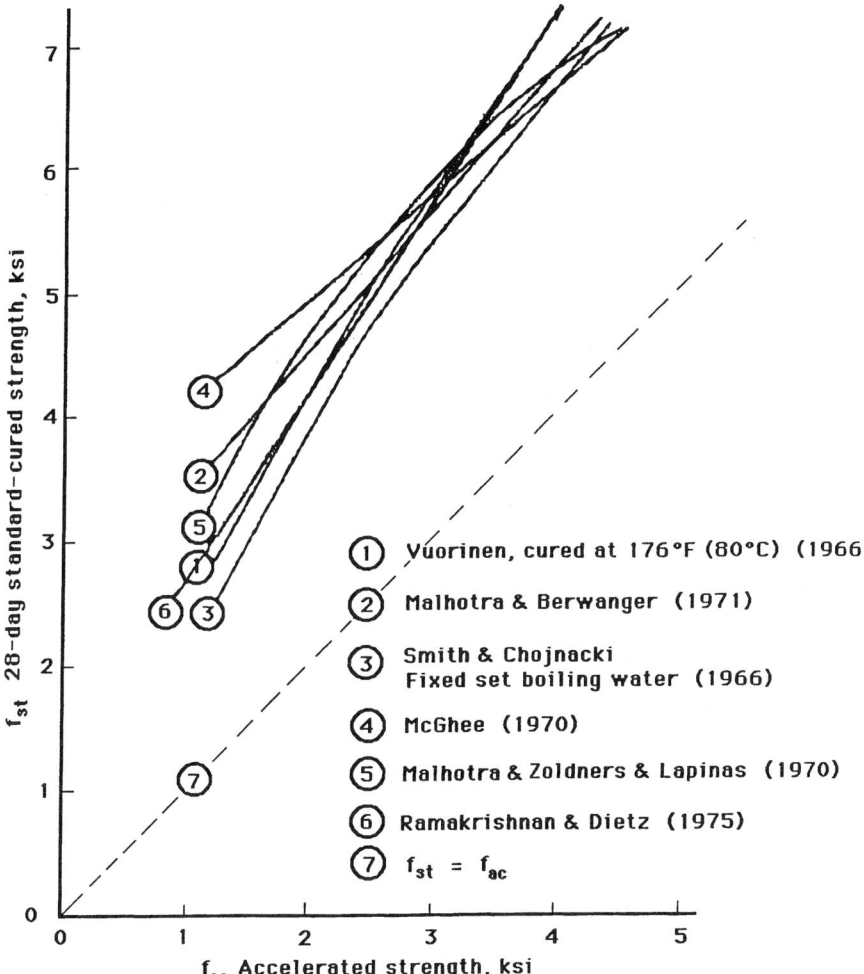

Figure 1.7 Recommended relationships between 1-day accelerated compressive strengths f_{ac} obtained by the boiling-water method and the corresponding normal 28-day strengths f_{st}. The average slope of the curves is approximately 1.4. 1 ksi = 6.90 MPa. (From Ramakrishnan and Dietz, 1978b. Copyright ACI. Reprinted with permission.)

wider range of strength, however, a curve, say a power function with a power of less than 1 (Fig. 1.9), fits better (Popovics and Pfeifer, 1992). The power function also has the theoretical advantage that it provides, correctly, zero for the predicted normal strength when the accelerated strength is zero, whereas the linear model provides an unrealistic predicted value at this point. It follows that in the case of a straight-line approximation, the slope of the line is steeper for low-strength concretes than for high-strength concretes. This is one reason why different investigators produced different linear formulas for calibration curves for the same accelerated testing method. Regardless of the form of the calibration curve, the f_{st}/f_{ac} ratio decreases with increasing concrete strength.

A refreshing novelty is that one can *derive* equations for the calibration curve from the strength versus water-cement ratio relationship instead of choosing the mathematical form of the curve arbitrarily. This is a more intelligent way of establishing a calibration curve because it is based on a fundamental law of concrete technology. Depending on what formulas we use for f_{st} versus w/c, different basic forms can be obtained for the calibration curve. This is illustrated below by two cases.

Example 1.1 The following calculation, based on Abrams' formula (Section 5.5), illustrates the derivation of a calibration curve for estimation of the normal 28-day strength from the strength obtained by an accelerated testing method.

A compressive strength experiment using standard methods (ASTM C192) was performed with a Type I portland cement and with different water-cement ratios. Abrams' formula fitted to the 28-day compressive strength results provided the equation

$$\log f_{st} = \log 11,417 - 0.7440 w/c$$

from which

$$w/c = \frac{\log 11,417 - \log f_{st}}{0.7440}$$

$$= 5.4539 - 1.3441 \log f_{st} \tag{1.1}$$

where f_{st} is in psi, w/c by mass, and the log is of 10 base. Specimens from the same concrete were also tested by accelerated testing procedure A (the warm-water method) of ASTM C684. From a graph similar to Figure 1.8, the relationship between the strength accelerated with procedure A and the water-cement ratio for this concrete is

$$\log f_{ac} = \log 24,567 - 1.8292 w/c \tag{1.2}$$

from which

$$w/c = \frac{\log 24,567 - \log f_{ac}}{1.8292}$$

$$= 2.400 - 0.5467 \log f_{ac} \tag{1.3}$$

But when the specimens come from the same concrete mixture, the water-cement ratios are the same and thus the right-hand sides of Eqs. 1.1 and 1.3 are equal. From this equality f_{st} can be expressed as

$$\log f_{st} = 2.272 + 0.4067 \log f_{ac} \tag{1.4}$$

that is,

$$f_{st} = 187 f_{ac}^{0.4067} \tag{1.5}$$

Thus, in this case the relationship between the normal 28-day strength and the accelerated strength is a power function. Equation 1.5 shows good agreement with the formula in Figure 1.9, considering the difference in the strength units.

Another kind of equation for calibration curves, with *two* independent variables, can also be derived from the strength versus water-cement ratio relationship (S. Popovics and Pfeifer, 1992). One of the independent variables is the accelerated strength, the other can be any other characteristic of the concrete that is related to the normal 28-day strength. Three such characteristics have been tried out experimentally as the second independent variable: the cement content, the water-cement ratio, and the normal 1-day strength. Out of these the inclusion of the water-cement ratio appeared to be the best. For instance, for the warm-water curing method of procedure A, the following equation provided a very good fit to the normal 28-day strength:

$$f_{st} = 4159 + 0.82 f_{ac} - 0.2593 w/c \tag{1.6}$$

where f_{st} = normal 28-day compressive strength of concrete, psi
$\quad\ \ f_{ac}$ = strength obtained with the warm-water curing method of procedure A, psi
$\quad w/c$ = water-cement ratio, by mass

The goodness of fit of Eq. 1.6 is indicated by $R^2 = 0.91$ for 16 different mixtures prepared with Type I, II, or III portland cement. This is a significantly better fit than that shown in Figure 1.8 for the three cement types without inclusion of the water-cement ratio. Equation 1.6 is derived in Example 1.2.

Example 1.2   A different form is obtained for the calibration curve if the starting equation is other than Abrams' formula. For instance, it is shown in Section 5.8 that the following formula describes the strength versus water-cement ratio relationship within wider limits of cement content than that using Abrams' formula (S. Popovics, 1990a):

$$f = a + bw/c + gc \tag{1.7}$$

where c is the cement content. The other symbols are the same as those in Example 1.1.

If the coefficients of this equation are known for the normal 28-day strength f_{st}, as well as for the accelerated strength f_{ac}, the equation of the pertinent calibration curve can be obtained as shown in Example 1.1: by expressing the water-cement ratios from the two equations and equating them. The result is

$$f_{st} = a_0 + a_1 f_{ac} + a_2 c \qquad (1.8)$$

That is, the calibration curve here is a straight line, the position of which, however, also depends on the cement content of the concrete. Note that the term c in Eq. 1.8 can be replaced by w/c with an appropriate modification of the parameters, the result of which is Eq. 1.6.

Application of Accelerated Curing

The results from accelerated curing can be used for practical purposes either by comparing them directly to appropriately specified strength values, as ASTM C684-95 points out; for estimation of the value of the normal 28-day strength; or for estimation of the batch-to batch variation of the normal 28-day strength, with a pertinent calibration curve. Although many such conversion factors, formulas, and calibration curves have been proposed, the most reliable way to establish a factor for use in converting from the results obtained with accelerated curing to the result of the normal 28-day strength is by trial mixes. Strength specimens are made from a batch of concrete with the intended materials and composition, then some of the specimens are cured in the accelerated manner for a specified number of hours, with the remainder cured in the normal manner for 28 days. Then the two are tested in the usual manner and the two sets of test results are compared. As long as the concrete ingredients and composition remain practically the same, this calibration curve or conversion factor will remain valid for the accelerated curing method used. If, however, there is any change, it is advisable to perform new trial mixes. Cements of different types may require different conversion factors (Fig. 1.8). What is more, it has also been shown that the conversion factor can be different for different brands of the same type of portland cement (S. Popovics, 1981a), possibly because of differences in the alkali contents of the various brands (Wills, 1978). Also, significantly different accelerated results have been obtained with concretes of the same nominal composition and test method in different laboratories, again perhaps due to differences in the cement used.

The duration of heat curing can be shorter at higher temperatures. Low curing temperatures have the advantage that the hydration products are similar to those produced at normal temperatures. However, the resulting low strength at early ages cannot reflect any possible lack of strength in the aggregate. In

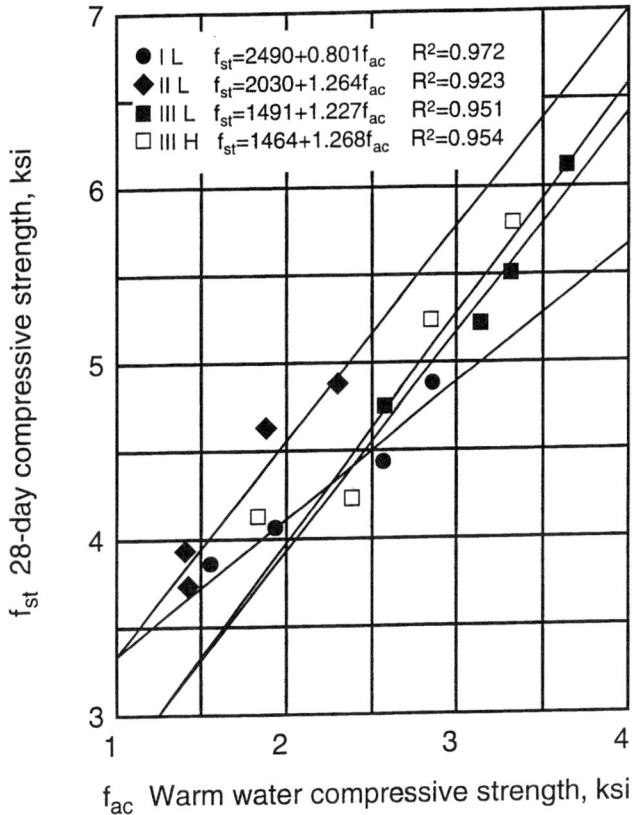

Figure 1.8 Graph of 28-day compressive strength f_{st} versus warm water compressive strength f_{ac}. The Roman numbers represent the cement type. Cement content: L, 500 lb/cu yd (300 kg/m³); H, 750 lb/cu yd (450 kg/m³). 1 ksi = 6.90 MPa. (From S. Popovics and Pfeifer, 1992.)

other words, a weak aggregate may be able to produce an accelerated strength of, say, 2000 psi (14 MPa) at an age of 24 hours but not the 4000-psi (28-MPa) strength that the same concrete is required to have with normal curing at an age of 28 days. Although the aggregate strength can be checked with separate tests, such as an additional accelerated strength test with concrete of lower water-cement ratio, it seems desirable to have accelerated curing methods that produce a relatively high percentage of the normal 28-day control strength. Higher curing temperatures may fulfill this goal, but this produces various problems, one of which may be the increased safety requirements for dealing with boiling water or other ambience of high temperature. Another, more fundamental problem is that the hydration products and their strength-developing properties obtained at high temperatures differ from those obtained at normal temperatures; the higher the curing temperature, the greater the differences. The practical consequence of this is that factors which cause changes in the 28-day normal strength can be over-, or underemphasized in the accelerated strengths. This is particularly true for high-temperature auto-clave curing; therefore, this curing has been shown unsatisfactory for general use for early determination of the strength potential (S. Popovics, 1992; Blaine et al., 1968b).

When the concrete strength specimens are to be submitted to elevated temperatures at an early age, such as in procedure B, the specimens are kept at normal temperature for several hours immediately after casting. After this they are heated up gradually by an external source of heat, such as hot air, steam, or water at atmospheric pressure. The specimens are kept at elevated temperature for a number of hours, followed by a cooling-down period, after which the specimens can be tested in the usual manner. As a rule, such tests provide a meaningful strength result approximately 24 to 48 hours after cast-ing. The type of mold used may have a significant effect when the boiling water method is used. In at least one test series, casting in steel molds resulted in accelerated concrete strengths 10 to 15% higher than in paper molds and 3 to 6% higher than in plastic molds (Ramakrishnan and Dietz, 1975, 1978).

In the *fixed set method* (Smith and Chojnacki, 1963), curing in boiling water starts at the final Proctor set. *Infrared radiation* (Spektor, 1956; Dani-low, 1960; Horimatsu, 1956) and *electrical heating* methods through wires (Horimatsu, 1956; Nikkanen, 1966; Martinet, 1963) or through the fresh con-crete itself (Ichiki, 1956) have also been tried as heat sources, but they have never become popular. Accelerated strength testing using the boiling-water method has also been extended to early splitting and flexural strength deter-minations (Villarreal Rivera, 1975; Malhotra, 1978; Ghosh et al., 1978).

Autogenous curing (procedure C) is special because this is the only method that does not use external heating for strength acceleration. It utilizes the heat generated by exothermic reactions during the hydration of cement (S. Popov-ics, 1992). Since the heat generated by a Type I portland cement is approx-imately 75 cal/g, the temperature rise in a concrete of 1:6 mix proportion by mass and 0.6 water-cement ratio by mass under adiabatic curing conditions is about 55°F (31°C) above the starting temperature 24 hours after casting. At 48 hours the temperature rise is 72°F (40°C) (Smith and Tiede, 1967). This means that placing a concrete specimen inside a well-insulated container immediately after its preparation is finished is alone sufficient to provide accelerated strength development comparable to that achieved by applying external heat. This method has the advantage of simplicity in procedure and equipment. The disadvantage is that it needs more than 48 hours, which is about twice as long as required by procedures A and B. Also, both the total amount and the rate of early heat development, thus the acceleration of strength development, can be influenced by a number of factors, such as the type and quantity of the cement, water-cement ratio, original temperature of the concrete, and presence of admixture (Ramakrishnan and Dietz, 1978; Bi-saillon, 1978). Thus the accelerated strength versus 28-day normal strength ratio can vary within wide limits for different concretes (Philleo, 1980).

A typical example for the relationship between the 2-day accelerated strengths obtained by the autogenous curing method and the corresponding normal 28-day strengths is presented in Figure 1.9. Here again, under other circumstances, different calibration curves (i.e., relationships) may be valid, as shown in Figure 1.10.

Application for Checking the Batch-to-Batch Variation of the Strength of a Concrete

An interesting feature of the presented calibration curves is that although they are different even for a given method of accelerated curing, their slopes are more or less identical. This approximately common (average) slope for the boiling-water method represents a 1400-psi (10-MPa) increase in the normal 28-day compressive strength per 1000 psi f_{ac} with good approximation in all series in Figure 1.7. So it seems that there is better agreement between *changes* in the accelerated strengths and changes in the normal strengths obtained by various investigators on various concretes than between these strengths of a concrete. Thus, perhaps the most reliable practical application of the accelerated curing methods is for the estimation of the batch-to-batch variation in strength of the concrete.

The more or less identical slopes of the various calibration curves in Figures 1.7 and 1.10 also seem to indicate that the boiling-water method cannot reliably distinguish changes in the normal 28-day compressive strength below 5% of the average strength, and that the autogenous curing method cannot distinguish similar changes below 4% since the maximum precision of accelerated curing methods is 3.6%.

Accelerated Methods Other Than ASTM Methods

The use of external heat sources for accelerated determination of control strengths began more than 60 years ago (Gerend, 1927; Patch, 1933). Since then a wide variety of accelerated test methods have been recommended by various investigators. The highest temperature to be applied is usually that of a boiling-water bath (Malhotra and Zoldners, 1975a; 1969, Sanchez-Trejo and Flores-Castro, 1978; Akroyd, 1961a; Lapinas, 1968). When lower temperatures are used, the curing ambience is water or steam, although dry hot air has also been used with sealed molds (King, 1955; Vuorinen, 1966). The autogenous curing method for accelerated testing was developed by Smith and Tiede (Smith and Tiede, 1967). More recent experiments with such methods have also been reported (Bisaillon et al., 1975; Bisaillon, 1978; Bickley, 1978). Computer programs for the development of numerical relationships for accelerated versus normal design strengths are available (Campbell, 1983).

A pragmatic disadvantages of all three ASTM methods is that overtime and week-end testing may be necessary. Therefore, modifications of these methods have been used in some instances to make tests conform with normal working hours. For such cases, that is, when the delay times and/or the heat

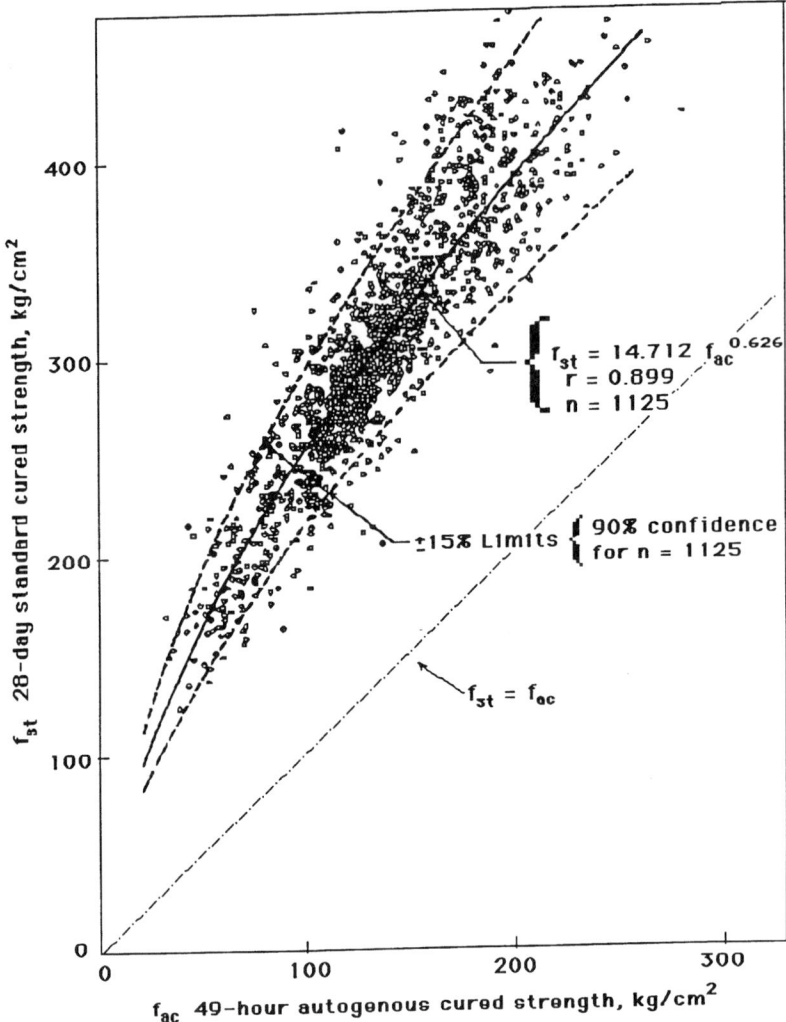

Figure 1.9 Experimental data of the relationship between 2-day accelerated compressive strengths obtained with autogenous curing and the corresponding normal 28-day strengths for a group of concretes using several brands of Type II cement. 1 kg/cm^2 = 14.2 psi = 0.098 MPa. (From Mena-Ferrer, 1978. Copyright ACI. Reprinted with permission.)

treatment and/or the age of testing are not kept constant, the strength of concrete under standard wet curing for any age (or maturity) can still be estimated from the early accelerated strength and an ASTM formula (C918-93) with the help of the maturity concept (Section 1.3), as follows (Hudson and Steele, 1971):

$$f_M = f_m + b(\log M - \log m) \tag{1.9}$$

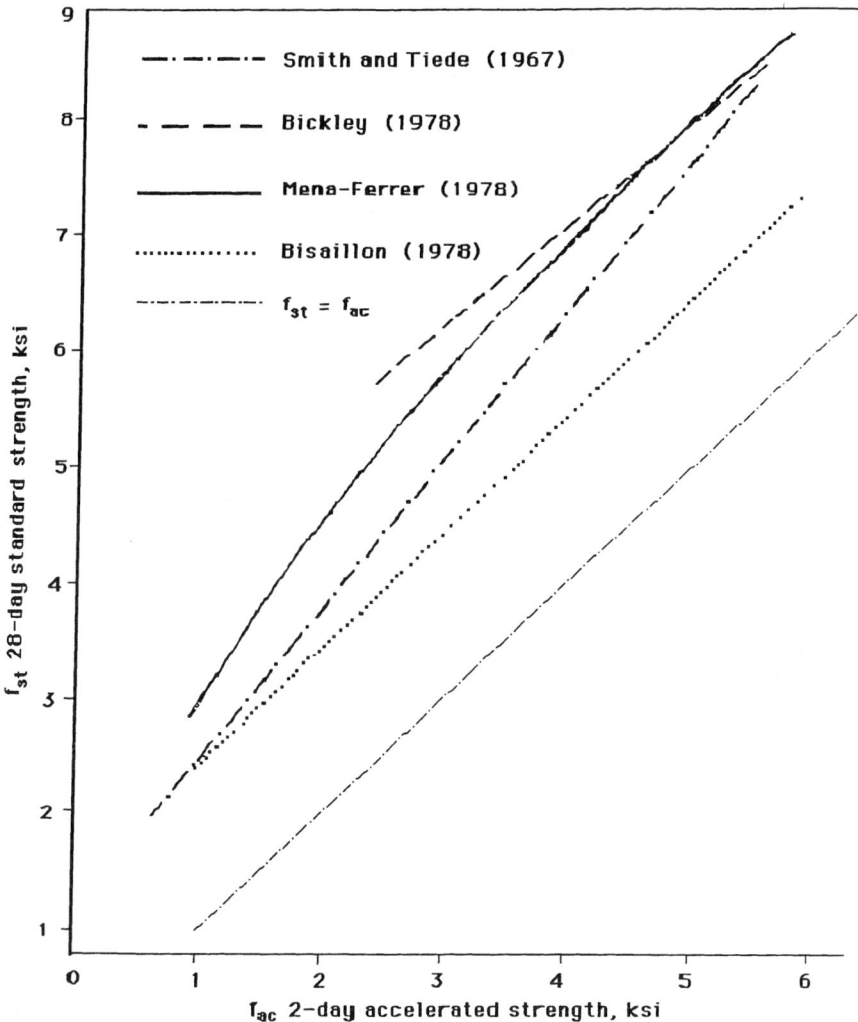

Figure 1.10 Recommended relationships between 2-day accelerated compressive strengths obtained with autogenous curing f_{ac} and the corresponding normal 28-day strengths f_{st}. 1 ksi = 6.90 MPa.

where f_M = predicted potential concrete strength after standard wet curing at maturity M corresponding to a desired age, psi (or MPa)

f_m = measured early compressive strength at maturity m, psi (or MPa)

M = degree-hours of maturity of specimens under standard wet curing up to the age for which the strength estimation is made, such as $73.4 \times 28 \times 24 \approx 50{,}000°F \times hr$

> m = degree-hours of maturity of specimens at the time, or early test after accelerated curing, such as accelerated curing temperature $\times$ age in hours + m'
>
> m' and b = empirical factors that are a function of the composition of the concrete and admixtures, the nature and duration of the heat treatment, the delay periods before and after the heat treatment, and so on; $m' = 0$ when the curing temperature is the standard 73.4°F (23°C)

For three portland cements used for paving purposes, four cement contents ranging from 376 to 682 lb/cu yd (222 to 402 kg/m³), and for an accelerated curing with almost-boiling water, Hudson and Steel obtained values of $m' = 1840$°F $\times$ hr. The value of b increased from 1900 for the low cement content to 2210 for the high cement content for one of their cements. This means that Eq. 1.9 gives misleadingly high strengths if the concrete is deficient in cement of amount unknown to the calculator, for instance, due to a mistake in batching. This overestimate occurs despite the resulting low value of f_m. Note that the values of b and m' are also a function of the type and even the brand of the cement used, and the type of heat treatment applied. Thus the appropriate values should be established statistically for specific cases by a designed series of trial mixes (Hudson and Steele, 1971). The details of such a method, that is, using Eq. 1.9 for the estimation of later age strengths from early age compression test results, are specified in ASTM C918-93 (Smith, 1984).

Comparison of Methods

On the basis of comparative tests performed on the three ASTM methods, Ramakrishnan and Dietz (1975, 1978) and Bauer and Olivan (1978) prefer the boiling-water method because it is simple, does not require extra temperature control, is less sensitive to changes encountered on projects, shows satisfactory correlation with normal 28-day control strength, and is quick. On the other hand, Abdun-Nur (1978) prefers those accelerated methods that use lower temperatures, close to the 100°F (37°C), especially autogenous curing, because these do not have safety hazards, and the hydration products are closer to those developed at normal temperatures. Philleo (1980) agrees that the autogenous curing method is the most convenient but believes that it is also the least reliable. Procedure A is technically good for both Type I and II cements with or without pozzolan addition (Lamond, 1979); therefore, it was adopted by the U.S. Corps of Engineers for control on major civil works projects (K. Mather, 1986). In the British Standard BS 1881 also, a modified version of the warm curing method has been accepted. It should be noted, however, that this method requires a thermostatically controlled water bath and a testing machine at the job site. Wills feels that when the accelerated testing is limited to a single laboratory–procedure–materials set of conditions, it results in an assessment of concrete quality at 1 or 2 days of age as reliable as that obtained after 28 days of normal moist curing (Wills, 1978). Thus which accelerated testing method is to be used to best advantage appears to depend on the circumstances prevailing on the construction site. The fixed set method has not become popular because of the additional complication of needing time-of-set determinations.

In brief, the use of 1- or 2-day strengths obtained by accelerated curing is not quite as simple as the 28-day strength for the estimation of control strength or the strength-developing capacity of concretes. Once the target strength has been established for an early accelerated strength for a given set of circumstances, 1- or 2-day accelerated strengths can be used for direct evaluation of the control strength of a concrete just as well as the normal 28-day strengths (Lamond, 1983). To put it another way, once a conversion factor or calibration curve has been established for an accelerated strength for a given set of circumstances (e.g., by the method specified in ASTM C918-93), the normal 28-day strength, and especially, batch-to-batch variation in this strength, can be estimated reliably. However, the magnitude of accelerated strengths may depend not only on the strength potential of the concrete but also, to varying degrees, on the nature of the concrete-making materials used, the concrete composition, the original temperature of the concrete, and the accelerated curing method. Therefore, if there is any change in the circumstances (i.e., in the concrete-making materials, etc.), it is prudent to recheck the target strength and conversion factor, respectively, for the new conditions by trial mixes. Nevertheless, the intelligent use of accelerated strength tests should serve well for control testing (Malhotra, 1981). Indeed, at an increasing rate, concrete producers and contractors are depending on such tests to forestall complications that occur with substandard 28-day strength results.

Prediction of Strength Potential from Early Strengths of Normal-Cured Specimens

Although it is possible to test a standard wet-cured cylinder or beam at an early age, say 1 day, this and its relation to the normal 28-day strength is very strongly influenced by the type and fineness of the cement used as well as the composition of the concrete. In addition, the measured 1-day strength is such a small fraction of the strength at 28 days that any small error, even the normal variability of the strength test, usually makes the 1-day strength an unreliable parameter for a judgment of the strength potential or control strength of a concrete. Thus only under special conditions may an early strength be used for estimating the strength potential of concrete. Such conditions are (1) when the 1-day strength is determined in such a manner that the result is highly reliable, and (2) when the composition of the concrete is kept unchanged. Prediction of the 28-day strength from the 7- or 14-day strength is more reliable (Hudson and Steele, 1971; Date and Schnomeier, 1985), but the practical significance of such prediction is minimal.

On the other hand, the fact that concrete specimens are tested without heat treatment is economically advantageous because it can accomplish savings in time and equipment costs in many cases. Therefore, methods have been proposed to improve the applicability of the early strength obtained under standard wet-curing methods for estimation of the strength potential of concrete.

One attempt to make this method more suitable for practical applications is based on the *maturity* concept (Section 1.3). A method is described in ASTM C1074-93. Equation 1.9 could also be used for prediction of the normal 28-day strength since the maturity method is quite good at temperatures around the standard 73.4°F (23°C), in which case $m' = 0$. However, this formula is unsafe for concretes deficient in cement, as has been pointed out. Another formula for the prediction of a normal 28-day compressive strength of concrete from an early strength test of standard wet curing is (Hudson and Steele, 1975)

$$f'_c = a \frac{f_m^b}{m^c} \tag{1.10}$$

where f'_c = estimated normal 28-day compressive strength of concrete, psi (or MPa)

f_m = compressive strength of specimens after standard wet curing tested at an early age and having maturity m, psi (or MPa)

m = degree-hours of maturity at the time of the early tests, °F × hr (or °C × h)

a, b, and c = empirical constants which are independent of the cement content, age, and maturity at testing within wide limits but may be a function of the type and brand of the cement used, water-cement ratio, admixtures, units used, etc.

For a Type I cement and strengths expressed in psi, Hudson and Steele (1975) established values of $a = 965$, $b = 0.75$ and $c = 0.5$ for Eq. 1.10. For other conditions these parameters should be determined from a series of trial mixes. They also found that strengths predicted by Eq. 1.10 from early strength tests of standard wet-cured specimens were as accurate as those based on specimens heat-treated by a $3\frac{1}{2}$ hour immersion in an almost boiling water, regardless at what age or maturity the early strength specimens were tested. Other formulas are also available for the concrete strength versus maturity relation (Carino, 1991a).

Another, different approach for generalization of the prediction of the compressive strength of concrete for various ages from the early strength of specimens under standard wet curing is provided by the formula (S. Popovics, 1981a):

$$f_t = f_j \frac{100 - C_3 e^{b_1 t} - (100 - C_3) e^{-b_2 t}}{100 - C_3 e^{b_1 j} - (100 - C_3) e^{-b_2 j}} \qquad (1.11)$$

where f_t = estimated normal strength of concrete at the age of t days, psi (or MPa)

t = age of concrete for which the prediction is required, days

f_j = experimentally obtained strength of concrete specimens after standard wet curing at the early age of j days, psi (or MPa)

C_3 = C_3S content of the cement, percent

b_1 and b_2 = rate parameters that are independent of the C_3S and C_2S contents of the cement and the strength as well as the age of the concrete but may be a function of the C_3A content and fineness of the cement, curing temperature, minor constituents, water-cement ratio, admixtures, type of strengths, test method, and any other factor that influences the strength development, 1/day

Figure 1.11 compares the measured values and corresponding values calculated from the measured 1-day normal strengths (i.e., $j = 1$) with Eq. 1.11

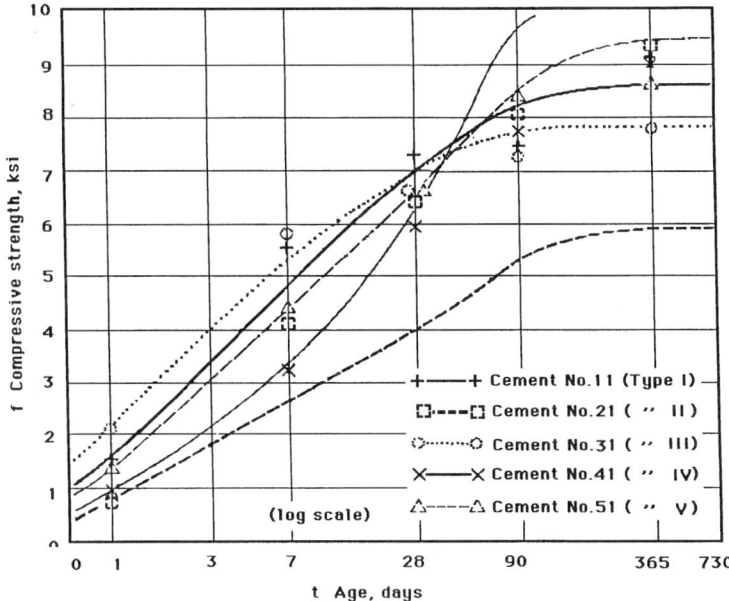

Figure 1.11 Prediction of compressive strengths of concretes from the 1-day measured strength of normal curing for various cement types. Points represent experimental results reported by Klieger; lines represent Eq. 1.11. 1 ksi = 6.90 MPa. (From S. Popovics, 1981a. Copyright ACI. Reprinted with permission.)

for the compressive strength of 6-bag/cu yd = 564-lb/cu yd (335 kg/m³) concretes made with the five standard types of portland cement. Another comparison using standard mortar and $j = 3$ days is presented in Chapter 3. Statistical analysis of the goodness of fit, calculations of the b_1 and b_2 parameters, and other details for this method of prediction are also given in that chapter. It can be seen, however, even from this abbreviated presentation, that the goodness of fit of Eq. 1.11 is quite good except for cement 21, the probable reason for which is that something was wrong with the experimentally obtained results of this cement. This shows again that the reliability of prediction of the strength potential of concrete from early tests of standard wet-cured specimens depends largely on the use of correct and uniform procedures of making, curing, and testing the specimens.

Although Eq. 1.11 also contains two parameters (b_1 and b_2) that should be determined by trial mixes, it is quite general because it is valid for all commercially available portland cements and for ages from 1 day through 1 year. A disadvantage of this formula is that not only should the early strength be measured with special care but determination of the compound composition and fineness of the cement used should be accurate enough.

Assurance of the Needed Strength Potential by Assuring the Proper Concrete Composition

While it is advantageous to obtain the control strength of a concrete at the age of not later than 48 hours instead of 7 or 28 days, it would be far better to know the strength potential before the concrete is placed in the forms, or even before it is discharged for delivery (Mather, 1976). This would eliminate the placement of substandard concrete in the structure, which is important since even 24 hours (i.e., the minimum time needed for accelerated strength testing) is more than enough to allow the concrete to set and harden in the form. Immediate establishment of the strength potential of a batch of concrete would also alert the concrete producer to find the source of the irregularity in strength and eliminate it promptly: for instance, by changing the composition of the concrete or checking the batching and mixing procedures.

The only method that has been tried out for such "immediate" estimation of the strength potential of a concrete batch is based on the relationship between the strength and composition of a concrete. When good cement, aggregate(s), admixture(s), and water have been used, then according to this method, the required control strength of a concrete is assured if its actual composition is close enough to its specified or intended composition and if the concrete is well compacted. Consequently, it is enough to check the composition and air content of the batch to be sure that the strength potential of this concrete is satisfactory.

The most important factor in the concrete composition from the point of view of strength potential is the water-cement ratio and then the air content (Chapter 5). At present there is no reliable test method for direct measurement of the water-cement ratio, but there are methods for the determination of the cement content and water content of the batch separately, from which the water-cement ratio can be calculated with limited reliability. In an ideal case the batching equipment would control itself from feedbacks and correct or reject the inadequate batch before discharge.

Mather (1976) has gone one step further by stating that there are already batching units available so reliable that one may know with sufficient confidence the amounts of each size of aggregate, cement, water, and each admixture in the mixer. This implies that no further tests of concrete is needed except for the determination of air content to have adequate confidence that the strength requirement will be met.

Practical experience with estimation of the strength potential of concrete batch from its experimentally established composition by "rapid" methods has generally been fair (Howdyshell, 1975; Bickley and Mukherjee, 1979). For instance, Howdyshell reported that the cement and water contents determined by the Kelly–Vail technique when used in conjunction with an air content test have estimated well the strength potential of a concrete batch (Howdyshell, 1978). On the other hand, a comprehensive examination of the most popular seven rapid test methods which determine either the cement or the water content of fresh portland cement concrete concluded that:

1. None of the test methods investigated could determine either the cement content or water content without sizable bias.
2. None of the test methods investigated were better than any of the other methods investigated for determination of the cement and water content.
3. The presence of mineral or chemical admixtures increased the bias of all the test methods investigated for every concrete mixture (Tom and Magoun, 1986). In other words, the rapid estimate is not as accurate as the results of the ASTM methods for accelerated strength testing, but if it is selected properly, it is sufficiently accurate to be meaningful for many practical purposes, and most important, the result is obtained in 15 minutes instead of 24 hours. Another advantage of this rapid method is that important properties other than the compressive strength, such as the durability, can also be appraised from the knowledge of the actual composition and air content of the concrete.

Note, however, that analysis of a fresh concrete may produce occasionally misleading results. Such a case is, for instance, when the analysis cannot clearly distinguish between cement and finely divided mineral admixture in the fresh concrete and, say, fly ash was batched mistakenly for the cement or a significant portion of it. The strength potential of such concrete is poor, of course, despite the fact that the analysis of the concrete composition may indicate an adequate amount of cement. Therefore, in important constructions Philleo would prefer the performance of both the analysis of the composition of the fresh concrete and, as an extra safety measure, the accelerated testing of strength.

The strength potential of a concrete for a given age is also influenced by factors other than the water-cement ratio. Some of these effects can be expressed mathematically (Malhotra, 1963; S. Popovics, 1967b); nevertheless, the best way to establish the needed composition of concrete for a specified control strength is again by trial mixes. Whenever there is any significant change in the concrete composition, performance of new trial mixes is highly recommended (Mena-Ferrer, 1978).

1.4   INDIRECT DETERMINATION OF CONCRETE STRENGTH

In-Place Testing of Concrete Strength

Due to the difficulties of direct determination or estimation of the compressive strength of concrete in structures (Section 1.2), considerable efforts have been made over the past 40 years to develop pertinent other methods, called in-place or in-situ tests (Zoldner and Soles, 1984; ACI, 1988a). These methods are nondestructive, or practically nondestructive; that is, such a test does not destroy the concrete, at least not to the extent that would damage the structure. It is apparent, however, that these "test" the compressive strength indirectly. They measure certain, more easily measurable properties other than the compressive strength of concrete in place; then from the results obtained an estimate of its compressive (or other) strength may be obtained (Gaede, 1941). That is, the power of a nondestructive test method for estimating the compressive strength of a concrete depends (1) on the accuracy of the measurement, and (2) on the reliability and limits of validity of the empirical relationship between the measurement and the compressive strength of the concrete. The unfortunate state of the art is that this power is not enough; that is, none of the presently known nondestructive methods can yield concrete strengths with a satisfactory reliability. For instance, according to the National Science Foundation (1993), the state of the art of in-situ strength assessment "is primitive and unreliable, prompting conservative, costly decisions." The specific reason for this is that the effects of various factors (composition, moisture content, etc.) on the concrete strength differ from the effects of the same factors on the measurements obtained by nondestructive methods. Consequently, the general predictability of the compressive (or other) strength from the measurements is uncertain.

However, these methods are suitable for determination of the relative strength of the concrete, strength changes in various spots of the same structure, or relative strengths in different structures, as long as the concrete composition, age, and curing are essentially the same. In other words, they provide inexpensive means of checking concrete uniformity (Tomsett, 1981). Due to the complex nature of concrete, the analysis and interpretation of the measurements are not easy. They must be carried out by a specialist in the applied technique, although many of these nondestructive methods are quick and simple enough to be performed in the field by a technician. The report by ACI Committee 228 (1996) is a good guide to assist the engineer in planning, conducting and interpreting the results of in-place tests, including statistical analysis of the data and development of calibration curves.

The above-mentioned uncertainties clearly show that none of the presently known nondestructive methods can be regarded as a substitute for standard compressive tests but rather as convenient methods for (1) comparing a concrete to another but essentially similar concrete, (2) determining the uniformity of concrete in the structure, especially the weak spots for further investigation if necessary, and (3) monitoring changes in strength with time at a given point of a structure. Under well-controlled conditions, they may also be used for determination of the safe removal time for forms or the earliest time at which posttensioning may take place.

When this kind of information is not enough and strength estimates are needed, an empirical correlation, that is, a *calibration curve,* should be established for a concrete between its strength and the result of the nondestructive test in question. Although manufacturers do provide calibration curves with their in-situ testing equipment, these curves do not appear satisfactory for general use. That is, various concretes typically require different curves. Thus the accuracy of strength estimation without the availability of the specific calibration curve is not good enough for most practical purposes. It will be shown, however, that more reliable estimates can be obtained with nondestructive methods for determination of the *variations* of the strength because the slopes of the calibration curves for various concretes show better agreement than the curves themselves.

Tests that utilize concrete cured under standard laboratory conditions for the development of a calibration curve may be at variance with field tests performed on concrete with a dry surface and a formed or troweled finish. Note also that the variability of all in-situ strength results is higher than that of the standard cylinder compression test (Carette and Malhotra, 1984).

In-place tests can be used to estimate concrete strength during construction, so that, for example, that operations can be performed safely. They can also be used to estimate concrete strength during the evaluation of existing structures. There are differences between the approaches for these two applications, although these are usually minor. For example, for new construction, a calibration curve is usually established empirically in a laboratory. For existing structures, the calibration curve is usually established by performing in-place tests at selected locations in the concrete under investigation and determining the strength of cores taken from adjacent locations (ACI, 1996).

We describe below the principles of selection of locations of nondestructive tests and coring for condition evaluation of concrete structures as well description of several nondestructive test methods. Details of these methods are presented in British Standard BS 6089 and elsewhere in the literature (ACI,

1968; 1996; Malhotra, 1976, 1984, 1991a; Whitehurst, 1966; RILEM, 1954; Borjan, 1981; Manning, 1985; Uzhpolevitchius, 1982; Luzhin et al., 1985; Agbabian and Masri, 1988, Teodoru, 1989; Nasser and AL-Manaseer, 1987; Galan, 1990; Aktan 1995; Schickert and Wihhenhauser, 1995). Keep in mind that careful and intelligent visual inspection must supplement all nondestructive testing.

Selection of Test Locations for In-Place Testing of Concrete Strength

The location of in-place tests in new construction is usually preplanned but not preplanned in existing structures. In both cases sufficient test locations should be provided to enable adequate estimation of the concrete strength. Informative data for the number of locations and number of tests per location are given in ACI (1996) and in ASTM E105 and E122.

In the case of *new construction,* access should be provided to the fresh concrete. This is usually done by cutting a hole in the mold wall. In selecting test locations, critical points should be established in the structures, although in certain cases, such as slabs, the test locations are distributed in a regular pattern. The selection of test locations in an *existing structure* is more complex. Here a good sampling plan is quite helpful. ASTM C823, "Practice for Examination and Sampling of Hardened Concrete in Constructions," and BS 6089, "Guide to assessment of concrete strength in existing structures," provide guidelines for this.

It has been proposed that where assessment of the load-carrying capacity of an *existing structure* is the principal aim, the tests should be performed, ideally, at points where likely minimum actual concrete strength and maximum stress from loads coincide (Bungey, 1982; Concrete Society, 1976; ACI, 1982). Under practical conditions, however, this ideal goal can rarely be achieved; besides, there may be considerations other than the residual strength of concrete in the structure that can influence the selection of test locations.

One of such additional considerations is in many cases determination of the cause and extent of the deterioration of concrete in a structure. The success of any repair or rehabilitation work depends first on the accuracy with which the cause and extent of the damage have been established. Only after a specific conclusion as to the cause and extent of the damage has been reached can a rational selection be made among alternative repair or rehabilitation strategies (Stowe and Thornton, 1984).

Keeping this in mind, the factors that may control selection of the locations of tests for condition evaluation of concrete structures can be divided into two classes, technical and practical (S. Popovics and J. S. Popovics, 1996a). *Technical* considerations are as follows:

1. The locations should be distributed in such a manner that a qualified person could form a reliable opinion from the test results concerning the condition of the *whole structure* or *structural element* in question. This evaluation includes:

- The nature and cause(s) of the damage
- The extent of the damage
- The locations and dimensions of the damaged as well as undamaged areas
- The strength and uniformity of the undamaged portions of the concrete

2. The goals mentioned under item 1 can be achieved if cores are taken both from certain damaged and undamaged portions of the structure. Test results of the *undamaged* portions are needed for judging:
   - The integrity and serviceability of the structure
   - The nature and extent of damage of the concrete throughout the structure
   - The uniformity of the sound concrete throughout the structure

3. The *damaged* (or sound) portions of the structure may be identified before testing by:
   - Visual inspection
   - Signs of seepage through the concrete
   - Sounding
   - A combination of these

4. Locations for strength determinations should be ideally at points where:
   - The concrete strength is probably the minimum regardless of the stresses.
   - The stresses are probably the maximum.
   - The concrete appears to be sound.

5. Ideally, tests should also be performed at points where the ratio estimated stress to estimated residual strength of concrete is highest. However, in many cases simultaneous estimation of stresses and strengths at various locations is quite uncertain; thus this criterion is applicable mostly in simple structures or structural elements.

6. Only a few nondestructive methods can be used for establishment of the cause and nature of the damage in the concrete. If this establishment is needed, it may be advantageous to take cores from the structure. Cores taken for strength determination are usually suitable for this purpose, too. In special cases locations should be selected where the damage seems particularly informative.

7. The same can be said when the thickness of the damaged outside layer or inside core of a concrete should be determined.

8. It should also be considered in the location selection that the upper portion of a wall or column regularly has lower strength than the lower portion of the same structural element.

Selections of test locations may also be controlled by *practical* considerations. It is not always possible, at least not economically, to select the test locations in the practice that fulfill all the technical requirements discussed above. Practical factors that control the test locations are access and safety.

Limitation by *access* means that the horizontal and/or vertical position of a location in or on the concrete structure and/or the inclination of a surface may be such that obtaining a usable measurement at that point is practically impossible under the prevailing conditions.

Limitation by *safety* considerations means not so much the safety of the personnel during testing but rather the potential damage the test may produce to the structure that may endanger the stability of the structure. This limitation is especially important in core taking. For instance, if a structural member selected on the basis of technical considerations for coring is slender, core cutting may impair future performance. In such a case cores should be taken at the nearest practical location. Also, cutting through reinforcing bar(s) should be avoided for several reasons, including safety. Use of a covermeter to locate reinforcement before cutting can eliminate this difficulty. When the purpose of core taking is the establishment of a valid calibration curve, ACI recommends selecting at least six to nine test locations and five replicate readings for this purpose (ACI, 1994).

In conclusion, the selection of test locations for condition evaluation of concrete structures is an important activity. If it is not done properly, the results obtained from the tests will not provide the information needed for the reliable evaluation of the condition of a concrete structure, regardless of how carefully the actual testing is performed otherwise. In other words, nondestructive tests may provide useful information about (1) the strength and uniformity of the sound concrete, and (2) locations, extent, and size of deterioration of the concrete throughout the structure but only if the core locations are selected properly.

Surface Hardness and Rebound Methods

Malhotra and Carette (1980a) characterize *surface hardness methods* as the indentation types that consist essentially of impacting the concrete surface in a specified manner and measuring the size of indentation. Such methods have long been used and standardized for metals (Brinell: ASTM E10; Rockwell: ASTM E18; Vickers: ASTM E92).

Although it is known that the results of the surface hardness methods are controlled primarily by the plastic deformability of the hardened cement paste on the surface (Novgordsky, 1973), there is little theoretical relationship between the strength of concrete and its hardness so measured. Nevertheless, within limits, an empirical correlation (i.e., a calibration curve) can be established for each concrete between strengths and data obtained from hardness tests (Facaoaru, 1976). The accuracy of such strength estimates is within 20 to 30% (RILEM, 1965), which can be improved when the mix proportion, type of coarse aggregate, age, and moisture condition of the concrete under test are kept essentially unchanged. On the other hand, if the concrete properties near the tested surface differ significantly from the properties in deeper portions (due to moisture, carbonation, damaged surface, etc.), the error of estimates can be greater. Therefore, it is important to consider the appearance of the tested concrete surface in the interpretation of the hardness results.

The standard deviation of the individual hardness results about the average is quite large and represents the combination of the nonuniformity of the concrete surface tested and the testing errors. This includes the variance resulting from the circumstance, whether the hammer hit a coarse aggregate particle, or a spot of mortar, or a pore. Therefore, the greater the maximum particle size and/or lower the cement content, the greater is the variation in readings. Thus, under special testing conditions, such calculations as the standard deviation or coefficient of variation can characterize the concrete uniformity and/or the magnitude of testing error. As a result of this and the ease and low price of the performance, usually quite a few (up to 20) readings are taken at short distances from one another, and the average of these is used for an estimation of strength.

Malhotra and Carette (1980a) describe three of these methods:

1. The Williams testing pistol, used since 1936, consists of a metal ball used as an indentor. The diameter of the impression made by the ball is measured; from this the curved surface area of the indentation is calculated, which is related, inversely, to the concrete strength.
2. The Frank spring hammer is controlled by a spring whose top can be fitted with balls of different diameters. The diameter and depth of indentation are measured and correlated with the compressive strength of concrete. Similar equipment known as the Zorn hammer has been standardized in Germany as DIN 4240.
3. The Einbeck pendulum hammer is at the end of an arm that is installed on a leg. The leg is held horizontally against the concrete surface under test, and the pendulum is released to strike the concrete with its spherical head. The diameter and depth of the resulting indentation are measured and correlated with the compressive strength of concrete. This hammer should be used on vertical surfaces only.

Another pertinent method, the Poldi–Waitzman hammer, utilizes a modified form of the Poldi hammer developed for metals. Here the impact energy does not have to be uniform or specified. Thus a common claw hammer can be used to produce the impact. However, two steel balls are applied: a smaller one that makes the impression in a steel specimen as a control, and simultaneously, a larger one for the concrete. The hardness of concrete is calculated as

$$K = K_s \frac{d^2}{D^2} \tag{1.12}$$

where K = hardness of concrete
  K_s = hardness of the steel control
  d = diameter of the impression in the steel
  D = diameter of the impression in the concrete

Here again the concrete strength can be estimated from K by an empirical relationship developed for the concrete in question. Further details concerning various indentation devices have been described in the technical literature (Gaede, 1952, 1954a, 1954b, 1957; Vassitch, 1954), including DIN 4240.

Among the *rebound methods,* and probably among all the methods for in-place testing of concrete strength, the various types of Schmidt hammer (Schmidt, 1954) are the most popular. This is due not so much to its accuracy but rather, to the ease of using it and to the low price of the equipment. The principle of the rebound methods is similar to the principle of the surface hardness methods. The equipment is a steel mass (hammer) with a smooth, flat end fixed to a spring in a tubular housing. The unit is usually less than 10 in. (250 mm) long. When the spring is pushed against the surface to be tested, it retracts completely, which causes the hammer to strike the concrete. Then the hammer rebounds, taking a rider with it along a linear glide scale. The rider can be held in position to allow a reading to be taken. This reading, which represents the magnitude of the rebound, is called the *rebound number.* Grinding may be necessary to provide a smooth concrete surface for running the test, and prewetting is suggested as a means of minimizing the effects of drying and carbonation. Other details of the test method are specified in ASTM C805 and in BS 1881, Part 202. The U.S. standard contains recommendations for the selection of test surfaces, interpretation of results, and details of the test report. It also states that the precision of the single-specimen, single-operator machine is 2.5 units. The Schmidt hammer has also been manufactured in a waterproof housing, which makes it applicable for underwater testing. The relationship between the rebound numbers obtained by two different types of Schmidt hammer is not necessarily linear (Borjan, 1981).

The Schmidt hammer method is based on the principle of elastic impact (Gaede and Schmidt, 1964; Akashi and Amaski, 1984). The kinetic energy of the hammer originates from its spring and is, at the time of the impact,

$$E_k = 0.5ch^2 \qquad (1.13)$$

where E_k = kinetic energy of the hammer
  c = spring constant
  h = length change of the spring

After impact the hammer rebounds to h'; that is, the kinetic energy is used up by the extension of the spring. Therefore, the rebound energy is

$$E'_k = 0.5ch'^2 \tag{1.14}$$

and the relative loss of energy is

$$\frac{E_k - E'_k}{E_k} = \frac{h^2 - h'^2}{h^2} = 1 - k^2 \tag{1.15}$$

where $k = h'/h$. In the case of completely elastic deformations $k = 1$; for completely inelastic deformations, $k = 0$. The rebound number R then is

$$R = 100k \tag{1.16}$$

Although there seems to be no advantage in taking more than one reading on a spot, it may be noted that the rebound number generally increases with successive repetitions of the test on the same spot (Keiller, 1982). Note also that the rebound number is influenced by the direction of impact because the gravity force on the hammer is added vectorially to the spring force. Correction factors for different impact directions are provided by the manufacturers.

The rebound number is influenced primarily by the elastic characteristics of the surface layer of about 1 in. (25 mm) of the concrete (Gaede and Schmidt, 1964). Whereas there are theoretical, although approximate numerical relationships between strengths and elastic properties of certain idealized materials (Nicholls, 1976; Akashi and Amaski, 1984), these relationships are not applicable to concrete. The main reason for this is that the, say, modulus of elasticity of a concrete is controlled primarily by the modulus of elasticity of the aggregate, but its strength is not. Therefore, such theoretical relationships serve only as a basis for the rule of thumb that concretes with higher modulus of elasticity, that is, with higher rebound number, are expected to be stronger. It has also been noticed that dry and/or carbonated concretes give higher rebound numbers than wet and/or noncarbonated concretes of the same compressive strength (Petersen and Stall, 1955). Troweled surfaces also provide higher rebound numbers than screeded or formed finishes. Nevertheless, within limits, an empirical quantitative correlation can be established for each concrete between strengths and data obtained by the rebound test (Facaoaru, 1976).

An example is shown in Figure 1.12. It can be seen that the accuracy of the strength estimation in this case (i.e., under strict laboratory conditions) is about $\pm 15\%$ of the established calibration curve. However, the probable accuracy of estimating concrete strength in a structure is about the same as that of the surface hardness methods ($\pm 25\%$), with the same limitations (Mitchell and Hoagland, 1961a; Petersen and Stoll, 1955; Greene, 1954; Grieb, 1958; Boundy and Hondros, 1964; Malhotra, 1987; Khajuria and Balaguru, 1994); so are the origin and size of the testing errors. Since, as in the case of the

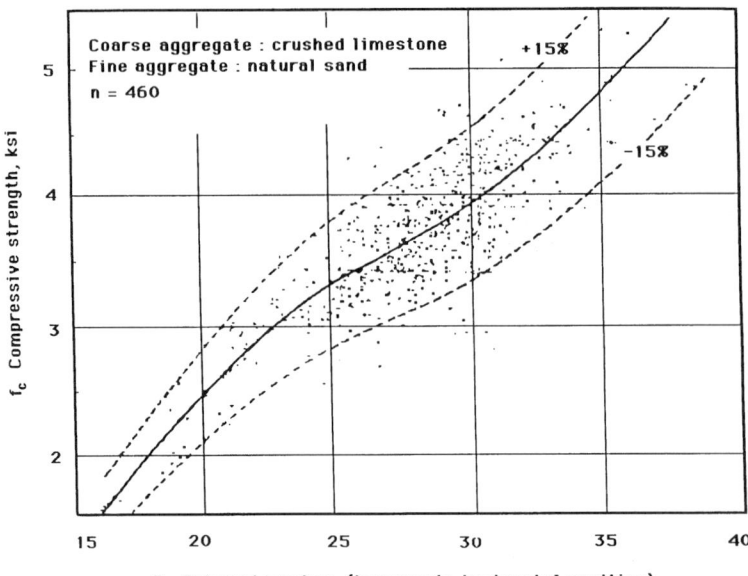

Figure 1.12 Relationship between compressive strength and rebound number for limestone aggregate concrete obtained with a type N2 hammer. 1 ksi = 6.90 MPa. (From Zoldners, 1957. Copyright ACI. Reprinted with permission.)

surface hardness methods, a calibration curve should be established for each concrete, the usual lack of a reliable calibration curve for the concrete in a structure can be the source of significant error in the strength estimates. This is demonstrated in Figure 1.13.

Another example is the following formula by Borjan (1981), used when no reliable supplementary data (composition, etc.) are available:

$$(f_{cu})_5 = -2.159 + 1.805 \log R + 0.345(\log R)^2 \qquad (1.17)$$

where $(f_{cu})_5$ = limit value of cube strength below which only 5% of the strength values are expected to fall, MPa

R = rebound number of a Type N Schmidt hammer held horizontally

When supplementary data are available or for other probability levels, only the first member of Eq. 1.17 will change. Therefore, the graphs of this equation form a family of parallel lines in a semilog system of coordinates. Jenkins also noticed that although different calibration curves are valid for the rebound hammer for various concretes (Fig. 1.13), the slopes of these curves are more or less the same, approximately 250 psi (1.7 MPa) to a unit change of rebound number when the hammer is of Type N2 (Jenkins, 1985). This means two

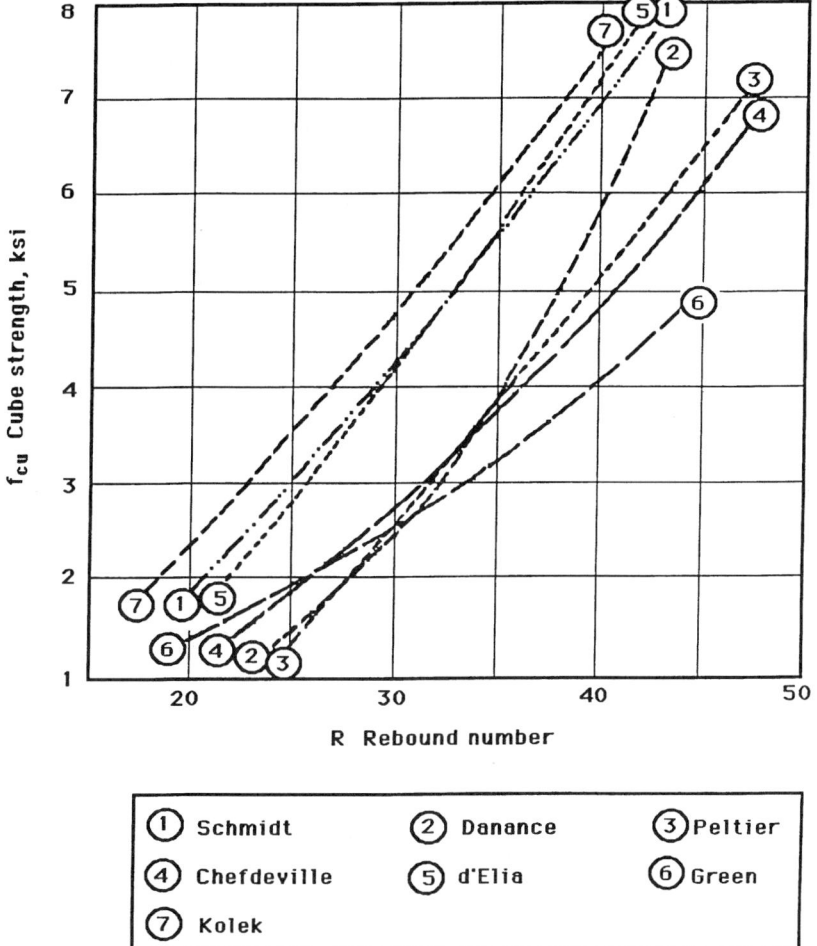

Figure 1.13 Calibration curves obtained by different experimenters. Curves 1, 2, 3, 4, and 7 were obtained with a type N2 hammer; curve 6 was obtained with a type N1 hammer. 1 ksi = 6.90 MPa. The average slope of the curves for the N2 hammer is approximately 250 psi (1.7 MPa) per rebound number. 1 ksi = 6.90 MPa. (From Kolek, 1958. Copyright Thomas Telford Publishing. Reprinted with permission.)

things: (1) if a reliable calibration curve is not available for the concrete in question, *strength differences* can be determined better with a rebound hammer than strengths, and (2) the N2 hammer cannot distinguish reliably strength variations of less than approximately 600 psi (4.1 MPa) since the maximum precision of the method is 2.5 × rebound number.

Other typical examples are presented under "Combined Methods" in Section 1.5. Incidentally, the available experimental evidence is contradictory whether valid correlation can be made even between the rebound number and modulus of elasticity of the concrete (Mitchell, and Hoagland, 1961; Petersen and Stall, 1955).

Rebound measurement can be supplemented by hardness tests, such as the Brinell (Kolek, 1958) or Mohs' method, for the improvement of the strength estimate. However, it is not clear yet if these improvements are large enough to justify the additional work.

The impact energy of the testing hammer should be checked not later than after each 200 measurements. Pendulum hammers, such as the Einbeck hammer mentioned previously, can be modified so that a single test can provide both the surface hardness and the rebound number (Pohl, 1969a).

Penetration Resistance and Pulling Methods

The basic principle of *penetration resistance* methods is to drive a metal probe of well-defined characteristics with high energy into the hardened concrete, determine the depth of penetration, and estimate the concrete strength from this depth (Malhotra and Carette, 1991). The first attempt to use such a method probably took place in 1934 in the Soviet Union, as reported by Skramtayev (Skramtayev, 1938).

The best known penetration method in North America is the Windsor probe (Kopf, 1969, 1981). This consists essentially of a powder-activated gun as the driver, a hardened alloy probe, load cartridges, and a depth gage for measuring the probe penetration. The gun fires the probe into the concrete, the exposed length of the probe is measured, and from this the concrete strength is estimated with the help of a pertinent calibration curve. The probe is relatively inexpensive and readily portable. Details of the method are specified in ASTM C803. Examples for relationships between the result of the Windsor probe and the compressive strength of the concrete are presented in Figure 1.14. Other penetration methods published earlier (Voellmy, 1954; Gaede, 1965) have not been accepted in the practice.

The penetration depth of a probe is influenced strongly by the hardness of the aggregate used as well as the maximum particle size and cement content whereas the concrete strength is not influenced as much. Therefore, there is no theoretically justified relationship between the penetration depth and concrete strength, only empirical correlations. These correlations should be established for each concrete. Supplementary hardness measurements, such as the Mohs hardness test on the concrete surface, may improve the accuracy of the penetration methods. Nevertheless, published data show that the variation in Windsor test results is larger than the variation in direct compressive strength on companion test specimens (Arni, 1972; Malhotra, 1974; Gaynor, 1969).

The standard Windsor probe apparatus, when new, is certified to deliver 575 ft-lb (780 J) of energy when shooting a probe and its driving head. This represents a muzzle velocity of 600 ft/sec (183 m/s), which is suitable for medium- and high-strength concretes. For concretes weaker than about 3000

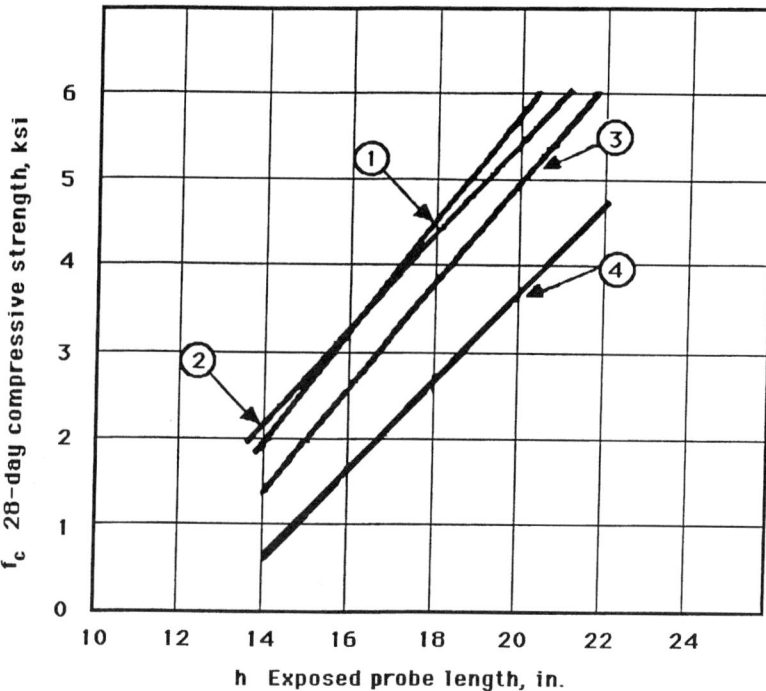

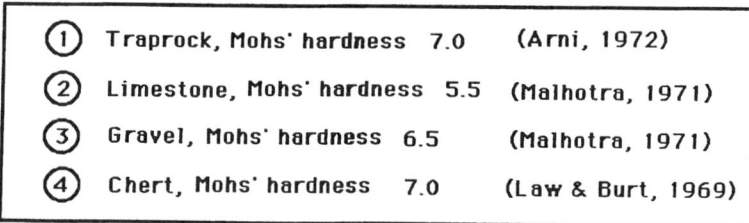

①	Traprock, Mohs' hardness 7.0	(Arni, 1972)
②	Limestone, Mohs' hardness 5.5	(Malhotra, 1971)
③	Gravel, Mohs' hardness 6.5	(Malhotra, 1971)
④	Chert, Mohs' hardness 7.0	(Law & Burt, 1969)

Figure 1.14 Relationship between exposed Windsor probe length and 28-day compressive strength of 6 × 12-in. (150 × 300-mm) cylinders obtained by various investigators. The average slope of the lines is approximately 5 ksi/in. 1 ksi = 6.90 MPa. 1 in. = 25.4 mm. (From Malhotra, 1976. Copyright ACI. Reprinted with permission.)

psi (about 20 MPa), a low-power test may be used. There are many factors that may affect muzzle velocity, one of which is the cleanliness of the gun barrel. Reduced muzzle velocity results in smaller penetration, falsely indicating a higher concrete strength. Testing errors of this sort would produce either a systematic error or a greater variance in the correlation of penetration to strength, or both.

Although Figure 1.14 demonstrates considerable differences between the calibration curves by various investigators, the slopes of these lines are more or less the same: approximately 5000 psi (35 MPa) per 1-in. (25.4-mm) variation of exposed probe length (Swamy and Ali, 1984b; Jenkins, 1985; Nasser and El-Manaseer, 1987; Bungey, 1981). This indicates a better agreement between *changes* in the penetration depth and changes in the concrete strength obtained by various investigators on various concretes than between the penetrations and strengths. Thus the most reliable application of the penetration test seems to be for estimation of the place-to-place variation of the concrete strength in a structure. It has also been suggested that the strength estimate obtained by the standard Windsor probe test can be considered as the upper limit of the actual strength of a typical concrete in the structure, whereas the standard core strength (Section 1.2) is the lower limit (Lee et al., 1992).

The precision statement in ASTM C803-82 says, among others, that for concrete with a 1-in. (25.4-mm) maximum size aggregate there is a significant difference in the strengths if the measured exposed length of two different sets of three probes is greater than 0.16 in. (4.1 mm). Considering the approximately common slopes of the calibration curves, this means that the Windsor probe cannot detect reliably differences less than 800 psi (5.5 MPa) in concrete strengths.

Pullout tests measure the force required to pull out a steel insert from the hardened concrete with a manually operated tension ram. The insert has an enlarged end which has been cast in the concrete during placing (Fig. 1.15). The pulled insert produces a complex state of stresses in the hardened concrete where shear stresses appear to be dominant. When these stresses exhaust the resistance of the concrete, a cone of concrete is removed at failure, the generating lines running approximately 30 to 35° to the direction of pull. The compressive strength of the concrete is then estimated from the pullout forces (Stone and Clarke, 1983; Stone and Giza, 1985; Carino, 1991b). The first attempt to use such a method was probably reported by Skramtayev (1938).

The pullout method is one of the few "nondestructive" methods that directly measures concrete strength in place. Thus its results are less affected by the properties of the aggregate than are the rebound number, penetration depth, or results of the sonic methods. However, even if the stresses were defined precisely, which they are not, it would be difficult to develop a sound theoretical relationship between this strength and the uniaxial compressive strength of the concrete (Stone and Carine 1984). In addition, the number of variables would be high. For instance, Malhotra and Carette (1980a) report that Kierkegaard and Hansen noticed an increase in the pullout force–compressive strength ratio with an increase in the maximum particle size of the aggregate. This may be the result of increased interlocking of the aggregate particles. Therefore, empirical relationships should be developed for each concrete for the estimation of compressive strength from the pullout force, although the differences between calibration curves here are less. Examples for such relationships are shown in Figure 1.16 and in the literature (Swamy and Ali, 1984; Khoo, 1984). The effects of age and temperature on the magnitude of the pullout force may be estimated with the maturity concept (Dilly and Ledbetter, 1984; Parsons and Naik, 1984, 1985).

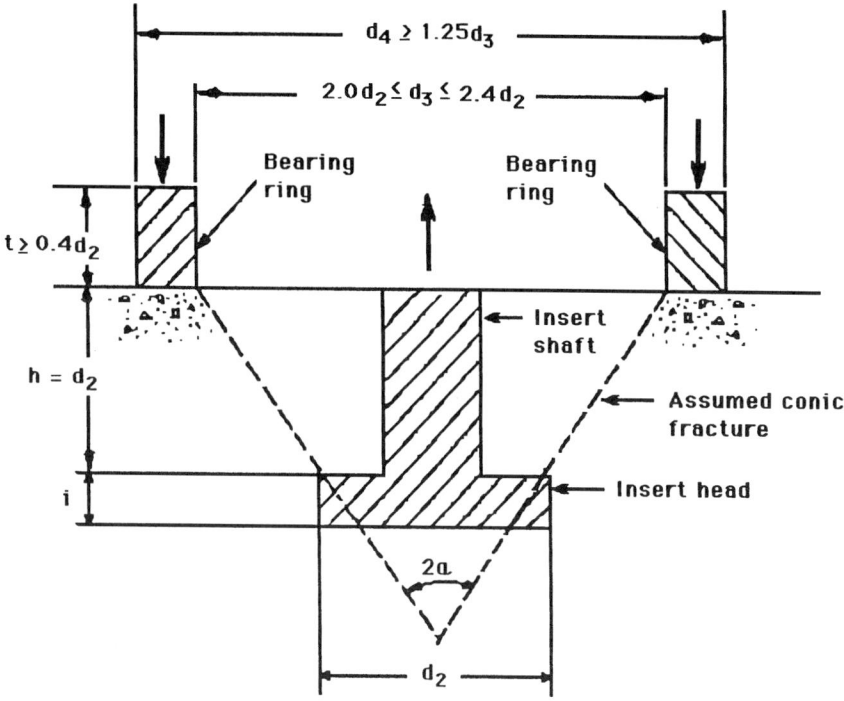

Figure 1.15 Schematic of pullout test as given in ASTM C900. (Copyright ASTM. Reprinted with permission.)

Two pullout tests have gained some popularity. One was promoted by Richards (1977) and specified in detail in ASTM C900. Here the ratio of the pullout strength (force divided by the surface area of the conic frostrum) to the compressive strength is in the range 0.2 to 0.25. The other is the Lok-Test, developed in Denmark (Kierkegaard-Hansen, 1975; Petersen, 1984). This test is similar to the ASTM (Richards) test, but the equipment is more compact, sophisticated, and hence, more expensive. Details are specified in Danish standard DS 411.

A variation of the pullout test is the *pulloff test.* Here a circular steel probe is bonded to the surface of the hardened concrete by means of an epoxy adhesive (Long and Murray, 1981, 1984; Keiller, 1985). After hardening of the adhesive, a slowly increasing tensile force is applied to the probe. As the tensile strength of the bond is greater than that of the concrete, the latter will eventually fail in tension near the surface. From the area of the probe and the force applied at failure, it is possible to calculate a nominal tensile strength for the concrete in the place of the break. From this and a calibration curve preestablished for the prevailing circumstances, the compressive strength of the concrete can be estimated. The result of the pulloff test is characteristic only of the condition of the concrete surface. This is a disadvantage when

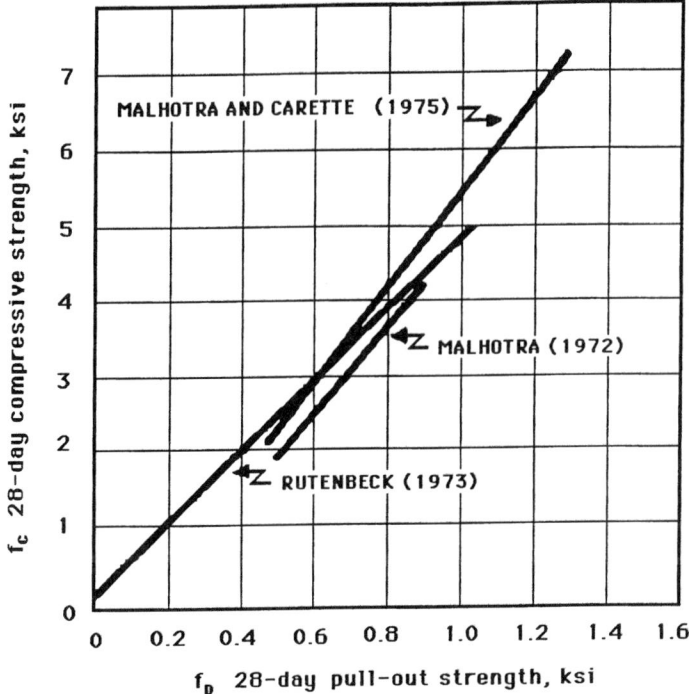

Figure 1.16 Relationship between pullout and 28-day compressive strength of 6 by 12-in. (150 300-mm) cylinders as obtained by various investigators. 1 ksi = 6.90 MPa. (From Malhotra and Carette, 1980b. Copyright ACI. Reprinted with permission.)

the strength of concrete in the structure is to be estimated since the properties of the concrete surface may, and frequently do, differ significantly from the properties of concrete inside. On the other hand, this test may provide some information about the durability of the concrete surface since it is likely that a low pulloff strength is a warning signal concerning low surface durability.

Another pertinent test was recommended by Tassios and Demiris (1968). Here again the pullout force is measured and correlated with the compressive strength of the concrete, but the probe is a specified nail driven into the surface of the hardened concrete by a gun. That is, this test is based on the friction between the probe and the hardened concrete. Incidentally, it is perhaps even simpler to test the friction by torque instead of using a pullout force (Pohl 1969b). A pullout test based on the measurement of friction between a steel insert and the hardened concrete was developed by the Building Research Establishment in England (Keiller, 1985; Chabowski and Bryden-Smith, 1979, 1980).

Neither the ASTM test nor the Lok-Test is, strictly speaking, nondestructive because they both damage the surface of the concrete and this requires repairs. Note, however, that if a pullout force equivalent to a given minimum strength is applied without failure, it may be assumed that the minimum required strength has been reached for the in-situ concrete; thus the structural unit need not be stressed to failure. Another, more important disadvantage of these two methods is that unlike the other in-situ tests discussed, these are limited to use in new construction. This is so because neither of them can be used in an existing structure unless the steel inserts were installed during construction. But even if they were, it is not certain that the test can be performed on every critical spot, since the criticality of certain points may become apparent only after hardening. Attempts have been made to eliminate these two disadvantages. Some of these are discussed in the literature (Bungey, 1982; Malhotra and Carette, 1980a; Petersen, 1984; Mailhot et al., 1979).

On the other hand, pullout tests are simple and quick, and the coefficient of variation of the results is about the same as that of the standard compressive strength (Bickley, 1983, 1984), although higher variations have also been reported (Murray and Long, 1987). Probably because they represent concrete strength, the results show better correlation with the compressive strength than do the nondestructive mechanical methods discussed earlier (Malhotra, 1975b, 1976). For instance, correlation coefficients of 0.97 and 0.99 for normal-weight concretes have been obtained from curve fitting of pullout and compression test results. When the reliability of the calibration curve is less certain, the estimate of the compressive strength may be off more.

Another "nondestructive" method that tests concrete strength directly in situ is the *break-off method* (ASTM C1150). One form of this consists of the determination of flexural strength in a plane parallel to and a certain distance from the surface of the hardened concrete (Carlsson, 1983; Carlsson et al., 1984; Johansen, 1979; Dahl-Jorgensen and Johansen, 1984; Di Maio et al., 1995). For this purpose, tubular disposable forms are inserted in the fresh concrete during construction. When testing in the hardened state, the inserts are removed and the concrete core is broken off at the bottom in flexure by applying a force to the top at right angles to the axis of the core. The test method is simple and rapid, although it may not be easy to insert the tubes in concretes of stiff consistency. The available limited experimental data show good correlations of break-off strength with flexural (Yener and Chen, 1985) and compressive strengths (Barker and Ramirez, 1988), although the within-test variation is high. Otherwise the test has the same limitations as the two major pullout tests. Also, with approximately the same effort, one could obtain cores with the push-out method (Section 1.2), which can be used more conveniently for the establishment of the compressive strength of concrete in the structure. The break-off method is not recommended for concrete having a nominal maximum particle size greater than 1 in. (25 mm).

Another form of the break-off principle developed in the Soviet Union for testing concrete is shown in Figure 1.17 (Novgordsky, 1973). The hole in the concrete can be formed during the construction or drilled out later. The concrete strength can be estimated from the force needed to produce break-off after the establishment of a calibration curve.

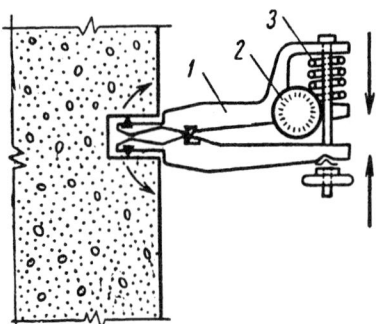

Figure 1.17   Soviet break-off method. (From Novgordsky, 1973.)

Other Nondestructive Test Methods

There are additional nondestructive methods for the *detection of defects and irregularities* in concrete. Most of these defects affect the load-carrying capacity of concrete structures.

1. The first step in most condition surveys of structures is a determination of the type and extent of the deterioration, if there is any, by a careful *visual inspection.* Guidelines for such inspection are available (Stowe and Thornton, 1984; Transportation Research Board, 1979; ACI, 1983a; S. Popovics and McDonald, 1989a).

2. *Infrared thermography* has been found to be capable of detecting delaminations in bridge decks and in other concrete elements exposed directly to sunshine. The method is based on the observation that in periods of heating, the surface temperature of delaminated spots on concrete is higher than that of the surrounding concrete (ASTM D4788-8). At night, when there is usually a loss of heat from the concrete to the air, the surface of the delamination is cooler than the average temperature of the solid concrete. These differences surface temperatures can be detected by sensitive infrared measuring systems (Manning and Holt, 1980; Manning, 1985, 1986; Weil, 1991; Luthi et al., 1995; Storozhenko and Meshkov, 1995).

3. *Low-power, high-frequency pulse radar* can also be used for the detection of deterioration inside concrete pavements or bridge decks (Park et al., 1995). Pulses of radio-frequency energy are directed into the concrete, a portion of which is reflected from any interface and displayed on an oscilloscope. An interface is any discontinuity or differing dielectric in concrete, such as delamination or a crack (Alongi et al., 1982; Cantor and Kneether, 1982; Cantor, 1984; Clemena, 1991; Maser, 1994; Tharmabala et al., 1994; Halabe et al., 1993). The biggest advantage of radar over thermography is that it is almost independent of weather conditions.

4. Electrical methods used in the field are currently limited to resistance and potential measurements (Lauer, 1991). One of the first applications of *electrical resistance* testing was a method for estimating of the permeability of bridge deck seal coats (Spellman and Stratfull, 1971). Details of this test are specified in ASTM D3633. Resistivity measurements have also been used to investigate the corrosion of steel in concrete. All resistance measurements of concrete must be made with an ac meter to eliminate polarization effects. To be precise, such measurements should be termed *impedance* rather than resistance because of the capacitance effects of the concrete (Manning, 1985). Resistivity is normally measured by the four-electrode method. Pavement thickness as well as moisture content determinations have also been attempted by electrical resistivity test (Malhotra, 1976; Moore, 1968). *Potential measurements* are used to determine in which section of a concrete structure the reinforcement is likely to be corroding (Okada et al., 1984; Tamura and Yoshida, 1984). When steel corrodes in concrete, a potential difference exists between the anodic half-cell areas and the cathodic half-cell areas along the reinforcing bar which can be measured. If the potential measurement is less negative than -0.20 V, the probability of corrosion is small; if it is more negative than -0.35 V, the probability of corrosion in high. Details of such a test method are specified in ASTM C876.

5. X-ray and especially gamma-ray *radiography* can be used for the detection of variations in consolidation of concrete members up to about 18 in. (450 mm) thick, location of reinforcement, cracks, measurement of the extent of corrosion, concrete density, steel fiber content, and assessment of the quality of grouting in prestressing ducts (Malhotra, 1976; Pohl, 1964a; Forrester, 1957, 1959; Preiss and Newman, 1964; Preiss, 1965, 1966). The rationale of these test methods is that as the radiation passes through a material of variable density, more radiation is absorbed by the denser parts of the material. In radiography the passing portion of the radiation is recorded on photographic film. In radiometry, variations in the intensity of the passing radiation are observed by radiation detectors such as Geiger counters.

6. *Nuclear methods* are based on principles similar to those for x- ray and gamma radiations, that is, on nuclear absorption and scattering techniques (Malhotra, 1976; Pohl, 1964b; Mitchell, 1991). The chief application of nuclear methods is in the measurement of moisture content insitu by neutron radiation and the density of concrete by gamma radiation

7. The main application of *magnetic methods* in the testing of concrete structures is in determination of the positions of reinforcement and the depths of cover (S. Popovics and J. S. Popovics, 1991a; Alldred 1995). A procedure is standardized in BS 1881, Part 204. Several commercially available portable battery-operated devices have been designed for this purpose, known as *cover meters* or *pachometers*. In lightly reinforced members such as bridge decks, cover meters can measure cover thickness as well as steel bar diameter, to within 0.25 in. (6 mm). However, in heavily reinforced members such as

columns with spiral reinforcement, it is virtually impossible to obtain meaningful results. For the same reason, cover meters do not produce reliable results about steel fiber content in concrete (Umoto and Kobayashi, 1984).

8. The principal *chemical methods* applicable in the field to concrete structures are used to determine depth of carbonation and chloride ion content. The results of these methods are used to establish if the concrete cover still has enough capability of protecting the steel reinforcement. Note that interpretation of the chloride ion content is quite uncertain (Manning, 1985; Morrison et al., 1976a, 1976b). Laboratory chemical methods are also used for the determination of the *cement content* of a concrete sample taken from a structure. However, it is difficult to obtain reliable results for cement content under general conditions despite the fact that pertinent standard test methods are available in ASTM C85 (Bungey, 1982; Figg and Bowden, 1971). BS 1881, Part 6 also provides comprehensive guidance for many chemical tests.

9. *Petrographic analysis* is a detailed examination of the internal structure of hardened concrete, usually with an optical microscope (K. Mather, 1978). It may include identification of the mineral aggregates and paste–aggregate interface, and assessment of the structure and integrity of the paste itself, including the air content (ASTM C457 and C856). These examinations are useful for causes and mechanisms of deterioration in the concrete, such as freeze–thaw, sulfate attack, alkali–aggregate reactivity, and fire damage (Roper et al., 1964a). Mix proportions in the hardened concrete can also be determined by microscopic examinations (Polivka et al., 1958; Axon, 1962).

10. A special case of examination by optical microscope is the determination of the original *water-cement ratio* on samples of hardened concrete. The method is based on the relationship between increases in water-cement ratio and increased capillary porosity in a cement paste (Section 4.4). Thin sections prepared from the hardened concrete in question are treated with a suitable fluorescent dye that penetrates the capillary pores of the paste. The hue of the dye, controlled by the magnitude of capillary porosity in the thin section, is observed through an optical microscope and compared to a calibration scale of colors prepared for this purpose. From this the original water-cement ratio can be estimated (Walker, and Marshall, 1979; Walker, 1979; Thaulow, 1982), which, in turn, provides the strength of the concrete. A similar principle is applicable for the detection of micro cracks on the surface of concrete (Novgordsky, 1973).

11. A less reliable method for determination of the original *water content* is described in the BS 1881, Part 6. Here *all* the pores in the hardened concrete sample are filled up with carbon tetrachloride, whose volume is then accepted as the volume of *capillary* pores, that is, the volume of the mixing water in the concrete specimen, which is an obvious oversimplification and overestimation.

1.5 TEST METHODS BASED ON WAVE PROPAGATION (S. Popovics, 1986a)

Sonic Methods in General

Several types of nondestructive test methods have been developed for testing concrete strength based mostly on the principle of propagation of stress waves, such as the measurement of acoustic pulse velocities, seismic velocities, and ultrasonic pulse velocities (J. S. Popovics et al., 1993; S. Popovics and J. S. Popovics, 1997; J. S. Popovics and Achenbach 1996). The common underlying principle of these methods is the recognition that factors that increase the concrete strength usually increase the velocity of the acoustic wave propagations in the material as well. Other methods are based on the determination of resonance frequency or some other feature(s) of the pulses, such as attenuation. These are often collectively termed *sonic tests.* These methods are attractive because they are (1) nondestructive, (2) simple to use, (3) relatively inexpensive, and (4) applicable in the field for a variety of structures.

Although measurement of the velocity of waves propagating through pavement and soil layers began in Germany in the late 1930s (Moore, 1968), the first ultrasonic method for testing concrete was developed by Leslie and Cheesman (1949) and by Jones (1948). Since then a considerable amount of work has been reported on this topic (Zoldners and Soles, 1984; Malhotra, 1984; RILEM, 1954, 1965, 1966; Pohl, 1969b, 1970 J. S. Popovics, 1996a; J. S. Popovics and Rose, 1996; S. Popovics and J. S. Popovics, 1998), resulting in standard test methods such as ASTM C597, BS 4408:1974, and others (Whitehurst, 1966; S. Popovics et al., 1995; Komlos et al., 1996).

Principles of Ultrasonics (J. S. Popovics, 1990a)

The word *ultrasonic* describes a vibration or stress wave propagation with a frequency above the human audible range, which is above 20 kHz. The same principles hold for ultrasonic wave propagation that hold for sound wave propagation; that is, ultrasonic stress waves can be reflected, refracted, and attenuated. The following formula is also valid for ultrasonic waves:

$$v = f\lambda \tag{1.18}$$

where v = wave (phase) velocity
  f = frequency of the wave
  λ = wavelength

Long ago, it was established, both theoretically and experimentally, that the velocity of a monochromatic pure longitudinal wave, called *longitudinal pulse velocity* or *compression* or *l wave velocity,* propagating in three dimensions in an infinite homogeneous elastic medium is controlled by certain mechanical properties of the material, as follows:

$$v_1 = \left[\frac{E}{d} \frac{1 - \mu}{(1 + \mu)(1 - 2\mu)} \right]^{0.5} \tag{1.19}$$

$$= \left(\frac{K + \frac{4}{3}G}{d} \right)^{0.5} \tag{1.20}$$

where v_l = longitudinal wave velocity
$\quad\quad E$ = modulus of elasticity
$\quad\quad d$ = mass density
$\quad\quad \mu$ = Poisson's ratio
$\quad\quad K$ = bulk modulus
$\quad\quad G$ = shear modulus

When the pulse propagates only in *one direction,* such as in rods, Eq. 1.19 simplifies to $v_l = (E/d)^{0.5}$ (Komlos et al., 1996).

The v_s velocity of a monochromatic pure *transverse wave* or *shear wave* can be shown to be

$$v_s = \left(\frac{G}{d} \right)^{0.5} \tag{1.21}$$

Therefore, the ratio of v_l to v_s is a function of Poisson's ratio only in such materials. The velocity of *surface waves,* or *Rayleigh waves,* v_R, is

$$v_R = \sqrt{\frac{0.87 + 1.12\mu}{1 + \mu}}\, v_s \tag{1.22}$$

that is, slightly less than the velocity of pure shear waves, and depends on Poisson's ratio (Timoshenko and Goodier, 1951). Note that the three velocities are not independent of each other. For instance, if v_l and v_s are known, v_R can be calculated.

The velocities represented by Eqs. 1.19 through 1.22 are referred to as *phase velocities* of the wave motion. If a specific phase velocity depends on the frequency of the wave, the wave is said to exhibit *dispersion.* If a traveling-wave group such as an ultrasonic pulse contains several frequencies, the wave will travel at *group velocity.* In a nondispersive medium, phase velocity equals group velocity; in a dispersive medium such as concrete, the group velocity will also be influenced by the frequencies, thus will not be equal to the phase velocity (Fung, 1965; Kolsky, 1963; Towne, 1967).

Another characteristic of the behavior of pulses in materials is the *damping* or *attenuation* of the pulses. This is signal reduction or loss as the pulse propagates. No standardized method exists to measure ultrasonic attenuation in concrete. Usually, a pulser-receiver and a set of longitudinal wave transducers of specific frequency are used in through-transmission mode. The signal loss is obtained by oscilloscope measurement of time-domain amplitudes before and after transmission through the material. The signal loss may be calculated as follows:

$$SL = 20 \log \frac{A_2}{A_1} \qquad (1.23)$$

where SL = signal loss for a specific frequency, dB
   A_1 = amplitude of the input signal
   A_2 = amplitude of the output signal

An *attenuation coefficient* can also be calculated for a specific material and frequency by dividing the signal loss by the material path distance traveled by the ultrasound wave. The unit of the attenuation coefficient is usually dB /mm (Rose and Goldberg, 1979).

Measurement of Pulse Velocity

The *ultrasonic pulse velocity* has been the most widely accepted vibrational method for assessing the quality of concrete in a structure, although other features of an ultrasonic pulse may also be applicable (J. S. Popovics, 1990b). The reason for using ultrasound is that adequate nonultrasonic (i.e., acoustic) waves in the audible range require large sound generators. Also, the higher the frequency, the smaller the defects that can be detected in the material. The rule of thumb is that if a defect is significantly smaller than the ultrasonic wavelength, the wave cannot detect it (Rose and Goldberg, 1979; Wells 1967). On the other hand, the acoustic energy loss (attenuation), already high in concrete, increases with higher frequencies. Also, ultrasonic pulses, having passed through concrete, carry noise due to scattering of the stress waves among the aggregate particles. The intensity of the noise increases with the frequency; thus in the case of high frequencies, it can mask meaningful signals. Therefore, there is a certain frequency that is optimum for a material. This optimum frequency is the highest permitted by the prevailing circumstances to penetrate the material adequately, usually between 20 and 200 kHz for concrete. This is much less than the range recommended for metals, which is due to the composite and viscoelastic nature of the concrete.

Ultrasonic stress waves are usually generated using a vibrating crystal. The crystal is shocked with an impulse voltage and then converts the electrical energy into a mechanical vibration of ultrasonic frequency. The reverse of this effect is utilized when the crystal receives an ultrasonic wave. This phenomenon is known as the *piezoelectric effect*. The piezoelectric crystal is housed in a protective casing, and the unit is called a *transducer*. Selection of the transducers is very important because they control the characteristics, such as peak frequency, bandwidth, and focal length, of the ultrasound sent through the material. The pulser-receiver unit generates and receives uniform impulse voltages that drive the transducers (Rose and Jeong, 1984). In some cases it may be necessary to study visually the waveform of a stress wave. To this end an oscilloscope can be used. If desired, the oscilloscope can be connected to an analog–digital converter, which can be used to save the waveform so that it can be analyzed by appropriate computer software.

Because air is highly attenuating to sound, the vibration cannot be transferred effectively with the usual procedures from the transducer to the material unless the transducer is in direct contact with the material. To decrease loss in signal, a layer of a viscous couplant such as petroleum jelly is placed between the material and the vibrating transducer.

There are several ultrasonic pulse velocity units commercially available for testing concrete (Malhotra, 1976; Whitehurst, 1951). According to Clifton and Anderson (1981), one of them, the ultrasonic concrete tester, has a testing range of about 50 ft (15 m), whereas two others, the Soniscope and Pundit, can be used to test concrete having a thickness up to about 80 ft (25 m). The differences are due to their respective operating frequencies, which are 150, 50, and 50 kHz. Since the frequency of seismic (i.e., mechanically generated) waves is lower (it falls in the audible range), their testing range is wider. The upper limit of the path length depends on surface conditions as well as on the characteristics of the interior concrete under investigation. Modern ultrasonic testing instruments with a single operator can measure pulse velocity with an accuracy of within 1% under laboratory conditions (S. Popovics et al., 1990). Jenkins found that under field conditions the average coefficient of variation of the pulse velocity data for similar concretes was 2.2%. That is, a significant difference in pulse velocity measurements may be 6.2% (Jenkins, 1985).

The basic procedure is to measure the *travel time* of an ultrasonic pulse, generated by a transducer, passing through concrete (Krautkramer and Krautkramer, 1983; S. Popovics et al., 1995; Komlos et al., 1996). The pulse is picked up by a receiver transducer and the time of travel is measured electronically. The straight distance between the two transducers along the travel path divided by the travel time gives the average velocity of wave propagation, that is, the *pulse velocity*. Usually, *longitudinal waves* are used in testing concrete, although other waves, such as *shear waves,* and *guided waves,* such as *surface* or *Rayleigh waves,* have also been tried (S. Popovics and J. S. Popovics, 1991b). Longitudinal waves are waves that are transmitted by particles vibrating parallel to the direction of propagation. Shear waves are transmitted by particles vibrating perpendicular to the direction of propagation. Surface waves travel along the boundary, that is, the free external surface of the material. The depth of penetration is on the order of one wavelength thickness.

Longitudinal waves can be measured in a more reliable manner than other waves because (1) the velocity of the longitudinal waves is the greatest, (2) their energy loss is the smallest, (3) their amplitude is the largest, and (4) it is relatively simple to separate them from the other wave types. It is usually assumed that the pulse velocity is independent of both pulse frequency and the stress conditions in the concrete. It has been shown that the first of these simplifying assumptions is true only within narrow limits (S. Popovics, 1990b) but the second assumption is valid within practical stress limits (S. Popovics and J. S. Popovics, 1991c).

It is recommended by ASTM C597-83 that for best results the transducers located directly opposite each other on the concrete surfaces. This is called the *through-transmission* mode (Fig. 1.18). However, because the effective beam width of the transducers is wide, transit times can be measured across corners of a structure but with some loss of sensitivity and accuracy (Manning, 1985). Pulse measurements on the same surface are also possible in principle since the pulse energy can reflect (echo) from the opposite face or defect to the original surface for measurement (McDowell and Millett, 1984; Krause et al., 1995). Such a method is the ultrasonic *pulse-echo technique,* in which a transducer sends an ultrasonic wave into the material and the reflected wave is received by the same transducer. The *pitch-catch technique* is similar to the pulse-echo technique except that two separate transducers are used, and the sending transducer is separated from the receiving one by a

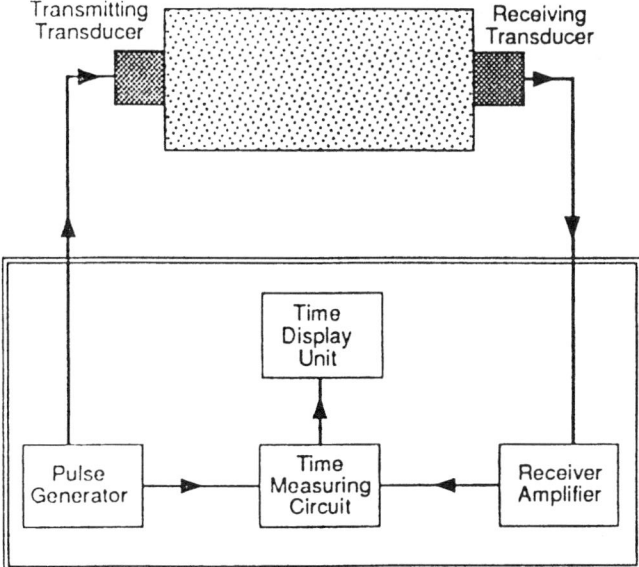

Figure 1.18 Schematic diagram of pulse velocity testing circuit according to ASTM C597-83 (reapproved 1991). (Copyright ASTM. Reprinted with permission.)

certain distance on the same surface (Krautkramer and Krautkramer, 1983). The two latter procedures have been used much more frequently for ultrasonic testing of metals than for concrete.

Since the pulse velocity in steel could be up to double that in concrete, especially in poor concrete, pulse velocity measurements in the vicinity of the reinforcing steel in concrete may be higher than in nonreinforced concrete of the same composition (Chung, 1978a). The quantitative estimation of this effect is difficult because not only the direction of the bars but also their quantity and diameters can influence significantly the measured velocity of pulse (Bungey, 1984; J. S. Popovics et al., 1992). Therefore, where possible, measurements in close proximity to steel should be avoided, especially when two-way steel or another complex bar arrangement is present.

If a wave encounters a crack or void, it will be diffracted around the discontinuity. This will increase the internal distance the wave must pass between the transducers, and consequently, its travel time. Therefore, for a given concrete and a given distance between transducers, the travel time of a wave will be less in a region where the concrete is sound than where the concrete is more porous and/or contains cracks. This simple principle, however, does not lend itself adequately to practical strength determinations for concrete. The discussion of the fundamental principles will show why this is so.

Ultrasound and Elastic Constants

The *elastic constants* of a *homogeneous* elastic material can be estimated from ultrasonic measurements. For instance, the modulus of elasticity can be calculated directly from Eq. 1.19 with the known pulse velocity and the other variables of the formula. A similar formula for the calculation of E is (K. Mather, 1981)

$$E = \frac{dv_s^2(3v_l^2 - 4v_s^2)}{v_l^2 - v_s^2} \tag{1.24}$$

where the symbols are as defined for Eqs. 1.19 and 1.21. Equations 1.19 and 1.24 do not provide reliable E values for *composite* materials, including a large group of concretes. When the paste and aggregate differ significantly in elastic properties, which is the case with most structural concretes at ages younger than 28 days, Eqs. 1.19 and 1.24 are misapplied, and the results are likely to be much higher than the true values. These formulas may apply only at later ages, say, after 2 years, when the elasticity of the cement paste approaches that of the aggregate, that is, when the concrete approaches elastic homogeneity (Beauzee 1954, Philleo, 1955; S. Popovics and J. S. Popovics, 1994). This is the reason for the recommendation to determine the practically important dynamic modulus of elasticity or the dynamic shear modulus by the resonance frequency method (ASTM C215). Note, however, that there are empirical relationships for the calculation of the resonance modulus E_{res} from longitudinal pulse velocity v measurement. One of these calculates the E_{res} from the E_{pulse} obtained from Eq. 1.19 or 1.24 for normal-weight structural concretes, as follows:

$$E_{res} = 1.18 E_{pulse} - 1.667 \times 10^6 \qquad (1.25)$$

where the moduli are in psi (1000 psi = 6.90 MPa) Another empirical formula is

$$E_{res} = Lv_l^n \qquad (1.26)$$

where v_l is the longitudinal pulse velocity and L and n are experimental parameters. For homogeneous materials, $n = 2$ from Eq. 1.19; for composite materials, $n > 2$. Pertinent data by Klieger (1957) provided $n = 2.83$ for a variety of concretes of various ages, whereas data by Babic (1987) produced $n = 2.25$ for different cement-stabilized soils, regardless of the age and composition of the mixtures within wide practical limits. Thus it appears that the following equation approximates well the E_{res} versus E_{pulse} relationship:

$$E_{res} = Lv_l^{2.5} \qquad (1.27)$$

The determination of longitudinal as well as transverse velocity of ultrasound in a homogeneous elastic material makes it possible to calculate other elastic constants approximately. Some of these formulas are [ASTM D2845–83 and E494–75(1985)]

$$\mu = \frac{1 - 2(v_s/v_l)^2}{2[1 - (v_s/v_l)^2]} \qquad (1.28)$$

$$G = dv_s^2 \qquad (1.29)$$

$$K = d(v_l^2 - 1.333v_s^2) \qquad (1.30)$$

and

where the symbols are as in Eqs. 1.10 through 1.27. Further discussion is presented in Section 6.3.

Ultrasound and Strength

Concrete is unique in that it is practically the only major material in which strength determination is attempted by ultrasound. Ultrasound is not a good tool for testing strength since there is no theoretically sound, direct relationship between longitudinal (bulk) pulse velocity, or some other feature of the ultrasound, and material strength. Nevertheless, the acute need for nondestructive determination of the in-situ strength of concrete, as well as the present lack of an adequate test method for this, forced the engineers to use ultrasound by using a loophole. The conceptual justification of this is twofold: one is the relationship existing for homogeneous materials between longitudinal pulse velocity and modulus of elasticity; the other is a relationship between the modulus of elasticity and the strength of the material, as will be discussed later. Thus, with the combination of these two relationships, and with a little license, the ultrasonic wave velocity may be correlated to concrete strength.

The relationship between modulus of elasticity E and strength f_o of a material exists, at least conceptually, since both are controlled by the bonds between atoms and molecules; that is, the two properties are related. An approximate textbook formula for this is

$$f_o = \left(\frac{E\gamma}{a_o}\right)^{0.5} \tag{1.31}$$

where γ = half of the surface energy
$\quad\;\; a_o$ = interatomic distance

Although measured strengths are regularly much lower than those calculated from Eq. 1.31, the discrepancy is attributed to the presence of flaws in the material, not to the lack of a sound relationship between E and f_o. It may seem, therefore, that there is a well-defined relationship between pulse velocity and strength of a material which can expressed mathematically with the combination of Eqs. 1.19 and 1.31 and a modifying factor for the consideration of flaws, as follows:

$$f = v_l \left[\frac{dg}{a_o\gamma}\frac{(1 + \mu)(1 - 2\mu)}{1 - \mu}\right]^{0.5} \tag{1.32}$$

$$= b_o v_l \tag{1.33}$$

where b_o is a materials parameter determined by calibration for a specific concrete. Note that according to Eq. 1.33, the strength is a *linear* function of the pulse velocity, which, however, is *not* supported by observations.

Equation 1.33 is but one of many possible models for the the f versus v relationship. Using other initial formulas, different relationships can be obtained for the concrete strength versus pulse velocity relationship. For instance, if the widely used empirical formula

$$E = af^{0.5} \tag{1.34}$$

by ACI is substituted into Eq. 1.19, the result is

$$f = c_o v_l^4 \tag{1.35}$$

where c_o is again a materials parameter. Note that according to Eq. 1.35, the concrete strength is a *fourth-order* function of the pulse velocity. Although this formula has been proposed by several researchers (Pohl, 1969a) and is better than Eq. 1.33, it is still not supported adequately by experimental results. Neither are any other formulas for the concrete strength versus pulse velocity based on a single longitudinal pulse velocity measurement, as shown in the next section, despite the theoretical soundness of the initial equations on which the derived formulas are based. But why? Why is it that a longitudinal pulse velocity measurement cannot provide the strength estimate of concretes with any acceptable degree of accuracy? The answer is because Eqs. 1.19 and 1.31 were obtained from the assumption that the material is homogeneous and linearly elastic, which concrete is not. Concrete is a highly composite viscoelastic material, and to disregard this, that is, to pretend that concrete is homogeneous and linearly elastic, is apparently a nonpermissible oversimplification from the standpoint of the strength versus pulse velocity relationship (Rose et al., 1994). This does not mean that there is no practical relationship between the strength and pulse velocity of a concrete but rather that such relationship is more sophisticated than Eq. 1.33 or 1.35. In other words, when an estimation of the concrete strength is attempted from a single pulse velocity measurement, as has been done during the past 40 years because of the need and for want of something better, this implies that the concrete is considered as a quasihomogeneous material. Thus the resulting strength estimate is necessarily unreliable.

Further reasoning along the same line is the following (S. Popovics and J. S. Popovics, 1992):

1. Approximately two-thirds of the volume of a typical structural concrete consists of aggregate. Therefore, the pulse velocity measured in the standard manner is controlled by the quality and quantity of aggregate in the concrete. The strength of the same concrete, however, is controlled by the matrix portion in the concrete, which is cement paste with air pores. Thus, trying to predict the concrete strength from a standard pulse velocity measurement is futile. Unfortunately, there is no test method at present for direct measurement of the pulse velocity in cement paste when it is part of the concrete. It is possible to calculate this velocity from certain assumptions if the exact composition of the concrete is known as well as the pulse velocities in the concrete and in the fine and coarse aggregate particles (Chung and Law, 1983). However, these data are almost never known accurately enough in practical cases of testing.

2. Some of the difficulties and uncertainties are due to the relative insensitivity of the pulse velocity to porosity. That is, the pulse velocity in a concrete is affected by the porosity in the cement paste far less than some of the

other concrete properties, especially the compressive strength, as was demonstrated by Kaplan (1960a) (Fig. 5.40). The reason for this is that loads produce stress concentrations in concrete around cracks and voids (see Section 2.4). Since the average stress field is much higher under compressive load than under ultrasonic pulse propagation, the compressive strength is reduced much more by porosity than the pulse velocity (S. Popovics, 182 1989). This insensitivity of the pulse velocity has the consequence that changes in the porosity in the cement paste of a given concrete can be large enough to cause significant changes in the compressive strength, yet they can be too small to be noticed by pulse velocity measurements. Thus it is no wonder that the attempt to estimate the concrete strength from the measured pulse velocity has been less than satisfactory.

3. The relationship between the elastic properties of a hardened concrete and its strength, and even more so, that between pulse velocity and its strength, is influenced by a number of factors. For instance, the higher the cement content (i.e., the lower the aggregate content) of a concrete, the higher its compressive strength related to a given longitudinal wave velocity (Jones and Gatfield, 1955). Additional such variables are the age of the concrete, moisture condition, type of aggregate, amount and location of reinforcement, and extreme temperatures (Whitehurst, 1951; Rayleigh, 1945; Talaber et al., 1979; Kaplan, 1959b, 1960b; Sturrup et al., 1984; Teodoru, 1986). When all these factors are kept constant, an empirical relationship, such as a calibration curve or formula, can be developed experimentally for each concrete for strength estimation. Note that the cement type, including air-entraining cements, does not seem to influence the strength versus pulse velocity relationship.

Quantitative Relationships Between Pulse Velocity and Concrete Strength

Several efforts have been made for the development of a mathematical model formula for the relationship between concrete strength and velocity of ultrasonic pulses, primarily longitudinal (compression) waves. Examples are Eqs. 1.33 and 1.35. The contradiction between these two equations, as well as the fact that neither of these formulas is supported by experimental results (S. Popovics, 1990b), clearly shows the futility of obtaining a satisfactory theoretical single-variable formula for f versus v. However, since such a relationship is needed for practical estimation of concrete strength from pulse velocity measurements, empirical formulas have been developed by statistical means (ACI Committee 228, 1996; Jones and Facaoaru, 1968). The graphical representative of such a formula is called a *calibration curve*.

A frequently used empirical formula for the pulse velocity versus strength relationship is

$$f = ae^{-bv_l} \tag{1.36}$$

which is a simplified form of a better, more general relationship,

$$f = ae^{-bv_l} + c \qquad (1.37)$$

where $\quad f$ = concrete strength
$\qquad v_l$ = pulse velocity
$a, b,$ and c = empirical constants depending on the type of strength,
composition, air content, as well as curing of the concrete

For instance, Borjan (1981) recommends the following formula for the case when the aggregate is quartz:

$$\log(f_{cu})_5 = 2.407 - 6.8 \times 10^{-4}(5760 - v_l) \qquad (1.38)$$

where $(f_{cu})_5$ = limit value of cube strength below which only 5 percent of the strength values are expected to fall, MPa
$\qquad v_l$ = velocity of the ultrasound in the concrete, m/s

The logarithm is of base 10.

When supplementary data are available or for other probability levels, only the 6.8×10^{-4} factor will change in Eq. 1.38. Therefore, graphs of this equation form a family of straight lines converging in one point in a log f_{cu} versus v_l system of coordinates. Other typical examples for such relationship are presented later in the section "Combined Methods." Similar relationships are offered by Pohl (1969a).

As long as the conditions establishing the parameters remain unchanged, such formulas can predict the compressive strength of the concrete within ±20% accuracy (Malhotra 1980a). If, however, the conditions (concrete composition, curing, etc.) change or are unknown, the concrete strength cannot be estimated from pulse velocity measurements with any acceptable degree of accuracy (Pohl, 1975; Jones and Facaoaru, 1968a). This statement is illustrated by Figure 1.19, where the strength estimate related to, say, 4 km/s (13,120 ft/sec) velocity ranges from 5 MPa (725 psi) to almost 30 MPa (4350 psi). Another example is Parker's experiment, in which he correlated pulse velocities and compressive strengths for concretes made with only one type of aggregate but containing different cements and a variety of admixtures (Parker, 1953). When the compressive strength estimated from pulse velocity was 4440 psi (30.7 MPa), the lower confidence limit for this value at the 95% level was 2100 psi (14.5 MPa). Such a strength estimate is not much more meaningful than the rule of thumb that a pulse velocity above 15,000 ft/sec (4570 m/s) represents an excellent concrete, and below 7000 ft/sec (2130 m/s) a very poor concrete.

Byfors attempted to reduce two of the interfering factors, the maximum particle size and air content, by introduction of a corrected pulse velocity (Byfors, 1979):

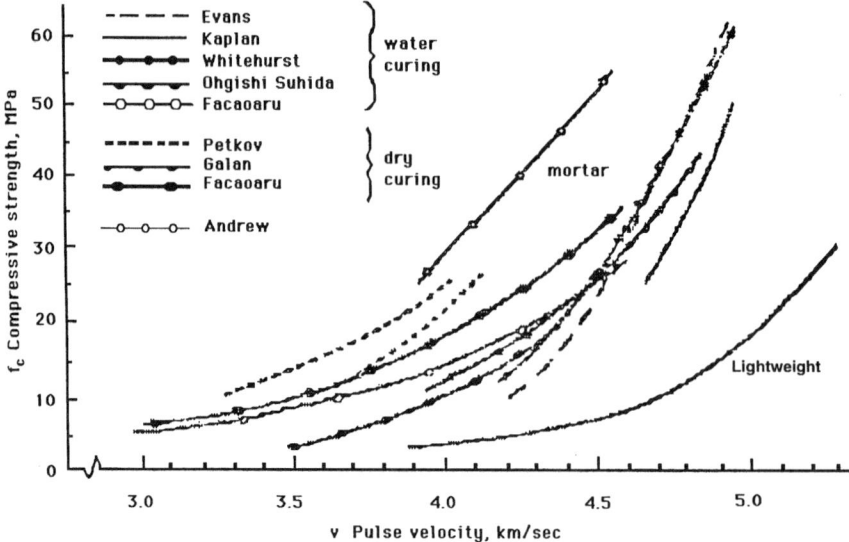

Figure 1.19 Relationship between compressive strength and pulse velocity of concrete as recommended by various authors. 1 MPa = 145 psi. 1 km/s = 3281 ft/sec. (From RILEM, 1969. Copyright RILEM. Reprinted with permission.)

$$v_{cor} = v_l + 0.05(a - 2) + 0.97 \log \frac{32}{D}$$

$$= v_l + 0.05(a - 2) + 0.42 \ln \frac{32}{D} \qquad (1.39)$$

where v_{cor} = corrected pulse velocity, km/s
v_l = measured pulse velocity, km/s
a = air content, %
D = maximum particle size, mm

The logarithm log is of base 10 and ln is the natural log.

The utilization of v_{cor} for estimation of concrete strength is demonstrated in Figure 1.20. For the sake of convenience the two fitted straight lines in the figure can be substituted for by the graph of the single equation

$$f_c = (1.2 \times 10^{-3})10^{0.8565v_{cor}} + 0.055$$

$$= 1.2 \times 10^{-3} \exp(1.972v_{cor}) + 0.055 \qquad (1.40)$$

where f_c is in MPa.

Other experiments have also indicated that an increase in the maximum particle size or in the quantity of aggregate, especially of the coarse particles,

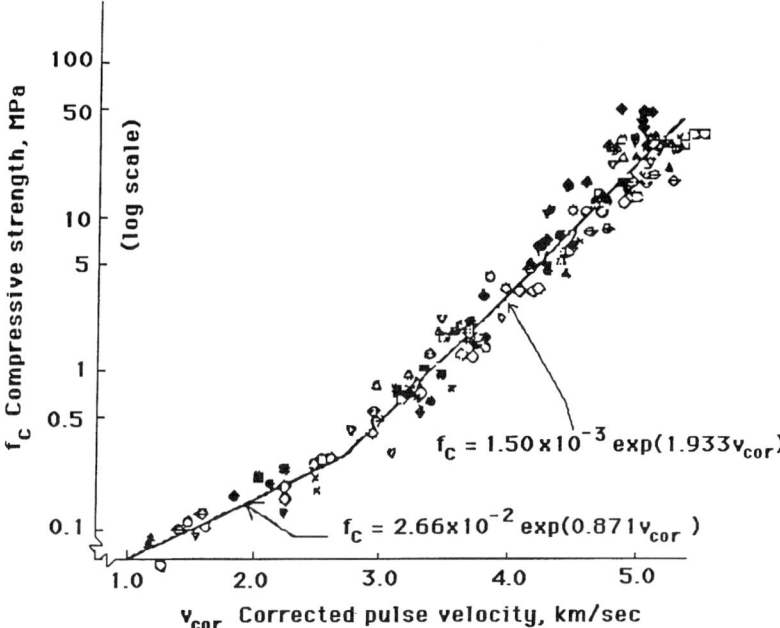

Figure 1.20 Relationship between compressive strength and pulse velocity corrected for air content and maximum particle size according to Eq. 11.39. Age: from 5 h to 28 days; w/c = 0.40, 0.58, 1.00 by mass; Cement: Type I, three different brands; cement content: from 165 to 620 kg/m³ (275 to 1030 lb/cu yd); D = 4, 8, 16, 32 mm; a = from 1.8 to 9.2%; curing: moist at 20°C (67°F); specimen: 100 × 100 × 400-mm (4 × 4 × 16-in.) prism. 1 MPa = 145 psi, 1 km/s = 3281 ft/sec. (From Byfors, 1979.)

increases the pulse velocity even when the concrete strength is kept unchanged (Tomsett, 1980).

An attempt to describe the effect of moisture content on pulse velocity is the following (Talaber et al., 1979):

$$\ln \frac{f_1}{f_2} = k f_1 (v_1 - v_2) \tag{1.41}$$

where f_1 and v_1 = compressive strength and pulse velocity, respectively, of concrete after standard wet curing

f_2 and v_2 = compressive strength and pulse velocity, respectively, of the same concrete after nonstandard, drier curing

k = empirical parameter for the given curing condition, representing the degree of drying

The logarithm is of base e.

k = 0.019 MPa⁻¹ was obtained in England for standard water-cured and air-cured cubes, respectively. Others also reported improved estimates of concrete strengths by the use of Eq. 1.41 (Keiller, 1982; Swamy and Al-Hamed, 1984). The effect of age has also been investigated (Andersen and Nerenst, 1952). In addition, in reinforced structural elements when the location and depth of the reinforcing steel are unknown, pulse velocity tests might provide misleading results.

More of the interfering factors are eliminated when the pulse velocity measurements are used for estimation of the *changes* in concrete strength. Such is the case, for instance, when pulse velocity measurements are performed under identical testing conditions on the same specimen or in the same point of a structure repeatedly at various ages to estimate, say, the deterioration of the concrete caused by freezing and thawing. Should the fluctuation of the uniformity of a concrete mass be expressed in terms of strength rather than simply by the changes in pulse velocity, the use of relative strength values is the reasonable approach. Such relative strength values can be obtained easily. For instance, if Eq. 1.35 were reliable enough for the estimation of concrete strength, a good formula for the relative strength, f_{rel}, would be (Pohl, 1969a)

$$f_{rel} = \frac{f_2}{f_1} = \left(\frac{v_2}{v_1}\right)^4 = (v_{rel})^4 \qquad (1.42)$$

where the subscripts 1 and 2 indicate values related to two different points (specimens) at the same age, or two different ages at the same point (specimen).

Similar formulas derived from the curves of Figure 1.21 for another group of concrete are (S. Popovics, 1973, 1975)

$$f_{crel} = (v_{rel})^{8.3} \qquad (1.43)$$

and $\qquad\qquad\qquad\qquad\qquad\qquad\qquad\qquad\qquad\qquad\qquad\qquad$ (1.44)

$$f_{frel} = (v_{rel})^{6.7}$$

where f_{crel} and f_{frel} represent the relative compressive and flexural strengths, respectively. Equation 1.44 is compared to a corresponding experimentally obtained curve (Kaplan, 1960a) in the upper part of Figure 1.21.

An even simpler approach is based on the observation that the calibration curves of different concretes are more or less parallel. For instance, the slopes of the various curves in Figure 1.19 are equal to 1000 to 1500 psi (approximately 7 to 10 MPa) per 1000-ft/sec (305-m/s) change in pulse velocity within the range 11,000 to 15,000 ft/sec (3.5 to 4.5 km/s). That is, many investigators obtained strength increase of approximately 1000 psi (approximately 7 MPa) for every increase of 1000 ft/sec (305 m/sec) in pulse velocity in this range, although their concretes and calibration curves were quite different. Similar observation was reported by Jenkins (1985).

The approximate sameness of the slope values also seems to indicate that pulse velocity measurements in their present forms cannot detect reliably

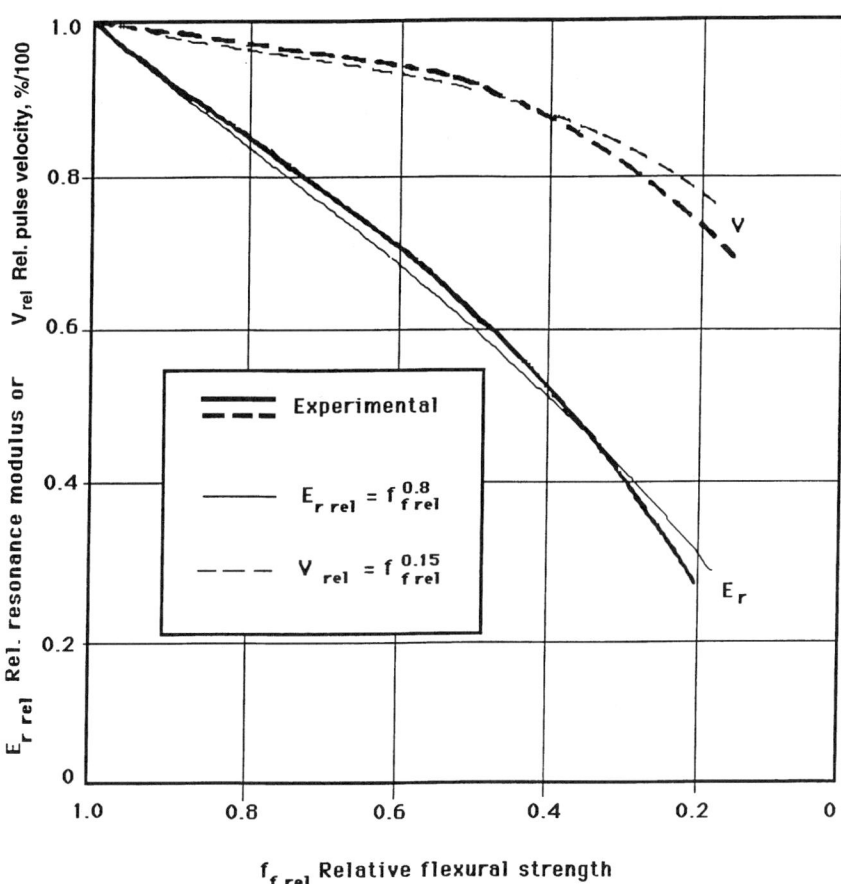

Figure 1.21   Relationship between relative values of flexural strength and pulse velocity as well as between relative values of flexural strength and resonance modulus of elasticity. (From S. Popovics, 1973, 1975. Copyright ACI. Reprinted with permission.)

strength differences smaller than approximately 900 psi (approximately 6 MPa) in structural concretes since the minimum significant difference in pulse velocity measurements is approximately 6.2%.

Conversion from the relative strengths to the actual in-situ strengths expressed in a stress unit can be done by taking cores from the structure and testing them for calibration. Examples of the use of pulse velocity measurements as well as the relative strength values of Eq. 1.42 that illustrate the uniformity of concrete in a block are presented in Figure 1.22.

Combined Methods

The idea seemed promising that the accuracy of the nondestructive determination of concrete strength may be improved by performing two or more

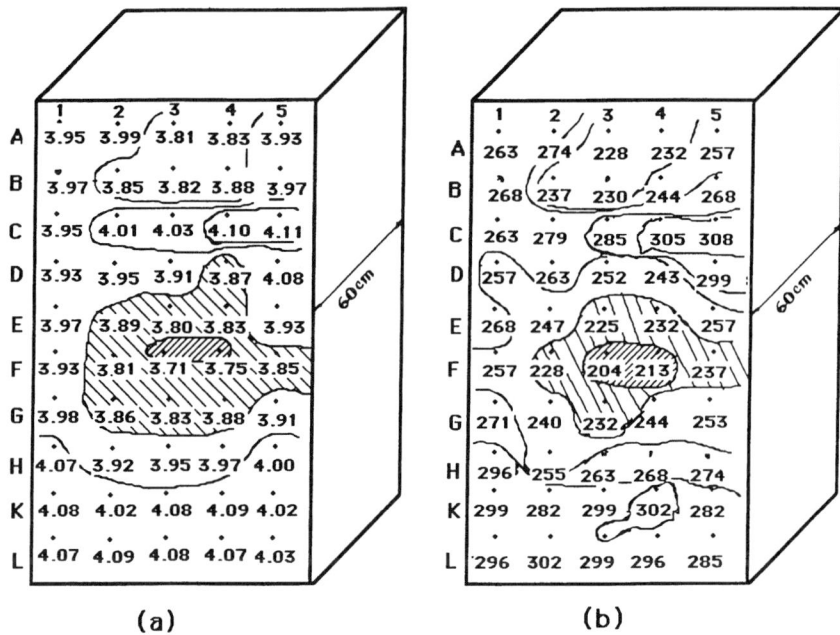

Figure 1.22 Nonuniformity of a concrete block by ultrasonic testing. (a) Pulse velocities in km/s. The lines represent equal velocities. (b) Compressive strengths in kg/cm². The lines represent equal strengths. 1 cm = 10 mm = 0.394 in. 1 km/s = 3281 ft/sec. 1 kg/cm² = 14.22 psi. (From Pohl, 1969a.)

different tests at the same time on the same specimen, or on the same location of a structure, and estimating the concrete strength from all results (RILEM Technical Committee, 1993). One such a combined method, use of the Windsor probe and the core test together, was mentioned earlier. In this case the strength estimate obtained by the Windsor probe can be considered as the upper limit and the standard core strength as the lower limit of the actual strength of a typical concrete in the structure (Lee et al., 1992).

Usually, however, the term *combined methods* means something else in nondestructive testing. Namely, the results of the various tests performed at the same time are combined in a suitable multivariable formula for the estimation of the concrete strength in the structure more accurately than from the individual results (Facaoaru et al., 1968; Facaoaru 1984; Skramtayev and Leschschinski, 1966a, 1966b; Tanigawa et al., 1980, 1984; Teodoru, 1988). The mathematical basis of this premise is that the accuracy of an approximation with one variable (simple correlation) can usually be improved by the introduction of a suitable second independent variable (multiple correlation). Several combinations are discussed below which show that strength estimates from these multiple measurements are not much more reliable than those from single measurements. It is recommended by Pohl (1969a) for the selection of nondestructive test methods for combination that the individual methods should measure different characteristics of the concrete. For instance, if the result of one of the methods is influenced primarily by the elastic properties of concrete, a suitable supplementary method could be another test that is controlled primarily by plastic properties. In practical terms this means that the accuracy of a hammer test method for strength estimation might be improved by measuring both the height of the rebound and the size of the indentation produced by the impact. Nevertheless, the true measure of the suitability of any combination of test methods is how much more accurate the estimation of concrete strength is from the combined measurements than from any of the single measurements. Combinations of rebound or penetration measurements with hardness determined with the Mohs scale were mentioned earlier in this section. Other combinations that have been tried include pullout or torque and ultrasonic pulse velocity tests (Pohl, 1969b), the pulse velocity method with measurement of the damping constant of the concrete; and the pulse velocity method with pulse attenuation measurement (Galan, 1967, 1990; Kesler and Higuchi, 1953). However, the latter two are laboratory techniques.

Density and moisture determinations may also supplement usefully other nondestructive measurements for strength estimation. The following formula by Brunarski as well as Wassmann and Eckhardt illustrates this combination (Pohl, 1969a):

$$f_c = 0.0145 w^{6.7} v_l^4 \qquad\qquad (1.45)$$

where f_c = compressive strength, kg/cm^2
$\quad\quad w$ = unit weight of the concrete, g/cm^3
$\quad\quad v_l$ = longitudinal pulse velocity, km/s

A similar combination is recommended by Osinski (Osinski, 1974, 1979) and others (Jones and Facaoaru, 1969).

The most popular combination has been the ultrasonic pulse velocity method along with the rebound hammer (Facaoarou and Stamate, 1969; Jones and Facaoaru, 1968a, 1969; Samarin and Meynink, 1981; Hamarin and Dhir, 1984) for estimation of the in-situ strength. That is, ultrasonic pulse velocity measurements are made on the concrete specimen or on in-situ concrete, and the rebound number is measured, for instance, by the Schmidt hammer. The two measurements are then substituted into an empirical formula obtained by multiple regression analysis to estimate the compressive strength. Malhotra and Carette (1980a) presented the following three examples for such formulas and for illustration of the improvement of the strength estimate resulting from the combined methods.

1. Samarin and Meynink

Test	Formula	Correlation Coefficient	Eq. No.
Rebound hammer	$f_c = A_1 R + A_2$	0.92	(1.46)
Ultrasonic pulse velocity	$f_c = A_3 v_l^4 + A_4$	0.87	(1.47)
Combined method	$f_c = A_5 R + A_6 v_l^4 + A_7$	0.95	(1.48)

where

f_c = estimated compressive strength of concrete
R = measured rebound number
v_l = measured longitudinal pulse velocity

A_1 through A_7 = empirical parameters to be obtained by regression analysis

2. Bellander (based on 221 measurements for each test)

Test	Formula	Correlation Coefficient	Eq. No.
Rebound hammer	$f_c = 0.00093 R^3 + 13.1$	0.92	(1.49)
Ultrasonic pulse velocity	$\ln f_c = 0.882 v_l - 0.259$	0.63	(1.50)
Combined method	$f_c = 0.00082 R^3 + 11.03 v_l - 32.7$	0.93	(1.51)

where ln is the natural logarithm, f_c is in MPa, and v_l is in km/s. (The conversion factors are: 1 MPa = 145 psi, 1 km/s = 3281 ft/sec.) The other symbols are the same as in Eqs. 1.46 through 1.48. Note that the numerical values of the parameters of these equations may change with any change in the composition of the concrete and in the curing and testing conditions.

3. Malhotra

Test	Formula	Correlation Coefficient	Eq. No.
Rebound hammer	$f_c = A_1 R + A_2$	0.92	(1.52)
Ultrasonic pulse velocity	$f_c = A_3 e^{A_4 v_l}$	0.93	(1.53)
Combined method	$\log f_c = A_5 \log R + A_6 v_l + A_7$	0.94	(1.54)

where the symbols are the same
as in Eqs. 1.46 through 1.48.

As an additional example, Pohl (1969a) quotes Slachta in connection with
good paving concretes:

$$f_c = 0.022R^{1.44}v_l^{2.63} \tag{1.55}$$

where the symbols are the same as in Eqs. 1.46 through 1.48. In this equation
the units are: f_c in kg/cm^2; v_l in km/s; and R in scale reading. The accuracy
of this equation is reported as $\pm 25\%$. The triple combination of rebound
hammer, pulse velocity, and pullout tests has also been proposed (Malhotra
and Carette, 1980a).

Several statistical methods are available for the development of such multivariable empirical formulas for the estimation of concrete strength from
combinations of the individual measurements. The most popular method is
curve fitting using *multivariable regression.* This is how most of the equations
in this section were obtained. The advantage of this method is that it is easy
to see through the correlation coefficients which test method contributes significantly to an improvement in the strength estimation and which one is
negligible. Another possibility is to use the statistical method called *discri-
minant analysis* (Van de Geer 1971). This can also show quantitatively the
extent to which each test method contributes to the strength estimation. Another method assigns an individual relative weight to each strength estimate
calculated from the individual tests, and sums these weighted strengths. Each
relative weight is inversely proportional to the variance of the results of the
test method in question. Another method is recommended by Skramtayev and
Leschschinski (1966a). This also calculates the most reliable strength estimate
from the individual strength estimates and variances of the test results obtained by two or more nondestructive test methods, but the use of the variances is different.

Incidentally, certain combined methods may contribute to an improvement
in the *uniformity testing* of concrete. For instance, if the strength estimate
obtained from a rebound test differs significantly from the strength estimate
obtained by ultrasonic pulse velocity on the same concrete, this may be an
indication that the quality of concrete on the surface differs from the quality
on the inside.

In any case, the above-discussed results of the combination of rebound and
pulse velocity tests suggest that the combined performance of both tests contributes little to the increased degree of accuracy of the estimation of the
compressive strength of concrete.

Further Applications of Sonic Methods

There are procedures based on sonic tests that can provide information about
certain technically important properties of concrete other than in-situ strength.

Since, however, these are also related to the load-carrying capacity of concrete structures, a brief description is presented below.

1. The pulse velocity technique has been used successfully for many homogeneous materials for the *detection and measurement of flaws, such as voids and cracks.;* therefore, efforts have been made to use it also in concrete (S. Popovics, 1986a; Strurrup, 1959; S. Popovics and J. S. Popovics, 1997). The idea is based on the observation that the pulse should pass around the end of the flaw; consequently, the pulse velocity calculated on the basis of a straight-line path will appear to be reduced suddenly on the spot of a flaw. The difference between this reduced pulse velocity and pulse velocity without the flaw can be used for calculation of the flaw depth. Since, however, the wavelength of the ultrasonic pulses usually applied in concrete is quite large, the method cannot detect cracks and voids smaller than about an 1 in. (25 mm). This insensitivity to small cracks has been shown experimentally by measuring pulse velocity in concrete specimens under increasing load, that is, under progressive cracking (S. Popovics, 1990b).

Note, however, that even the detection of large, hidden cracks is doubtful because reductions in pulse velocity can also be caused by the presence of a zone of low-strength concrete rather than the presence of a crack. Also, the orientation of the defects as well as the presence of moisture in the concrete can produce sizable errors in the measurements. The situation is complicated further if more than one crack is present in the mass of concrete between the transducers. That is, this method could be used for crack detection only when the concrete has a highly uniform quality and the pulse velocity is known to high accuracy in the uncracked concrete. Since this is not the case in most damaged (cracked) concrete structures, any attempt to use pulse velocity measurements for crack detection and characterization is futile, and may be misleading. A similar principle is applicable for the determination of the thickness of concrete and asphalt layers (Bungey, 1982) and for the characterization of reinforcement in concrete (J. S. Popovics et al., 1992) with the pulse velocity technique.

2. The *resonance frequency test* was developed by Powers in 1938 (Powers, 1938). The method, described in ASTM C215-91, covers measurement of the fundamental transverse, longitudinal, and torsional frequencies of concrete prisms and cylinders for the purpose of calculating dynamic (resonance) modulus of elasticity, dynamic (resonance) shear modulus, and dynamic (resonance) Poisson's ratio (Malhotra, 1991c). The fundamental resonant frequencies can be determined using one of two alternative procedures: the forced resonance method or the impact resonance method. In the *forced resonance method,* a supported specimen is forced to vibrate by an electromechanical driving unit. The specimen response is monitored by a lightweight pickup unit on the specimen. The driving frequency is varied until the measured specimen response reaches a maximum amplitude. The value of the

frequency causing maximum response is the resonant frequency of the specimen. The fundamental frequencies for the three different modes of vibration are obtained by proper location of the driver and the pickup unit.

In the *impact resonance method,* a supported specimen is struck with a small impactor and the specimen response is measured by a lightweight accelerometer on the specimen. The output of the accelerometer is recorded. The fundamental frequency of vibration is determined by using digital signal processing methods or counting zero crossings in the recorded waveform. The fundamental frequencies for the three different modes of vibration are obtained by proper location of the impact point and the accelerometer. This method has the advantage that it can generate pulses of very high intensity. The same testing procedure should be used for all specimens of an associated series. This method is essentially a laboratory method, intended primarily for detecting significant changes in the resonance modulus of elasticity of concrete specimens that are undergoing exposure to weathering or other types of potentially deteriorating influence.

The fundamental resonance frequency is controlled by the modulus of elasticity of the concrete more intensively than the pulse velocity (S. Popovics, 1970a; Jones and Facaoaru, 1969). Thus it can provide better indirect information about concrete strength (Pohl, 1962). In other words, a reduction of the resonance frequency is a better indicator of deterioration and strength reduction in a concrete specimen than is a reduction in pulse velocity. Note, however, that different computed values of resonance modulus of elasticity, or strength, may result from widely different resonance frequencies of specimens of sizes and shapes of the same concrete. Therefore, comparison of results from different specimens should be made with caution. Also, Jones reported that for wet concrete there was no appreciable difference in the resonance (dynamic) modulus of elasticity determined from the flexural and longitudinal modes of vibration despite the difference in the frequencies. The resonance shear modulus and the damping coefficient also showed no appreciable change with frequency. When the concrete was allowed to *dry,* however, the modulus of elasticity calculated from the flexural vibrations was lower than that calculated from longitudinal vibrations (Jones, 1957). Unfortunately, the resonance frequency method is rarely suitable for field use, mainly because there is no method at this time for the determination of the effect of geometry of and the reinforcement in the tested elements on the results. Details of the resonance test methods are given in several specifications, such as ASTM C215, and BS 1881, Part 5.

3. There is a mechanical version of the *pulse-echo method,* called the *impact-echo method.* This is an analysis of reflected pulse traces to detect defects inside the concrete. A single mechanical blow is applied to the concrete surface and the out-of plane component of the resulting transient displacement on the surface of the structure, at a point near the impact site (within a few centimeters), is monitored with a piezoelectric, broadband, dis-

placement transducer. The captured time-domain signal is then transformed into the frequency domain with the Fourier transform technique. Study of the frequency values of peaks in the magnitude spectrum enables detection of reflectors (i.e., defects), such as weak spots, cracks, and voids, within the concrete for the region immediate to the impact (Bungey, 1982; Carino, 1984 Carino and Sansalone, 1992; Sansalone and Carino, 1988, 1989, 1991, 1995; J. S. Popovics et al., 1995). The interpretation of the frequency peak values with the existing analytical method depends on the cross-sectional form of the structure being tested. For a large plate of thickness b and bulk longitudinal wave velocity v_l one frequency peak is obtained, the value of which is given by $f_{plate} = v_l/2b$; this has been well established in the literature through laboratory and field tests. For structures such as long cylindrical columns, rectangular columns, and hollow cylindrical rods, multiple-frequency peaks are expected (J. S. Popovics and Rose, 1996). This existing analysis approach for these, however, relies on *empirical* adjustments to the established formulation for platelike structures in order to obtain results for cross-sectional forms such as solid cylindrical rods, and hollow cylindrical rods; thus its validity is quite limited. Also, the analysis of the data is quite complicated and at present still ambiguous (J. S. Popovics and Rose, 1993). It has been proposed that an analysis approach based on the theory of *elastic guided wave propagation* would be more suitable for impact echo data than the existing analysis approach. It would result in more powerful signal excitation and capture schemes, superior analysis of data, and therefore, improved flow detection (J. S. Popovics, 1993, 1994; J. S. Popovics and Rose, 1994).

4. Ultrasonic techniques have also been used for the determination of *thickness of concrete pavements* (Golis, 1968; Mailer, 1972). Some of the methods require the determination of wave velocity in the concrete as well as the longitudinal resonance frequency of the slab when vibrated in its thickness dimension (Muenow, 1963; Takabayashi and Ishida, 1984; Jones, 1949, 1955; Jones and Mayhew, 1966; S. Popovics and J. S. Popovics, 1997; J. S. Popovics and Achenbach, 1996).

5. The property of a material causing free vibration in a specimen to decrease in amplitude is called *attenuation* or *damping capacity.* One of its numerical characteristics is the *logarithmic decrement.* Most of the damping in concrete is due to absorption and scattering from aggregate particles, pores, and cracks. Thus, it occurs in the matrix, with some on the paste–aggregate interfaces and a little in the aggregate (Swamy, 1971; Suaris and Fernando, 1987a, 1987b; Akashi, 1960; Landis and Shah, 1995). A significant increase in attenuation could indicate an increase in porosity, voids, a cracked region, or weaknesses in the paste–aggregate or paste–reinforcement interfaces. Nevertheless, attenuation to locate *defects* in concrete has attracted only limited attention, although the technique has been applied successfully to the characterization of other engineering materials (Adler et al., 1986; Martin, 1976; Stone and Clarke, 1975). The presence of air voids in dry specimens

contributes little to the damping capacity, but moisture in the pores causes major damping. Thus, damping characteristics may also be used in the laboratory to detect variations of *moisture content.* Attempts to correlate attenuation to concrete *strength* have been unsuccessful, as the method produces less reliable strength estimates than the standard pulse velocity method. However, as was mentioned earlier, a small improvement can be obtained by the use of pulse velocity and attenuation results in combination for estimation of the concrete strength (Kesler and Higuchi, 1953; Galan, 1990).

In brief, the presently available nondestructive test methods do not provide reliable estimates for the in-situ strength of concrete either individually or in combination without a specific calibration curve developed for that concrete. Even the best ones may produce unacceptable errors. The greater the maximum particle size and the lower the cement content, the more uncertain the estimated concrete strength. Therefore, they must not be used as a substitute for standard direct strength tests. However, when performed along with direct tests, they can reduce the cost of testing. These nondestructive methods are much more suitable for checking the concrete *uniformity* in various points of a structure at the same age, including the estimation of *strength differences,* or for *monitoring* changes in the concrete strength in the same point of the structure at various ages.

As far as the selection of the nondestructive test method is concerned that is most suitable for a given purpose, this writer's subjective opinion is the following:

1. For determination of the safe *form-removal time,* the standard *pullout test* seems the best because its results have the best reproducibility, its calibration curve is least sensitive to the aggregate properties, and the requirement that the steel inserts be installed during construction does not cause any difficulty here. In addition, modification(s) of an *ultrasonic pulse velocity* method also seems useful for form-removal-time determination.

2. For in-situ estimation of the *compressive strength* of concrete, the standard *rebound hammer* seems best because of the ease and economy of its application in the field and the relatively good reproducibility of its results. This recommendation assumes that the surface properties of the tested concrete are the same as the properties inside the concrete. Also, the standard *pullout tests* have shown relatively good results for estimation of the compressive strength.

3. If there are reasons to suspect that the surface properties do not represent the properties of the entire concrete mass, the standard *ultrasonic pulse velocity test* seems best.

4. For the in-situ estimation of the *flexural strength* of concrete, the *break-off method* seems best because here the testing conditions are closest to the standard testing of the flexural strength.

1.6   VARIABILITY OF COMPRESSIVE STRENGTH

Variability Sources

Since the failure of concrete under load is essentially a random process (Section 2.4), even macroscopically identical concrete specimens would not be likely to fail at exactly the same stress. It is, therefore, reasonable to follow S. Walker's suggestion to use statistical methods for the treatment of variability, that is, for the estimation and specification of concrete strength (Walker, 1955). Nowadays excellent statistical methods and computer softwares are available for evaluation of concrete properties, especially strength (Balaguru and Ramakrishnan, 1987a).

The overall variability in concrete strengths, as well as in many other properties, can be divided into two portions:

1. The variability in strength caused by differences in the macroscopic features (composition, etc.) of the concrete specimens
2. The variability in strength caused by the random nature of concrete failure as well as by outside sources

The first group of sources of variability includes inherent variability of each of the ingredients entering into the mixture as well as the variabilities in the processes of batching, placing, and compacting. The second group includes variability from sampling, curing, testing, and so on. The principal sources of strength variation are shown in Table 1.4. A similar but more detailed list was presented earlier by Mercer (1951). The numerical characterization of the variability is discussed elsewhere (S. Popovics, 1982b). It is important to realize that uniform strength results are not necessarily accurate strength results.

Individual Values and Averages

The distribution of *individual strengths* of comparable concrete specimens is usually assumed normal, although logarithmic-normal distribution has also been recommended (Soroka, 1968; Torrens, 1978; Hindo and Bergstrom, 1985). Both theory and experience have shown, however, that the *average strength* of a set of nominally identical, carefully produced concrete cylinders or cubes usually follows the normal distribution with an acceptable approxi-

TABLE 1.4 Principal sources of strength variation

Variations in the Properties of Concrete	Discrepancies in Testing Methods

Changes in water-cement ratio
  Poor control of water
  Excessive variation of moisture in
    aggregate
  Retempering
	Improper sampling procedures

Variations in water requirement
  Aggregate grading, absorption,
    particle shape
  Cement and admixture properties
    Poor quality molds
    Delivery time and temperature
	Variations due to fabrication
	    techniques
	  Handling and curing of newly
	    made cylinders
	  Air content

Variations in characteristics and
  proportions of ingredients
  Aggregates
  Cement
  Pozzolans
  Admixtures
	Changes in curing
	  Temperature variation
	  Variable moisture
	  Delays in bringing cylinders to the
	    laboratory

Variations in mixing, transporting,
  placing, and compaction
	Poor testing procedures
	  Cylinder capping
	  Compression tests

Variations in temperture and curing	

Source: ACI (1977). (Copyright ACI. Reprinted with permission.)

mation (Novgorodsky, 1973; Mathews and Metcalf, 1969; Rusch et al., 1969; Balaguru and Ramakrisknan, 1987b).

A simple way to check visually how well a distribution approximates normality is to plot the cumulative occurrences of the variable on a scale of the normal probability integral. The better the plotted values approximate a straight line in this system, the closer is the distribution of the variable to normal distribution (Fig. 1.23). Another example for the distribution of concrete strengths is presented in Figure 1.24.

If the strengths follow the normal distribution adequately, it is possible to calculate a minimum expected strength for any percentage of results falling below this minimum from known average $\bar{x}$ as well as the standard deviation s or the coefficient of variations v. For instance, when the standard deviation is used to characterize the variability

$$x_m = \bar{x} - ts \tag{1.56}$$

When the coefficient of variation is used,

$$x_m = \bar{x}(1 - 0.01tv) \tag{1.57}$$

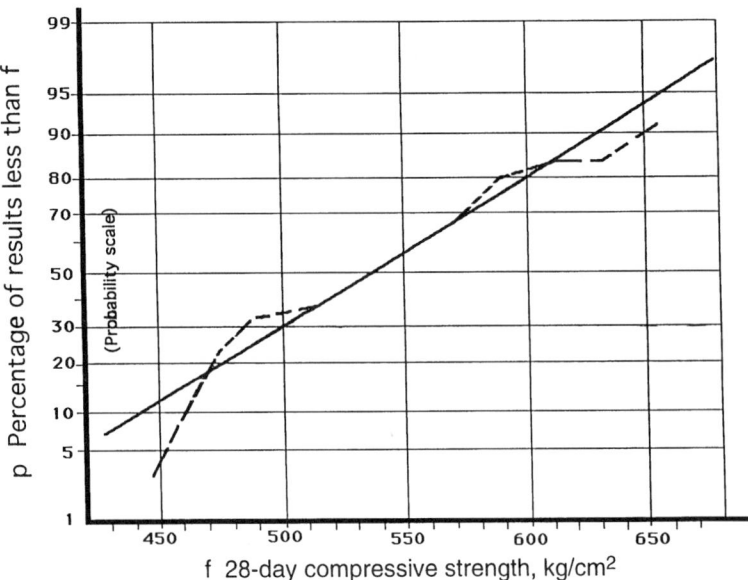

Figure 1.23 Comparison of the distribution of strengths of standard mortar cubes to normal distribution. Solid line, theoretical normal distribution; dashed line, actual distribution. Number of tests: 153. 1 kg/cm² = 142 psi = 0.098 MPa. (From S. Popovics and Ujhelyi, 1955.)

where x_m = lower limit of strength below which only a specified p percentage of the strength may fall

$\bar{x}$ = arithmetic average of the individual strength results

t = factor depending on the p percentage of results allowed to fall below x_m and on the sample size (see Table 1.5).

For very large samples ($n \approx \infty$) the t values are identical with the corresponding values of the normal distribution. It can be seen from Eqs. 1.56 and 1.57 that s and v have considerable influence on the value of x_m (Abdun-Nur, 1966).

PROBLEM 1.1 The average compressive strength of a large series of concrete cylinders is $\bar{x}$ = 3000 psi (20.7 MPa); the standard deviation of the individual results is s = 500 psi (3.45 MPa). What is the strength x_m below which only 1% of the individual strength results is expected to fall in the total volume of concrete?

Solution From Table 1.5 the value of the appropriate t is 2.326. Therefore, from Eq. 1.56,

$$x_m = 3000 - 2.326 \times 500 = 1835 \text{ psi}$$

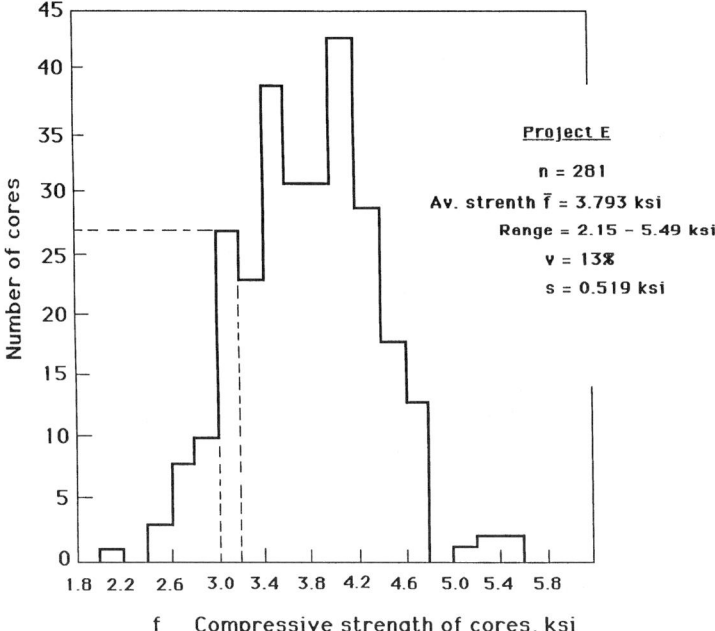

Figure 1.24 Distribution of the compressive strengths of concrete cores taken from one project. It shows, for example, that the number of cores having compressive strengths between 3.0 and 3.2 ksi (20.7 and 22.1 MPa) was 27. 1 ksi = 6.90 MPa. (From Newlon, 1966.)

In other words, the concrete in question is expected to provide compressive strengths greater than 1835 psi (12.7 MPa) 99 times out of 100.

The standard deviation of a set of sample averages drawn from the same large population is less than the standard deviation of the population of the individual values:

$$s_n = \frac{s}{\sqrt{n}} \tag{1.58}$$

where s_n = expected value of the standard deviation of sample averages
s = standard deviation of the population
n = number of specimens in the samples

The uncertainty of any average also decreases with increasing number of specimens tested under the same conditions. The limits of this uncertainty with respect to the unknown but true average of a large number of similar tests can be determined from Table 1.5 or by multiplying the computed sample deviation for the test group by a factor a which depends on the number

TABLE 1.5 Values of t^a

$100p$ Percentage of Tests Falling Within the Limits $\bar{x} \pm ts$

Number of Samples	Degrees of Freedom	50	60	70	80	90	95	98	99
		Chances of Falling Below Lower Limit							
		2.5 in 10	2 in 10	1.5 in 10	1 in 10	1 in 20	1 in 40	1 in 100	1 in 200
2	1	1.000	1.376	1.963	3.078	6.314	12.706	31.821	63.657
3	2	0.816	1.061	1.386	1.886	2.920	4.303	6.965	9.925
4	3	0.765	0.978	1.250	1.638	2.353	3.182	4.541	5.841
5	4	0.741	0.941	1.190	1.533	2.132	2.776	3.747	4.604
6	5	0.727	0.920	1.156	1.476	2.015	2.571	3.365	4.032
7	6	0.718	0.906	1.134	1.440	1.943	2.447	3.143	3.707
8	7	0.711	0.896	1.119	1.415	1.895	2.365	2.998	3.499
9	8	0.706	0.889	1.108	1.397	1.860	2.306	2.896	3.355
10	9	0.703	0.883	1.100	1.383	1.833	2.262	2.821	3.250
11	10	0.700	0.879	1.093	1.372	1.812	2.228	2.764	3.169
16	15	0.691	0.866	1.074	1.341	1.753	2.131	2.602	2.947
21	20	0.687	0.860	1.064	1.325	1.725	2.086	2.528	2.845
26	25	0.684	0.856	1.058	1.316	1.708	2.060	2.485	2.787
31	30	0.683	0.854	1.055	1.310	1.697	2.042	2.457	2.750
∞	∞	0.674	0.842	1.036	1.282	1.645	1.960	2.326	2.576

^{a}Values of t extracted from table originally produced by Fisher and Yates, *Statistical Tables for Biological Agriculture and Medical Research*.

of tests in the group and on the desired degree of probability for falling within the limits. This relationship is shown graphically in Figure 1.25. The $p = 0.9$ probability of falling within the limits (i.e., 9 times out of 10) is often adequate for engineering work, and $p = 0.99$ is generally regarded as near certainty (Troxell, 1968). The *probable error* is a special case of limits of the uncertainty for $p = 0.5$.

An important utilization of these statistical methods is the estimation of the the target strength required for proportioning.

PROBLEM 1.2   The average compressive strength of a series consisting of 25 concrete cylinders is 4215 psi (29.1 MPa). The standard deviation is known to be 420 psi (2.90 MPa) from previous experience. Calculate the *limits of uncertainty* for this average compressive strength so that the true average strength fall within these limits 9 times out of 10.

First Solution   From Figure 1.25, the value of a for $n = 25$ and $p = 0.9$ is 0.34. Therefore, the limits of uncertainty are $\pm 0.34 \times 420 = \pm 143$ psi (± 1.0 MPa), and the strength range within which there are 9 chances out of 10 that similar future averages will fall is 4215 ± 143 psi, or from 4072 to 4358 psi (28.1 to 30.1 MPa).

Second Solution   From Table 1.5 with interpolation, $t = 1.711$ for $n = 25$ and $p = 0.9$. Therefore, from Eq. 1.58 the limits of uncertainty for the average in question are

$$\pm 1.711 \times \frac{420}{\sqrt{25}} = \pm 143 \text{ psi (1 MPa)}$$

If the standard deviation is not known with adequate accuracy but rather, is calculated from the 25 available strength results, a correction factor should also be applied from Table 1.6.

TABLE 1.6   Multiple factor for correcting standard deviation based on fewer than 30 tests

Number of Tests	Multiplying Factor	Number of Tests	Multiplying Factor	Number of Tests	Multiplying Factor
10	1.36	17	1.14	24	1.05
11	1.31	18	1.12	25	1.04
12	1.27	19	1.11	26	1.03
13	1.24	20	1.09	27	1.02
14	1.21	21	1.08	28	1.02
15	1.18	22	1.07	29	1.01
16	1.16	23	1.06	30	1.00

Source: Philleo (1981). (Copyright ACI. Reprinted with permission.)

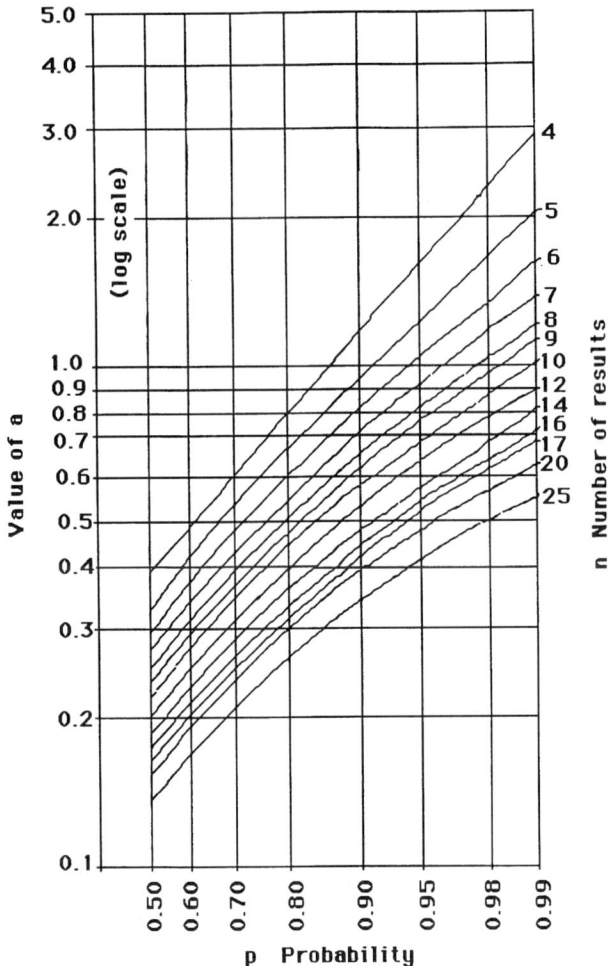

Figure 1.25 The factor a for calculating 50 to 99% confidence limits for averages of n test results. $a = t/\sqrt{n}$. (From ASTM, 1976. Copyright ASTM. Reprinted with permission.)

PROBLEM 1.3   Calculate the number of concrete cylinders needed for determination of the average compressive strength with an accuracy such that at least 98 times out of 100 ($p = 0.98$) the average of the test group is within 5% of the true average, that is, the average of a very large but otherwise similar group.

Solution   If the coefficient of variation can be assumed as 8.8%, the value of a is

$$a = \frac{5}{8.8} = 0.568$$

It can be seen from Figure 1.25 that the intersection of the values of $p = 0.98$ and $a = 0.568$ indicates a need for 20 specimens to meet the specified accuracy requirement.

Note, however, that a little caution is appropriate here. It may be financially beneficial for the concrete producer not to base determination of the target strength or the needed sample size solely on probabilities obtained from prior information and sampling information but also on the producer's risk, including the engineer's judgment of the existing economic situation. The reason for this is that the replacement of a substandard concrete on one job may cost more than replacement of the same concrete volume on another job. It has been shown that such refinement of the sample size can be done through the use of Bayes' theorem of information theory (Webster, 1971).

Logarithmic-Normal Distribution

In the case of logarithmic-normal distribution the logarithms of the random variable follow the normal distribution. Therefore, the minimum expected strengths for such distribution can be calculated from transformation of Eq. 11.56 (Hindo and Bergstrom, 1985):

$$\ln x_m = \overline{\ln x} - t s_{\ln x} \tag{1.59}$$

$$x_m = \exp(\overline{\ln x} - t s_{\ln x}) \tag{1.60}$$

where $\overline{\ln x}$ and $s_{\ln x}$ are the arithmetic average and standard deviation, respectively, of the logarithms of the individual strength results. That is,

$$\overline{\ln x} = \frac{1}{n} \sum \ln x \tag{1.61}$$

$$s_{\ln x} = \left[\frac{\sum (\ln x)^2 - n(\overline{\ln x})^2}{n - 1} \right]^{0.5} \tag{1.62}$$

ln is the natural logarithm.

Effect of Concrete Strength on Variability

Ideally, the method of measuring the variation of the individual results about the average should be affected only by the degree of control exercised in the production and application of the concrete, and in particular, it should be independent of the average strength of the concrete.

There are conflicting opinions based on conflicting experimental evidence as to whether the standard deviation or the coefficient of variation fulfills this requirement. For instance, Himsworth, among others, has the opinion based on Figure 1.26 that the standard deviation of the compressive strength results is practically independent of the average strength (Himsworth, 1954). Neville (1959) has formed the view that under laboratory conditions the standard deviation is proportional to the average strength of the concrete (Fig. 1.27); that is, the coefficient of variation is independent of the average strength. Murdock suggested that up to a certain strength (3000 psi in Fig. 1.28) the coefficient of variation may be considered independent of the average strength; at higher strengths, the standard deviation may be so considered. Although Erntroy's experimental data also support the latter approximation, he rejected any simple relationship between concrete strength and variability, and instead, recommended the use of an empirical linear relationship between the two water-cement ratios that are needed to provide the $\bar{x}$ and x_m strengths, respectively (McIntosh, 1964).

Neville also found that the standard deviation of comparable concretes increased in proportion to the strength as the strength of the concrete in-

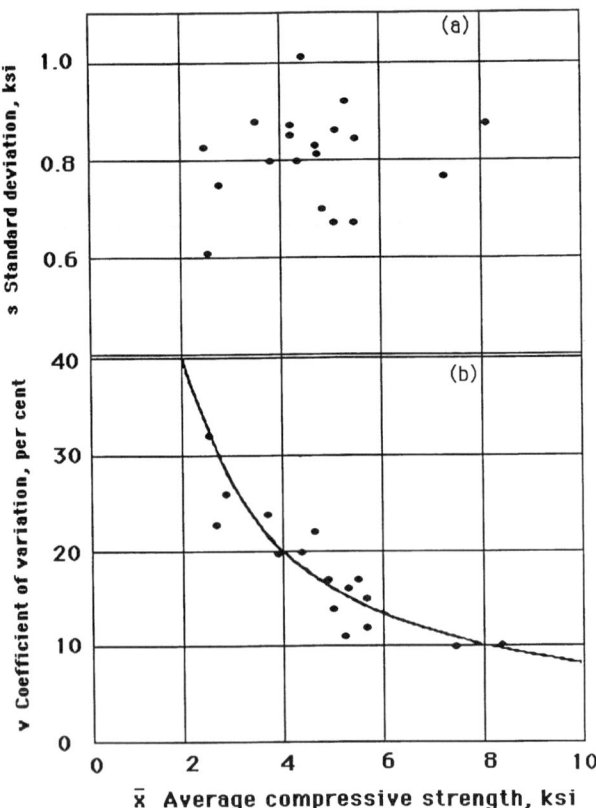

Figure 1.26 Standard deviation and coefficient of variation of compressive strengths as a function of the average compressive strength under field conditions. 1 ksi = 6.90 MPa. (From Himsworth, 1954. Copyright Thomas Telford Publishing. Reprinted with permission.)

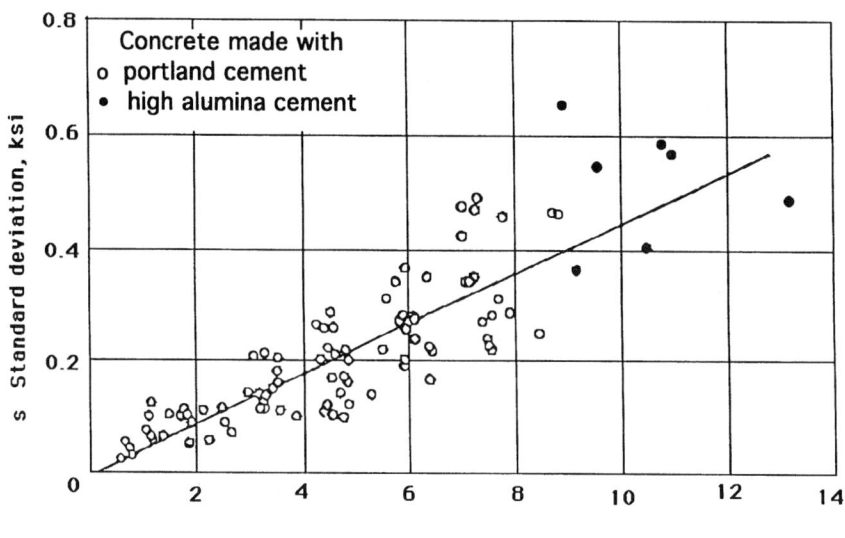

Figure 1.27 Standard deviation of compressive strengths as a function of the average compressive strength of laboratory specimens. 1 ksi = 6.90 MPa. (From Neville, 1959. Copyright Thomas Telford Publishing. Reprinted with permission.)

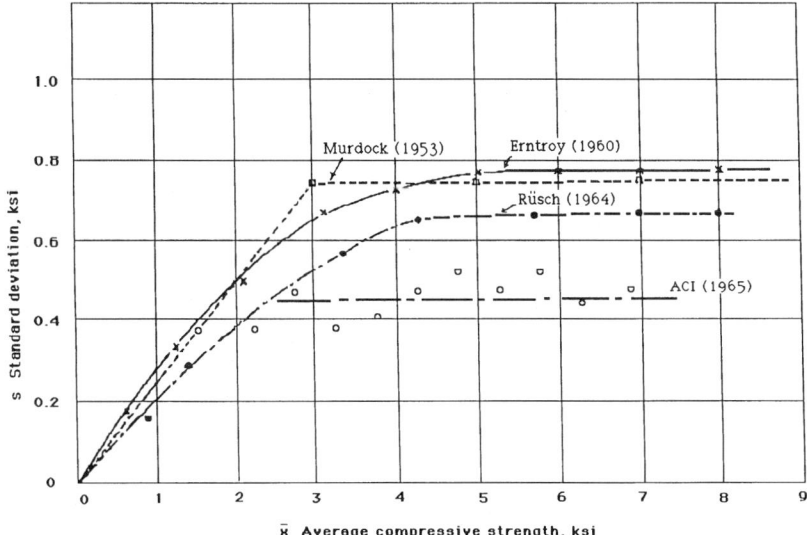

Figure 1.28 Relationship of standard deviation and average compressive strength. 1 ksi ≈ 6.90 MPa. (From Newlon, 1966.)

creased with age. Data by Rusch et al. (1969), however, show that this is not true for high-strength concretes.

The probable reason for these contradictions is that in certain cases the degree of control of the concretes compared was the same for the entire strength range, while in other cases the control of concretes of higher strength was stricter, which reduced the variation of strength results. The first type of case is typical of laboratory conditions, where, therefore, the coefficient of variation is likely to be more independent of the average concrete strength than the standard deviation. The second type of case is typical of field conditions where, therefore, the standard deviation is likely to be less dependent. This could be the reason that more and more specifications have been adopting the standard deviation for the practical characterization of the variability of compressive strength (Soroka, 1971; S. Popovics, 1979; Teychenne, 1973; RILEM/CEB/CIB/FIP, 1975). ACI Standard 214 and DIN 1084 also recommend the standard deviation for the characterization of the overall variability of strength, although ACI also states that the coefficient of variation is more suitable for the within-batch variations.

The variations in concrete strength depend primarily on the degree of control applied during the making and using of fresh concrete (Abdun-Nur, 1966; Smith, 1971). It appears from the huge amount of published data that for concretes for usual structural or paving purposes in a single project an overall standard deviation of 500 to 550 psi (3.5 to 3.8 MPa, 35 to 40 kg/cm^2) or an overall coefficient of variation of 10% is an achievable value under well-controlled field conditions. This variability can be reduced to about half under strict laboratory conditions, but it will be doubled or even tripled under poor control (ACI, 1977).

It follows from the demonstrated variability of concrete strength that caution should be exercised when conclusions are to be drawn from small apparent differences in concrete strength, particularly when the number of specimens tested is small. Halstead, for instance, calculated (Halstead, 1969) that (1) if two sets of three cubes from a single batch of six were used and three cubes are kept for control purposes, the average strength of the treated cubes (kept at differing temperature, etc.) must differ from the average strength of the controls by at least 7.5% for the effect of the treatment to be regarded as significant with 95% confidence; and (2) if a control set of three cubes is made without treatment and another set of three cubes is made from another batch with the treatment (with different cement, admixture, etc.), the average strength of the two sets must differ by at least 16% at 3 days, 12.5% at 7 days, 11% at 28 days, or 10% at 56 days for the treatment to be regarded as having had a significant effect. These percentages are valid only for the level of testing variability produced in Halstead's experiments.

Components of the Variability of Compressive Strength

Although information considering the total variation of compressive strength results is useful and can be applied to design specification limits as well as form a view of the overall uniformity of the concrete, it is of little worth to the problem of finding the sources of the variations and reducing them in concrete construction. The basic need is to isolate the factors affecting variability, that is, to break down the total variation into meaningful components.

This breakdown is done by special statistical calculations, called *analysis of variance,* based on results of samples taken appropriately (Brown, 1966). The advantage of using the s^2 variance instead of the s standard deviation is that the total variance is equal to the sum of the component variances, as illustrated by Eq. 1.63. Three methods of breakdown are discussed.

1. The first such attempts estimated the effects of the variations in the qualities and quantities of the concrete components. It was found, for instance, that the primary such factor is fluctuation of the water-cement ratio. According to other evaluations, variation in the strength-producing capability of the cement contributes not more than 10% of the total variation in concrete strength (Troxell et al., 1968), although higher values have also been reported (Erntroy, 1960; Wright, 1958), especially for high-strength concretes.

2. Another approach is to establish (a) the strength variation within a mass of concrete that makes up a single batch (*within-batch variation*); and (b) the variation from batch to batch (*between-batch variation*). The following relationship exists:

$$s^2_{total} = s^2_{within} + s^2_{between} \qquad (1.63)$$

Because many of the factors, such as mixing time, batching weights, and so on, are identical, or at least more identical, for a single batch, the within-batch variation is expected to be less than half of that between batches under normal conditions (Table 1.7). On the other hand, if some of the factors are producing nonuniform distribution of the components in the concrete, the within-batch variation may be large. Furthermore, it is apparent that the standard deviation is apt to be larger in each subsequent case of the following sequence (Newlon, 1966): within batch; between batches on the same job; between jobs within the same region or from a single source; and between large regions or countries.

Since the within-batch and between-batch variations are the most important for construction engineers and inspectors, ASTM has accepted this characterization for the precision of the various standardized test methods. For example, the "Standard Method of Making, Accelerated Curing, and Testing of Concrete Compression Test Specimens" (C684-95) contains the following precision statement:

1. The *single-laboratory coefficient of variation* has been determined as 3.6% for a pair of cylinders cast from the same batch, as used in procedures A, B, and C. Therefore, results of two properly conducted strength tests by

TABLE 1.7 Between-batch and within-batch variances obtained in a laboratory[a]

Cement	Strength as Average of All Cubes (MPa)				Total Variation[b] (MPa)				Between-Batch Variation[b] (MPa)				Within-Batch Variaiton[b] (MPa)				Total Coefficient of Variation (%)			
									Age at Test (days)											
	3	7	28	56	3	7	28	56	3	7	28	56	3	7	28	56	3	7	28	56
A	18.46	26.34	37.21	42.02	1.19 (38)	1.44 (39)	2.00 (41)	2.23 (41)	1.63 (18)	1.99 (19)	2.68 (20)	2.89 (20)	0.48 (19)	0.52 (20)	0.99 (21)	1.30 (21)	6.5	5.5	5.4	5.3
B	17.26	23.81	36.18	41.53	0.97 (39)	1.01 (35)	1.37 (27)	1.31 (23)	1.33 (19)	1.28 (17)	1.32 (13)	1.30 (11)	0.42 (20)	0.68 (18)	1.41 (14)	1.32 (12)	5.6	4.3	3.8	3.2
C	19.46	26.61	35.78	40.07	1.27 (45)	1.21 (49)	1.41 (47)	1.59 (47)	1.73 (22)	1.42 (24)	1.75 (23)	2.02 (23)	0.54 (23)	0.97 (25)	0.96 (24)	1.02 (24)	6.5	4.5	3.9	4.0
D	19.77	28.87	40.98	45.71	1.43 (31)	1.50 (31)	1.86 (31)	2.34 (31)	1.83 (15)	2.00 (15)	2.23 (15)	3.16 (15)	0.90 (16)	0.76 (16)	1.44 (16)	1.12 (16)	7.2	5.2	4.5	5.1
E	16.26	23.40	36.95	43.98	1.15 (74)	1.33 (77)	1.79 (41)	1.61 (35)	1.92 (24)	2.10 (25)	2.76 (13)	2.29 (11)	0.43 (50)	0.70 (52)	1.06 (28)	1.17 (24)	7.1	5.7	4.8	3.7

Source: Halstead (1969).

[a]1 MPa = 145 psi.

[b]The figures in parentheses are the degrees of freedom associated with each value.

the same laboratory on two individual cylinders made with the same materials should not differ more than 10% of their average.

2. The *single-laboratory, multiday coefficient of variation* has been determined as 8.7% for the average of pairs of cylinders cast from single batches mixed on two days, as used in procedures A, B, and C. Therefore, results of two properly conducted strength tests each consisting of the average of two cylinders from the same batch made in the same laboratory with the same materials should not differ by more than 25% of their average.

3. The third approach is to separate the components of variability resulting from the differences in the macroscopic features of the concrete specimens from those that result from outside sources. An example for this two-way breakdown is provided by Erntroy (1960). Also, separation of the materials variance from the testing variance within the total variance of strength results is possible (S. Popovics and Ujhelyi, 1955; S. Popovics, 1982b). Another example, for three components, is given for the variability components of compressive strength in Tables 1.8 and 1.9. Here the s_{total}^2 variance is broken down into three components: testing variance, sampling variance, and materials variance. It can be seen from these tables that the variations caused by testing errors form a considerable portion of the total variation of the compressive strength determined by the standard method in the field. Similar statistical evaluations showed the same tendency (Teychenne, 1973; David, 1967).

As pointed out earlier, if concrete samples were taken and tested from various portions of a batch as was done for the measurements resulting in Table 1.7, except for sampling and testing errors, the specimens would reflect the within-batch variation. In actual practice, specifications for sampling concrete, such as AASHTO T141 and ASTM C172, either require that the sample be taken from the center portion of the batch, or that the sample be remixed before making test specimens. Therefore, the use of multiple specimens taken from a single batch by the standard method does not furnish within-batch variation but measures primarily the combination of sampling and testing errors. In other words, the variation obtained with such specimens as within-batch variation is actually testing and sampling variation (Newlon, 1966). Further details concerning pertinent methods and formulas are available in most textbooks on statistics and in publications related specifically to the variability of concrete (Cordon, 1966; Neville and Kennedy, 1964; McIntosh, 1963; ASTM, 1950, 1964).

1.7 JUDGMENT OF ACCEPTABILITY OF CONCRETE BASED ON STRENGTH

Producer's Risk and Consumer's Risk

An important application of the statistical concepts discussed previously is in the judgment of the acceptability of a concrete. The rational approach to this

TABLE 1.8 Components of overall variation in compressive strength of concrete: I[a]

				28-Day Compressive Strength Variations					
Project	Ave., x (psi)	Overall Standard Deviation, σ_0 (psi)	Overall Coefficient of Variation, v_0 (%)	Standard Deviation, Testing, σ_t (psi)	Coefficient of Variation, Testing, v_t (%)	Standard Deviation, Sampling, σ_s (psi)	Coefficient of Variation, Sampling, v_s (%)	Standard Deviation, Materials, σ_m (psi)	Coefficient of Variation, Materials, v_m (%)
Structural Concrete									
1	4235	425	10.0	170	4.0	236	5.6	310	7.2
2	4420	482	10.9	323	7.3	39	0	360	8.1
Paving Concrete									
1	4675	545	11.7	377	8.1	91	0	386	8.3
2	3755	420	11.2	322	8.5	42	0	270	7.1
3	3720	575	15.5	318	8.5	—	0	495	13.3
4	4760	467	9.8	200	4.2	34	0	420	8.8
5	4688	773	16.5	585	12.5	—	0	545	11.7

Source: Baker and McMahon (1969).

[a] 1 psi = 0.0069 MPa.

TABLE 1.9 Components of overall variation of compressive strength of concrete: II[a]

Project	Overall Variance, σ_0^2 (psi)^2	%	Testing Variance, σ_t^2 (psi)^2	%	Sampling Variance, σ_s^2 (psi)^2	%	Materials Variance, σ_m^2 (psi)^2	%
					28-Day Compressive Strength Variances			
			Structural Concrete					
1	180,630	100	28,900	16.0	55,690	30.8	96,100	53.2
2	232,324	100	104,329	44.9	1,521	0.6	129,600	55.8
			Paving Concrete					
1	297,025	100	142,120	47.8	8,281	2.8	148,996	50.2
2	176,400	100	103,684	58.8	1,764	1.0	72,900	41.3
3	330,625	100	101,124	30.6	—	—	245,025	74.1
4	218,089	100	40,000	18.3	1,156	0.5	176,400	81.0
5	597,529	100	342,225	57.2	—	—	297,025	49.7

[a] 1 psi = 0.0069 MPa.

is by a statistical evaluation of a group of results obtained from tests on specimens made from samples of the concrete in question (Anon., 1966). Two such methods are illustrated below using the compressive strength as the measured property. To the reader of this section it should already be obvious that regardless of what statistical method or concrete property is used for judging the acceptability of the concrete, the judgment is inevitably subject to risk of mistaken decision. On the one hand, the *producer's risk* is that concrete of acceptable quality may nevertheless be judged unacceptable occasionally on the evidence of valid, but pessimistic test results. On the other hand, the *consumer's risk* is that a concrete of quality below the specified level may nevertheless be judged acceptable on the evidence of valid, but optimistic test results. It is desirable that in acceptance criteria both risks be minimized within the limits of practicality.

Operating Characteristic Curves

Both the producer's and consumer's risk vary in magnitude with variation in the quality of the concrete and with the number of specimens involved. Given a particular compliance criterion, the risks can be evaluated in terms of the probability P_a that the compliance criterion will indicate acceptance of a particular quality of concrete (Koufopoulos, 1982). Variation of the risk with variation in the true quality of the concrete is best shown diagrammatically by plotting P_a against the true quality, measured by the fraction p of the distribution of strengths defective to the specified strength, f_c'. Such diagrams are called *operating characteristic* (O-C) *curves* (RILEM/CEB/CIB/FIB, 1975; Taerwe, 1983; Bonzel and Manns, 1970; Tait, 1981; Snell and Rutledge, 1982; Guedes and Souza, 1978). It has been shown on this basis that the acceptance criteria in the ACI Building Code 318-83, Section 7.5, provides lower risks to the concrete producer than to the consumer (Chung, 1978b). When both P_a and p are plotted on a scale of the normal probability integral, most of the commonly used compliance criteria give straight-line O-C curves, as shown in Figure 1.29.

Several compliance functions and corresponding O-C curves can be developed; Eqs. 1.56 and 1.57 are typical examples. The general form of such a compliance function $Z(x)$ is

$$Z(x) = \bar{x}_n - \lambda s_n \tag{1.64}$$

where $\bar{x}_n$ = arithmetic average of n individual strength measurements
  s_n = standard deviation of the n individual strength measurements
  λ = factor dependent on n and on the degree of assurance required that the acceptance decision will not be mistaken

The compliance criterion then is

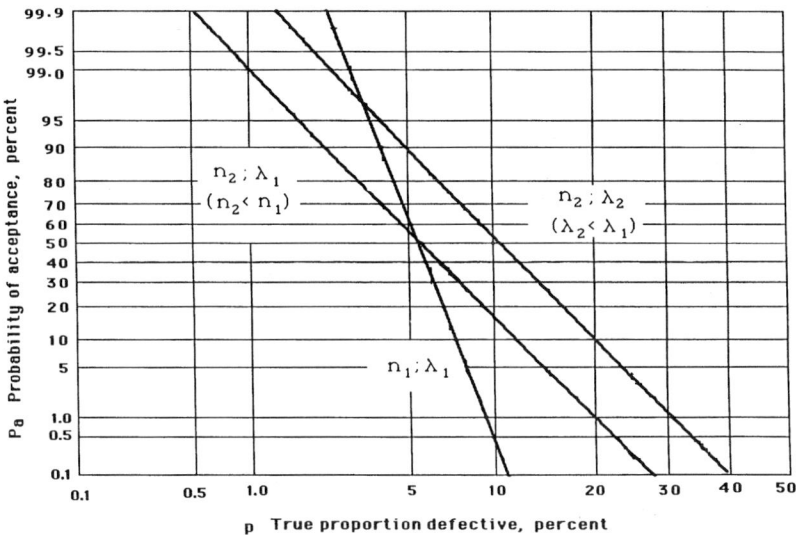

Figure 1.29 Typical O-C curves for compliance criteria of the form $\bar{x}_n - \lambda s_n \geq f'_c$ with parameters n and λ. (Copyright RILEM. Reprinted with permission.)

$$\text{accept if} \qquad Z(x) \geq f'_c \qquad (1.65)$$

When the sample size is large ($n \geq 30$) or the standard deviation of the strength results is known with adequate reliability from prior experience, Eq. 1.64 will become Eq. 1.56 (i.e., $\lambda = t$); otherwise, $\lambda \neq t$ because of the need to apply the modifying factors listed in Table 1.6. Consideration of structural safety suggests that an upper boundary for P_a should be kept in the region of 5% defectives. Otherwise, presently, only tentative recommendations are available for boundaries of O-C curves. An example mentioned by RILEM is presented in Figure 1.30.

Control Chart Analysis

The second tool for judging the acceptability of the quality of a concrete is the control chart analysis simplified here specifically for concrete control. The purpose of this analysis is again to detect if the quality of a concrete in question represented by a property such as compressive strength is above the specified minimum with an acceptable risk. Such charts also enable the engineer to analyze trends in the quality of concrete and develop a course of improvement, if necessary. Methods are well established for the setting up of control charts (ACI, 1977; ASTM Committee E11, 1976; Barros et al., 1983; Lambotte, 1985).

The most frequently used statistical characteristic for concrete quality is the individual or average compressive strength. Values of the strength of suc-

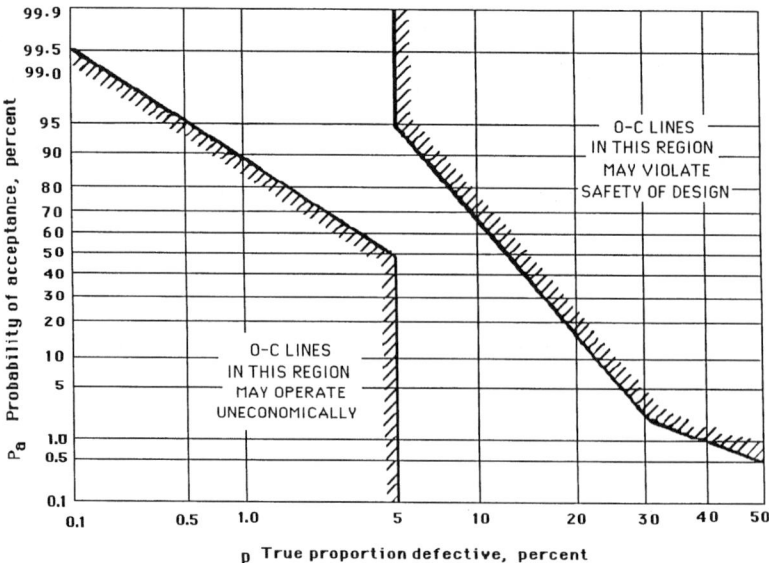

Figure 1.30 Boundaries for O-C curves for compliance criteria shown in Figure 1.29, proposed tentatively by RILEM. (From RILEM/CEB/CIB/FIP Joint Committee, 1975. Copyright RILEM. Reprinted with permission.)

cessive samples, each containing an equal number of specimens, are plotted as ordinates against a scale of abscissa giving the numerical sequence of samples. The specified strength, f'_c, and the required average strength, called the *target strength* of proportioning, f_{cr}, are also plotted. These appear as lines parallel to the abscissa (Fig. 1.31). The value of f_{cr} is established from the allowable number of strength values below the specified strengths f'_c (S. Popovics, 1982b). Points that fall below the specified strength indicate that something is probably wrong with the batch represented due either to lower intrinsic strengths, higher variability than anticipated, poor sampling, or faulty testing.

Another useful kind of control chart utilizes the moving-average concept (Cordon, 1979). Figure 1.31 also shows a control chart for the moving average of any five consecutive tests. More specifically, the first five tests (each consisting in our case of two companion cylinders) are averaged for the first point. For the next point, the sixth test is included in the average and the first test is deleted, and so on. The advantage of this method is that occasional low strength values will not cause the concrete to fail these specifications. When, however, low tests occur consecutively, which is the dangerous case, the moving-average chart will reveal this by showing that the concrete fails to meet specifications. This chart is valuable in indicating trends and will show the influence of seasonal changes, changes in materials, and so on (ACI, 1977). The number of tests averaged to plot moving averages with an appro-

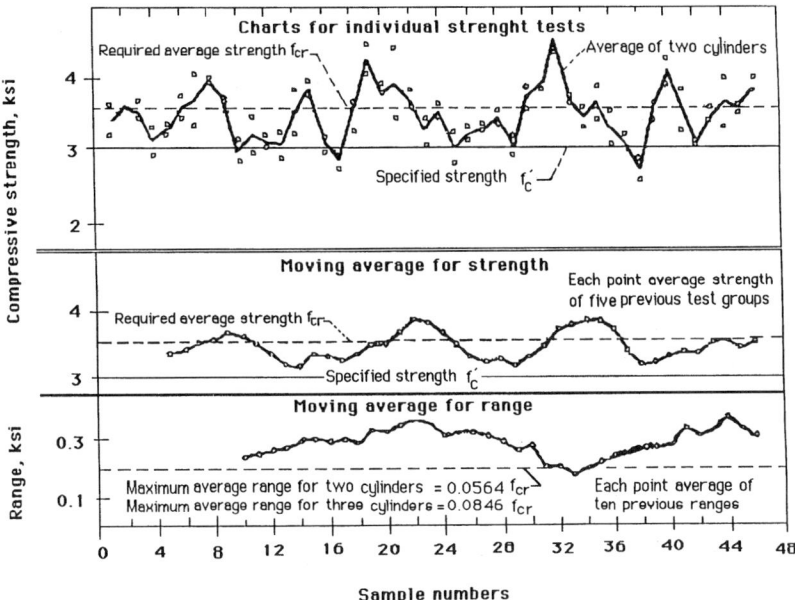

Figure 1.31 Three types of control charts for concrete. 1 ksi = 6.90 MPa. (From ACI, 1977. Copyright ACI. Reprinted with permission.)

priate lower limit can be varied to suit each job. For instance, when the ACI Building Code 318 applies, a chart averaging three previous tests is appropriate for moving average since the code requires that the frequency of an average of three consecutive tests below f'_c should not exceed 1 in 100.

The control charts described above are most commonly used in North America. In the United Kingdom, cumulative sum (*cusum*) control charts are now widely used. A characteristic feature of these charts is that the points plotted on the chart contain information from all the observations up to and including the plotted point. To construct a cusum chart, the cumulative sum of the algebraic differences between each result and the assumed (or specified) value is plotted as successive results are obtained. The slope over any part of the cusum chart is then proportional to the difference between the actual and assumed values. The horizontal plot indicates that the actual results are the same as the specified value. A rise in the plot indicates that the average value is greater than the specified value, while a fall indicates that the average values are less than the specified values. It is claimed that it is easier to visualize changes in the concrete strengths in this chart than in the control charts recommended by ACI. The reason for this is that the cusum chart presents only the surplus or deficiency in strength; thus small changes in the average appear as quite different slopes, whereas in the control charts the strengths are in full size (Mindess and Young, 1981).

Control charts or cusum charts may be prepared for ranges or standard deviations of the test results (Fig. 1.31). If the average range or standard deviation is greater than an upper limit, this indicates poor sampling and/or faulty testing procedures.

Concrete Strengths Below Specifications

The consequences of a failure of a concrete to meet strength requirements vary greatly. It is a reasonable premise that a deviation below specified strength levels will reduce the service life or increase required maintenance, and therefore the producer of the concrete should be penalized in proportion to that loss in the value. Naturally, below some strength levels the concrete will be unacceptable and must be removed from the structure.

Under traditional specifications, engineering judgment was relied on to solve such problems. O-C curves or control charts provide a better basis for answering such questions more objectively. In practical terms, however, the problem is still unsolved to express the financial equivalent of a low concrete strength quantitatively; that is, (1) to establish a reasonable scale for reduced payments for strengths less than some specific levels; and (2) to determine the strength level below which the concrete must be removed. ASTM C94-94 has no adequately spelled out penalty clause for ready-mixed concretes. A number of gradual penalty functions have been recommended (La Course et al., 1983), but they are all more or less arbitrary. Often, the concrete producer is asked to accept reduced payments based on the costs of concrete in place, which is often three to eight times the price he receives at the end of the chute. For instance, if a producer quotes $30/cu yd ($39.24/m³) on concrete that becomes worth $200/cu yd ($261.61/m³) when placed in the structure, even a small pay reduction of only 5% would mean a loss to the producer of 5% of the in-place cost, or $10/cu yd ($13.08/m³), a very undesirable result from the producer's standpoint. Although an increase in the sample size may reduce the producer's risk in certain cases (see Problem 1.3), at present the only reliable protection for the concrete producer against such loss is to proportion the concrete at a target strength that is higher than the one related statistically to the strength specified by the designer (Weed, 1979). A method for the selection of the most economical extra strength margin was illustrated elsewhere (S. Popovics, 1982b).

Once it has been decided that the concrete strengths failed to meet specifications, an investigation should be undertaken in every case where the structure has any importance. The National Ready Mixed Concrete Association recommends the following gradual approach (NRMCA, 1979):

Step 1 Verify the accuracy of testing of the concrete, including sampling, preparation of specimens, curing, capping, testing machine, breaking, and so on. If the testing is found to have deviated from the standard methods, it may be possible to terminate the investigation at this point by declaring the measured low strengths invalid. Otherwise, it is necessary to continue the investigation with step 2 or step 3 or both.

Step 2 Compare structural requirements with the measured strength of concrete. It is possible that the part of structure in which the substandard concrete was used does not actually require the full specified strength. The structural engineer is to decide whether or not the measured low strength of the concrete specimens actually represents a possible impairment of the load-carrying capacity or service life of the structure.

Step 3 Estimate the strength of the concrete in the structure by nondestructive tests(s). Such tests may indicate quite convincingly whether or not the concrete being questioned differs appreciably from concrete judged acceptable elsewhere in the structure. If evaluation by such methods fails to alleviate concern over structural adequacy, step 4 may be necessary.

Step 4 Drill concrete cores from the questioned structural element and test them for compressive strength according to ASTM C42 (Martin and Junces, 1982). The core strengths are usually less than the strength specified for molded cylinders (Section 1.2). Thus according to ACI Building Code 318, the concrete should be considered acceptable if the average strength of three cores is at least 85% of that specified, with no core less than 75%. The structural engineers should examine cases where the core strengths fail to meet these requirements to determine if there is cause for concern over structural adequacy. If the answer is affirmative, step 5 is necessary provided that the significance of the structure justifies it. Otherwise, step 6 should be taken.

Step 5 As a last resort, load tests may be required, performed according ACI Building Code, to check the load-carrying capacity of the structural members whose strength is seriously in doubt (Javor, 1978; RILEM, 1978; RILEM Tentative Recommendation, 1978, 1981). Generally, such tests are suited only to flexural members.

Step 6 In those rare cases where a structural element fails the tests discussed above, or where structural analysis of untestable members indicates inadequacy, appropriate corrective measures must be taken. The alternatives are reduction of the load rating on the faulty structural element, augmentation of the construction to bring its load-carrying capacity up to original expectation, or replacement of the unacceptable element.

More specific acceptance criteria, based on in-place testing, are given in the ACI Committee 318-89 Standard, Section 5.6.

2

OTHER CONCRETE STRENGTHS

2.1 FLEXURAL STRENGTH (S. Popovics, 1970b)

Significance

In the practice, flexural strength is most commonly utilized in beams and slabs. Here, the stresses are caused by flexural loads, temperature changes, uneven shrinkage, and moisture changes. To avoid undue cracking in the concrete of such structures, flexural strength is of special importance despite its low magnitude.

Testing of the Flexural Strength

The flexural strength, or flexural tensile strength, or bending strength, or modulus of rupture of concrete is determined with simple rectangular beams without any reinforcement submitted to increasing bending until failure occurs. The simple and inexpensive nature of the testing procedure makes it appealing to use the flexural test for checking the quality of concrete on the construction site. This can be done either directly or by estimating the compressive strength from the measured flexural strength. There is considerable literature, dating back to 1904, by Talbot (Talbot, 1904), covering flexure test methods.

Specimens for the determination of the standard *control flexural strength* and for the *design flexural strength* (Section 1.1) can be made in the laboratory or in the field, as described in ASTM C192 and ASTM C31, respectively, or in other specifications (RILEM, 1975b). The molding procedure in these cases is essentially similar. The beams should have a minimum cross-sectional dimension at least three times the maximum nominal size of the coarse aggregate in the concrete. The length should be at least 2 in. (50 mm) greater than three times the depth of the beam as tested. A commonly used specimen is 6 × 6 in. (150 × 150 mm) in cross section, 21 in. (525 mm) long, that is tested on a span of 18 in. (450 mm).

The sample of fresh concrete is placed in serviceable beam molds, usually in two equal layers. The criteria for determining whether cylinders should be rodded or vibrated also apply to beams. In the case of rodding, usually one stroke should be applied for each 2 sq in. (1300 mm²) of concrete surface using a $\frac{5}{8}$-in. (9.5-mm) steel rod. The curing methods for beams are similar to the curing prescribed for cylinders except that all the beams should be stored in lime water at around 23°C for at least 20 to 24 hours before testing. That is, the curing method depends mainly on whether the control or the design flexural strength is requested. The factors governing the needed number of beams, testing age, and so on, are similar to those that apply to compression specimens.

The standard formula, based on the validity of Hooke's law, for calculation of the flexural strength of a beam is the following:

$$f_{fl} = \frac{M}{I} \frac{d}{2} \qquad (2.1)$$

where f_{fl} = flexural strength
M = bending moment, assumed to cause the rupture
I = moment of inertia of the cross section
d = depth of the beam

It can be seen that f_{fl} represents a calculated maximum tensile stress in the bottom fiber of the test beam. If the material in question does not follow Hooke's law, the f_{fl} value calculated by Eq. 2.1 is not the actual flexural strength but only a fictitious, greater-than-true value.

Beams should be tested in moist conditions under either center-point loading (ASTM C293), or third-point loading (ASTM C78). The first method is for small specimens and is not an alternative for the second. Tests indicate the following order of decreasing magnitude of the flexural strength measured:

1. *Center loading* with the M moment computed at the center, in which case f_{fl} = 1.5Pl/bd^2 (P is the ultimate load applied, l the span length, b the width of the beam, and d the depth of the beam).
2. *Center loading* with the moment computed at the point of fracture, in which case f_{fl} = 3Py/bd^2, y being the distance between the line of fracture and the nearest support measured on the bottom surface of the beam.

3. *Third-point loading,* in which case $f_{fl} = Pl/bd^2$, provided that the fracture occurs within the central one-third of the beam.

The loading rate specified in ASTM C78-94 and C293-94 for flexural specimens is such that the increase in external fiber stress is between 0.86 and 1.21 MPa/min (125 and 175 psi/min) when calculated in accordance with Eq. 2.1.

The third-point loading gives about 10% lower strengths than the center-point loading (Meyer, 1963a; Wright, 1952) because the maximum moment M is distributed over a greater length of the beam. Flexural strength is very sensitive to the moisture content and distribution in the specimen. For instance, drying for only 30 minutes caused in a test series an average reduction of 8% in the measured flexural strengths (Walker and Bloem, 1957a).

Beam specimens for the determination of design strength can be sawed out from the finished structure (ASTM C42). The small amount of experimental evidence concerning the strength relationship between sawed and molded specimens seems to indicate again that for the same curing conditions, sawed specimens produce measured flexural strength less, in an experimental series about 25% less (Walker and Bloem, 1957b), than those secured with comparable molded specimens.

The flexural strength of a concrete calculated by Eq. 2.1 is always greater than its actual tensile strength (Platts and Kirchner, 1971). The main reason for this difference is that the use of Eq. 2.1 assumes the applicability of Hooke's law of stress-strain proportionality. Nevertheless, values of f_{fl} are suitable for comparison of resistances of various mixtures against tensile stresses, and their use is convenient because of the simplicity of the test.

A theoretically interesting variation of the traditional flexural method is when perforated beams are tested (Roup and Fillmore, 1961; Ladanyi and Nguyen, 1968). A cylindrical transverse hole in the middle of the beam modifies the stress distribution around the minimum cross section. Formulas are available for calculation of these stresses (Imbert 1970; Evans and Marathe, 1968), including the tensile strength. Results show, however, that this method cannot furnish the direct tensile strength of the concrete either without detailed knowledge on the stress-strain behavior of the material up to fracture. Among the other indirect test methods (Section 1.4), the break-off method is probably the most suitable at present for the estimation of the *in-situ flexural strength* (Yener and Chen, 1985).

Relationship Between Flexural and Compressive Strengths

Because of all these factors, it is only natural that the relationship between the compressive strength and flexural strength of a concrete is influenced by these and a number of other factors, such as fineness of the cement. This is illustrated by the variety of approximations shown in Table 2.1 that were recommended for the f_{fl}/f_{co} flexural strength–compressive strength ratio by

TABLE 2.1 Approximations for the relationship between flexural and compressive strengths

f_{fl}/f_{co}	Authority[a]	Remarks
0.09 to 0.12[b]	Walz and Wischers	$f_{co} = f_c$ structural lightweight concrete
0.1 to 0.3[b]	Kaplan	$f_{co} = f_{cm}$
0.11 to 0.20[b]	Bonzel	f_{cu} varies from 600 to 100 kg/cm²; aggregate is sand and gravel
0.112 to 0.23[b]	Gonnerman and Shuman	$f_{co} = f_c$
0.12 to 0.22[b]	Kesler	$f_{co} = f_c$
0.125 to 0.20[b]	Ros	$f_{co} = f_p$
0.125 to 0.20[b]	Graf	$f_{co} = f_{cu}$
0.13 to 0.25[b]	Kenis	$f_{co} = f_c$
0.13 to 0.25[b]	Bonzel	f_{cu} varies from 600 to 100 kg/cm²; aggregate is crushed stone
0.13 to 0.20[b]	Walker and Bloem	f_c varies from 6500 to 1500 psi; with aggregates of different maximum sizes
0.13 to 0.17	Walz	f_{cu} = about 460 kg/cm² at the age of 28 days; with different coarse aggregates
0.154 to 0.289[b]	Akazawa	$f_{co} = f_c$
$0.14 - 0.0001f_{cu}$	Palotas	f_{cu} is in kg/cm²
$0.29 - 0.000032f_c$	Abrams	f_c is in psi
$1.15/f_{cu}$	Williams	f_{cu} is in psi; lightweight concrete, drying
$5/\sqrt{f_c}$ to $11/\sqrt{f_c}$	ACI Committee 435	f_c is in psi
$7.5/\sqrt{f_c}$ to $12/\sqrt{f_c}$	ACI Committee 435	f_c is in psi
$8/\sqrt{f_{cu}}$	Short and Kinniburgh	f_{cu} is in psi; structural lightweight concrete
$9.2/\sqrt{f_{cu}}$	Road Research Lab.	f_{cu} is in psi

TABLE 2.1 (Continued)

f_{ft}/f_{co}	Authority[a]	Remarks
$0.75 l^{1/3}\sqrt{f_{cu}}$	Palotas	f_{cu} is in kg/cm²; average values
$1/f_{cu}^{0.3}$ to $1/f_{cu}^{0.4}$	Hummel	Ratio increases with angularity of particles
$2.793/f_{cm}^{0.37}$	Sen	f_{cm} is in psi
$0.09 + 50/f_{cu}$	Williams	f_{cu} is in psi
$0.11 + 1.7/f_{cu}$	Komlos	f_{cu} is in MPa
$0.16 + 12/f_{cu}$	Ujhelyi	f_{cu} is in kg/cm²; for lightweight aggregate concrete
$3000/(4f_c + 12,000)$	Sozen et al.	f_c is in psi
$g(t)(2.5)^{w/c}$	Popovics	w/c = water-cement ratio by weight; $g(t)$ is a parameter that is dependent on age, type of aggregate, curing and testing methods
$0.15\sqrt{u/f_{co}^{0.3}}$	Popovics	f_c is in psi, u unit weight of concrete is in lb/cu ft; for lightweight and normal weight concretes[c]
$a_1\sqrt{u/f_{co}}$	ACI Committee 209	a_1 can vary between 0.6 and 1.0

Source: S. Popovics (1967a). (Copyright TRB. Reprinted with permission.)

[a]References for the authors listed in this column can be found in S. Popovics (1967a).

[b]The higher the concrete strength, the lower becomes the ratio. 1 ksi = 6.90 MPa. 1 kg/cm² = 0.098 MPa.

[c]See Figure 2.2

various investigators (S. Popovics, 1967a). Figure 2.1 presents a graphical comparison for some of the formulas.

Within a single test series of a concrete, the compressive strength can be estimated from the result of a flexural test with an accuracy of about ±1000 psi (7 MPa). The reliability of a general, overall relationship is, of course, poorer. Similar conclusions can be drawn from a report by Saul (1960). These data show that the flexural strength of a high-strength concrete is about 10% of its compressive strength, but it may increase up to 30% with a decrease in strength. However, the f_{fl}/f_{co} ratio is also influenced to a high degree by the age and composition of the concrete, by the aggregate type, as well as by the curing and testing conditions. Walz pointed out, for instance, that at higher strengths the flexural strength of structural lightweight concretes is only about 70 to 90% of the flexural strength of normal concretes of the same compressive strength (Walz and Wischers, 1965). This is also indicated by Table 2.1 and Figure 2.2a. The difference in the flexural strengths of normal and lightweight concretes of identical compressive strength may be attributed to the difference in the total porosity of the concrete (Section 6.4). It may also be due to their differing modes of failure in flexure. According to one opinion, in a normal-weight concrete, the failure occurs primarily as a result of a breakdown of the bond between the hardened cement paste and the surface of the aggregate; whereas in a lightweight concrete the fracture is caused mainly by the weakness of the aggregate particles (Bache and Nepper-Christensen, 1968). However, if the phenomenon is rephrased, stating that the compressive strengths of structural lightweight concretes are about 10 to 30% higher than the compressive strengths of normal-weight concretes of the same flexural strengths, a better explanation offers itself. It is more likely that the greater deformability of the lightweight aggregate particles reduces the stress concentrations in the concrete around aggregate particles. This reduction is more effective in compression than in flexure because of the higher stress field during compression testing (S. Popovics, 1987a).

The correlation is better when compressive and flexural strengths of standard cement mortars are compared because here the number of variables is reduced by the standardized circumstances. This has been demonstrated both for U.S. standard cement mortars (Lynn and Palmer, 1961) and for German mortars (Harig, 1966; S. Popovics and Ujhelyi, 1954, 1955).

The main reason for the enlarged uncertainty of the f_{fl} versus f_{co} relations seems that the flexural strength is affected by the concrete composition, namely by the type and fineness of the cement, water-cement ratio, air content, type of mineral aggregate, and so on, as well as by the wetness of curing to different degrees than is the compressive strength (Walker and Bloem, 1956; Bonzel, 1964; Neville, 1981; Fulton and Davis, 1961; Jones and Kaplan, 1957; Shacklock and Keene, 1959; S. Popovics, 1967b). For instance, fundamental is the observation that a decrease in concrete porosity affects the compressive strength more than the flexural strength (Fig. 5.40). This explains that the f_{fl}/f_{co} ratio decreases both with increasing age and with decreasing

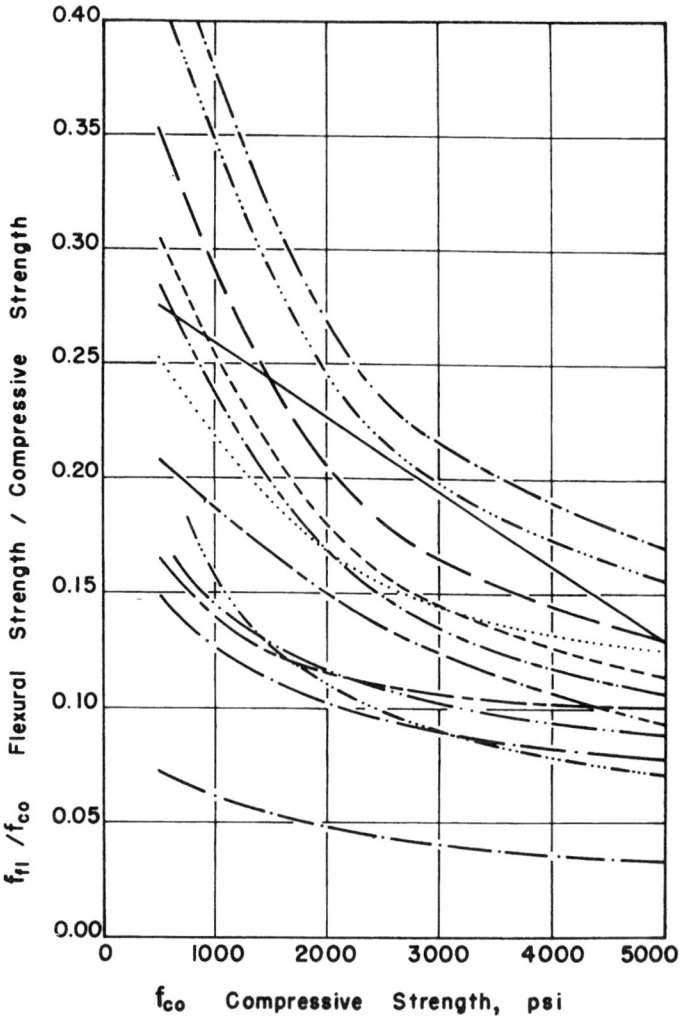

Figure 2.1 Various recommendations for the flexural strength–compressive strength ratio in terms of the compressive strength. 1 ksi = 6.90 Mpa. ____, Abrams (1922): f_{fl}/f_{co} = 0.29 − $0.000032f_{co}$. ____ ____, Road Research Lab. (1951): $9.2/\sqrt{f_{cu}}$. – – – – – –, Short & Kinniburgh (1963): $8.0/\sqrt{f_{cu}}$. ____ . ____, Hummel (1959): $1/f_{cu}^{0.3}$ to $1/f_{cu}^{0.4}$. ...., Sen (1961): $2.793/f_{cu}^{0.37}$. ____ .. ____, Williams (1962): $1.15/f_{cu}^{0.3}$. ____ _ ____, Williams (1962): $0.09 + 50/f_{cm}$. ____ . _ . ____, ACI (normal concrete) (1966): $7.5/\sqrt{f_c}$ to $12/\sqrt{f_c}$. ____ ... ____, ACI (lightweight concrete) (1966): $5/\sqrt{f_c}$ to $11/\sqrt{f_c}$. ____ _ . _ ____, Sozen et al. (1959): $3000/(4f_c + 12{,}000)$: f values are in psi (1 ksi = 6.90 MPa). (From S. Popovics, 1970b. Copyright ASME. Reprinted with permission.)

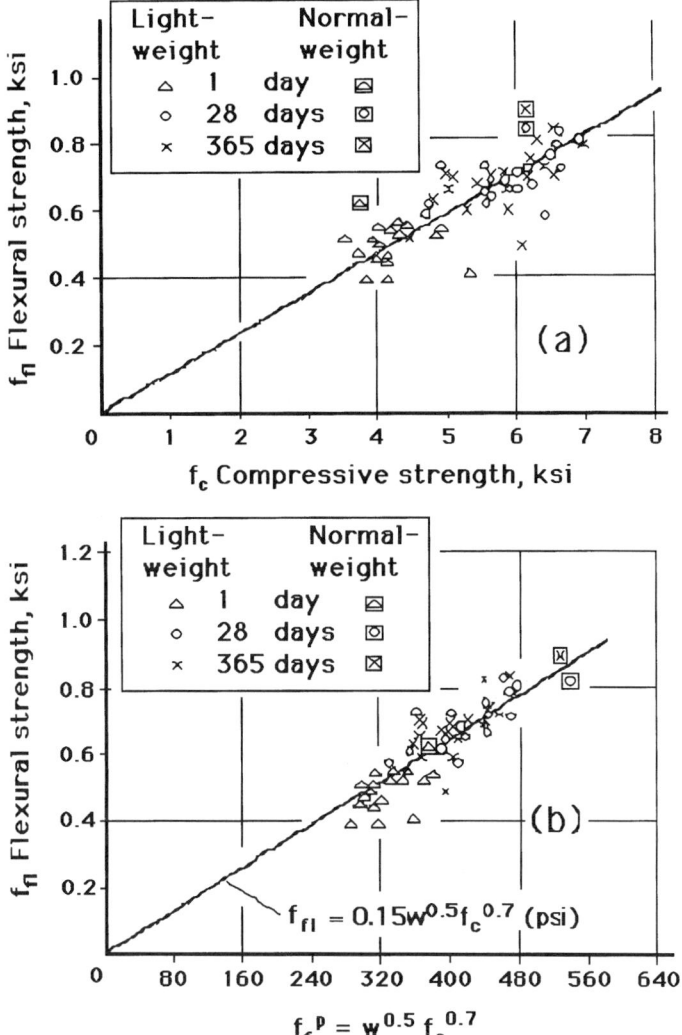

Figure 2.2 (a) Experimental f_{fl} flexural strengths plotted against related f_c compressive strengths as reported by Reichard. (b) Comparison of the experimental f_{fl} flexural strengths of (a) to f_f^p calculated from Eq. 2.8. 1 ksi = 6.90 MPa. (From S. Popovics, 1975, 1977. Copyright RILEM. Reprinted with permission.)

water-cement ratio (Fig. 2.3), that is, with increasing strength, although differences in the complete stress-strain curves in compression and flexure may also contribute to this decrease (Welch, 1966). The decrease of the f_{fl}/f_{co} ratio with age is also influenced by the fact that the development of the tensile strength of concrete, thus that of the flexural strength as well, is faster, therefore stops sooner (Fig. 3.8) than development of the compressive strength of

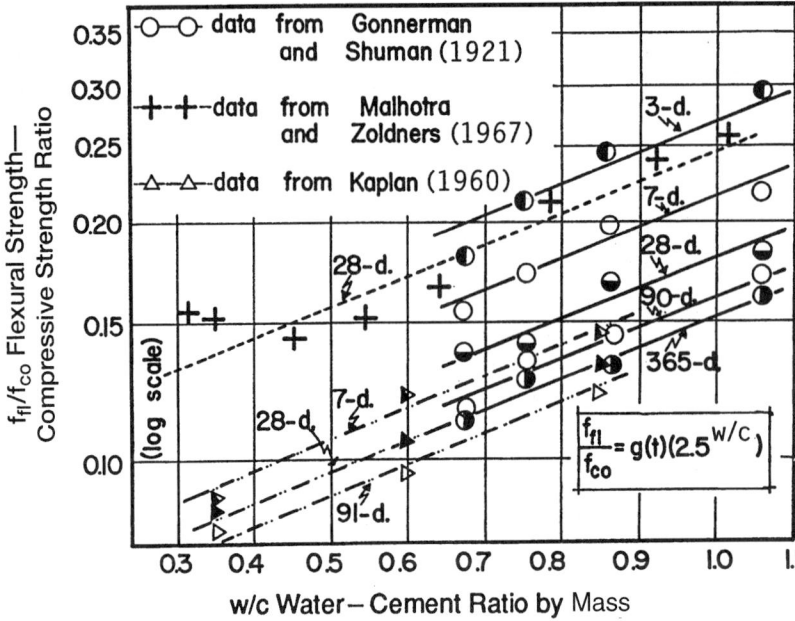

Figure 2.3 Flexural strength–compressive strength ratio at various ages as a function of the water-cement ratio. (From S. Popovics, 1969a. Copyright ASTM. Reprinted with permission.)

the same concrete (Wright, 1952; Walker and Bloem, 1957b; S. Popovics, 1967c). In addition, the effects of the size and shape of the specimen as well as the effects of curing and testing methods are considerably higher on the values obtained for the flexural strength than on those obtained for the compressive strength. It is enough to mention here the significant increase in the compressive strength caused by sudden drying (S. Popovics, 1986b), in contradistinction to the reduction in flexural strength caused by drying or sudden cooling of the concrete specimen (Shacklock and Keener, 1959; Graf et al., 1960; Dutron, 1962).

Some of these effects can be expressed numerically by approximate, mostly empirical formulas. Abrams was probably the first to propose a formula, a parabola, for the relationship between the compressive and flexural strengths (Abrams, 1922). Since then, numerous formulas have been recommended by other investigators. The simplest is a linear approach:

$$f_{fl} = b_0 + b_1 f_{co} \qquad (2.2)$$

where b_0 and b_1 are empirical factors that are independent of the strength but depend on the applied units as well as on the composition of concrete, curing and testing conditions, and other factors. Such a linear relationship is supported by experimental data within restricted limits, such as the range of structural concretes. In addition, Eq. 2.2 is supported by the well-documented fact (Table 2.1) (Meyer, 1963a; Lynn and Palmer, 1961; Bonzel, 1964; Fulton, 1964; Price, 1951; Rothfuchs, 1962; Malhotra, 1967; Jindal, 1964) that the f_{fl}/f_{co} ratio decreases as the concrete strength increases because from Eq. 2.2,

$$\frac{f_{fl}}{f_{co}} = \frac{b_0}{f_{co}} + b_1 \qquad (2.3)$$

Many other nonlinear relationships have been recommended for the f_{fl} versus f_{co} relationship or for the f_{fl}/f_{co} ratio on an empirical basis. Some of the latter formulas are presented in Table 2.1. Additional formulas can be derived from application of Abrams' formula for the compressive strength versus water-cement ratio and the flexural strength versus water-cement ratio relationships (Chapter 5). The significance of such formulas is not so much that they provide the flexural strength–compressive strength ratio more accurately than the other formulas in Table 2.1 but rather, that they show how certain factors influence this ratio. For instance, the ratio of Eqs. 3.21 and 3.20 provides the following formula, which shows the effect of the water-cement ratio and cement fineness on the flexural strength–compressive strength ratio of concrete:

$$\frac{f_{fl}}{f_{co}} = AB^{w/c} + \sqrt{\frac{S_0}{S_s}} \times comp \qquad (2.4)$$

where w/c = water-cement ratio
S_s and S_0 = specific surface of the cement in question and that of an average Type I portland cement, respectively
comp = parameter that is characteristic of the compound composition, that is, the type of cement
A and B = parameters similar to those in Eq. 5.6; when the water-cement ratio is expressed by mass, $A = 0.085$ and $B \approx 2.5$ are valid for a large group of commercially available portland cements

The effect of air content is illustrated by the following formula (Example 5.7):

$$\frac{f_{fl}}{f_{co}} = AB^{w/c} \times 10^{\alpha a} \qquad (2.5)$$

where a is the air content of the concrete expressed in percent and α is a parameter.

The following formula derived from Eqs. 5.36 through 5.39 shows the effects of age at testing, water-cement ratio, and their interaction on the f_{fl} versus f_{co} relationship (S. Popovics, 1967b):

$$\frac{f_{fl}}{f_{co}} = b \left(\frac{46 + 6t}{15 + 2.5t} \right)^{w/c} \tag{2.6}$$

$$\approx b \left(2.45 + \frac{1.3}{t} \right)^{w/c} \tag{2.7}$$

where b = empirical parameter that usually varies between 0.05 and 0.1, depending on the type of aggregate, curing and testing conditions, etc.

t = age at testing, days

w/c = water-cement ratio by mass

The effect of lightweight aggregate on the f_{fl} versus f_{co} relationship can be illustrated by the following formula (S. Popovics, 1973, 1975, 1977):

$$f_{fl} = bu^{0.5}f_{co}^{0.7} \tag{2.8}$$

where b = empirical parameter, which equals approximately 0.15 when the tests are performed under the standard conditions and the compressive strength is the standard cylinder strength in psi ($= 0.0069$ MPa)

u = standard density of the concrete, lb/cu ft ($= 16$ kg/m^3)

Equation 2.8 is presented in Figures 2.2b and 6.11. These formulas are valid for structural concretes not younger than 3 days. The f_{fl}/f_{co} ratio represented by Eq. 2.8 is given on the bottom of Table 2.1.

It follows from the above that flexural strength of plain concrete may be used for checking the compressive strength on the construction site only if the same concrete-making materials are used continuously, and the entire testing procedure, including pouring, curing, testing, and so on, is kept unchanged (Walker and Bloem, 1957b). In such a case, a reliable value for parameter b in Eqs. 2.6 through 2.8 can be obtained from a couple of trial mixes, and the formulas can be used for practical purposes.

Variability of Flexural Strength

Study of the variability in flexural strength has been lagging behind. It is clear, however, that the flexural strength has a relatively greater fluctuation than the compressive strength (Kaplan, 1960b). Under laboratory conditions, one may expect an overall coefficient of variation of about 5% or more for the flexural strength, while the *within-batch* and *between-batch* coefficients can be about 4 and 3%, respectively (Malhotra and Zoldners, 1967). This variability is, of course, much greater under field conditions. For instance, a report by the Michigan State Highway Department shows 12 to 15% overall coefficient of variation and 5 to 7% within-batch coefficient of variation in flexure (Michigan Department of State Highways, 1966).

It has also been documented that the standard deviation of both within-batch tests and between-batch tests appear to be independent of the magnitude of the flexural strength (Walker and Bloem, 1956). In addition, an analysis of laboratory flexural strengths ranging between 500 and 700 psi (3.5 and 4.9 MPa) at the age of 28 days from 145 concrete mixtures revealed the following (Greer, 1983):

1. The within-batch standard deviation of flexural strengths for individual batches ranged from 0 to 75.3 psi (0 to 0.52 MPa). The average value for the 28-day strength test was 33.5 psi (0.23 MPa).

2. The within-batch standard deviation of 40 psi (0.28 MPa) encompassed 80% of the batches; the value of 55 psi (0.38 MPa) encompassed 95% of the batches.

3. Any given set of three 28-day beams could differ from the average strength of all beams by 20 psi (0.14 MPa) or more as much as 14% of the time.

4. The between-batch standard deviation of flexural strengths ranged from 0 to 59.4 psi (0 to 0.41 MPa). The average value for the 28-day strengths was 27 psi (0.19 MPa).

5. The between-batch standard deviation of 30 psi (0.21 MPa) encompassed 80% of the batches; the value of 40 psi (0.28 MPa) encompassed 95% of the batches.

6. The average between-batch standard deviation of 27 psi (0.19 MPa) would indicate that the difference between the average of three beams from each of two batches of the same mixture could range up to approximately 75 psi (0.52 MPa) approximately 95% of the time based on ASTM precision limit calculations.

Note that the average between-batch variation above is less than the average within-batch variation. This is due to the use of averages for between-batch calculations.

According to ASTM C293-94, the single-operator coefficient of variation has been found to be 4.4%. Therefore, results of two properly conducted tests by the same operator on beams made from the same batch sample should not differ from each other by more than 12%. The multilaboratory coefficient of variation has been found to be 5.3%. Therefore, results of two different laboratories on beams made from the same batch sample should not differ from each other by more than 15%.

2.2   **TENSILE STRENGTH** (S. Popovics, 1970b)

Significance

As demonstrated in Section 2.1, the flexural strength of a concrete is really a calculated tensile strength. Nevertheless, it is customary to recognize *tensile strength* separately. This recognition is justified because there are major differences between flexural and tensile strength determinations in (1) how the stresses are produced, (2) how the stresses are distributed, and (3) how the strengths are calculated.

Tensile stresses in structures are caused by more-or-less uniform shrinkage or drying and by temperature changes. The case of uniaxial tension from loads is rarely encountered in practice, and even in laboratory tests it can be obtained only with care. However, significant principal tensile stresses may be associated with multiaxial states of stress, including earthquake conditions.

In addition to its practical significance, the tensile strength of concrete has a fundamental role in the fracture mechanism of hardened concrete. Generally, almost any kind of fracture in concrete occurs through cracking caused by tensile stresses. This means that concrete fraction is essentially a tensile failure regardless of whether the fracture is caused by compression (or other loading), freezing, or by other factors. The type of loading controls only the character of the cracking through which the strengths and other mechanical properties of the concrete are influenced. For instance, the difference between crack propagation under compressive load and that under tensile load is the reason that the concrete tensile strength is about one-tenth of the compressive strength (see Section 2.4).

The various kinds of specimens for determination of the tensile strength of concrete should be molded and cured essentially in a way similar to that used for flexural strength specimens. Several different methods have been recommended for determination of the tensile strength of concrete. The more important tests are presented below.

Direct Tension Test Method

The direct way to determine the tensile strength of concrete or mortar is similar to the tensile strength tests of many metals and many other materials; that is, a specimen—briquette, long cylinder, or prism—is submitted to a uniformly distributed increasing axial tension load in a suitable testing machine until the specimen breaks into two parts. The strength is expressed as the ultimate tension load per cross-sectional area, usually in psi, MPa, or kg/cm². Talbot was probably the first in the United States to report such tests (Talbot, 1904). There is no standard American test method for direct determination of the tensile strength of concrete. RILEM has a pertinent recommended method (CPC 7-1975).

Gripping of the specimen is generally achieved either by truncated cones or by steel reinforcement embedded into concrete specimens. The difficulty associated with this classic form of direct tension test is that it is burdened with misalignment and clamping stresses. Thus the stress concentrations, bending, and/or torsion created during testing produce reductions of unknown magnitude in the measured strength, thereby making the reliability of these strength results highly questionable. Because of the stresses introduced due to gripping, there is a tendency for the specimens to break near the ends. This problem is often overcome by reducing the section of the central portion of the strength specimen (briquette). Similarly, the method in which metal pulling pieces are glued with epoxy (Komlos, 1967–1969; Heilmann et al., 1969; Hughes and Chapman, 1965; Spetla and Kadlechek, 1966) or fixed with grout to the end of the tension specimens eliminates stresses caused by gripping. These, however, offer no final solution for the eccentricity problem (Elvery and Haroun, 1968; Malhotra, 1966b).

Tests of extremely young concrete are frequently performed horizontally to avoid the influence of the specimen's own weight. In such cases, however, the friction between the specimen and the supporting base may disturb the results. This friction can be eliminated by using a mercury base (Orr and Haigh, 1971). In another technique the tensile load is applied purely by friction (Gonnerman and Shuman, 1928; Johnston and Sidwell, 1968; Ward and Newman, 1970). The advantage of this method is that simple specimens can be used.

A promising modification of the direct tension test was proposed by Todd (1955). This method consists of molding a concrete cylinder with an installed steel tube or reinforcing bar through its center upon which strain gages are mounted (Fig. 2.4). This composite specimen is placed in a testing machine and the steel tube or bar is pulled in uniaxial tension until concrete tension failure occurs. The load carried by the concrete is the difference between the total load and the load carried by the steel calculated from the strain gage readings. The method is also described by Pincus and Gesund (1965) and in a similar test method by L'Hermite (1955). Ledbetter and Thompson (1965) modified the original Todd method by applying a tube whose surface has special deformations to aid in bonding the concrete to the steel. Todd's method eliminates a large part of the problem of both clamping and misalignment. However, this test is slow and requires skilled operators as well as the use of relatively sophisticated equipment and technique. An increase in the size of the specimen as well as drying reduce the tensile strength (Spetla and Kadlecek, 1966; Malhotra, 1970a; Kadlecek and Spetla, 1967). Reliable numbers concerning the magnitudes of these reductions are not available.

A concrete specimen submitted to uniaxial tension fails ordinarily along a more-or-less plane surface that is approximately perpendicular to the direction of loading. The major part of the coarse aggregate particles usually pull out of this surface; that is, the number of broken particles is small. This indicates that the interface between coarse aggregate and mortar (i.e., the bond) is the weakest link in the concrete.

The overall as well as the within-batch and between-batch variabilities of the carefully measured direct tensile strength of concrete appear to be the

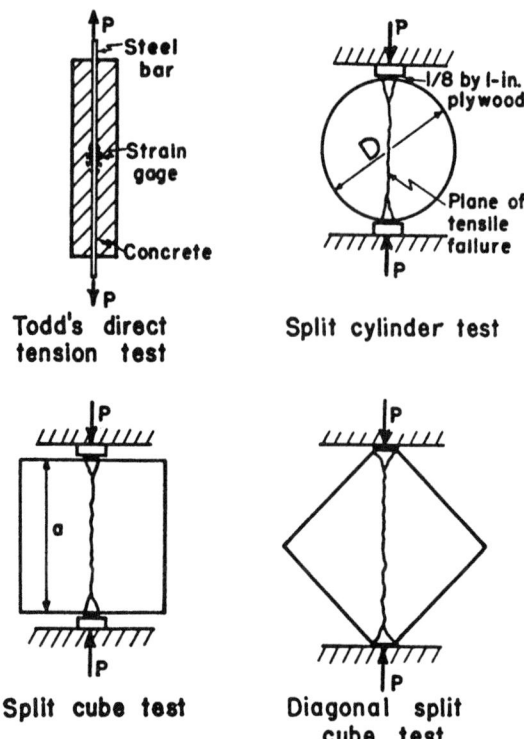

Figure 2.4 Direct and indirect methods for determination of the tensile strength of concrete. (From S. Popovics, 1970b. Copyright ASME. Reprinted with permission.)

same as those of the flexural strength (Orr and Haigh, 1971; Gonnerman and Shuman 1928; Ward and Cook, 1969; Komlos, 1970). Because of the problems associated with direct tension tests, other methods have been recommended for *indirect* estimation of the tensile strength of concrete. One should not expect from these methods to provide the "true" tensile strength. Nevertheless, tensile strengths obtained indirectly are suitable for comparison of various mixtures, and their use is convenient because of the simplicity of the tests.

Concrete *cores* taken from a structure can also be tested in direct tension. Raphael (1984) reports, however, that such core strengths were about half the tensile strength of comparable laboratory specimens. He attributes this to differences in the curing conditions of the two types of specimens.

Splitting Tension Test

This method has become the most popular test for tensile strength determination. The splitting test was developed in Brazil in 1943 by Carneiro and Barcellos (1949, 1953) and, independently, in Japan by Akazawa (1953).

Originally, a cylindrical specimen was recommended, and this is standardized in ASTM C496, but the method has been applied successfully on cubes (Berthier, 1951; RILEM, 1975c; Nilsson, 1961; Bossi, 1964; Davies and Bose, 1968), as Figure 12.4 shows, on beam portions remaining from flexural testing (Sell, 1963; Ramakrishnan et al., 1967a; Krishnaswamy, 1969) and on elliptical cylinders as well (Krishnaswamy, 1969; Brisbane, 1963). Among these perhaps the diagonal split cube test is the least suitable (Fig. 2.4) The methods can be supplemented to provide also a stress–strain diagram (Davies, 1968). In the splitting test a vertical *compressive* load is applied, uniformly distributed as a line on two opposing generatrices of the specimen, as shown in Figure 2.4. The load is increased until failure takes place by splitting the specimen along the vertical diameter.

The compressive load applied on the generatrices creates compressive and tensile stresses in the specimen. The maximum stresses are located on the elements located in the internal vertical surface between the loaded generatrices. Here the *vertical* stresses are compressive: the *horizontal* stresses are also compressive near the loaded generatrices but become *tensile* soon as they move away from the external surface so that an almost uniform tensile stress field exists over about 80% of the vertical plane (Fig. 2.5).

In the case of a *cylindrical* specimen, the maximum tensile stress in this vertical plane is (L'Hermite, 1955; Timoshenko and Goodier, 1951; Wright, 1955; Mitchell, 1961; Ramesh, 1960)

$$\sigma_{tm} = \frac{2P}{\pi LD} \tag{2.9}$$

and the compressive stresses are

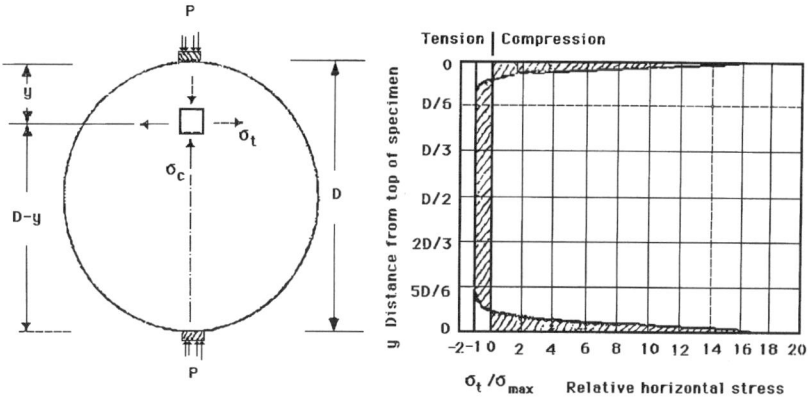

Figure 2.5 Distribution of horizontal stresses in the internal surface between the loaded generatrices of a cylinder under splitting test. The load P is transmitted over a strip of width $D/12$. (From Wright, 1955. Copyright Thomas Telford Publishing. Reprinted with permission.)

$$\sigma_c = \frac{2P}{\pi LD}\left[\frac{D^2}{y(D-y)} - 1\right] \tag{2.10}$$

$$= \sigma_{tm}\left[\frac{D^2}{y(D-y)} - 1\right] \tag{2.11}$$

where σ_{tm} = maximum tensile stress
σ_c = compressive stress
P = load applied
L = length of the cylinder
D = diameter of the cylinder
y and $D - y$ = vertical distances of the element from the two loaded
generatrices

Equations 2.9 through 2.11 show that a state of biaxial stress exists in the specimen during the splitting test where the compressive stresses are larger than the tensile stresses. Nevertheless, the failure takes place in tension provided that the compressive strength of the material is greater than three times the tensile strength. The form of the failure is separation along the vertical plane. Thus the cylinder splitting tensile strength f_{sp} of a concrete is calculated from the formula

$$f_{sp} = \frac{2P_{max}}{\pi LD} \tag{2.12}$$

where P_{max} is the ultimate load applied. The other symbols are as in Eq. 2.9. The tensile strength from cube or prism splitting can be calculated by formulas similar to Eq. 2.12 except that the pertinent nominal area of the plane of failure (i.e., a^2 in the split cube test of Fig. 2.4) should be substituted for DL (Komlos, 1967–1969; Bonzel, 1965; Sen and Desayi, 1962; Palotas and Halasz, 1963; Iyengar, 1963). The splitting strength of a cube specimen tested diagonally is about 80% of that predicted by Eq. 2.12. These findings have been reconfirmed by use of the finite-element method (Davies and Bowes, 1968).

In contradistinction to concrete failure under direct tensile load, in the splitting test the major part of the coarse aggregate particles is usually broken along the surface of failure. This may be due to the fact that the tensile stresses have their high values in the splitting specimen within a narrow strip along the central vertical plane. In any case, the strength of the aggregate particles appears to have a sizable influence on the splitting strength of the concrete. For instance, an increase in the maximum particle size causes a greater increase in the splitting strength than in the direct tensile strength (Hannant et al., 1973). Also, when the tensile strength of a cement paste has been increased, for instance by epoxy modification, the splitting strength of the concrete containing this cement paste shows less increase than the flexural or direct tensile strength (S. Popovics, 1984).

The formulas above are based on the assumption that Hooke's law holds true up to the failure in the tested specimen and that a state of plane stress exists. In the case of concrete, neither assumption is valid. It has been shown, for instance, that splitting strengths calculated from certain assumptions of the theory of plasticity are somewhat smaller, whereas those calculated from other pertinent assumptions are somewhat higher than the values calculated by Eq. 2.12 (Seefried et al., 1967; Franca and Pincus, 1969). It was also shown that the arithmetic average value of the solutions derived from the upper and lower bound theorems is identical to the solution derived using Eq. 2.12 (Chen and Chang, 1978). In addition, it is believed that disregard of plastic deformations is not a decisive weakness of Eq. 2.12 (Peltier, 1954) mainly because the tensile stresses have no local concentrations or peak values along the central vertical plane. This has been shown not only by the theory of elasticity but also by photoelastic studies (RILEM, 1975b; N. B. Mitchell, 1961; Rudnick et al., 1963). Nevertheless, applicability of the splitting test for low-strength materials such as several-hours-old concrete is doubtful because such specimens suffer considerable compression and squeeze during the test, which alter the distribution of stresses (Frydman, 1964; Bynum et al., 1971). Two additional simplifications in calculation of the splitting strength using Eq. 2.12 are that the effect of the compressive stresses on the failure of the splitting specimen is omitted and that the concrete is assumed to be a homogeneous material rather than a composite of paste and aggregate or an anisotropic substance (Van Cauwelaert and Eckmann, 1994).

Equations 2.10 and 2.11 also show that the compressive stress is, theoretically, infinitely large in the specimen immediately under load (i.e., when $y = 0$) if the loads act on an infinitely narrow line. Such a very high stress would distort the result of the splitting test. Therefore, in practice, soft strips are placed between the specimen and the plates of the testing machine to reduce the load intensity (Oehlers and Johnson, 1981). This also reduces the maximum values of compressive stresses in the specimen. If, however, the width of the strips is less than $0.1D$, the effect on the tensile stresses is negligible (Ramakrishnan et al., 1967b). ASTM C496-90 specifies that two bearing strips of $\frac{1}{8}$-in. (3-mm)-thick plywood approximately 1 in. (25 mm) wide and as long or slightly longer than that of the specimen should be provided for each splitting specimen. The strip material may have a noticeable effect on the splitting strength measured (Sell, 1963; Spooner, 1969). When the splitting specimens contain pieces of reinforcement located perpendicular to the plane of loading, such as cores from bridge decks, the ultimate splitting load increases linearly with the amount of reinforcement (Kirtschig and Dulgeroglu, 1966).

The greatest advantage of the splitting test is its simplicity and the fact that the standard compression test specimen and testing machine can be used.

Another advantage is that failure in this test, unlike that in the flexural test, starts inside the concrete; therefore, the test result is not overly influenced by the surface condition of the specimen, such as moisture or temperature (Grieb and Werner, 1962), or by minor irregularities in the testing, such as slight eccentricity (McNeely and Lash, 1963; Ivey and Buth, 1966). A further advantage is that in the case of shortage in specimens, the portions of a specimen remaining from the splitting test can be used for additional determination of the compressive strength. It was found that the compressive strength of split halves of concrete cylinders is about 85% of the value obtained with comparable specimens that have not been split (Poijarvi and Syrjala, 1965; Nilsson, 1962; S. Popovics, 1961).

A disadvantage of this method is that the tensile stresses have their high values in the splitting specimen within a narrow strip. Thus the effects of the weak links, as well as the effect of the composite nature of the concrete, prevail less in the splitting test than in the direct tension test or in the concrete of a structure. This also means that the splitting strength of a concrete can be influenced more by the aggregate properties and less by the properties of the cement paste than can the actual tensile or flexural strength of the concrete exhibited in the structure. Considering also the theoretical simplifications usually applied, it is clear that the f_{sp} is, again, a more-or-less fictitious value.

The splitting strength of concrete *cores* was found to be practically the same as that obtained on comparable laboratory specimens despite the different curing (Raphael, 1984). The variability in splitting strength again appears to be about the same as that of the flexural strength (Komlos, 1970). The overall coefficient of variation of splitting strengths obtained by various investigators varies within 2.5 and 10% (Sell, 1963; Ramesh 1960; Ramakrishnan et al., 1967b). The usual minimum value of the within-batch coefficient is around 4 to 5% (Malhotra and Zoldners, 1967b; Ward and Cook, 1969).

Ring Tension Test and Other Methods

1. After a few preceding efforts (Roper and Bryden, 1964; Graf, 1923; Pogany, 1937; Sedlacek and Halden, 1962), Malhotra reintroduced another technique, the *ring test,* for determination of the tensile strength of concrete (Malhotra et al., 1966b). In this method, hydrostatic pressure is applied radially against the inside periphery of a concrete ring specimen. The resulting tensile stresses developed in the specimen are determined from the ultimate pressure by the pertinent equations of the theory of elasticity for thick-walled cylinders. If the ratio of the radius of the ring to its wall thickness is high enough (around 10 or higher), the ring tensile strength can be well approximated as (Malhotra, 1970b)

$$f_r = \frac{PD}{2t}$$ (2.13)

where f_r = ring tensile strength
  P = ultimate hydrostatic pressure
  D = average diameter of the ring
  t = wall thickness of the ring

The ring tension test has certain merits. The equipment is simple and portable, thus the testing can be carried out even at a construction site. Also, the entire volume of the ring specimen is subjected to tensile stresses without clamping and misalignment stresses. On the other hand, the drawbacks of this method are that special specimens are needed, large aggregate sizes require large ring that may be cumbersome, and the f_r ring tensile strength calculated is again a somewhat fictitious value because it is based on the validity of Hooke's law. The overall as well as the within- and between-batch coefficients of the variation of the ring tensile strength appear to be slightly greater than those of the flexural strength (Malhotra and Zoldners, 1967; Malhotra et al., 1968a).

2. The *double punch test* is a variant of the splitting test. Here the splitting load acts in a pair of points instead of a pair of lines (Chen 1970; Chen and Colgrove, 1974). The specimen is a 6 × 6-in. (153 × 153-mm) concrete cylinder placed vertically between the loading platens of a testing machine. It is compressed by two 1.5-in. (38-mm)-diameter steel punches 1 in. (25 mm) thick placed concentrically on the top and bottom surfaces of the cylinder. This load develops tensile stresses inside the specimens which split the cylinder across many vertical diametral planes. The average tensile stress over all of the cracked diametral planes can be calculated from the ultimate load and geometry of the specimen:

$$f_{dp} = \frac{p}{\pi(1.2LD - r^2)}$$ (2.14)

where f_{dp} = tensile strength calculated from the double punch test
  r = radius of the punch

The other symbols are identical with the symbols of Eq. 2.12.

These calculated f_{dp} tensile strengths show good one-to-one correlation with the tensile strengths obtained by splitting cylinders (Chen and Yuan 1980; Elices and Planas, 1982).

3. A combination of the splitting tension test and the double punch test is the *point-load core test* (Robins, 1980, 1984; Richardson, 1989). It is similar to the splitting test except that point load and support are applied in two

opposing points on the curved surface of a cylindrical specimen instead of line load and support. The point-load index characteristic of the tensile strength is defined as the maximum load divided by the square of diameter of the cylinder. The same test, with little extra preparation, is applicable to specimens with shapes other than cylindrical, including irregular shapes. The point-load index is calculated as the maximum load divided by the square of the distance between the loading and supporting points. However, experimental results revealed that size and shape effects were very pronounced when lumps of irregular shapes were tested with point loading.

4. Tensile stresses can be produced in suitable specimens by *centrifugal force*. The method has been tried out on hardened cement pastes (Roper and Bryden, 1964). The limited amount of data showed a relatively small coefficient of variation of the experimental results.

5. *Gas pressure* can also be used on concrete specimens to produce uniaxial tensile stresses. If the specimen is a cylinder, it is inserted into a cylindrical steel jacket with seals at each end, and gas pressure (nitrogen) is applied to the bare curved surface. The pressure is increased at a specified rate until failure occurs, usually by the formation of a single cleavage plane transverse to the axis of the specimen (Clayton, 1978). Cores may also be used for this purpose.

6. Another indirect test for determination of the tensile strength of concrete consists of *testing frame* specimens between the compression platens of a testing machine, where the geometry of a frame is such that one of its members is in tension (Desayi and Veerappan, 1972). Comparative tests provided about 80 to 90% cylinder splitting strengths of comparable structural concretes.

7. The tensile strength can be tested by using a very rigid, yet simple machine in which the load is generated by *thermal contraction of a sturdy aluminum column* (Kirchner and Rishel, 1971).

Relations Between Various Tensile Strengths

Table 2.2 summarizes the results of several investigations concerning the relationship between the f_{sp} *splitting* and f_t *direct tensile strengths*. The data clearly show that the splitting strength is usually greater than the direct tensile strength of the same concrete, although the opposite has also been reported (Raphael, 1984; Malhotra, 1967; Desayi, 1969; Sen, 1955; Kesai et al., 1972). The following discussion throws some light at the source of this contradiction.

As mentioned, the measured value of tensile strength is reduced more by loading eccentricity, stress peaks, and so on, than is the measured value of splitting strength. But apart from this, several simplifications can be mentioned as reasons why the f_{sp} value calculated by the usual method may be greater than f_t, although the failure of concrete in the splitting test is con-

TABLE 2.2 Approximations of the relationship between direct tensile and splitting strengths[a]

f_t/f_{sp}	Authority[b]	Remarks
0.68	Wright	$f_{sp} = 405$ psi at the age of 28 days
0.69	Baus and Campus	Cores taken from a 14-year-old concrete beam
0.75	Bonzel	Average based on the findings of several investigators
0.85	Laboratorio Central de Ensayo de Materiales de Construction	$f_{sp} = 14.1$ kg/cm^2 at the age of 7 days
0.9	Pincus and Gesund	f_{sp} varies from 150 to 450 psi
1.0	Newman	Gravel and crushed stone concretes; maximum particle size is $\frac{3}{4}$ in. f_{sp} varies from 300 to 600 psi
1.28	Newman	Mortar and lightweight concrete; f_{sp} varies from 230 to 500 psi
0.41 to 0.75	Ramesh and Chopra	Cement mortar; f_{sp} varies from 65 to 14 kg/cm^2
0.52 to 0.81	Kadlecek and Spetla	f_{sp} varies from 17 to 28 kg/cm^2
0.55 to 0.93	Campus et al.	Aggregates with different surface textures
0.62 to 0.68	Ledbetter and Thompson	Structural lightweight concrete; f_{sp} varies from 100 to 500 psi
0.69 to 0.96	Ali et al.	Mortars of 1:1 and 1:2 mix proportions; f_{sp} varies from 316 to 618 psi
0.72 to 1.0	Rusch and Hilsdorf	f_{sp} varies from 14 to 33 kg/cm^2
0.86 to 0.94	Malhotra and Zoldners	Maximum particle size is $\frac{3}{8}$ in.
$0.86 + 63/f_{sp}$	Ward	Normal-weight concrete; f_{sp} is in psi
$1.06 - 0.23/f_{sp}$	Komlos	f_{sp} is in MPa; f_{sp} varies from 1.5 to 4 MPa

Source: S. Popovics (1967a). (Copyright TRB. Reprinted with permission.)
[a]1 ksi = 6.90 MPa; 1 kg/cm^2 = 0.098 MPa.
[b]References for the authors listed in this column can be found in S. Popovics (1967a).

trolled by the tensile strength. The concrete does not follow Hooke's law; therefore, use of this law in calculation of the splitting strength increases the apparent value of f_{sp}. Also, in the direct tensile test the concrete can fail anywhere in the length of the specimen; in the splitting test, however, the failure occurs in, or near, the central vertical plane of the specimen, which strongly reduces the number of weak links in the zone of maximum stresses. Finally, restraint due to friction between the splitting specimen and the plates of the testing machine may increase the apparent value of f_{sp}. On the other hand, one would expect from the biaxial stress condition in the splitting specimen that the value of f_{sp} as calculated by the usual method is somewhat smaller than f_t (L'Hermite, 1955).

The numerical effects of these simplifications on the measured value of f_{sp} may depend on the composition and age of the concrete, curing and testing of the specimen, and so on, thus they may be partially responsible for the large variation in f_t/f_{sp} values that is demonstrated in Table 2.2. Another frequently overlooked source of this variation is the composite character of the structure of concrete (Soshiroda, 1972). Since the failure surface of concrete specimens subjected to a splitting test usually contains more broken aggregate particles than the failure surface caused by direct tension, it appears that the bond between the cement paste and aggregate particles has less influence and the strength of the aggregate particles more influence on the splitting strength of concrete than on the direct tensile strength. Accordingly, the f_t/f_{sp} ratio may be influenced by the type of mineral aggregates. For instance, one would expect that the f_t/f_{sp} ratio (and the f_{fl}/f_{sp} ratio as well) is higher for structural lightweight concretes, or that an increase in the maximum size of the aggregate reduces the splitting strength of concrete less than its tensile strength. Experimental evidence appears to support these expectations (Ward, 1964; Walker and Bloem, 1960; Newman, 1964). A change in the porosity of the cement paste and concrete may also affect the f_t/f_{sp} ratio (S. Popovics, 1967b, 1969a, 1969b; Hannant et al., 1973). For instance, this ratio seems to decrease somewhat with an increase in tensile strength (Ward and Newman, 1970; Bonzel and Manns, 1970).

In brief, one can always suspect that a low f_t/f_{sp} ratio is the result of some unnoticed irregularity in the tensile test. Apart from faulty load measurement, no experimental error can increase the measured value of f_t over its true value. On the other hand, crushed aggregates with small maximum particle size, or certain lightweight aggregates, may produce f_t strengths higher than their splitting strengths.

Limited experience with the *ring tension test* appears to indicate that the ring tensile strength is slightly higher than the comparable value obtained in the splitting test.

Relations Between Flexural and Tensile Strengths

Results by several investigators concerning the relationship of the *flexural strength* to *direct tensile strength* are summarized in Table 2.3. Relationships concerning the *splitting strength* are illustrated in Figures 2.6 and 2.7. These data show that the f_{fl} flexural strength calculated in the usual way is higher than either the comparable direct tensile or the splitting strength of the same concrete. Otherwise, experimental data indicate good correlation between the flexural strength and direct tensile strength or splitting strength within a single test series. More specifically, it is a fair approximation that the f_t versus f_{fl}

TABLE 2.3 Approximations for the relationship between direct tensile and flexural strengths[a]

f_t/f_{fl}	Authority[b]	Remarks
0.45	Wright	$f_{fl} = 605$ psi at the age of 28 days
0.48	Graf	Wet curing
0.6	Hamada	
0.6	L'Hermite	
0.6	Ros	
0.70	Pincus and Gesund	Average value; f_{fl} varies from 200 to 500 psi
0.744	Raphael	Calculated value supported by experimental data by others
0.37 to 0.56	Walz	
0.37 to 0.55	Kadlecek and Spetla	f_{fl} varies from 23.2 to 42.0 kg/cm²
0.38 to 0.57	Campus et al.	Aggregates with different surface textures
0.48 to 0.63	Price	Valid from $f_t = 110$ to 630 psi
0.48 to 0.63	Gonnerman and Shuman	Valid from $f_{fl} = 230$ to 1010 psi
0.48 to 0.60	Malhotra and Zoldners	Maximum particle size is $\frac{3}{8}$ in.
0.52 to 0.70	Kaplan	f_{fl} varies from 550 to 850 psi
0.52 to 0.77	Hummel	
0.55 to 0.70	Dutron	With different aggregates
$0.41 + 0.41/f_{fl}$	Komlos	f_{fl} is in MPa; f_{fl} varies from 2.7 to 9.3 MPa
$0.63 + 61/f_{fl}$	Ward	f_{fl} is in psi; for normal-weight and lightweight structural concretes

Sources: S. Popovics (1967a). (Copyright TRB. Reprinted with permission.)
[a]1 ksi = 6.90 MPa; 1 kg/cm² = 0.098.
[b]References for the authors listed in this column can be found in S. Popovics (1967a).

relationship is linear within practical limits (Ward and Cook, 1969; Pincus and Gesund, 1965; Kaplan, 1963; Pineiro et al., 1970).

The f_{sp} versus f_{fl} relationship can also be considered as linear, as was shown by several authors (Jindal, 1964; Komlos, 1970; Akazawa, 1953; Grieb and Werner, 1962; Sen and Bharara, 1961; Narrow and Ullberg, 1963; Hanson, 1961). This is also demonstrated in Figure 2.6 by experimental data. By using these experimental values, the following approximate formula can be obtained for the f_{fl} versus f_{sp} relationship:

$$f_{fl} = 1.2f_{sp} + 100 \qquad (2.15)$$

where the stresses are in psi.

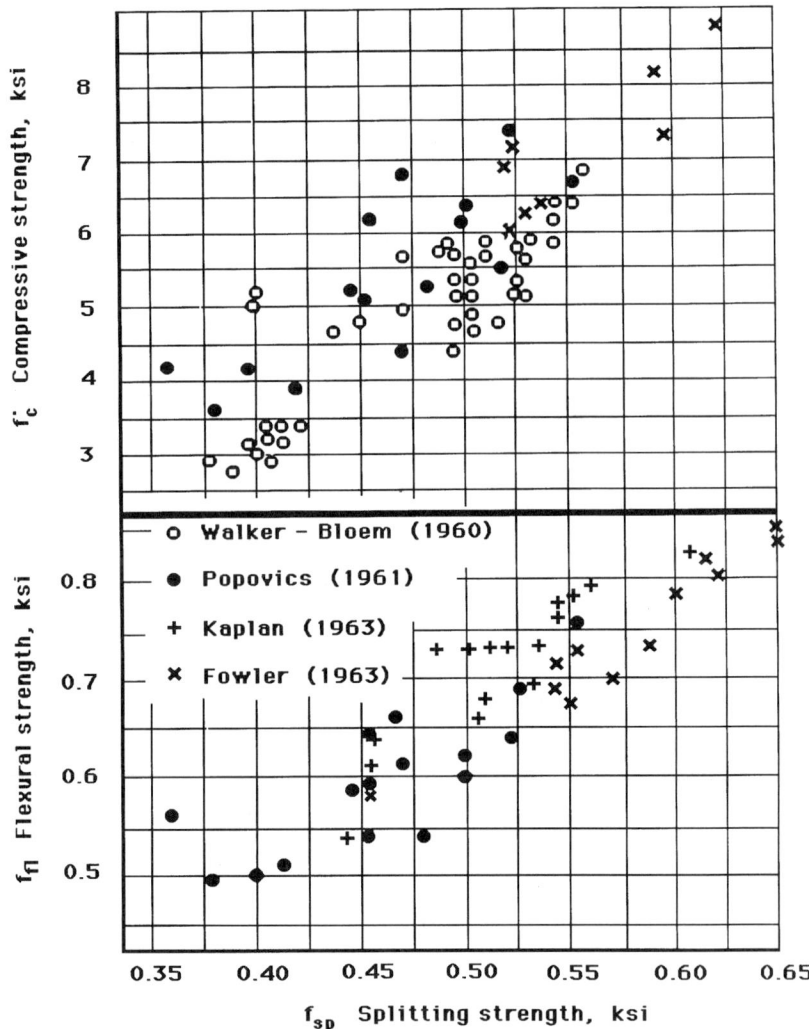

Figure 2.6   Empirical relations for compressive and flexural strengths versus splitting strength. (From S. Popovics, 1967a. Copyright TRB. Reprinted with permission.)

As indicated by Figure 2.7, Eq. 2.15 would not necessarily hold for tests carried out under different conditions and within a wider strength range. Nevertheless, the formula further demonstrates that the ratio of f_{sp}/f_{fl} increases as the concrete strength increases. This is in accordance with the observations of other authors (Table 2.4) (Bonzel, 1964; Malhotra and Zoldners, 1967; RILEM, 1975b; Palotas and Halasz, 1963).

The failure of plain concrete beam in the bending test is controlled by the tensile strength of the concrete. The propagation of microcracks has an im-

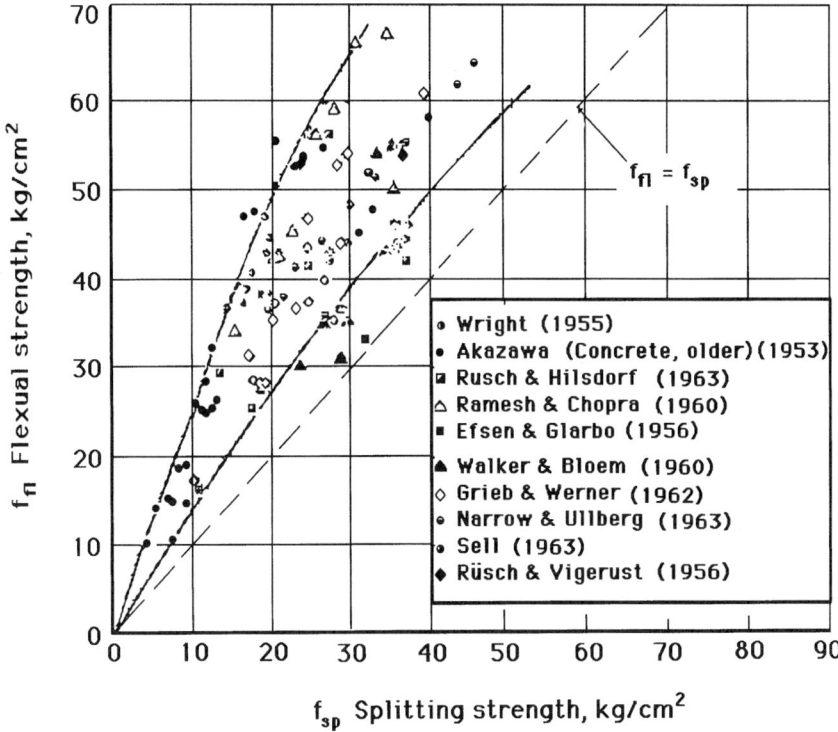

Figure 2.7 Experimental results obtained by various investigators for the relationship between splitting strength and flexural strength of concrete. 1 kg/cm² = 14.2 psi = 0.098 MPa. (From Bonzel, 1965. Copyright Beton Verlag. Reprinted with permission.)

portant role in this (Yoshimoto et al., 1972). A simplified explanation is that as soon as the tensile stress reaches a value in the bottom fiber of the beam which the section cannot resist, cracking and failure occur. Nevertheless, the flexural strength of concrete calculated by Eq. 2.1 is always greater than its actual tensile strength (Platts and Kirchner, 1971).

L'Hermite (1955) attacked this problem from two different theoretical directions. When he applied the difference between the elastic energy of the fissure propagation in the direct tensile test and that in the bending test with third-point loading, he obtained a limit value of 0.575 for the ratio of f_t/f_{fl}. On the other hand, when he used the difference between the average deformations in the two tests, he obtained 0.5 for the limit of this ratio. Tucker (1945) showed that at the failure of nonuniformly stressed specimens local stresses may be present that may exceed the tensile strength of the same material as determined by uniformly distributed loading. He found that according to a statistical theory of direct tensile strength, the flexural strength may be up to 90% greater than the direct tensile strength. Another and perhaps the main reason for the difference between f_{fl} and f_t seems to be the appli-

TABLE 2.4 Comparison between splitting and flexural strengths[a]

f_{sp}/f_{fl}	Authority[b]	Remarks
0.63	Sell	f_{fl} = about 45 kg/cm² at the age of 14 days
0.66	McNeely and Lash	f_{fl} = 690 psi; coarse aggregate is crushed gravel
0.67	Wright	f_{sp} = 405 psi at the age of 28 days
0.8	Sen and Bharara	f_{fl} varies from 214 to 630 psi; tested at various ages
0.39 to 0.74	Akazawa	Recommended average: 0.47
0.45 to 0.53	Ramesh and Chopra	f_{fl} varies from 34 to 68 kg/cm²
0.50 to 0.80	Grieb and Werner	f_{fl} varies from 250 to 790 psi; sand and gravel with 1½ in. maximum size
0.51 to 0.78	Grieb and Werner	f_{fl} varies from 350 to 955 psi; crushed stone with 1½ in. maximum size
0.57 to 0.88	Grieb and Werner	f_{fl} varies from 430 to 750 psi; lightweight aggregate concrete
0.65 to 0.89	Grieb and Werner	f_{fl} varies from 640 to 840 psi; crushed stone with 1 in. maximum size
0.55 to 0.71	Narrow and Ullberg	f_{fl} varies from 550 to 850 psi; with different aggregates
0.62 to 0.90	Walker and Bloem	f_{fl} varies from 800 to 300 psi; with aggregates of different maximum sizes
0.63 to 0.83	Rusch and Vigerust	f_{fl} about 45 kg/cm²; ratio decreases with decreasing strength
0.67 to 0.91	Efsen and Glarbo	f_{fl} varies from 16 to 42 kg/cm²; ratio decreases with decreasing strength
0.72 to 0.77	Kaplan	f_{fl} varies from 850 to 550 psi
$0.6 + 100/f_{fl}$	Popovics	f_{fl} varies from 490 to 750 psi; with different aggregates

Source: S. Popovics (1967a). (Copyright TRB. Reprinted with permission.)
[a]1 ksi = 6.90 MPa; 1 kg/cm² = 0.098 MPa: 1 in. = 25.4 mm.
[b]References for the authors listed in this column can be found in S. Popovics (1967a).

cation of Hooke's law in the usual calculation of the flexural strength of concrete. Since concrete deformations do not follow Hooke's law, the assumption of linear tensile stress distribution throughout the depth of the beam results in fictitious tensile stress values which are higher than the actual stresses in the beam. For instance, the flexural strength calculated from the hypothetical stress distribution that the material is rigid-plastic in tension and proportional in compression is two-thirds of the flexural strength when calculated from the assumption of linear stress distribution (Spetla and Kadlecek, 1966). Pincus and Gesund (1965) also state that Blakey and Beresford (1953) reported a similar factor of 0.735 using second- and third-degree parabolas for the stress distribution in the beam as compared to linear distribution. Raphael obtained practically the same number by using a rectangular diagram for the tensile stresses (Raphael, 1984). The importance of the stress distribution with respect to the calculated flexural strength of concrete has also been demonstrated experimentally by Blackman et al. (1958).

Relations Between Compressive and Tensile Strengths

Correlations between the *compressive strength* and *direct tensile strength* follow the tendency discussed earlier in connection with flexural strength (Saul, 1960; Malhotra and Zoldners, 1967; Ward 1969; Ledbetter and Thompson, 1965; Raphael, 1984; Schuman and Tucker, 1943; Duriez and Arrambide, 1961; Hedstrom, 1966; Campus, 1955). At extremely early ages, however, the f_t/f_{co} ratios exhibit unexpected fluctuations (Kesai et al., 1972). Similarly, there is a correlation between the *compressive* and *splitting strengths* of concrete, as demonstrated in Figure 2.6 by using experimental data of several investigators. Other data also support such a correlation (Fig. 2.8) (Malhotra and Zoldners, 1967; Akazawa, 1953; Mitchell, 1961; Ramesh, 1960; Bonzel,

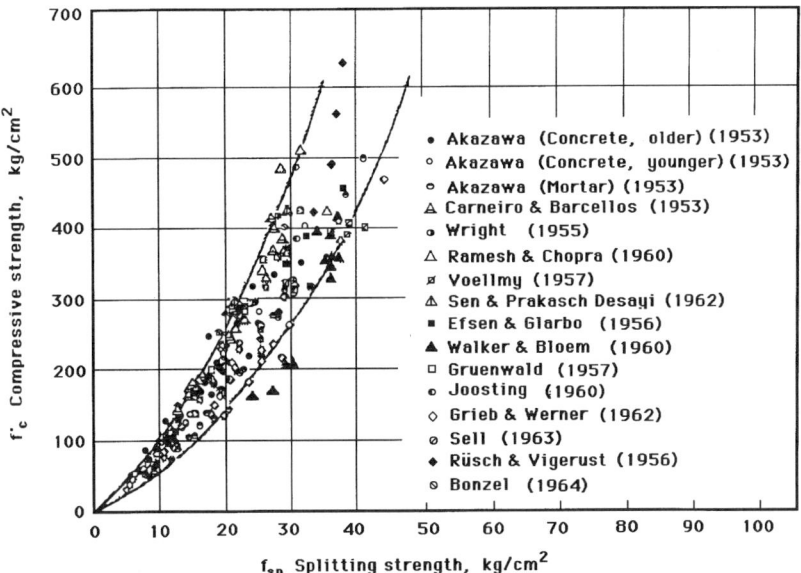

Figure 2.8 Experimental results obtained by various investigators for the relationship between splitting strength and compressive strength of concrete. 1 kg/cm² = 14.2 psi = 0.098 MPa. (From Bonzel, 1965. Copyright Beton Verlag. Reprinted with permission.)

1965; Grieb and Werner, 1962; Ivey and Buth, 1966; Sen, 1955; Sen and Bharara, 1961; Hanson, 1961; Short and Kinnikurgh, 1963; Campus et al., 1966; Deutsche Bundesbahn, 1962; Abeles, 1960; Avram et al., 1981; Narayanan, 1961; Chandrashekhara and Krishnaswamy, 1964; Gyengo, 1959). Bonzel (1965), for example, presents empirical formulas by several investigators for the relationship between the compressive and splitting strengths of cement mortars and concretes. The general form of these formulas is

$$f_{sp} = C f_{co}^{n} \tag{2.16}$$

In the case of cylindrical specimens the value of C varies from 0.71 to 0.74, depending on the composition of the specimen and the test method used (Carino and Lew, 1982). These formulas as well as numerous other experimental data show that this relationship is curve linear because the ratio of tensile strength to compressive strength decreases with increasing concrete strength (Price, 1951; Grieb and Werner, 1962; Duriez and Arrambide, 1961; Gyengo, 1959; Thaulow, 1957). It may be mentioned that the f_{sp}/f_{co} ratio is less for structural lightweight concretes than for normal-weight concretes of identical compressive strength (Ward, 1964; Hanson, 1961). For example, ACI Committee 209 recommends the following formula (ACI, 1971):

$$f_{t} = 0.33\sqrt{u f_{co}} \tag{2.17}$$

The reader may recall that a similar phenomenon was observed earlier in connection with the f_{fl}/f_{co} ratio.

2.3   OTHER CONCRETE STRENGTHS

As has been emphasized, other concrete strengths are usually affected by the same factors as compressive or tensile strengths, but the degree of the effects is usually different. Therefore, there again exist correlations between torsion, shear, and other strengths and compressive, tensile, and other strengths of a concrete, but these correlations are influenced significantly by the composition, applied test methods, and other circumstances of the testing.

Torsion Strength

It is difficult to talk about the torsion strength of concrete with any satisfactory precision because the "strength" calculated from the ultimate load depends to a large extent not only on the shape of the specimen but also on whether the formula used was derived from an assumption of elastic behavior of the concrete or from some other behavior. Nevertheless, when the failure of the concrete is produced by torsional load, it is customary to talk about torsion strength.

The only case that is satisfactory, theoretically, is when the torsion specimen is a circular tube with relatively thin walls. In this case, the failure takes place in pure shear. Otherwise, a relationship is expected between the torsion strength and tensile strength of a concrete because the failure in such torsion appears to be a tensile failure. This was first proposed in the United States probably by Andersen in a paper and in subsequent discussions (Andersen, 1935). However, even this relationship is strongly dependent on the size and shape of the torsion specimens, as shown in Table 2.5, and in addition, on the method used for calculation of the torsion strength (Cowan, 1953; Zia 1961; Navaratnaraja, 1968). The torsion strength determined on circular cylinders and calculated on the basis of rigid-plastic stress distribution (plastic theory) is 75% of the torsion strength calculated from a linear stress distribution (elastic theory) (Timoshenko, 1957). Therefore, the magnitude of the pure tensile strength of a concrete is expected to be less than its comparable torsion strength calculated from a linear stress distribution (Pincus and Gesund, 1965). Experimental data support this statement conclusively (Table 2.5). Incidentally, the torsion strength of concrete has an increasing practical significance because the failure of prestressed concrete in torsion is essentially similar to that of plain concrete (Zia, 1961).

Table 2.5 summarizes experimental data for the torsional strength of concrete by several investigators. In addition, two related experiments can be mentioned. Iyengar et al. (1963) found that torsion strengths of various concretes were just a little higher than their comparable splitting strengths, which varied from 174 to 488 psi (1.20 to 3.37 MPa) in their series. Unfortunately, it is not reported in their paper how the torsion strength was calculated. Also, numerous specimens of square cross sections and of various compositions were tested at various ages by Rudeloff as quoted by Graf et al. (1960); the results approximated the value of 0.43 for the ratio of direct tensile strength to torsion strength where the latter was calculated presumably by the elastic theory. The values of direct tensile strength varied from 100 to 240 psi (0.69 to 1.66 MPa) in this test series.

Marshall is of the opinion that the elastic approach is not the correct one for calculation of the torsion strength of concrete since the shear stresses obtained from this theory are in some cases more than twice the value of the direct tensile strengths. This opinion is not unexpected because the concrete cannot be considered as a linearly elastic material, especially at high stress levels. Moreover, if concrete behaved elastically under torsion, the failure would begin at the midpoint of the longer side of a specimen with a rectangular section and gradually spread to the remainder of the section. Careful observation showed, however, that such failure never occurred; the process of failing is sudden cracking of the entire section, tending to show a uniform stress distribution (Marshall, 1964). Nevertheless, it is conceivable that the

TABLE 2.5   Comparison between torsion and direct tensile strengths[a]

Authority[b]	Section Tested	Torsion Strength (psi)		Direct Tensile Strength (psi)
		Plastic Theory	Elastic Theory	
Graf and Morsch	Circle	151	202	164
Graf	Circle	Not given	265	164
	Circular ring	Not given	196	
Bach and Bauman	Circle	269	364	264
	Circular ring	269	244	264
	Square	269	433	264
	2:1 Rectangle	269	463	264
Bach and Graf	Circle	261	350	Not given
	Square	265	425	Not given
	2:1 Rectangle	263	450	Not given
Marshall and	Circle	243	325	282
Tembe	Square	318	508	282
	1.5:1 Rectangle	287	472	282
	1.7:1 Rectangle	313	530	282
	2:1 Rectangle	302	516	282
	3:1 Rectangle	349	580	282
Turner and	Square	247 and 264	395 and 422	301
Davies	2:1 Rectangle	309 and 223	528 and 381	219
Young et al.	Square	336	539	Not given
	1.5:1 Rectangle	278	467	Not given
	2:1 Rectangle	352	602	Not given
Anderson	Square	336	539	Not given
	1.25:1 Rectangle	306	505	Not given
	1.5:1 Rectangle	342	575	Not given
Anderson	Square	400	640	Not given
	1.25:1 Rectangle	369	609	Not given
	1.5:1 Rectangle	392	658	Not given

Source: S. Popovics (1967a). (Copyright TRB. Reprinted with permission.)
[a]1 ksi = 6.90 MPa.
[b]References for the authors listed in this column can be found in S. Popovics (1967a).

elastic theory is more applicable for calculation of the torsional stresses in high-strength concretes, especially when the specimen is a hollow cylinder, since the plasticity of concrete decreases with increasing strength.

Due to the uncertainties of the classical methods, Hsu proposed a new approach. This is based on the assumption that plain concrete beams subjected to pure torsion actually fail by bending (Hsu, 1968a, 1968b, 1968c). Expressing this mathematically, he obtained the following formulas for rectangular beams:

$$T_u = \frac{x^2 y}{3} (0.85 f_{fl}) \tag{2.18}$$

$$\approx 6(x^2 + 10)y \sqrt[3]{f_c'} \tag{2.19}$$

where T_u = ultimate torque, in.-lb
$\quad\quad\, x$ = size of the shorter size of the beam, in.
$\quad\quad\, y$ = size of the longer side, in.
$\quad\quad f_{fl}$ = standard flexural strength of the concrete, psi
$\quad\quad f_c'$ = standard cylinder compressive strength of the concrete, psi

   The failure torque of T- and L-sections of plain concrete can be calculated by dividing these sections into stem and flange sections, each of which is rectangular. The strength of the entire section of the beam then is taken as the sum of the strengths of stem and flange calculated by Eq. 2.18 or 2.19. As Figure 2.9 shows, experimental data support Eq. 2.19 favorably.

Shear Strength

There are several test methods for determination of the shear strength of a concrete each providing different results. One such test is the *direct shear test*, where a plain concrete beam or slab of a short span is subjected to a load applied very close to the supports. This method does not provide pure shear because sizable normal stresses are also produced in the specimen. The direct shear strength f_{dsh} of the concrete can be calculated from the ultimate load and the dimensions of the specimen in the usual way. Graf et al. (1960) applied this method on concretes where the cube compressive strength f_{cu} varied from 130 to 310 kg/cm^2 (12.74 to 30.38 MPa) and in another series of concretes where the f_{fl} flexural strengths varied from 18 to 43 kg/cm^2 (1.76 to 4.21 MPa). He found that within these strength limits the f_{dsh}/f_{cu} ratio was 0.23 and the f_{dsh}/f_{fl} ratio was 1.6, with good approximation. An empirical formula was recommended by Palotas (1960) for the relationship between direct shear strength and the cube strength of a concrete:

$$\frac{f_{fdsh}}{f_{cu}} = 0.25 - \frac{f_{cu}}{8000} \tag{2.20}$$

where f_{cu} is expressed in kg/cm^2 (= 0.098 MPa).
   Another popular test method for the determination of the shear strength of a concrete is to load reinforced beams without web reinforcement for bending up to shear failure; the shear resistance f_{sh} of the beam at a diagonal (i.e., the *shear cracking strength* of the concrete) can be calculated from the ultimate load and the characteristics of the specimen (i.e., dimensions, reinforcement, etc.) in the usual way. This shear strength is not identical with the direct

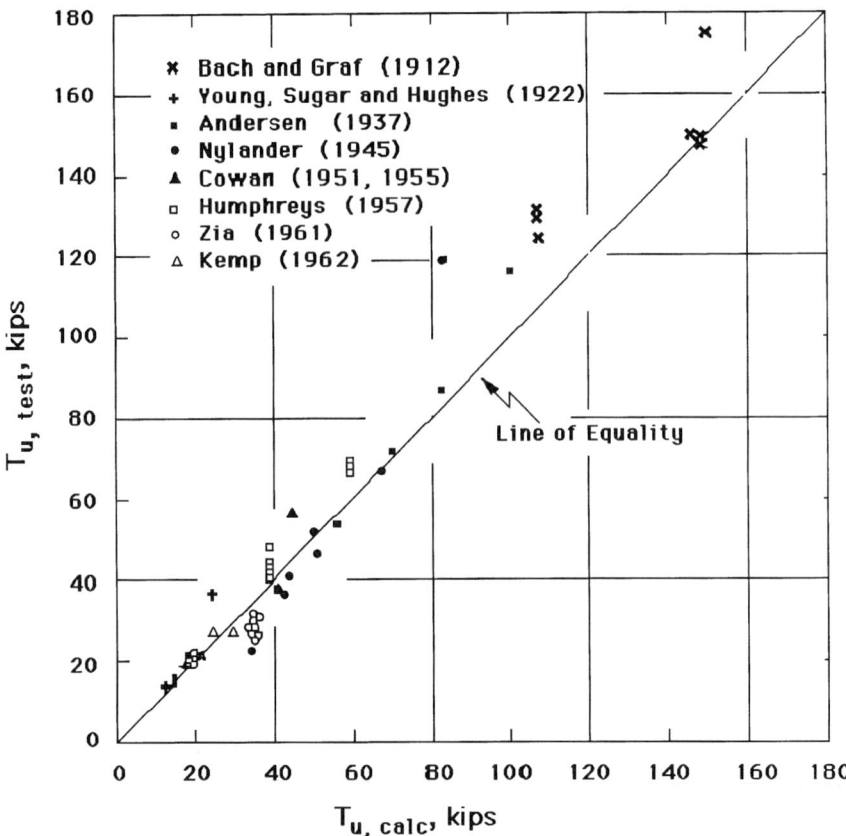

Figure 2.9 Comparison of experimental values of ultimate torque and corresponding values calculated by Eq. 2.19. (From Hsu, 1968. Copyright ACI. Reprinted with permission.)

shear strength and is not pure shear either, because here the failures are actually of a tensile nature on a plane different from that on which the maximum shear stress occurs. Viest and others (Viest, 1959; Hognestad et al., 1964; Bresler and MeGregor, 1967) found a correlation between the shear cracking strength f_{sh} and $\sqrt{f_c'}$, whereas Hanson (1961) found a correlation between f_{sh} and f_{sp}. In both cases the proportionality factor is a function of the characteristics of the beam. In accordance with this, the ACI Building Code contains a formula that provides the f_{sh} shear resistance of a concrete as a linear function of $\sqrt{f_c'}$ (ACI, 1983b). On the other hand, recent experimental evidence appears to indicate that the f_{sh} shear strength varies just slightly with the compressive strength of the concrete, at least for underreinforced beams. A typical example is the recommendation by Evans and Dongre (1963). The empirical formulas presented for the relationship between the shear strength f_{sh} and cylinder strength f_c of a concrete are for normal concrete,

$$f_{sh} = 0.04f_c + 100 \qquad (2.21)$$

and for structural lightweight concrete,

$$f_{sh} = 0.03f_c + 75 \qquad (2.22)$$

where all strengths are expressed in psi (= 0.0069 MPa). Although these formulas are valid, strictly speaking, only for a certain type of beam, other investigators have also published experimental results and/or formulas that show a similar tendency for normal concrete (Ferguson and Thompson, 1953; Kani, 1966; Bazant and Sun, 1987), for structural lightweight concrete (Nelson and Frei, 1958), or for both (Short and Kinniburgh, 1963; Taylor and Broms, 1964).

The strength of concrete in *pure shear* f_{psh} (i.e., when no normal stresses are present on the plane of failure) can be obtained by subjecting a long, thin hollow cylinder to torsion away from the ends (Reeves, 1962). The strengths obtained by this method are shear strengths comparable to biaxial strength results with tensile and compressive stresses of equal magnitude. On this basis, Kesler estimated that the pure shear strength is approximately 20% of the compressive strength of the concrete (Kesler, 1966). On a similar basis, Sen (1955) recommended the relation

$$f_{psh} = 0.608f_{cu}^{0.737} \qquad (2.23)$$

and

$$f_{psh} = 0.689f_c^{0.741} \qquad (2.24)$$

where f_{cu} is the cube strength and f_c is the cylinder strength in psi.

Comparative experiments show again that the shear strength f_{sh} of a structural lightweight concrete is less, in general, than that of a normal-weight concrete of identical compressive strength. This difference can be as high as 40%, depending on the type of lightweight aggregate used (Short and Kinniburgh, 1963; Evans and Dongre, 1963; Taylor and Broms, 1964; Hanson, 1958).

Bond and Impact Strengths

Several different types of specimens and test procedures may be employed in measuring the strength of concrete *bond to reinforcement*. These include the pullout and pushout tests, the embedded bar tensile test (ASTM C234-91a), the split-bar method (Jiang et al., 1984), and various types of flexural specimens. The bond strength is usually calculated as the ultimate load in the reinforcing bar divided by the normal embedded surface area of the bar, although the actual stress distribution is complex (Tassios and Koroneons,

1984). The flexural test is probably preferred in that bond is of prime concern in bending. The pullout test is very simple to perform but has the disadvantage of producing a confining stress in the concrete in the region where bond may be most critical.

Different test methods provide different results. For instance, results of pushout tests are usually unrealistically high, due to the diameter increase of the bar under compression. The difference between bond values obtained from pullouts and beams are due primarily to the different crack patterns that develop in the two types of specimens. Although longitudinal cracks develop along the reinforcing bars (as the stresses increase and slip becomes appreciable) in both beams and pullouts, there is a basic difference in the cracking phenomena. In pullout specimens, the concrete is in compression, while in the beam specimens both the concrete and reinforcement are in tension. As the steel stresses increase, transverse cracks develop in the beams, and each new transverse crack tends to initiate a new longitudinal crack. Transverse cracks are entirely absent in pullout specimens and cracking is confined to longitudinal splitting (Lutz, 1978).

Despite these differences, most of these tests can, for design purposes, provide a comparison of bonds of different concretes but not bond strength. The reason for this is that although the bond strength is usually improved by an increase in the compressive strength of the concrete (Price, 1951; Jindal, 1964; Lutz, 1970), especially by an increase in the cement content (Hsu and Slate, 1963), it is usually more dependent on the surface characteristics of the reinforcement than on the properties of concrete. More specifically, the mechanisms of bond failure in the plain and deformed reinforcement differ considerably (Balazs, 1991). For the bond to plain-surface bars, there is a fairly large initial contribution from adhesion which is completely lost on the second and subsequent pulls. The bond in deformed bars decreases gradually as the concrete around the deformations is fractured on repeated pulling (Lutz and Gergely, 1967).

Equally important is the *bond of the cement paste or mortar portion of the concrete to the coarse aggregate particles.* This bond carries the load from the matrix to the aggregate particles, without which there would not be sizable concrete strength. The rule of thumb is that crushed stone aggregates develop stronger bond to the paste than gravel particles with smooth surface, and smaller particles develop stronger bond than larger particles. In any case, this bond (i.e., the paste–aggregate interface) is the weakest link in the composite structure of hardened concrete (Section 4.5). The direct consequence of this is that the failure of a concrete under load starts with crack development and propagation on this interface. Therefore, if this bond is weakened further, for instance, by a clay coating on the aggregate particle, this can reduce the concrete strength drastically, especially the flexural and tensile strengths. Efforts to develop a coating that *strengthens* the bond between cement paste and aggregate particles have remained unsuccessful (S. Popovics, 1987b).

Excessive bleeding is harmful for the bond because the rising water may be arrested on the underside of the horizontal elements of the reinforcement and aggregate particles, reducing the contact. The same can be the result of poor compaction of concrete. The bond strength also decreases strongly with increasing temperature which probably is due to the difference in thermal coefficients of expansion between cement paste and other solids.

There are several methods for testing the *impact strength* of concrete. Some of these use repeated hammer blows, such as the standard German method (DIN 52107); others use pendulum arrangements (Monack, 1971); and so on. The impact strength measured depends on the test method used (Dinsdale and Wilkinson, 1958; Zielinsky et al., 1981a, 1981b). Other factors that influence the impact strength of concrete are (1) the type of the ingredients of concrete, especially the aggregate; (2) the concrete composition, especially the water-cement ratio and cement content; (3) curing conditions; and (4) age.

These factors, of course, also influence the static strength but usually not to the same extent. Thus the impact strength–static strength ratio is not a constant value (Sandhu, 1963). It is not surprising, then, that various investigators arrived at conflicting conclusions concerning these ratios. For example, Glanville et al. (1938a) concluded that the impact strength of a concrete may be assumed to be between one-half and three-fourths of its static cube strength. Others also found correlation between impact and compressive strengths (Fulton and Davis, 1961). On the other hand, Green (1958) states that Dutron believed the impact strength of a concrete to be connected with its tensile strength, and that Passov and Framm found no relationship at all. Results of impact tests by Walz are strongly influenced by the angularity and surface texture of the coarse aggregate used, but they do not show good relation with either the compressive or flexural strengths of the concrete (Walz, 1939). More recent Belgian experiments with various cements indicate certain correlations between the impact strength and the compressive and flexural strengths, respectively, although not for the splitting strength. This correlation exists up to the age of 1 year but only when the specimens were cured in 100% relative humidity until testing. Comparable specimens cured in 50% relative humidity did not show any of these correlations, due to the fact that the impact strength of these specimens declined at longer ages while the other strength values increased (Dutron, 1962).

As pointed out earlier, the primary reason for these conflicting opinions is that the impact strength is affected by various factors to a greater extent than are any of the static strengths. For example, under certain circumstances, such as dry curing and brittle and/or smooth aggregate, the concrete becomes more brittle with increasing static strength; thus the resistance to impact falls. It also appears probable that the magnitude of bond between the cement paste and aggregate has a more pronounced effect on the impact strength of concrete than on its static strengths. Even more to the point are findings by Dahms (1968, 1969) which showed that raising the water-cement ratio from 0.4 to 0.8 by weight produced a 42% reduction in the compressive strength, a 35% reduction in the splitting strength, and about a 95% (!) reduction in the impact strength at the age of 28 days. He also found that while the compressive strength at age of 7 days was about 65% and the splitting strength about 70% of the comparable 90-day static strengths, the 7-day impact strength was only about 5% of the 90-day static strength. Just how many of his findings can be generalized for other concretes and for concretes tested under different circumstances is open to question. Nevertheless, the warning is loud and clear: Impact tests performed under different conditions (age, water-cement ratio, aggregate, test method, etc.) may easily produce seemingly contradicting results. Thus one may expect a correlation between the impact and static strengths of a concrete only if the composition of the concrete remains unchanged and the specimens are tested at the same age in the comparative test series.

Finally, Atchley and Furr (1967) noticed that the stress caused by impact testing reaches a maximum before the strain. This may indicate that like metals, concrete has a delay time for the initiation of plastic deformation.

Phenomenological Relationships

The relationships between various types of concrete strengths discussed so far have been empirical, obtained by curve fitting or derived from empirical relationships. It is also possible, however, to obtain similar relationships from certain theoretical or phenomenological considerations. For example, by assuming the validity of a straight line for Mohr's envelope (Mohr, 1911) (Fig. 2.10 and Section 2.5), a direct shear strength and torsion strength can be calculated. This shear strength, f_{dsh}, is represented by point C in Figure 2.10 and is equivalent to the *cohesion* as this term is used in soil mechanics; it is half of the *geometric average* of the compressive and tensile strengths:

$$f_{dsh} = 0.5\sqrt{f_{co}f_t} \qquad (2.25)$$

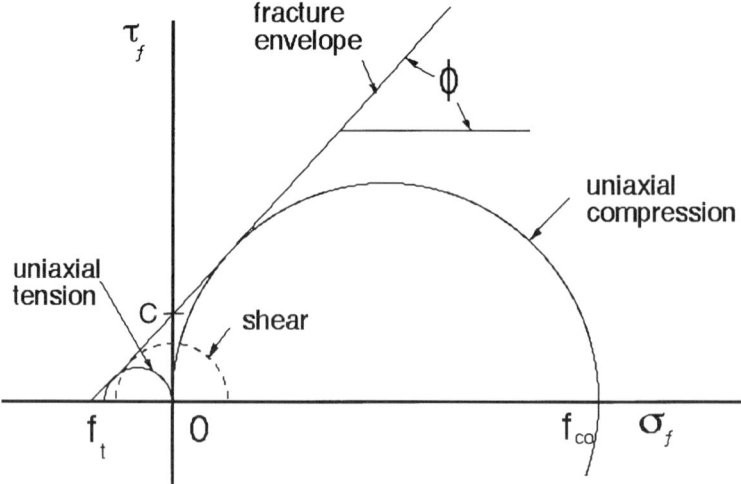

Figure 2.10 Straight line as a Mohr fracture envelope (see also Fig. 2.18).

The strength, here called torsion strength, f_{tor}, is represented by the dashed-line circle in Figure 2.10. It is really pure shear strength. The name *torsion* originates from the fact that it is usually determined by *torsion test* of a concrete tube. It is half of the harmonic average of the compressive and tensile strengths:

$$\frac{1}{f_{tor}} = \frac{1}{f_t} + \frac{1}{f_{co}} \tag{2.26}$$

Although the Mohr criterion of fracture is not valid strictly for concrete and the assumed envelope is curved, Eqs. 2.25 and 2.26 provide approximate values that are at least qualitatively valid. For example, Eq. 2.26 implies that the greater the compressive strength relative to the tensile strength of a material, the closer is the torsion strength to the direct tensile strength. By using the rule of thumb that the compressive strength of a concrete is 10 times as high as the tensile strength, Eqs. 2.25 and 2.26 provide the following approximate values for the direct shear strength and the torsion strength, respectively, of a concrete:

$$f_{dsh} = 0.16f_{co} = 1.6f_t \tag{2.27}$$

and

$$f_{tor} = 0.09f_{co} = 0.91f_t \tag{2.28}$$

Other fracture criteria can provide different relationships (Cowan, 1953; Zia, 1961; Stassi D'Alia, 1963).

Using the same simplifying assumptions as above, one can also obtain formulas for the relationships between various strengths with the help of the *coefficient of internal friction* of hardened concrete. This coefficient is shown as the angle of inclination, ϕ, of the straight line for the Mohr's envelope (Fig. 2.10). For example,

$$f_{tor} = f_{co} \frac{1 - \sin \phi}{2} \tag{2.29}$$

$$f_{dsh} = \frac{f_{tor}}{\cos \phi} = f_{co} \frac{1 - \sin \phi}{2 \cos \phi} \tag{2.30}$$

and

$$f_t = f_{co} \frac{1 - \sin \phi}{1 + \sin \phi} = f_{co} \frac{1}{\tan^2(45 + 0.5\phi)} \tag{2.31}$$

The oversimplifications applied in the underlying assumptions hurt the reliability of these formulas, too, especially Eq. 2.31. Nevertheless, they are interesting because they present the strength relationships from a different point of view. If one considers, for instance, that the value of ϕ decreases rapidly with increasing moisture content of the concrete (Akroyd, 1961b), Eqs. 2.29 to 2.31 seem to indicate that the relationships between various concrete strengths are also a function of the moisture content of the concrete.

2.4 FUNDAMENTALS OF THE FRACTURE MECHANISM OF CONCRETE (S. Popovics, 1969d)

The classical theories of fracture developed for brittle materials do not describe the failure of concrete under load adequately. It took a while to realize that concrete is more complex than an ideally brittle material; it is better to classify it as a quasibrittle material. This complicated the research on concrete fracture because linear fracture mechanics is not suitable for quasibrittle composites. So the development of an acceptable fracture mechanism for concrete had to wait until suitable models for nonlinear fracture mechanics were developed.

Significance

This section is concerned with the load-induced internal mechanical changes within a concrete specimen that lead to fracture. The primary change is the decreasing ability of the material to transfer stress during increased deformation. Resistance to these changes is called the *strength* of concrete from a phenomenological point of view. The subject is theoretical in nature, yet it is obvious that it has technical significance as well. A proper theory of fracture enables the engineer to control as well as utilize the concrete strengths more intelligently, thus more efficiently. Therefore, the fracture mechanics of concrete is discussed in this section, although only on an elementary level.

One would expect a theory of concrete fracture to provide quantitative predictions of the behavior of concrete under load. This encompasses the following items:

1. Numerical relationships between the strength and composition of a concrete, including the porosity (air content) of that concrete
2. Numerical relationships between the various concrete strengths, including the strengths under sustained, repeated, and multiaxial loads
3. Numerical relationships for the deformations of a concrete as a function of the composition and type of testing
4. Explanation of paradoxes of concrete fracture. For example, whereas the hardened cement paste and aggregate are each notch-sensitive, their mixture, the concrete, is not; or, a concrete in a drying condition has

higher compressive strength than the same concrete would under wetting, even when the average moisture contents are the same.

Since strength is the most sought after property of structural materials, wide-ranging research has been devoted to the various aspects of concrete strength. Unfortunately, the mechanism of fracture encompasses an enormous body of knowledge, which includes fundamental theories as well as experimentation and phenomenological work. Thus the attempts have been only partially successful in developing an adequate quantitative theory of fracture for any material, and even less successful for concrete in particular.

One source of the difficulties connected with the fracture mechanism of concrete is their highly heterogeneous composite nature: Aggregate particles act as rigid inclusions in a softer but still brittle matrix offering restraint to and resisting cracking and moisture move. This imparts greater stiffness to the matrix and reduces the dimensional instability, but the differential stiffness of the aggregate and matrix creates heterogeneity and weak interfacial contact zones (Swamy, 1972).

The fracture behavior of these materials is not fully understood in either its microscopic or macroscopic aspects, and the problem is still at the stage of description and classification of observed behavior. A fracture theory for the paste is a demanding problem in itself, since the structure of the hardened cement paste is very complex. For example, there are at least four different types of pores in the hardened cement paste: gel and capillary pores, pores from entrained air, and pores from incomplete consolidation. Each of these may have different effects on the fracture of concrete. Besides, any pore can contain either water or air or both. As a result of this heterogeneity, the primary reason for the failure of a concrete can be not only the weakness of the paste but also the weakness of the aggregate, that of the paste–aggregate interface, or any combination of these factors. The compressive failure of specimens under uniaxial loading is discussed in Section 1.2 from a phenomenological point of view.

Crack Initiation and Propagation in Concrete

It is a well-established fact that in most cases the failure of concrete under load takes place through progressive internal cracking, starting in the matrix portion, rather than abruptly (Jensen and Chatterji, 1996). In other words, the typical failure is the result of an essentially continuous material changing gradually to an essentially discontinuous one caused by internal tensile stresses. The conspicuous disintegration of a strength specimen under load is only the last stage of this process.

Brandtzaeg is probably the first to deduce the existence of internal cracking in concrete under load (Brandtzaeg, 1927; Richart et al., 1929). He found that the volume of his concrete specimens decreased under uniaxial compressive loading, roughly in proportion to the load applied, up to perhaps two-thirds of the ultimate load. He called the compressive stress corresponding to this load the *critical stress*. Above this critical stress, the material is further condensed by a load increase; however, the rate of volume change decreases up to a point where a small increase in load causes no corresponding change in volume. In his experiment this took place at about 80% of the ultimate load, but this value may decrease to as low as 50% with increasing age of the concrete (Beres, 1967, 1971), or with an increase in its aggregate content (Shah and Chandra, 1968). At even higher loads, the volume increases with an increase in load so that the apparent volume of the tested specimen at the ultimate load actually becomes larger by the application of compressive load than it was before the load was applied. This volume increase is also reflected by a radical increase (over 0.5) in the ratio of lateral to longitudinal strain, the apparent Poisson's ratio. From this, Brandtzaeg properly concluded that the bulging and eventual failure of the material result from a gradual development of internal tension-microcracking throughout the specimen parallel to the direction of the applied compressive stress. Total failure takes place when the cracks begin to form continuous patterns.

The gradual development of microcracking in concrete under loads less than ultimate is also confirmed by further observations for compressive, tensile, and other loadings. Jones, for example, shows (Jones, 1952; Jones and Kaplan, 1957) that for cubes in compression, the ultrasonic pulse velocity in the direction of loading remains constant as the load is increased to failure. In the transverse direction, however, there is a decrease in the pulse velocity at loads less than the ultimate, and further decreases in velocity occur as the load is increased to failure (Fig. 2.11). It is deduced from these results that the internal cracking starts at a fraction of the ultimate load and the cracks are oriented parallel to the direction of loading. According to other observations, and in contradistinction to concrete, the volume change of hardened cement paste under compression is one of continued decrease. From this it may be deduced that the apparent volume increase of mortars and concretes under load is related to the presence of aggregate particles (S. Popovics et al., 1990; S. Popovics and J. S. Popovics, 1991c).

X-ray photography and microscopy provide an even more detailed picture (Hsu et al., 1963; Slate and Olsefski, 1963). Accordingly, microcracks at the *interface* between coarse aggregate and mortar exist in concrete *before* it is subjected to any external load. These *bond cracks* seem chiefly to be a result of differential volume changes of matrix and aggregate during hydration and drying (Hsu, 1963; Slate and Meyers, 1969) as well as of bleeding and segregation (S. Popovics, 1982b). Their presence indicates that the interface (i.e., the bond) is the weakest link within the concrete. Incidentally, this is typical of most composite materials composed of constituents of different stiffness. The weakness of the interface has been proved directly (Hsu, 1963a), at least in normal-strength concretes where the hardened paste is weaker than the aggregate.

Above about 30% of the ultimate compressive load, these bond cracks begin to increase in length, width, and in number with increasing strain, while

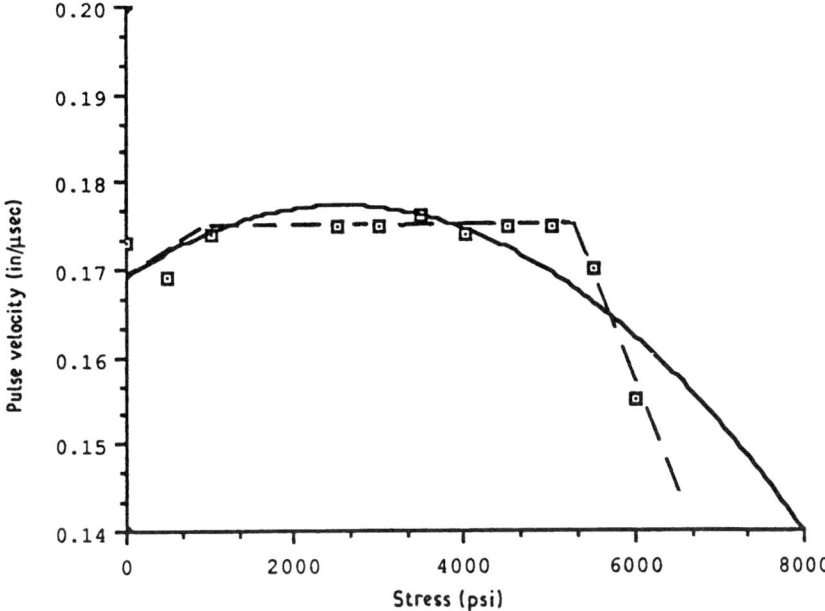

Figure 2.11 Typical compressive stress against transverse ultrasonic pulse velocity relationship for a concrete. The dashed line shows the conceived relationship represented by the experimental data (points); the solid line is a computer approximation. 1 ksi = 6.9 MPa, 1 in./μsec = 25,400 m/s. (From S. Popovics and J. S. Popovics, 1991c. Copyright RILEM. Reprinted with permission.)

cracks in the mortar remain negligible. In general, bond cracks develop first around the larger aggregate particles; this suggests that the strength of the comparable concretes should decrease with an increase in maximum particle size. At about 70 to 90% of the ultimate load, cracks, through the mortar and/or aggregate particles, begin to increase noticeably (Fig. 2.12) and begin to form a continuous crack pattern by bridging between nearby bond cracks.

When the continuous crack pattern is developed extensively, the load-carrying capacity of concrete decreases. The failure through aggregate particles is rare in normal-strength concretes made with conventional aggregates, indicating that the strength of such aggregates is higher than the strength of the hardened paste or of the interface. Inversely, cracks frequently penetrate weaker aggregates, such as certain lightweight particles as well as conventional aggregates in high-strength concretes rather than detour around them. Thus, the properties of the coarse aggregate have a significant influence on crack development and therefore on concrete strength and deformation; with larger sizes of the aggregate and/or greater aggregate contents in the concrete there is an increasing amount of interfacial cracking.

The evaluation of stress–strain diagrams of concretes leads to similar conclusions (Testa and Stubbs, 1977). The curvature (i.e., the deviation of the stress–strain diagram from linearity) is also due to microcracking (S. Popovics

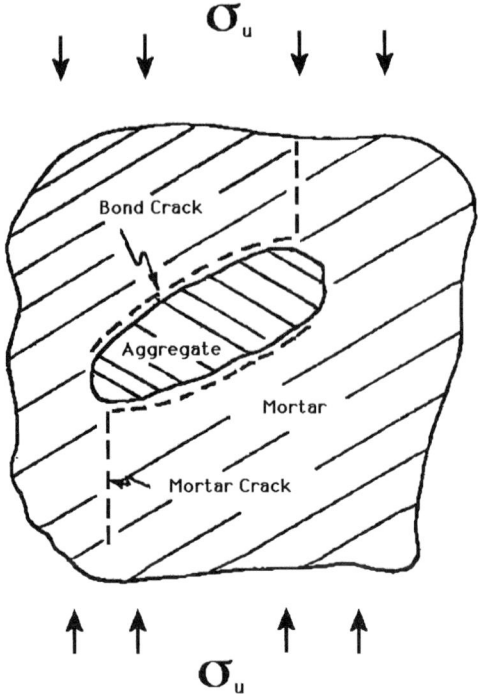

Figure 2.12 Crack initiation and propagation in concrete.

and J. S. Popovics, 1991b). In the case of short-time compression, this diagram is more curved for concrete than for the comparable mortar, which, in turn, has a more curved diagram than the comparable paste of the same water-cement ratio (Shah and Winter, 1966). Also, concretes with low strength (i.e., with low cement content) have a more curved stress–strain diagram than comparable concretes with high strengths. These points again demonstrate that the larger the size of the aggregate and/or the greater the aggregate content in the concrete, the greater the influence of the interfacial bond.

The existence of microcracks in concrete has been verified further with optical microscopes (Berg, 1950; Evans, 1946), electrical wire resistance strain gages (Blakey and Beresford, 1955), fluorescent-dye techniques (Jones, 1968), reflective photoelasticity (Swamy, 1972), surface cracks with laser interferometry, and internal cracks with acoustic emission monitoring the noise which results from the cracking (L'Hermite, 1955, 1962; Rusch, 1959). It has also been observed that the pattern of the crack propagation is influenced by the rate of loading. For instance, when the rate of stress increase is kept constant during testing, fewer cracks develop before failure than when the *strain rate* of the specimen is kept constant. Consequently, the failure of the concrete is abrupt in the first case and more gradual in the second. Figure 2.13 presents a general, comprehensive picture of the effects on various test results due to microcracking.

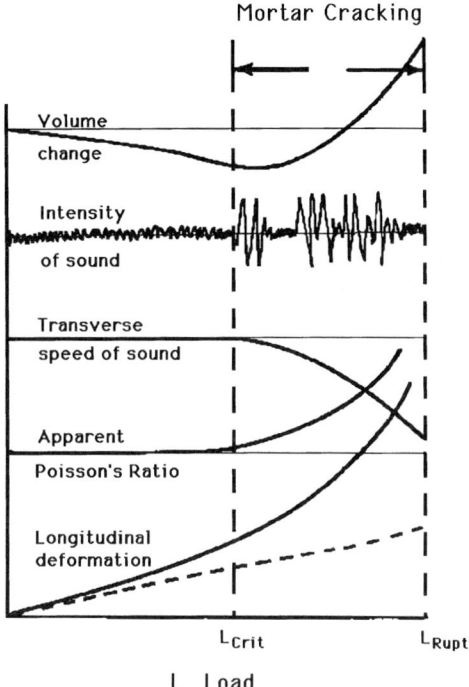

Figure 2.13 Effects of microcracking under uniaxial compression on certain properties of concrete.

Strength and Cracking of Concrete

After the cracks form a continuous pattern in the mortar part of the concrete by bridging bond cracks, the load-carrying capacity of the matrix becomes low. Thus it seems probable at this stage that a contribution to the ultimate compressive strength of the concrete is provided by internal friction developed on the surfaces of the newly created cracks and by the mechanical interlocking of the coarse aggregate particles. This contribution of internal friction can be demonstrated experimentally. For instance, saturation of concrete or mortar specimens with petroleum or a derivative thereof results in a reduction in the compressive strength of about 20%. Since petroleum derivatives do not attack concrete chemically, this strength reduction can be attributed to the lubrication of the rough surfaces. This assumption is further supported by the fact that the strength of the concrete is regained when the oil is expelled from the concrete by infrared drying (Steinbach, 1967; Weisz, 1960).

In brief, the strength of concrete originates from the following sources, more or less in the order of appearance:

1. Resistance to cracking on the interface of cement paste (or mortar) and (coarse) aggregate

2. Resistance to cracking in the hardened paste
3. Resistance to cracking in the (coarse) aggregate particles
4. Aggregate interlocking and friction between the broken internal surfaces

Stress Concentrations

The elastic solution of the notch problem by Inglis (1913) provides the first step in explaining the background of internal cracking of concrete under load. It shows that existing pores, voids, cracks, and so on (i.e., *flaws* in the material) can act as tensile stress concentrators, regardless of whether the load is under compression or tension. Thus stresses at the flaw tip could be much higher than the average stress (the applied load divided by the area of the specimen) and may eventually exceed the ultimate tensile strength of the material. More specifically, Inglis found that if the flaw is narrow, with the tip elliptical in shape, the stress concentration σ_m/σ_u in a linearly elastic noncracking high-strength (i.e., a hypothetical) material will be approximately (Fig. 2.14)

$$\frac{\sigma_m}{\sigma_u} = 2 \left(\frac{a}{r_0}\right)^{0.5} \tag{2.32}$$

where σ_m = maximum tensile stress at the elliptic tip of the flaw
σ_u = applied tensile stress far from the tip of the flaw, which is uniform and normal to the plane of the flaw
a = flaw depth
r_0 = minimum radius of curvature at the tip

Equation 2.32 shows that when r_0 approaches zero, the stress concentration increases to infinity. In other words, a material containing sharp flaws would have no strength at all. In actual materials, however, this is not the case. This is so partly because the atomic spacing sets a lower limit to r_0, and partly

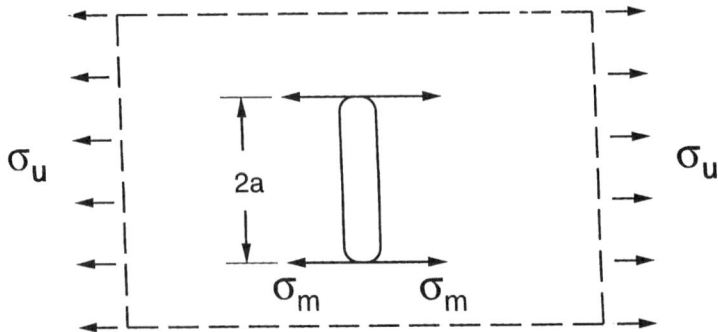

Figure 2.14   Stress concentration at an initial crack of 2a length.

because a major portion of the resistance to crack propagation comes not from the atomic bond strength but rather from the conservation of surface energy (Cottrell, 1964). Even so, stress concentrations at flaw tips can be very high. For instance, if the flaw depth is 10^{-2} in. (0.25 mm) and r_0 is 10^{-6} in. (25.4×10^{-6} mm), the stress concentration from Eq. 2.32 is 200; that is, the stress at the flaw tip is estimated to be 200 times higher than the uniform, remote average tensile stress. This means that a remote average tensile stress su as low as 10 psi (0.07 MPa) would produce a tensile stress concentration of 2000 psi (13.8 MPa). This stress, of course, is not reached in the concrete because the concrete cracks, resulting in a reduction of the stresses. In real brittle materials, the strength concentrations can reach the value predicted by Eq. 2.32 only under very low stresses, because at higher stresses the material cracks. It is important to note that *stress concentrations by flaws reduce the concrete strength much more than would be expected solely from the decrease caused in the load-carrying area.* This is discussed further in Chapter 5.

As will be shown in Chapter 4, a hardened cement paste contains many flaws in the form of gel pores, capillary pores, and other voids filled, fully or partially, with air or water. Therefore, the actual strength of a hardened paste or concrete is only a fraction of the theoretical strength which is estimated on the basis of the molecular cohesion of a perfectly flawless, homogeneous material. Note in connection with Eq. 2.32 that the size of the flaw and its shape and the angle between the direction of the applied stress and the direction of the flaw affect the magnitude of the stress concentration (Philleo, 1967; Grudemo, 1974).

A different kind of internal stress concentration is produced by the difference in the stiffness of the matrix and that of the aggregate particles (Swamy, 1980). External loads and internal stresses due to shrinkage, thermal or moisture gradient, and so on, produce a random, nonuniform strain distribution due to the wedging action of the aggregate particles (Fig. 2.15). The greater the difference between the elastic moduli of the aggregate and the matrix, the greater the strain concentration that typically occurs in the matrix and at interface. Also, the larger the aggregate size, the greater are the strain concentrations. The magnitude of this stress concentration is less than that around flaw tips but is still significant enough to contribute to the propagation of cracks (Brown and Mostaghel, 1967; Bremner and Holm, 1986).

Fracture Toughness

The resistance of a material to crack propagation is called *fracture toughness*. This can be measured for many brittle materials by testing the strength of specimens containing deliberately introduced flaws or sharp holes called *notches*. The stress concentration at the end of the notch reduces the strength, the extent of which is called *notch sensitivity*.

More specifically, the notch sensitivity of a concrete can be defined as the ratio of its flexural strength calculated from the net cross section of the notched beam to the flexural strength of a comparable but unnotched beam

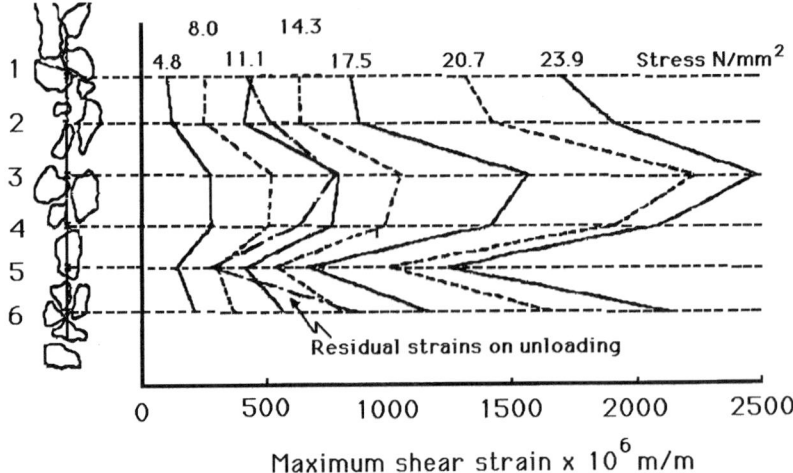

Figure 2.15 Typical strain distribution in the constituents of concrete under uniaxial compression. Aggregate, crushed basalt, D = 12 to 19 mm; sand:cement:coarse aggregate :: 1:2:3; w/c = 0.55 by mass. (From Swamy, 1980.)

(Shah and McGarry, 1971). Thus if a notched beam had the same fracture load in bending as a comparable unnotched beam having the same depth as the remaining depth beneath the notch of the notch beam, this ratio would have the size of unity. In a brittle material, this ratio is typically less than the unity. In a ductile material, relaxation of the stress concentration by plastic flow may give a ratio approaching unity. Thus this ratio can be used as a measure of the notch sensitivity of the material.

The Griffith Criterion

Stress concentrations have an important role in the fracture of concrete; nevertheless, this role is only a part of the theory. As pointed out in the preceding paragraph, the initial bond cracks have little influence on the load-bearing ability of the concrete under the usual short-time testing conditions. The reason for this is that they will not propagate unless the applied stress exceeds the tensile strength of the material. This is the fundamental criterion for crack propagation. An equivalent criterion is that a crack will start to propagate when the decrease in elastic strain energy becomes greater than the energy required to create new crack surfaces. The latter criterion was used by Griffith for the development of a theory for the progressive development of failure for homogeneous brittle materials (Griffith, 1920; Jaeger and Cook, 1978, Chap. 12; Tegart, 1967, Chap. 7). This is also applicable, at least qualitatively, for concrete. Assuming that all the elastic work involved in fracture goes into the creation of two new surfaces, each having a surface energy of T per unit area, that linear elasticity and plane stresses exist, and that Eq. 2.32 is valid,

Griffith obtained a necessary and sufficient condition for crack propagation from the requirement that the total free energy of the system be minimized. The numerical form is (Griffith, 1924)

$$\sigma_{u,m} = \left(\frac{2ET}{\pi a}\right)^{0.5}$$ (2.33)

where $\sigma_{u,m}$ = minimum σ_u uniform tensile stress that can cause crack propagation

E = modulus of elasticity

T = surface energy of the material per unit area

The other symbols are as in Eq. 2.32.

The term $\sigma(\pi a)^{0.5}$ is called the mode I *stress intensity factor in an infinite solid,* representing the crack-opening mode, and is usually designated by K_I. When this factor reaches some critical value, crack propagation occurs. This *critical* stress intensity factor is designated K_{IC} and is the numerical characteristic of the fracture toughness. K_{IC} values usually lie in the range 100 to 450 lb-(in.)$^{-3/2}$ [0.1 to 0.5 MN-(m)$^{-3/2}$] for hardened cement pastes, whereas for mature concretes they range from about 400 to 1300 lb-(in.)$^{-3/2}$ [0.45 to 1.40 MN-(m)$^{-3/2}$]. However, K_{IC} for the interface region seems to be much smaller, about 100 lb-(in.)$^{-3/2}$ [0.1 MN-(m)$^{-3/2}$].

An alternative way of describing the fracture of concrete is by the *strain energy release rate,* G_I. Here crack propagation occurs when G_I reaches the *critical rate* G_{IC}. It can be shown that for plane stress,

$$G_{IC} = \frac{K_{IC}^2}{E}$$ (2.34)

and for plane strain,

$$G_{IC} = \frac{K_{IC}^2}{E(1 - \mu^2)}$$ (2.35)

It can also be shown theoretically that

$$G_{IC} = 2T$$ (2.36)

However, values of T calculated from Eq. 2.36 are always much larger than the accepted value for the surface energy of a hardened cement paste.

Although Eq. 2.33 was developed for uniaxial tensile loading, the Griffith analysis has been extended to uniaxial compression and other loading types as well (Swamy, 1979). This extension is applicable to concrete, but only in a qualitative sense, as shown below.

According to the Griffith theory, when the stress concentration at the tip of a flaw exceeds a certain value, this crack will propagate; that is, new surfaces are created while elastic strain energy is relaxed near the crack. Therefore, the energy demand to create a new surface by further cracking increases with the growth of the crack (Glucklich, 1963). An initial crack length a_0 will begin to grow under a stress of σ_0 but will soon be arrested at a_1 length because of the increase in energy demand. The stress must be increased to σ_1 so that the crack can continue to grow to a length a_2. Observations indicate that during this stress increase there is a $\sigma' < \sigma_1$ stress that is usually sufficient to cause another crack to propagate. Thus the cracking is not in the form of a single crack but in the form of a zone of tortuous microcracks. This process increases the material resistance to cracking. Crack propagation will continue until the largest crack reaches a certain critical size a_{cr}. This part of the process is called *slow* crack propagation. Beyond this point, the rate of change of surface energy will be less than the rate of stored energy change. This simply means that the energy demand for creating new surfaces is now less than the existing strain energy; therefore, the crack will now grow spontaneously, with increasing speed, until specimen failure occurs. This process is called *rapid* crack propagation. It is likely that the critical crack causes a radical redistribution of stresses within the specimen, thus hastening the growth of other cracks, because the velocity of the rapid crack propagation may be very high, approaching sonic velocity (Anderson, 1959; Zaitsev and Wittman, 1978).

Experience has shown that slow crack propagation in brittle materials is much more extensive under compression than under tension. This gives explanation of why the compressive strength of these materials is so much greater than their tensile strength.

Comparison of Fracture Mechanism Under Tension and Compression

Apart from certain special types of loading, such as a group of direct shear tests or triaxial compression with high confining pressures to be discussed later, the mechanism of fracture in concrete is similar under different loading conditions, such as tension, compression, some combined stresses, or even under freezing or other deterioration. In each case, crack propagation leads to the failure of the concrete. The cause of stresses in the concrete controls only the *character* of the cracking through which the strengths and other mechanical properties are influenced. Therefore, the criterion of Griffith, qualitatively, can be extended to concrete fracture under these loading conditions. Attempts have also been made to model numerically certain aspects of crack propagation in concrete, but these have been only partially successful (Shah and Swartz, 1989). Specific aspects of concrete fracture under several loading types are discussed briefly below.

As has been shown, the failure of concrete under *uniaxial compression* is caused by lateral tensile stresses. Naturally, this is similar to the fracture mechanism produced by pure tension, although not identical to it. Three differences can be mentioned:

1. Cracks normal to the compressive stress cannot be pried open and will not propagate under any circumstances. In solids, however, cracks are oriented randomly. If the crack is at an angle to the applied compressive force, it is subjected to shear in addition to compression. There exists a critical length for spontaneous crack propagation in shear as there is in tension. If the shear stress associated with this critical length is computed in terms of the applied compressive stress, the latter turns out to be eight times larger than the stress computed from Eq. 2.33. Ratios greater than 8, sometimes as high as 50, have been reported for glass (Rosenthal and Asimow, 1971).
2. The crack propagation is arrested much more frequently under compression than under tension. For instance, experimental results (Kaplan, 1963) show that the ratio of stress at crack initiation to stress at ultimate failure in flexural tests varies within 0.7 and 0.8, depending on the composition of concrete. The same ratio in compressive tests varies within 0.45 and 0.50. Blakey (1952) reports an even greater difference. In other words, the complete failure of concrete in the compressive test is preceded by numerous microcracks while the rupture in the tensile test takes place after only a few cracks.
3. Perhaps not unrelatedly, the slow crack growth period is relatively shorter in concrete specimens under uniaxial tension than those under compression. This is because the stored-energy release rate increases rapidly with a small increase in the size of a crack in a tension field (Lott and Kesler, 1967).

These differences suggest that the compressive strength of a concrete is considerably greater than its tensile strength, and that the stress-strain diagram for a concrete under tension has less curvature than the corresponding diagram for compression. In addition, the nature of both differences is such that the pores in the concrete can accommodate the slow propagation of cracks to a greater extent under compression than under tension. Since rapid crack propagation is essentially independent of the porosity and is controlled by the inertia situation associated with the displacement around the tip of the crack and the stiffness of the testing machine, it is expected that an increase in porosity will weaken the compressive strength of a concrete relatively more than its tensile or flexural strengths. It will be shown later that experimental results support these speculations (S. Popovics, 1967b, 1969b). Fracture of concrete under torsion is also caused by crack propagation (Navaratnarajah, 1968).

Sustained and Repeated Loads

The principle of crack propagation is also applicable to failure of concrete under sustained (Desayi and Viswanatha, 1967) and repeated loads (Raithby, 1979; Antrim, 1967; Rascon Chavez, 1967; RILEM, 1966b). In these cases the propagation of cracks is helped by the extended time under load and the repetition of load, respectively. Consequently, the ultimate load in these cases is less than the ultimate load obtained with the usual short-time, static loading. The greater the induced stress, the sooner the failure takes place under the sustained load. Similarly, a greater load produces failure with a fewer number of repetitions.

Under *sustained load,* the concrete suffers much larger deformations (creep) than under short-term loading. This helps the crack propagation directly and indirectly by the redistribution of stresses within the concrete. In other words, cracks will propagate under sustained load despite the constant stress because the extended time period under the load will (1) create additional weak spots through the creep, and (2) lend more opportunity for the cracks to find the weakest paths for propagation.

The mechanism of weakening under *repeated load* is similar. Here, again, concrete suffers greater deformations than under standard static loading. This helps cracks grow under repeated cycles of loading even when the maximum load remains the same. Also, the loading and unloading branches of the stress-strain curve form a hysteresis loop. The area within this loop represents the irreversible energy of deformation, that is, the energy that propagates the cracks.

Brandtzaeg (1927) raises the question of whether a specimen will ultimately fail under a stress that is just sufficient to initiate propagation of mortar cracks if this stress acts long enough or if it is repeated frequently enough. The answer appears to be affirmative (Rusch, 1959).

Crack closure upon reduction or removal of load in concrete does not appear to be definitive (Swartz et al., 1988). With the increase in rate of loading, the measured strength of concrete is also increased mainly because the time available for the slow growth of cracks is shortened. When the loading rate is high, for instance in impact tests, the first crack, or first few cracks, control the failure process and show a certain similarity to tensile loading.

Multiaxial Loading

The load-bearing ability of a concrete specimen under multiaxial, or combined, loading, such as triaxial loading, may be either greater or smaller than that under uniaxial loading, depending on the testing condition. In general, if the arrangement of the combined loading is such that it hinders the crack propagation as compared to uniaxial testing, the load-bearing ability will also be increased (Krishnaswamy, 1968; Orowan, 1948–1949). The reverse is also true. For instance, concrete subjected to tensile stress in one direction and compressive stresses in the other two directions will have lower strength than the pure uniaxial compressive strength. Concrete submitted to triaxial compressive stresses, on the other hand, will have higher strength. Tests on a particular concrete using different stress paths in the triaxial path domain indicate that the results are path dependent (Kotsovos, 1979). More specific discussions of the combined strength of concrete are presented below, in principle, and quantitatively at the end of this section. Concrete specimens under biaxial compression fail by splitting of the cube or prism parallel to the unloaded surfaces, by a corner or edge failure, or by a combination of the two modes (Andenaes et al., 1977).

In the case of triaxial compression, two distinct modes of failure were observed for paste, mortar, and concrete specimens as illustrated in Figure 2.16 (Palaniswamy and Shah, 1974). Under low confining pressures, the failure mode is tensile and shows a splitting type of fracture, similar to that

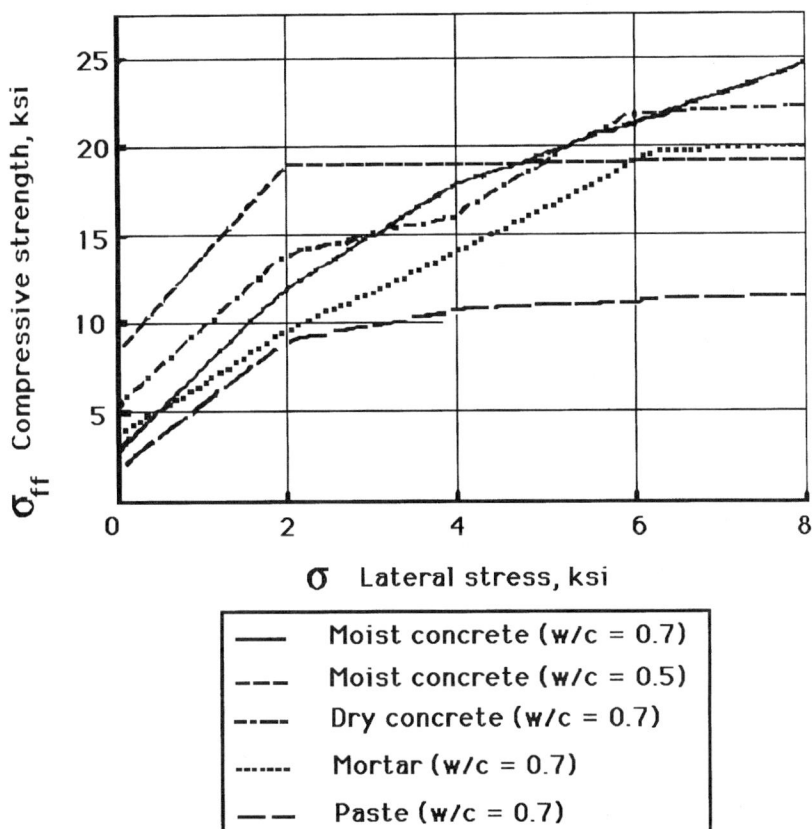

Figure 2.16 Ultimate compressive strength σ_{ff} of hardened cement paste, mortar, and concrete specimens as a function of the lateral stress $\sigma = \sigma_2 = \sigma_3$. (From Palaniswamy and Shaw, 1974. 1 ksi = 6.90 MPa. Copyright ASCE. Reprinted with permission.)

observed in uniaxial compression. This type of fracture is accompanied by relatively large axial compression strains and lateral tensile strains and significant internal microcracking; this was also indicated by a decrease in lateral pulse velocity (Fig. 2.17). In contrast, for relatively high lateral pressures the fracture mode is compressive, showing a crushing type of failure of the cement paste. Here the axial strains at failure are smaller, lateral strains are compressive, very little microcracking occurs at failure, and the specimens do not appear damaged. Similar observations have also been reported by several other investigators (Andenaes et al., 1977; Cowan, 1953; Berg, 1959). For the splitting type of fracture mode above, the maximum deviatoric or

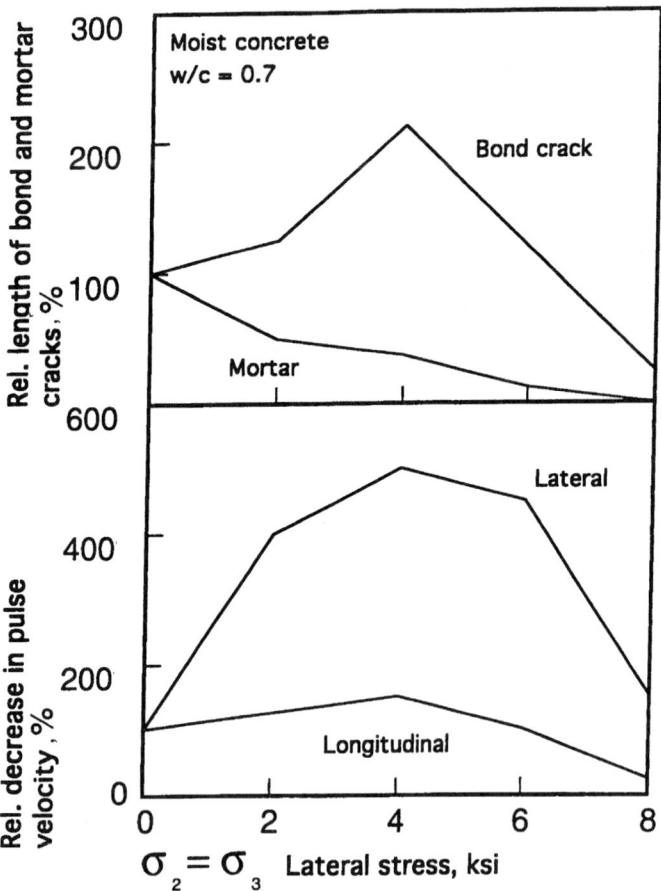

Figure 2.17 Effects of the magnitude of lateral pressure on the fracture mode. The relative values are percentages of the corresponding uniaxial case. The bond cracks at failure increased in this concrete specimen up to 4000 psi (27.6 MPa) lateral stress; above this lateral stress they gradually decreased so that there was little microcracking at failure. (From Palaniswamy and Shaw, 1974. Copyright ASCE. Reprinted with permission.)

distortional component $[= (2\sigma_1 - \sigma_2 - \sigma_3)/3]$ of the set of stresses at failure was found to be regularly greater than the hydrostatic or volumetric component $[= (\sigma_1 + \sigma_2 + \sigma_3)/3]$. The reverse was found to be true for the crushing type of fractures observed at high lateral pressures. Thus a reasonable explanation of this dual behavior of concrete is that the fracture at low lateral pressures is governed by bond failure as a result of shear by the predominant deviatoric component on the paste–aggregate interface. At high lateral pressures, the cement paste is crushed by the dominating hydrostatic component before the weaker deviatoric component could produce bond failure. This explanation also implies the possibility of a finite hydrostatic compressive strength for hardened cement paste specimens, the existence of which has been reported for other porous materials.

Since the majority of concrete structures are submitted to sustained, repeated, and combined loads of considerable sizes during the service life, the process of crack propagation discussed above may change the perspective of the strength problem. While the problem has been, until recently, to determine the strength of concrete at failure, it may become a question of at what load the mortar cracks begin propagation in the concrete.

Considering this, it appears that the Griffith criterion is most suitable for concrete among the available failure hypotheses because it provides a qualitatively correct picture of the fracture mechanism of concrete. The most serious difficulty with the *numerical* application of the Griffith criterion is the elusiveness of the E, T, and so on, values for concrete (Kaplan, 1961; Lott and Kesler, 1966). Introduction of a nonlinear fracture mechanics approach improves the situation but still does not provide all the answers (Shah and Quyang, 1993). Therefore, much work must be done before this, or any other method, can satisfy the majority of the requirements listed at the beginning of this chapter (Murrell, 1971). For practical engineering purposes, numerous empirical strength hypotheses have been devised, such as Mohr theory. Some of these are discussed later in this chapter.

Crack Propagation and Strain

Another approach, somewhat different from the Griffith criterion, is based on the observation that the initiation and propagation of cracking may be more dependent on the average *tensile strain* than on average stress. This approach explains the apparent strength increase (or decrease) under triaxial loading as the consequence of a reduction (or increase) in the tensile strains caused by this loading type. The previously mentioned strength reduction of concrete under sustained or repeated loads is explained again as a result of increased strains. Direct experimental data also support the importance of tensile strain in connection with the failure of concrete (Newman, 1968). For instance, Todd (1955) concludes from his experiments, and from others, that the maximum tensile strain concrete can withstand is on the order of 10^{-4}, that is, 100 μin./in. Pertinent experiments by Kaplan (1963) provide more specific information. He found that cracking starts in mortars under short-time loading when tensile strains reach a value of about 160 μin./in. regardless of whether the load is flexural, splitting, or compressive. With the addition of more and more coarse aggregate to the mixture (up to 55% of the volume), the cracking strains gradually reduce to about 80 μin./in., but they are still independent of the type of loading. These cracking strains are also independent of the two types of coarse aggregate and the two water-cement ratios used in his experiment. In contradistinction to strains, the calculated average tensile *stresses* at cracking are higher (1) in flexure than in splitting tests, (2) in limestone concretes than in comparable gravel concretes, and (3) in concretes of lower water-cement ratio than in those of higher water-cement ratio. Tensile strains at 95% of the ultimate loads are independent of the type of aggregate and water-cement ratio used; stresses are not. Using a somewhat different approach, Carino and Slate (1976) have done work that also supports the criterion of failure based on the concept of limiting tensile strains. Their results show that this limiting strain is a linear function of average normal stress at failure. The role of the maximum strain energy in fracture has also been investigated in connection with glasses (Anderson, 1959).

Statistical Aspects of Microcracking

The fact that there are so many flaws in a hardened concrete and the reasoning that crack propagation starts in the weakest of the randomly existing flaws in the concrete specimen make statistics a convenient tool for estimation of the global effects of these cracks. Indeed, a number of statistical theories have been offered, all of which are based on the assumptions that (1) flaws of various sizes are distributed randomly but with a constant density per unit volume throughout the specimen, (2) these flaws do not interfere with each other, and (3) the weakest flaw controls the strength of the specimen. A form of the latter assumption is the truism that "a chain is only as strong as its *weakest link.*" It is not difficult to see, however, that as far as the failure of concrete is concerned, this "chain" model is an oversimplification. A bundle of threads represents a more sophisticated model where the load is applied to the threads in parallel. Also, this model is more satisfactory because the bundle will not necessarily break under load that causes one, or even several, of the threads to break (Daniels, 1945). However, the mathematics of this bundle model is more complicated; since this model is also an oversimplification when compared to the real problem, it is not discussed further here.

Perhaps the most important application of the weakest-link principle is in the estimation of the *effect of the size of specimen on the measured strength* of concrete. The Griffith criterion implies that the measured strength should decrease as the area under maximum stress increases, because as this area increases, so does the number of flaws and consequently, the chance of finding a more damaging flaw.

2.5   PRACTICAL CRITERIA FOR FRACTURE

Difficulties in Application

The application of the Griffith theory to a group of brittle materials, such as glass (Phillips, 1965) or certain metals (Yokobory, 1965), has been proven to be useful. It is also useful to explain the fracture of concrete in *qualitative* terms. However, attempts to apply it *quantitatively* to concrete, or even to cement paste fracture, has been only partially successful (Chen, 1980a; Wittmann, 1983). For instance, the best that an extension of the Griffith criterion can do for biaxial stresses is to predict that the compressive strength of a concrete is eight times its tensile strength, regardless of the composition, age, strength, and so on. This is an acceptable approximation only for a small group of concretes. The same failure criterion provides even poorer approximation when the concrete is under compression–compression stresses (Swamy, 1979). Also, the observed directions of cracks in compression specimens are not in accord with the directions expected from the Griffith criterion.

The reason for these discrepancies is probably the high degree of compositeness of concrete. As a result, both the aggregate particles and the various pores embedded in the concrete may play double roles: (1) they may cause stress concentrations and initiate crack propagation, thus weakening the concrete; or (2) they may arrest the propagation of a crack, thus causing a local increase in strength. For instance, it has been shown that a bond crack will usually propagate first under load. As long as this crack develops around an aggregate particle, the energy demand for continued cracking is low. When it must force its way into the paste, the energy demand suddenly is increased. Another example is the case of a crack propagating in the paste and encountering a zone of higher strength, such as an aggregate particle, an unhydrated cement grain, or a location of more advanced hydration. If the crack penetrates this zone, the energy demand will increase. If the crack detours around this zone, the energy demand will increase again because the actual area of the newly formed surface is increased by the detour (Glucklich, 1963). In other words, Eq. 2.33 is invalid for the reliable calculation of concrete fracture due to the uncertainty of the values of E and T that should be substituted into the formula: those of the hardened cement paste, those of the aggregate, those of the aggregate–paste interface, or some combination of these cases (ACI, 1980a). Another problem is the uncertainty related to the value of T, as was mentioned in connection with Eq. 2.36. Also, tests of notched specimens have shown that although hardened cement paste specimens and stones are notch-sensitive to a certain extent, mortar and concrete specimens are not (Lloyd et al., 1968; Shah and McGarry, 1971; Swartz et al., 1988). With these limitations in application, it seems excusable that the fracture mechanics of concrete is not treated here to the full depth of the present state of the art. Besides, the full treatment of the subject is available in the literature (Shah, 1984; Shah and Swartz, 1989; Shah et al., 1995).

The need for a practical failure criterion for combined stresses exists, how-ever, for design and compliance purposes for concrete shells, pressure vessels, dams, and so on, since the strength of concrete under such stresses can differ considerably from the uniaxial compressive or tensile strengths. Thus, engi-neering methods have been offered for practical purposes that relate the failure condition to easily performed strength tests (Philleo, 1979; Newman, 1968, 1971). Several empirical or semiempirical fracture criteria are discussed briefly below.

Mohr's Criterion of Fracture

Mohr's is probably the oldest fraction criterion (Mohr, 1911) offered for con-crete. The essence of this is a pair of curved lines that are envelopes to the $\sigma_1-\sigma_3$ stress circles (Popov, 1976). Compressive stresses σ are considered as positive, tensile stresses as negative. Some of these circles can be constructed from the results of simple strength measurements, such as uniaxial tension, pure shear, and uniaxial compression, as shown in Figure 2.18. The smooth envelope around these circles then defines the fracture criterion for other stress conditions. More specifically, $\sigma_1-\sigma_3$ circles of combined stresses drawn tan-gent to this envelope give the condition of fracture at the point of tangency. If the stress circle for a given state of stress lies entirely within the Mohr envelopes, fracture will not take place in the concrete for that state of stress, according to Mohr's criterion.

Formulas for Mohr's envelope have also been recommended. The simplest of these is based on the assumption of a straight line (Fig. 2.10):

$$\tau_{ff} = \sigma_{ff} \tan \phi + c \qquad (2.37)$$

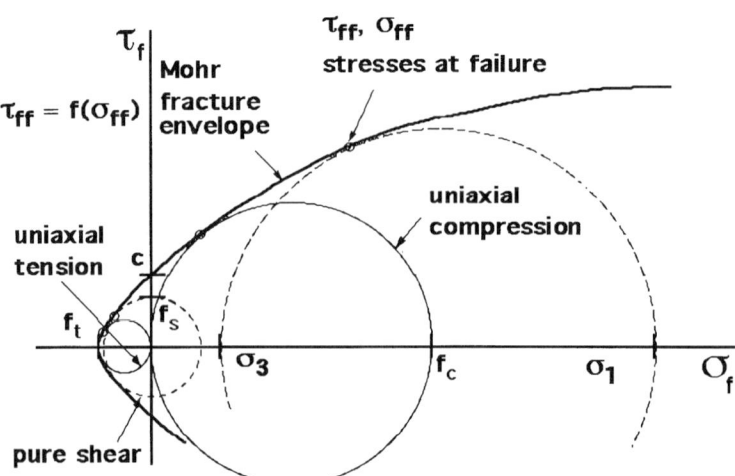

Figure 2.18   Mohr envelope as a criterion of fracture (see also Fig. 2.10).

Note that the coefficients have physical meaning: ϕ represents the coefficient of internal friction or the angle of shearing resistance of the hardened concrete, and c is the direct shear strength. Incidentally, Eq. 2.37 is equivalent to the Coulomb criterion of fracture. ASTM C801-81 (1986) contains formulas for calculation of the best-fit slope and intercept values in Eq. 2.37 for stress circles obtained experimentally, including 95% confidence limits. Other methods are also available (Lisle and Strom, 1982).

The next simplest mathematical case, that of a parabolic Mohr's envelope, can be derived from the Griffith theory of fracture (Murrell 1958; Jaeger and Cook, 1978). This envelope has the advantage that it uses the same formula, or curve, for compressive and tensile conditions.

Unfortunately, the experimentally established Mohr's envelopes for concrete are neither linear nor parabolic, but are slightly concave downward and somewhere between these two cases, as shown below. Several recommended envelopes are presented in Figure 2.19. The differences in these envelopes are due to differences in the testing conditions, specimen shape, moisture content, and so on.

The Mohr criterion implies that the fracture is independent of the magnitude of the σ_2 intermediate stress in the case of triaxial $\sigma_1 > \sigma_2 > \sigma_3$ stresses. Since experiments have shown that the intermediate stress *does* have an influence on the fracture of concrete under certain conditions, such as biaxial compression alone or together with tension (Theocaris and Prassianakis, 1974;

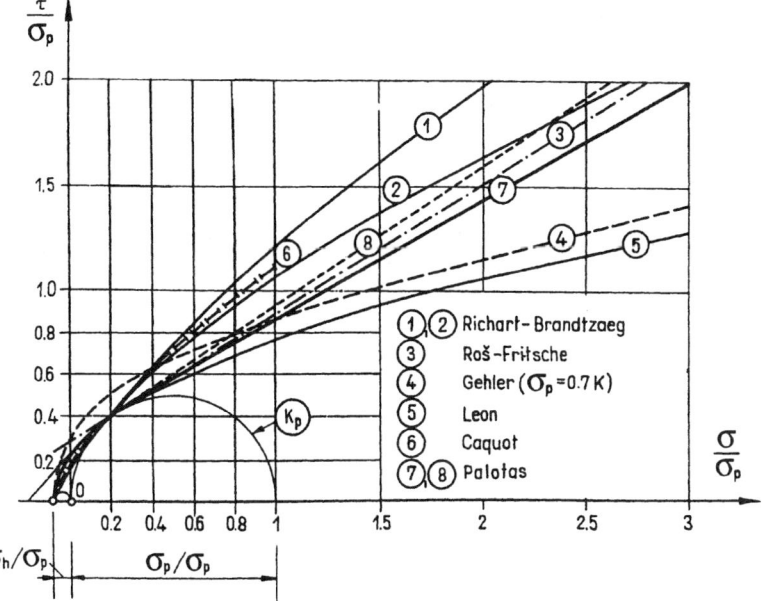

Figure 2.19 Recommended Mohr envelopes for concretes normalized for $f_c = 1$. (From Palotas, 1973.)

McHenry and Karni, 1958), the Mohr fracture criterion does not have a general applicability for the satisfactory prediction of the strength of concrete under multiaxial loads.

Applications of Mohr's Criterion

Despite the above-mentioned limitations, the Mohr fracture criterion can be used for numerical estimation of the strength-increasing effect in axial direction of a uniform lateral pressure on a specimen. The best known formula is

$$\sigma_1 = f_c + \sigma \tan^2(45 + 0.5\phi) \tag{2.38}$$

which results from the simplifying assumption of a straight line for Mohr's envelope. For $\phi = 35°$, which is a reasonable value for dry concretes, and $f_c = 3000$ psi (20.70 MPa), Eq. 2.38 takes the form:

$$\sigma_1 = 3000 + 3.7\sigma \tag{2.38a}$$

The graph of this equation is presented in Figure 2.20. Note that ϕ can be considerably smaller than 35° for wet concretes, due to the effect of the porewater pressure in the concrete (Akroyd, 1961b). It can also be seen from Figure 2.20 that the effect of σ on σ_1 is not linear in reality, indicating that the valid Mohr's envelope is not a straight line. The following empirical modification of Eq. 2.38a provides a better fit with experimental results (Murrell, 1971):

$$\sigma_1 = 3000 + 50\sigma^n \tag{2.39}$$

where n is an empirical parameter. This equation with $n = 0.72$ is also plotted in Figure 2.20. Note that according to this equation, approximately 30 psi (0.021 MPa) lateral pressure σ is enough to produce a 20% increase in the value of σ_1 and raise it from 3000 psi (20.70 MPa) to 3600 psi (24.84 MPa). Additional justification of Eq. 2.39 can be found in Figure 2.21 and in the literature (Hobbs, 1971).

Another empirical generalization of Eq. 2.38 is (Sokolov, 1978)

$$\sigma_1 = f_c + \frac{\sigma P b}{(f_c)^{0.5}} \tag{2.40}$$

where P is an experimental parameter depending on the testing conditions and b is a dimensional factor in (psi)$^{0.5}$ or (MPa)$^{0.5}$, respectively. For $f'_c = 3000$ psi (20.7 MPa) and $P = 200$, Eq. 2.40 becomes practically identical with Eq. 2.39. In addition, however, it also shows that the strength gain pro-

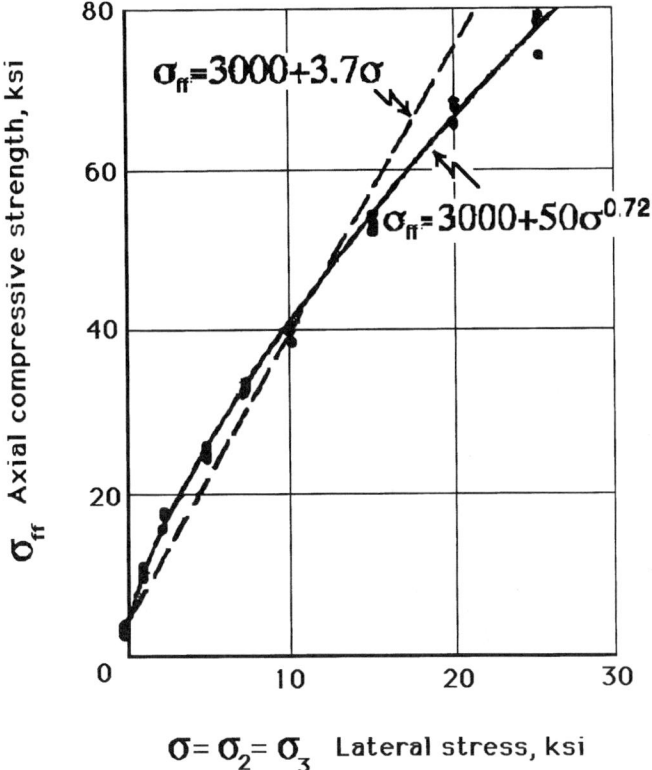

$$\sigma_{ff}=3000+3.7\sigma$$

$$\sigma_{ff}=3000+50\sigma^{0.72}$$

$$\sigma = \sigma_2 = \sigma_3 \quad \text{Lateral stress, ksi}$$

Figure 2.20   Effect of lateral stresses on the strength of 6 × 12-in. (150 × 300-mm) concrete cylinders. 1 ksi = 6.90 MPa. The experimental data were taken from Price (1951). (Copyright ACI. Reprinted with permission.)

duced by the lateral pressure is relatively less for stronger concretes than for weaker ones.

Three Other Fracture Criteria

1. Instead of representing the resistance of material in stress–space plots as in Figure 2.18, stress invariants, or the equivalent σ_{oct} and τ_{oct} octahedral stresses, can be used as the coordinate axes (Kupfer and Gerstle, 1973; Ottosen, 1977; Andenaes et al., 1977). Since all criteria for failure must be invariant with respect to change of axes, it is logical to express them in terms of stress invariants and to study the simplest invariant first. Accordingly, a material is assumed to fail when the octahedral shear stress reaches a critical value characteristic of the material (D.W. Murray, 1979). To make the fracture criteria valid for concretes of different compressive strengths, relative values

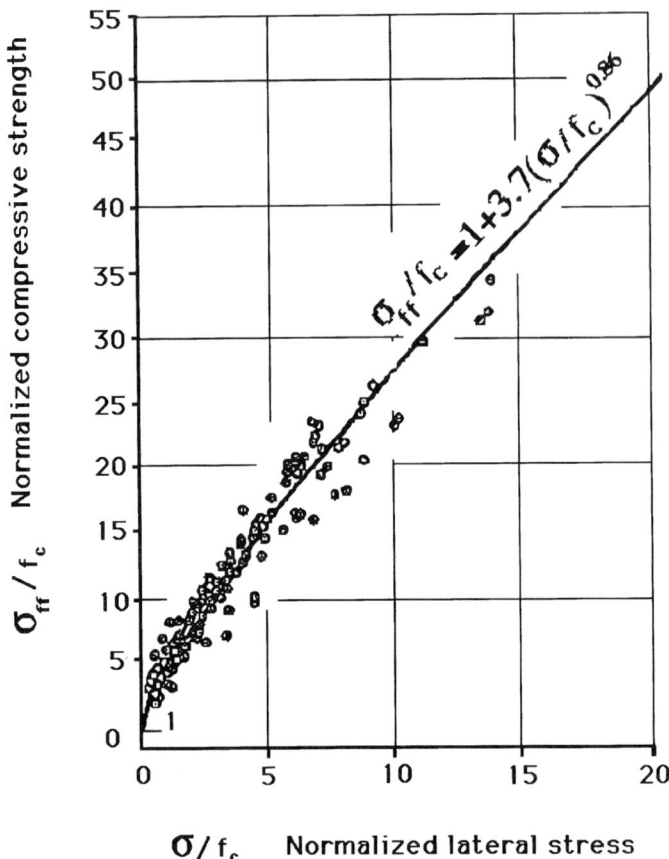

Figure 2.21 Ultimate compressive strengths σ_{ff} of various concrete mixes in triaxial compression plotted in terms of f_c and the minor $\sigma = \sigma_2 = \sigma_3$ principal stresses. (From Newman, 1971, Copyright Wiley. Reprinted with permission.)

are applied, which are usually obtained by dividing the octahedral stresses by the f_c uniaxial compressive strength of the concrete specimen. Using this approach, as well as the $\tau_{oct} = f(\sigma_{oct})$ fracture formulation by Nadai (1950), Bresler and Pister (1957, 1958) recommended the following linear relationship as a first fracture criterion:

$$-\frac{\tau_{oct}}{f_c} = \frac{a\sigma_{oct}}{f_c} + b \qquad (2.41)$$

where

$$\sigma_{oct} = \frac{\sigma_1 + \sigma_2 + \sigma_3}{3} \tag{2.42}$$

and $\qquad$ (2.43)

$$\tau_{oct} = \frac{[(\sigma_1 - \sigma_2)^2 + (\sigma_2 - \sigma_3)^2 + (\sigma_3 - \sigma_1)^2]^{0.5}}{3}$$

The failure surface corresponding to Eq. 2.42 is a circular cone with the line $\sigma_1 = \sigma_2 = \sigma_3$ as axis.

Although this equation represents many results satisfactorily, it is inadequate for some other stress combinations. Therefore, Bresler and Pister suggested the inclusion of additional invariants as a refinement:

$$\frac{\tau_{oct}}{f_c} = f_1 \frac{I_1}{f_c} + f_2 \frac{I_3}{f_c^3} \tag{2.44}$$

where I_1 and I_3 are stress invariants.

Although experimental results support this approach quite well (McHenry and Karni, 1958), perhaps because it involves all the principal stresses, the weakness of this method is that it provides differing formulas for biaxial and triaxial stresses.

2. The fracture criterion developed by Berg provides the relative value of the σ_1 principal stress that will just cause fracture in concrete (Berg, 1959). The formula is (Neville, 1973)

$$\frac{\sigma_1}{f_p} = \left(1 + 2Kn_3 + K^2n_3^2 - \frac{n_2^2}{c_1^2} - \frac{n_3^2}{c_1^3}\right)^{0.5} \tag{2.45}$$

where $\sigma_1 > \sigma_2 > \sigma_3$ = principal stresses
$\qquad\qquad\quad f_p$ = prism compressive strength of concrete
$\qquad\qquad\quad f_{cl}$ = cleavage strength which is approximately equal to the uniaxial tensile strength
$\qquad\qquad\quad \sigma_{cr}$ = uniaxial stress in concrete that initiates the crack propagation
$\qquad\qquad\quad n_2 = \sigma_2/f_{cl}$
$\qquad\qquad\quad n_3 = \sigma_3/f_{cl}$
$\qquad\qquad\quad c_1 = f_p/f_{cl}$
$\qquad\qquad\quad K = (f_p - \sigma_{cr})/f_p$

The f_p, f_{cl}, and σ_{cr} parameters can be determined by uniaxial compression and tension tests. Experimental data by Berg seem to support Eq. 2.45. The criterion does not apply, however, for cases when σ_2 and σ_3 have values that prevent transverse stress to reach the value of cleavage strength. The reason for this limitation is that under these conditions, the behavior of concrete is no longer brittle but plastic.

3. Figure 2.22 shows results of biaxial compression tests on three different concretes when the end restraint of the specimen was eliminated by the use of steel brush platens (Kupfer et al., 1969). It can be seen that:

(a) Under $\sigma_1 = \sigma_2$ biaxial compression the concrete strength is about 16% greater than the uniaxial compressive strength.

(b) The maximum biaxial strength seems to be about 25% higher than the uniaxial compressive strength and occurs approximately at $\sigma_1 = f_p$ (Darwin and Pecknold, 1977).

(c) The biaxial tensile strength is practically the same as the f_t uniaxial tensile strength and is approximately one-tenth of the uniaxial compressive strength.

(d) The general shape of the curves resembles the half-ellipse representing the criterion of failure based on the maximum distortion energy, but the experimental curves pass over and beyond the $\sigma_1/f_p = 1$ and $\sigma_2/f_p = 1$ points in the compression zone.

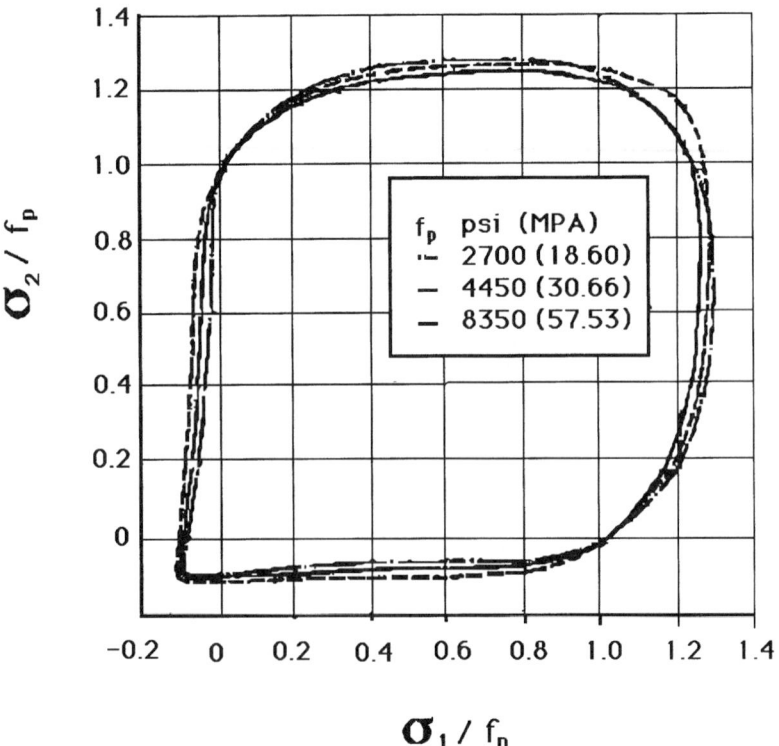

Figure 2.22 Strength of concrete under biaxial stress. The end restraint is eliminated. (From Kupfer et al., 1969. Copyright ACI. Reprinted with permission.)

(e) A good correlation with experimental data on concrete can be obtained by equations that approximate the curves in Figure 2.22 (Kupfer et al., 1969). For compression–compression stresses,

$$\left(\frac{\sigma_1}{f_p} + \frac{\sigma_2}{f_p}\right)^2 + \frac{\sigma_2}{f_p} + \frac{3.65 \ \sigma_1}{f_p} = 0 \qquad (2.46)$$

For compression–tension stresses,

$$\frac{\sigma_1}{f_t} = 1 + \frac{0.8\sigma_2}{f_p} \qquad (2.47)$$

For tension–tension stresses,

$$\sigma_1 = f_t = 0.64(f_p)^{2/3} = \text{const.} \qquad (2.48)$$

A somewhat similar modification of the maximum shearing stress criterion was recommended by Duguet in 1885 (Popov, 1976).

(f) The value of f_p hardly affects the curve. In other words, this biaxial fracture criterion is valid for a wide range of structural concretes.

Similar observations have been reported by other investigations (Andenaes et al., 1977; Vile, 1968; Okajima, 1972). Note, however, that the relative strength in any compression tension or biaxial tension stress combination decreases as the uniaxial compressive strength of the concrete increases. This is the consequence of the fact that the uniaxial tensile strength follows the same trend. A more detailed explanation for this is offered in Section 2.4.

Evaluation of the Various Fracture Criteria

Since the criterion curves obtained experimentally with biaxial tests on concrete pass over and beyond the $\sigma_1/f_p = 1$ and $\sigma_2/f_p = 1$ point (Fig. 2.22), the maximum distortion energy criterion, based on the f_p measured uniaxial compressive strength, provides an overly conservative prediction of the biaxial compressive strength of mortar and concrete. The same evaluation is valid for the Mohr criterion and for the maximum normal stress criterion, despite the fact that these criteria have been used successfully for many other materials. For instance, the maximum normal stress criterion appears to be particularly suitable for brittle cast iron (Murphy, 1964). In other words, concrete displays a peculiar behavior and, again, it is not clear why. Possible reasons are that the lateral pressure (1) delays the bond failure on the paste–aggregate interface, or (2) eliminates the effect of a group of weak links in the concrete by restricting the crack propagation to certain directions, or (3) closes certain cracks in the concrete and increases the friction between the broken internal surfaces.

3

STRENGTH DEVELOPMENT OF PORTLAND CEMENT

The strength of a hardening portland cement paste originates from the development of the hydration products, the major portion of which is in the form of a rigid gel, the *cement gel*. Both the chemistry and the internal structure of such a paste is of great complexity. Therefore, no adequate theoretical treatment of the relationship between strength and composition of portland cement has been developed for hardening cement pastes. Nevertheless, it appears reasonable to assume that two opposing factors have primary roles in the magnitude of the strength developed: interparticle bonds as the origin of strength, and the porosity of the paste as a strength-reducing factor. Note that these two factors are not independent of each other: new bonds are developed by new hydration products, which, simultaneously, reduce the porosity in the paste. Properties of portland cement and other physical and chemical factors (temperature, etc.) exert their influence on strength mainly indirectly through their effects on these two primary factors.

3.1 HYDRATION PROCESS OF PORTLAND CEMENT (S. Popovics, 1992)

Major Constituents of Portland Cement

Although most of the substances of which portland cements are composed contain three or more elements in a state of combination, one may introduce a considerable simplification into their study by regarding them as being built up of certain oxides.

Publication of Le Châtelier's research (Le Châtelier, 1887) started the recognition that the four constituents, or compounds, shown in the bottom part of Table 3.1, can be considered as the major constituents in the portland cement clinker. Most of these *clinker minerals* form crystals so small that the particles in the ground clinker contain more than one crystal. It should be noted that a microcrystalline glassy material can also be identified in the clinker, which is calcium and aluminates for the most part. The amount of the clinker materials (i.e., *the compound composition*) can also be calculated approximately from the data of oxide analysis. Bogue is the first to offer formulas for such computation (Bogue, 1929). These calculations are based on the assumption that certain ideal conditions exist in the kiln and during the cooling of the clinker, which in practice do not exist; thus they have inherent bias, especially in underestimating the C_3A content. Nevertheless, they are useful for many practical purposes. It is important, however, that the oxides form the proper clinker minerals in the cement, which can be assured by the correct manufacturing procedure.

TABLE 3.1 Names, compositions, and abbreviations of common oxides as well as major compounds in portland cement

Name	Composition	Molecular Weight	Abbreviation	Mineral Name
Lime	CaO	56	C	
Silica	SiO_2	60	S	Quartz
Alumina	Al_2O_3	102	A	
Iron	Fe_2O_3	160	F	
Water	H_2O	18	H	
Hydrated lime	$Ca(OH)_2$	74	CH	Portlandite
Sulfuric anhydrite	SO_3	80	S	
Anhydrite	$CaSO_4$	136	$\overline{CS}$	Anhydrite
Gypsum	$CaSO_4 \cdot 2H_2O$	172	$\overline{CS} \cdot H_2$	Gypsum
Magnesia	MgO	40	M	Periclase
Soda	Na_2O	62	N	
Potassa	K_2O	94	K	
Tricalcium silicate	$3CaO \cdot SiO_2$	228	C_3S	Alite
Dicalcium silicate (beta)	$2CaO \cdot SiO_2$	172	C_2S	Belite
Tricalcium aluminate	$3CaO \cdot Al_2O_3$	270	C_3A	
Tetracalcium aluminoferrite[a]	$4CaO \cdot Al_2O_3 \cdot Fe_2O_3$	486	C_4AF	Celite

Source: S. Popovics (1992). Copyright Noyes. Reprinted with permission.
[a]In reality, the iron-containing phase is a solid solution of variable composition. C_4AF seems to be its fair average composition.

In reality, the clinker minerals in a portland cement are not in the form of pure compounds. The calcium silicates, for instance, contain small amounts of alumina, magnesia, and possibly some other oxides. Since these impurities can influence the crystal forms and other properties of the compound, it is reasonable to call these compounds, as they occur in portland cement, by their names (alite, etc.), originated by Tornebohm. These names are also shown in Table 3.1.

Each of the clinker materials has important individual characteristics, as shown first by Bates. *Alite* paste hardens fast. It attains the greater part of its strength in a week, and a little increase occurs at longer ages. The heat development that is produced by reactions between alite and water (*heat of hydration*) is also rapid and intensive. The *belite* produces little strength until after several weeks but gains steadily in strength at later ages until it approaches the equality with alite. Also, the development of its heat of hydration is much slower. The two calcium silicates form regularly 70 to 80% of the portland cement clinker. *Tricalcium aluminate* alone attains very little strength, but in mixes with calcium silicates, it can produce much higher rates of strength development than what would follow from its small quantity. It is not clear yet how tricalcium aluminate contributes to the hardening and strength. In any case, the reaction of tricalcium aluminate with water is very rapid, causing a flash setting (i.e., *quick setting*) of the cement paste which is accompanied by a vigorous heat evolution. Gypsum in the amount of 3 to 6% is added to the clinker in the cement factory to slow down this reaction and control the time of setting as well as the strength development. Tricalcium aluminate of the clinker in large quantities also reduces the resistance of the hardened paste against sulfates and other chemical attacks. The *celite* hydrates rapidly, accompanied by marked heat evolution, but the reactions are less intensive than those of tricalcium aluminate. The ferric phase may contribute to the strength development of portland cement at later ages, although this mechanism is not clear. The presence of the ferrite phase gives portland cement its characteristic gray color. Portland cement without iron compounds has white color.

It is not known yet what causes the differences in reactivities of the clinker minerals, or, in general, why only a few compounds (C_3S, β-C_2S, etc.) act as hydraulic cements, whereas many others of broadly similar composition (γ-C_2S, etc.) do not. A possible explanation is that SiO_2 can form stable compound with only one CaO molecule at ordinary temperature. However, at high temperatures such as those required for cement manufacturing, silica can combine with two or even three lime molecules either because the coordination number of the calcium silicates changes with temperature (Lea, 1971) or because the coordination of Ca^{2+} changes from symmetric at ordinary temperature to increasingly irregular at higher temperatures (Taylor, 1971). The lime-rich compounds are unstable at ordinary temperature; therefore, their structures tend either to form a new polymorphic arrangement or to hydrate. In either case the calcium in the new structures is in its normal coordination.

Clinker minerals can be identified under a microscope by x-rays or other methods. Some of the compounds can form more than one type of crystal form, which can influence the hydration characteristics of the compound. For instance, in the case of the dicalcium silicate, only the beta form has cementing values under normal hardening conditions. Also, C_3A in cubic form is less vulnerable to sulfate attack than in orthorombic form (Mehta, 1980).

Minor Constituents

The term *minor* refers to the quantity of these constituents rather than to their importance. This is particularly true for the *magnesia* and *alkali oxides*. The magnesia is usually in an uncombined state in the portland cement. If it is present in an excessive quantity (about 5% or more), especially in crystalline form as periclase, it may cause the cement to be unsound, similar to the effect of free lime. The alkalies are usually combined with the major compounds. Yet they may react with active silicate of certain aggregates, causing extensive expansion and cracking. This *aggregate–alkali reaction* is not harmful when the total amount of alkali is under 0.6% by mass of the cement expressed as Na_2O equivalent. Such cements are called *low-alkali cements*. The N_2O equivalent is calculated as $Na_2O + 0.64K_2O$.

Types of Portland Cement

All portland cements have the same constituents. It is the differences in the relative proportions of these constituents that determine the individual type of cement. Currently, five main types of portland cement are recognized in the United States.

Type I, or, according to the British terminology, *ordinary* portland cement, is the general-purpose cement that is used when the special properties of the other four types are not required. Concrete blocks, floors, reinforced frames, beams, and slabs are typical examples. *Type II,* or *modified* cement, is a modification of Type I to increase resistance to sulfate attack and reduce heat evolution. It is used, for instance, in concrete pipes, pavements, and foundations. It is also popular for the production of high-strength concrete (Peterman, 1986). *Type III,* or *high early-strength* cement, is used when rapid strength development in concrete is essential, as in precast plants, winter concreting, and repairs. *Type IV,* or *low-heat,* and *Type V,* or *sulfate-resistant,* cements can be considered extreme and special cases of the Type II cement. Type IV has been used for massive dams and other large concrete structures to reduce the cracking tendencies resulting from accumulated heat of hydration. Type V cement has been used for structures where the concrete may be in contact with soils and groundwater containing larger amounts of sulfates. Such uses include canal linings, culverts, and foundations. Because of their special nature, Types IV and V are not commonly carried in stock and usually are made on special order. Type IV cement has not been used much in recent years because improved construction techniques have made the application of this slow-hardening cement unnecessary in mass concretes.

There are also Types I, II, and III cements modified by the addition of an air-entraining admixture during the manufacturing. These are called *air-entraining cements* and are designated as Types IA, IIA, and IIIA, respectively. The air-entraining cements are used mainly in structures where the concrete would be exposed to frost action. Note, however, that the development of automatic admixture dispensers greatly reduced the importance of air-entraining cements because it is not possible to control the air content in concrete at the desired level when they are used.

In addition, there are *blended cements* which are intimate and uniform mixtures of portland cement and a pozzolanic material. *Pozzolanic materials*, or pozzolans, are materials that in themselves possess little of no cementitious value but will, in finely divided form and in the presence of water, react chemically with calcium hydroxide at ordinary temperatures to form compounds possessing cementitious properties. The most important artificial pozzolan is the *fly ash* (pozzolan Class F according to ASTM C618) because it is used in the largest quantity. It is the finely divided residue resulting from the combustion of ground powdered coal (Clifton et al., 1977; Berry and Mahotra, 1980; Smith, 1978).

The general properties of the blended portland cements are similar to the properties of comparable (plain) portland cements (Mather, 1957; Klieger and Isberner, 1967; Kokobu, 1968; Kokobu and Yamada, 1974). The advantage of adding a pozzolanic material to the clinker is that it can combine chemically with the lime and alkalies liberated from the portland cement paste during the hardening period. Thus the resistance of concrete against certain chemical effects is improved and the expansion caused by excessive aggregate–alkali reaction is reduced. In addition, blended portland cements generally provide a lower heat of hydration and an improvement in the properties of fresh concrete compared with portland cement. On the other hand, a blended portland cement may give lower strengths at early ages, although the ultimate strength is not reduced when extended moisture curing is provided.

ASTM C595-95 recognizes five kinds of blended portland cement. These are *portland blast-furnace slag cement* (slag content is between 25 and 70% of the mass of portland blast-furnace slag cement), *portland-pozzolan cement* (pozzolan content is between 15 and 40 mass %), *slag cement* (slag content is at least 70 mass %), *pozzolan-modified portland cement* (pozzolan content is less than 15 mass %), and *slag-modified portland cement* (slag content is less than 25 mass %). They may come in air-entrained and non–air-entrained form.

Guidelines for the selection of cement for a given purpose have been published by ACI (1988). Further engineering aspects of blended cements and pozzolanic materials are discussed elsewhere (S. Popovics, 1968; Barton,

1965; Kramer, 1960; Faber et al., 1974; Turriziani, 1960; Malquori, 1960; Fifth International, 1969; Malhotra, 1983).

Hydration: Reactions Between Cement and Water

The reactions between portland cement and water are described briefly below. Further engineering aspects of the hydration have been presented in numerous publications, including Chapter 3 in S. Popovics (1992). When portland cement is mixed with a limited amount of water, the cement particles get dispersed in the aqueous phase. The result is cement paste, which displays considerable plasticity immediately after mixing and maintains this for a period of time called the *dormant period.* After awhile, 2 to 3 hours under normal conditions, however, the paste starts stiffening, less and less plasticity can be observed, finally all the plasticity is gone, and the paste becomes brittle, although it is still without any sizable strength. This stiffening process is called *setting.* The *gain of strength* (i.e., the *hardening process*) takes place subsequent to the setting (Fig. 3.1), that is, typically several hours after completion of the mixing.

Both the setting and the hardening of a portland cement paste are the results of a series of simultaneous and consecutive reactions between the water and

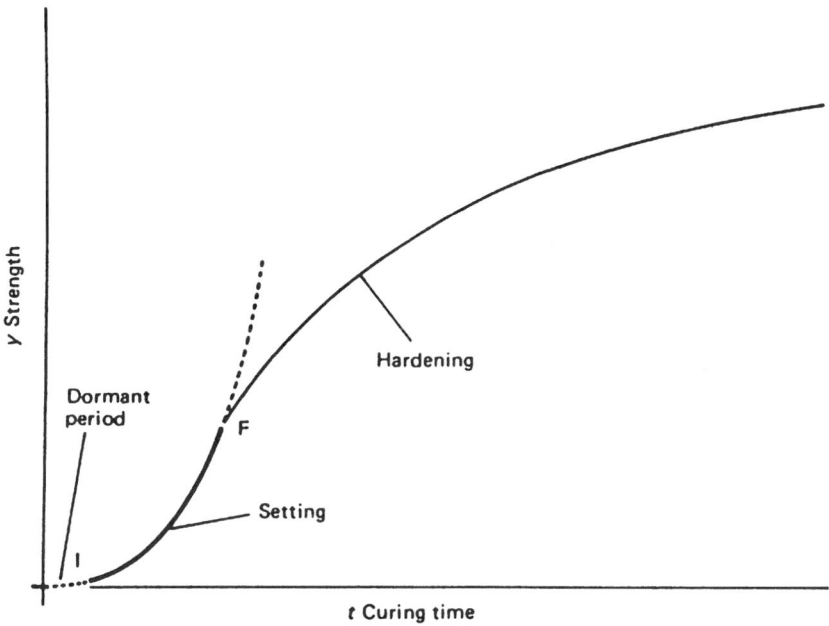

Figure 3.1   Illustration of the terms *dormant period, setting,* and *hardening.* I, initial setting; F, final setting. (Not to scale.)

the constituents of the cement. These reactions altogether are covered by the term *hydration* of portland cement. The results of the hydration are the hydration products that develop gradually with time and form finely structured porous solids called *cement gel.* This consists primarily of *calcium silicate hydrate* (CSH) *gel* in which larger crystals of other materials, most notably *calcium hydroxide* $Ca(OH)_2$, are present. The hydration products bind a portion of the water chemically. This water is not removed under standard drying conditions; therefore, it is called *nonevaporable* water.

The pores in the hardening paste are filled with liquid of high alkalinity resulting primarily from the alkalies in the cement. The pH of such liquids is typically over 13 (Diamond, 1975). So, strictly speaking, the hydration of portland cement is not the results of reactions between the portland cement compounds and water but rather between the cement and this alkaline liquid phase. Since the solubility of calcium hydroxide is reduced by the presence of alkali hydroxides, Lea speculates that they might also have influence on the hydration (Lea, 1971).

The two most important chemical reactions during the *early* period of the hydration are

- the reaction between the C_3A as well as the gypsum of the cement, producing *ettringite,* which is calcium aluminate trisulfate hydrate, and
- the hydration of alite in the cement with water, producing calcium silicate hydrate (CSH).

Most of these reactions are exothermic; that is, they are accompanied by heat development beginning when cement and water are first brought into contact. This heat is called *heat of hydration.*

After the completion of setting (i.e., during hardening), chemical reactions continue at a diminishing rate until one or more of the conditions necessary to the reactions will be lacking.

The chemical compounds found in the gel of hydrated cement are complex; most of them are impure in the sense that they contain elements not ordinarily given in their chemical formulas, and they do not have exactly the same composition when formed under differing conditions, especially with respect to temperature conditions and original water-cement ratio (Skalny et al., 1978). Nevertheless, it may be worthwhile to mention that examination of hardened cement pastes after long hydration provides approximate information concerning the final compositions and the related reactions of the hydrated compounds of portland cement. For instance, using the symbols of Table 3.1, we have

$$2C_3S + H_6 = C_3S_2H_3 + 3CH$$

where $C_3S_2H_3$ represents a CSH (tobermorite) gel, or

$$C_3A + H_{10} + \overline{CS} \cdot H_2 = C_3A \cdot \overline{CS} \cdot H_{12}$$

where the right side of the equation is calcium aluminate monosulphate hydrate (Brunauer and Copeland, 1964). This monosulphate phase can also develop from calcium aluminate trisulphate hydrate (ettringite) formed during the early stage of hydration (Mehta, 1993).

The main difference between the chemistry of hydration of tricalcium silicate and that of dicalcium silicate is that the former develops more CH. The compositions of the other hydrated calcium aluminates and ferrites are quite complex.

The hydration of portland cement results in the gradual development of a number of features of the cement paste that can be measured and used for the characterization of the extent of hydration. There are standard methods for some of these characteristics, such a strength and heat of hydration (ASTM C186). Two test methods may be mentioned for the chemical examination of the hydration products. One uses a *derivatograph.* This examines directly the calcium hydroxide developed gradually by the hydration, from which the chemical structures of the other hydration products can be estimated (Tamas, 1966). The other method is called *silylation,* which examines the structure of the silicate anions in mature pastes using the principle of structure retention by end-blocking the reactive hydroxyl groups. Then the volatile end-blocked trimethyl silyl esters of the silicate anions can be separated and analyzed by gas–liquid partition chromatography (Tamas, et al., 1976). There are also accepted methods for the measurement of chemical shrinkage, specific surface of the hydration products, and the nonevaporable water content.

The hydrations of the various cement compounds are not necessarily independent of each other. For instance, there is a considerable amount of experimental evidence presented in the literature (Hansen, 1970; Kuhl, 1961; Celani et al., 1968; Alexander et al., 1969, 1972; Bobrov, 1974; Copeland and Kantro, 1968; Schramli, 1978; and Verbeck, 1965) indicating that the hydration processes of the two calcium silicates are interacted by the C_3A. A possible mechanism is that the C_3A acts as a catalyst on the hydration of the silicates (S. Popovics, 1974, 1976a, 1980). A physical mechanism might be that C_3A makes the structure of the developing CSH gel somehow denser similarly to elevated temperatures (Section 3.6). In any case, because of the interaction, and because of the differences in the kinetics of the development of the various hydration features, the relationship between these features is usually not simple.

It is important to note that the calcium hydroxide is the weakest hydration product in the hardened cement paste from the standpoint of strength, and it also has the lowest chemical resistance among the hydration products. In other words, the spots of calcium hydroxide are usually the center of deterioration in concrete. The presence of calcium aluminate sulfate hydrates is also harmful, especially from the standpoint of sulfate resistance.

Mechanism of the Hydration Process

The hydration—that is, the reaction series between cement and water, more precisely the aqueous phase—starts on the surface of the cement particle similar to the corrosion of a metal. Then, with time, this surface of reaction moves gradually deeper into the interior of the cement particle (Fig. 3.2). On contact with the still unhydrated part of the cement particle, water reacts with it and/or dissolves a portion of it. This supersaturated solution diffuses out from the surface of reaction toward larger (capillary) pores through the very small (gel) pores of the solid shell of previously created hydration products around the unhydrated cement particle. The newest hydration products precipitate from the solution in air- or water-filled pores that are large enough to allow nucleation of a new solid phase and form a flocculation surface. This mechanism has been confirmed by microscopic observation (Williamson and Tewari, 1972), and even better by SEM (Uchikawa, 1987). Moreover, the hydrated material in portland cement pastes can be categorized as *inner* and *outer* product, according to whether it has been formed in space previously occupied by cement grains or by water, respectively. It is relatively easy to distinguish, by x-ray and scanning electron microscopy (SEM), the inner product formed from the larger cement grains, but impossible to determine whether other regions of the hydrated material have been formed in space

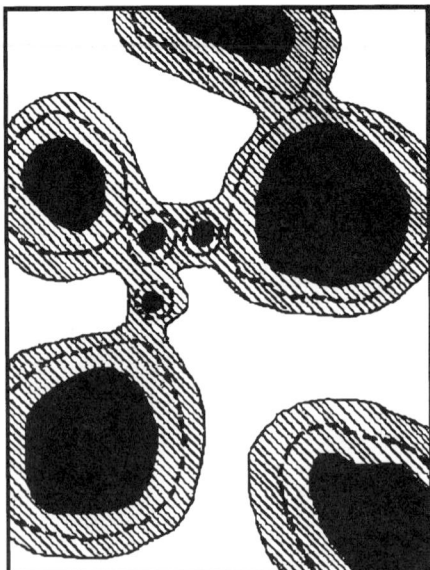

Figure 3.2 Two-dimensional schematic presentation of hydration process of portland cement particles in a compacted paste at an early age. Dashed line: original surface of the unhydrated particles; shaded area; hydration products, mostly cement gel; black area: remaining unhydrated portion of the particle; white area: capillary porosity filled with water. (Not to scale.)

previously occupied by water or in that previously occupied by smaller cement grains, the outlines of which have disappeared. (Taylor, 1993). The inner product has a more dense structure than the outer product, although their composition is more or less the same.

The continuation of hydration requires that the solid hydration product always contain a considerable amount of gel pores (Locher and Richartz, 1974). Note that the water in gel pores, (i.e., the gel water) is not capable of reacting with unhydrated cement, but the capillary water is.

The reaction between cement and water is accompanied by a diminution in the total volume of the paste. Called *chemical shrinkage,* this was probably observed first by Le Châtelier (Geiker, 1983). Measured values of the chemical shrinkage show good correlation with other characteristics of the hydration, such as nonevaporable water and compressive strength.

The strength development of a cement is complex because of the complexity of the hydration mechanism. Namely, it has been established that at the beginning, that is, when the portland cement and water are first brought into contact, intensive reactions take place for a brief period (Kantro et al., 1962). This is called the *zero stage* of strength development in this book because these reactions contribute little to strength. The zero stage covers primarily the dormant period and the setting. Shortly after this, the solid gel portion of the hydration products starts adhering to the surfaces of the cement particles in larger quantities, so the liquid phase of the cement paste has to diffuse through this increasing protective layer to reach the unhydrated portion of the cement for further hydration. In the early period of hydration this protective coating is thin, so the liquid phase can diffuse through it easily and quickly, providing more than enough water for all the reactions; therefore, the speed of the hydration and strength development is controlled by the reactivity of the cement compounds, that is, by the rates of reactions between the liquid phase and the cement compounds. This period is called the *first stage* of strength development. Later, however, after further hydration, the gel coating becomes so thick that the diffusion through it is slower than the slowest step in the chemical reaction proper. In other words, after this point just so much reaction can take place as the restricted amount of liquid phase permits. Therefore, the diffusion speed takes over control of the rates of hydration and strength development. This is called the *second stage.* The rates produced by the two control mechanisms are different.

The durations of the zero and first stages seem to depend on the overall rate of hydration. More specifically, when the hydration is fast (high curing temperature, high C_3S and C_3A contents, high fineness, etc.), the duration of the first stage may be a week or even less, whereas at slow hydration rates it may last for many months (Popovics, 1987c).

Since C_2S is far less reactive than C_3S, a thicker gel coating must build up on it, taking longer before diffusion becomes rate controlling (Brunauer 1964). The rates of chemical reactions strongly increase with the increase in paste temperature; the rate of a diffusion-controlled hydration process shows a small temperature dependence (Section 3.6). Therefore, the hydration of C_2S is expected to be more temperature dependent, and for a longer period, than the hydration of C_3S.

The mechanism of hydration presented implies that the hydration process, and therefore the strength development, stops when one of the following conditions occurs: (1) no more unhydrated cement is available for reactions, (2) not enough free water is available, or (3) diffusion can no longer take place. The first condition needs no further discussion. The third condition can exist because of lack of spaces of sufficiently large dimension left for nucleation of new solid hydration products, or because the coating is too impervious or too thick. Powers has calculated from the double volume of hydration products that the water-cement ratio in a paste must be at least somewhere between 0.35 and 0.40 by mass to provide enough space for all the hydration products that can be derived from the cement. It follows, therefore, that the more closely the cement particles are packed originally in the cement paste, that is, the lower the water-cement ratio, the sooner the setting and hardening process will start, resulting, for instance, in relatively high early strengths; but also, the reactions will decelerate more intensively and come to an end. It should be stressed, because it is not recognized generally, that for a high strength it is not necessary for all cement to hydrate. This is so because when the capillary porosity in the hardened cement paste has become negligible, the concrete has reached its maximum strength, regardless how much cement is still unhydrated. This is quite fortunate since complete hydration of the cement is rarely achieved in practice.

The second condition above is not as simple as it may appear. Since there is practically always more water in fresh concrete than is necessary for the hydration, the gradual loss of a portion of the free water does not harm the concrete strength. So the questions are:

- How much water loss is harmful?
- What is the earliest age when the water loss starts getting less harmful?

Due to the importance of these questions, the effect of drying on the hydration is discussed next.

Effect of Premature Drying on Strength (S. Popovics, 1986c)

For continuing hydration the presence of liquid phase is needed in a cement paste. The duration and magnitude of strength development, respectively, are longer and larger in a wet-cured concrete than in a concrete that is dry cured. If too much moisture is lost from the concrete too soon, the strength is damaged. Examples of strength reduction by premature drying are shown in Figure 3.3. Whether the drying is premature or not is measured not so much by

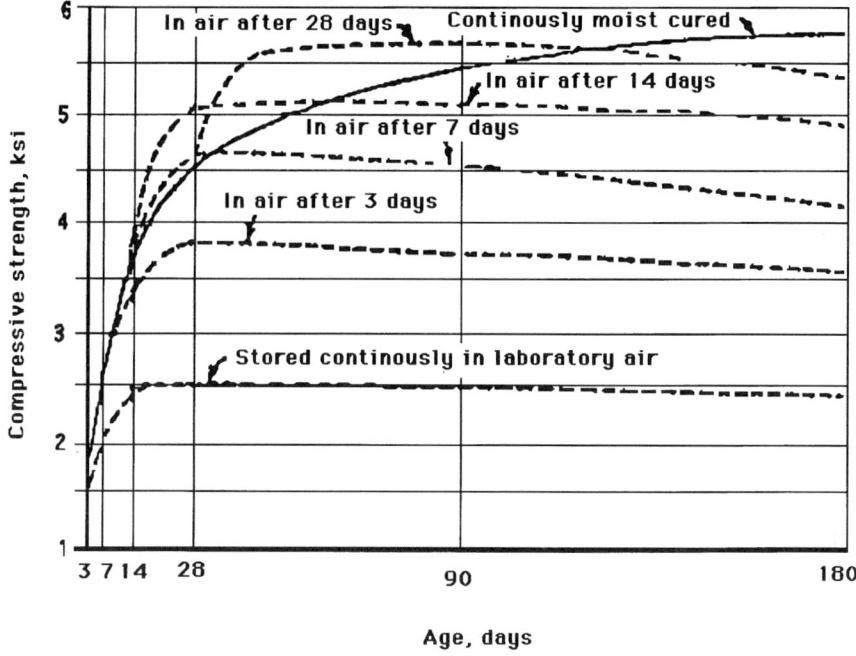

Figure 3.3 Compressive strength of concrete stored in laboratory air after previous moist curing. Water-cement ratio = 0.5 by mass; slump = 3.5 in (90 mm); cement content = 556 lb/cu yd (330 kg/m³); air content = 4%. 1 ksi = 6.90 MPa. (From U.S. Department of the Interior, 1981.)

the age of the specimen but rather by the degree of hydration in the cement paste, as demonstrated by an analysis of Klieger's (1958) data.

Klieger's results were obtained on 150-mm (6-in.) modified cube specimens (ASTM C116) that were cured in a fog room at various temperatures for 28 days. Subsequently, all the specimens were cured at 23°C (73°F), but with half of them still in the fog room, whereas the other half were cured at 50% relative humidity. When these specimens were broken at the age of 90 and 365 days, respectively, no significant differences were shown between the compressive strengths obtained after relative humidities of 100% and 50%, respectively, except for a few specimens. When it is noted that the exceptions are the specimens with Type I and II cements cured at low temperatures, especially −3.9°C (25°F), the explanation seems evident. Low-temperature curing slowed down hydration of the Type I and II cements to a large extent. Thus, even after 28-day wet curing, a large portion of the cement remained unhydrated in these specimens along with a large portion of free moisture. So, under curing in 50% relative humidity, this free moisture evaporated prematurely, leaving not enough moisture in the paste for adequate further hydration of the cement and for strength development. Since Type III cement develops strength faster at any temperature than the other types, its hydration develops enough during the first 28 days, even at $-3.9°C$ (25°F). So subsequent dry curing does not hurt the strength because the drying is not premature. The same is valid for Type I and II cements cured at temperatures higher than 12.8°C (40°F). This explanation is further supported by Klieger's data showing that concretes similar to those mentioned above, but made with the addition of 2% calcium chloride, suffered little strength reduction by dry curing at any temperature.

Two paradoxes should be mentioned. Assuming adequate hydration:

1. A concrete specimen tested in a saturated state produces lower strength than the same specimen would in a drier state.
2. The compressive strength is increased when the outside portion of the specimen contains less moisture than the inside, and vice versa. The effect of moisture distribution on the flexural strength is the opposite.

Details of these paradoxes are discussed in Section 3.8.

Fractional Rates of Hydration

In contrast to some of the statements above, certain earlier experimental results seemed to suggest (Powers, 1958) that the same products are formed at all stages of hydration of the portland cement; that is, the fractional rate of hydration of all compounds (alite, belite, etc.) in a given cement is the same. One can visualize this hydration mechanism by imagining that a cement particle is made up of layers of identical composition; upon contact with water, first all the compounds in the outside layer hydrate before the hydration in the next layer underneath can start; the hydration in the third layer can start only after completion of the hydration of all the compounds in the second layer, and so on. More recent, refined measurements, however, have disproved the assumption of equal fractional rate (Copeland and Kantro, 1964). The C_3A crystals in a portland cement particle hydrate more intensively than the alite crystals, which, in turn, hydrate more intensively than the belite component.

Soroka (1979) resolved the conflict between the two views, stating that at the first stage of hydration, that is, when the rates of chemical reactions control the rate of hydration, the compounds hydrate at their own individual rates. Later, however, when diffusion takes over control of the rate of hydration, the fractional rates of hydration of all compounds are the same in a given cement.

Measurements have also demonstrated that the rate, or fractional rate, of hydration of any component in cement is affected by the composition of the cement. For instance, the rates of hydration of C_3S and C_2S increase with increasing C_3A content, as discussed earlier. Thus, the hydration characteristics, including the strength and heat developments, of the pure individual cement compounds separately have limited value in the quantitative characterization of the hydration characteristics of a portland cement.

3.2 EFFECT OF CEMENT COMPOSITION ON STRENGTH DEVELOPMENT: LINEAR MATHEMATICAL MODELS

Problem Statement

The effects of compound composition on strength development were mentioned in Section 3.1 but only in qualitative terms. Here, and in the following section, mathematical *cement models* (formulas) are discussed that attempt to describe these apparent effects quantitatively.

The term *cement model* means a simplified, hypothetical cement in which many of the factors influencing the hydration and hardening of an actual cement are disregarded; the remaining few variables are combined in a form that can reproduce quantitatively one specific property, such as strength development. The mathematical form of the strength development of this hypothetical (or model) cement is called a *mathematical cement model.* The process of hydration and the mechanism of the strength development in the cement model do not not have to be identical with, or even similar to, the actual ones to make the model acceptable. The only criterion for this is whether the model can produce results that are supported adequately by experimental data. Nevertheless, frequently, although not always, the closer the values produced by the model are to the experimental values, the higher the probability that the two working mechanisms are similar.

Modeling is an important problem for cement manufacturers because reliable relationships between the fineness and composition of a portland cement and its properties help them produce cements of specified properties. It is also helpful for the concrete engineer in selecting of the proper cement for a given purpose.

As mentioned, the strength of a portland cement paste originates from the primary and secondary bonds as well as from the reduction of porosity in the hardening cement paste. Therefore, it is safe to say that the fineness and chemical composition of the cement affect the rate and magnitude of the developing strength only indirectly, through their effects on the quality and quantity of the developing hydration products. Since, however, the relationships between fineness and composition on the one hand and bonds and porosity on the other had not been established yet, except in a rudimentary attempt (Jons and Osbaeck, 1982), direct relationships between the composition of portland cement and its properties can be established only at the phenomenological level, through more-or-less empirical models. Nevertheless, such models do have a rational background since the compound composition of the cement and/or its fineness determines the rate and final extent of hydration and, consequently, the bonds and the porosity in the hardening cement paste.

Approach

The apparent effects of the compound composition and fineness of cement on many of the technically important properties of the cement paste, including the strength development, have been recognized for more than 70 years. The character of these effects has also been recognized even earlier in qualitative terms. For instance, the early strengths are influenced primarily by C_3S and C_3A, the late strength development, by C_2S. The next step was the attempt to establish numerical relationships between the compound composition of cement and its strength development (Bates and Klien, 1917; Woods et al., 1932). However, there were, and are, two main difficulties with these attempts: (1) finding the proper form for the formula (model), and (2) the fact that the compound composition as we consider it presently does not adequately characterize the strength-developing capability of a cement in many cases because the conditions of clinkering, cooling, presence of minor constituents, and so on, can also have presently unaccountable, yet significant interfering effects (Locher and Richartz, 1974).

Note, however, that these interfering effects can be reduced by formulating the cement model for relative strengths, for instance in the form of a ratio. If, for instance, a cement model based on the major constituents says that the strength developed by a portland cement at the age of 7 days under given circumstances is 65% of the strength of the same mixture at the age of 28 days, the ratio structure of such a model cancels out a large portion of the side interferences from the 65% value. Thus, models providing relative strengths are quite suitable for the quantitative representation of certain effects of the major cement constituents on the hydration and hardening. For most practical purposes, of course, strengths expressed in a stress unit are needed. Therefore, formulas are also available for such purposes with more or less reasonable reliability. The recently developed exponential models especially appear satisfactory. In the following sections several models are discussed for the hardening of portland cement pastes, mortars, and concretes in terms of compound composition and age. These are the original linear model and its modifications, the exponential models for using the first-stage mechanism for relative strength and for strength in stress units, the complete quadratic model, the truncated quadratic model, and the exponential model using three stages of strength development. Factors other than the composition of portland cement that can influence the strength development, such as temperature, curing, and test method, will be discussed subsequently.

Linear Model

The first attempt to establish a quantitative relationship between the compound composition of a cement and the strength it develops at various ages was the working hypothesis, or hardening model, stating that each cement compound contributes its intrinsic strength at a given age in proportion to the percentage of that compound present independently from the others. The mathematical form of this model for the four main compounds is linear:

$$f = \text{strength} = a(C_3S) + b(C_2S) + c(C_3A) + d(C_4AF) \qquad (3.1)$$

where the symbols in parentheses represent the calculated (Bogue) percentages by mass of the compounds, and a, b, c, and d are empirical coefficients (parameters) representing the contribution of 1% of the corresponding compound to the strength of the hardening mixture at a given age under the given circumstances (temperature, fineness, etc.).

This approach implies also that the following five assumptions are true:

1. Only these four compounds contribute to the strength development, provided that the fineness, curing and testing conditions, and SO_3 content are kept unchanged.
2. Each of these cement compounds develops its strength independently, that is, without any interaction with the other compounds.
3. The amounts of the four compounds are independent of each other in cements.
4. The air contents in all mortar or concrete specimens are the same.
5. The mechanism of the strength development is the same at early ages as at late ages.

A set of the coefficients in Eq. 3.1 for a given set of conditions can be determined by regression analysis from results of such strength test series where the compound composition of the cement is the sole variable; that is, where the fineness and gypsum content of the various tested cements, age, air content and composition of the strength specimens, curing and testing methods, and so on, are practically identical. Such a test series was examined by Gonnerman (1934), where, among others, standard Ottawa sand mortars were made with a variety of portland cements of differing compound composition and tested for compressive and tensile strengths at ages from 1 day to 2 years. From the strength results of 57 cements, 28 coefficients were calculated by the method of least squares for the compressive strengths of the mortars of 2.75 aggregate–cement ratio by mass for seven age groups, and other 28 parameters for the standard tensile strengths. These coefficients are shown in Table 3.2.

TABLE 3.2 Coefficients for Eq. 3.1[a,b]

Age	1 day	3 days	7 days	28 days	3 months	1 year	2 years
Compound							
	1:2.75 Standard Ottawa Sand Plastic Mortar Cubes of 2-in. for Compressive Strength (psi)						
C_3S	8.5 ± 0.40	27.4 ± 0.98	40.0 ± 1.47	48.8 ± 3.10	55.7 ± 3.67	61.8 ± 4.10	70.7 ± 4.05
C_2S	0.3 ± 0.37	−1.1 ± 0.91	−5.1 ± 0.91	19.1 ± 2.88	62.9 ± 3.41	80.6 ± 3.81	82.2 ± 4.13
C_3A	11.3 ± 1.11	24.1 ± 2.74	58.4 ± 4.11	100.1 ± 8.67	56.4 ± 10.2	85.6 ± 11.47	12.5 ± 11.43
C_4AF	−6.5 ± 1.26	−9.8 ± 3.12	−0.2 ± 4.68	30.8 ± 9.88	39.7 ± 11.71	39.6 ± 13.07	27.2 ± 13.12
	1:3 Standard Ottawa Sand Briquets for Tensile Strength (psi)						
C_3S	2.1 ± 0.14	3.6 ± 0.10	4.6 ± 0.21	5.0 ± 0.20	4.7 ± 0.17	4.6 ± 0.26	4.9 ± 0.23
C_2S	0.3 ± 0.13	0.8 ± 0.18	1.3 ± 0.19	3.8 ± 0.18	6.1 ± 0.16	6.4 ± 0.25	6.1 ± 0.24
C_3A	4.6 ± 0.39	6.3 ± 0.54	7.0 ± 0.57	7.1 ± 0.55	4.4 ± 0.48	2.1 ± 0.74	0.9 ± 0.65
C_4AF	0.4 ± 0.45	3.7 ± 0.62	3.5 ± 0.63	4.0 ± 0.63	4.0 ± 0.55	2.6 ± 0.84	2.2 ± 0.74

Source: Gonnerman (1934). (Copyright ASTM. Reprinted with permission.)

[a]The ± numbers represent the probable error for each coefficient. Curing of specimens: in moist room at normal temperatures up to 28 days, then water.

[b]If it should be desired to express the relation between composition and strength or any other property in terms of the *oxides*, coefficients for such equations may be readily obtained. Thus if a, b, c, and d are the coefficients of C_3S, C_2S, C_3A, and C_4AF, respectively, in the equation for strength for a particular type of specimen and test period, then strength $= (4.4710a − 3.3710b)CaO + (8.6024b − 7.6024a)SiO_2 + (1.0785b + 3.0432d − 1.4297a − 1.6920c)Fe_2O_3 + (5.0683b − 6.7187a + 2.6504c)Al_2O_3$; 1 in. = 25.4 mm; 1 psi = 0.0069 MPa.

Example 3.1 The composition of cement 24 in Gonnerman's investigation was $C_3S = 41\%$; $C_2S = 37\%$; $C_3A = 7\%$; $C_4AF = 12\%$. Thus, from Table 3.2 the estimated compressive strength of this cement at the age of 7 days in a 50-mm (2-in.) standard mortar cube is $f = 40.0 \times 41 - 5.1 \times 37 + 58.4 \times 7 - 0.2 \times 12 = 1860$ psi (12.83 MPa). Strengths for other ages calculated by Eq. 3.1 and the pertinent parameters of Table 3.2 for this cement are shown in Figure 3.4 along with the experimental values obtained by Gonnerman.

It should be noted that any change in the conditions concerning the cement (fineness, SO_3 content, etc.) or testing the strength (composition of the mortar, curing, etc.) would result in a different set of parameters.

Equation 3.1, in general, gives fair agreement between calculated and observed strengths with the appropriate parameters for cements of usual compositions. Nevertheless, this equation is objectionable for several reasons:

1. The *a, b, c* and *d* parameters are *not* equal to the intrinsic strengths of the four major cement compounds as they are measured on pure, indi-

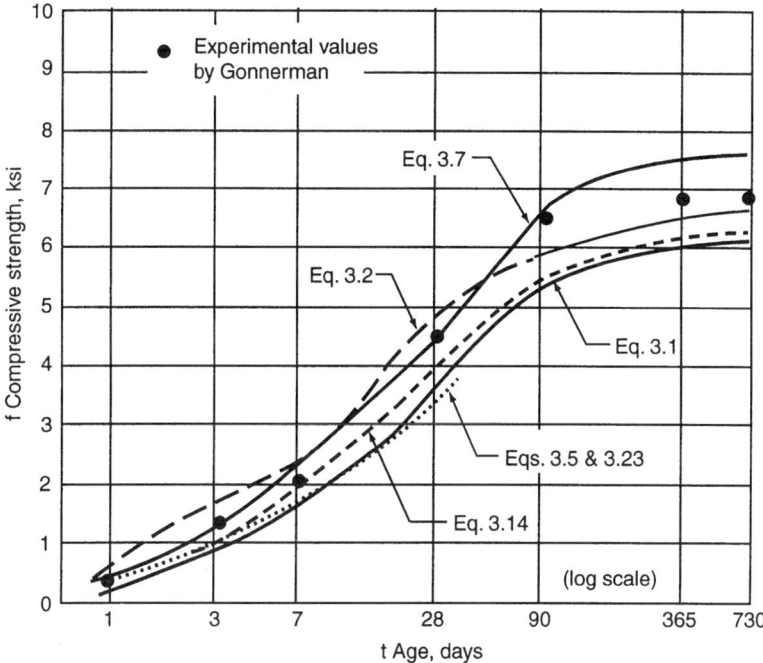

Figure 3.4 Comparison of compressive strengths estimated for the same cement. The specimens were 2-in. (50-mm) cubes of 1:2.75 Ottawa sand mortar made with cement 24 from the series of Gonnerman. Lines represent values calculated by formulas as marked; points represent experimental values. 1 ksi = 6.90 MPa. (From S. Popovics, 1980b Copyright Septima. Reprinted with permission.)

vidual compounds. This has been shown experimentally (Bogue and Lerch, 1934) and also by the fact that some of these parameters in Table 3.2 are negative, since there is no such thing as negative intrinsic strength.

2. Not only the four major compounds but also some of the other ingredients of portland cement (SO_3, alkalies, other minor constituents, etc.) as well as air content can influence the strength at any age.

3. The hydration and therefore the hardening process of the various compounds in a cement are not necessarily independent of each other. Therefore, a linear superposition cannot express adequately, say, the effect of C_3A on the strength development.

4. The magnitude of strength contributed by a compound at a given age depends on whether the strength development is in the first stage or in the second (Section 3.1). This, in turn, depends on the composition and fineness of the cement. Therefore, two coefficients may be needed for each age for each compound for a set of conditions.

5. The *amounts* of the four major compounds in a portland cement are not independent from each other either. In a large test series (Blaine et al., 1968a) C_3S was found strongly correlated with C_2S (correlation coefficient $r = 0.97$), and C_3A correlated with C_4AF ($r = 0.72$).

6. The number of the needed empirical coefficients for Eq. 3.1 is very high, as implied in Table 3.2.

7. In the case of an unusual cement composition, (high or low C_3A content, etc.) Eq. 3.1 may provide values that differ significantly from the strengths obtained experimentally.

Modifications of the Original Linear Model

Several attempts have been made to eliminate or reduce these objections above while keeping the linearity of the model. Blaine and his co-workers (1968a), for example, attempted to improve the linear formulas by introducing a large number of additional variables for the description of strength development. Their formulas are based on their strength tests of more than eighty different portland cements in Ottawa sand mortars. They selected 15 characteristics of the cement composition, including the four major compounds of Eq. 3.1 and the minor constituents as well the fineness of the cement and the air content of the mortar strength specimens but disregarded the free-lime content; they ran a linear regression analysis to obtain the coefficients for these 17 independent variables for estimation of the mortar strengths at various ages, tested these coefficients statistically; kept only those that were found significant or at least close to be significant statistically, and set up several groups of linear formulas for the compressive strength of 50-mm (2-in.) Ottawa sand mortars with and without the inclusion of the minor constituents. The group of formulas that is comparable to Eq. 3.1 (50-mm cubes of standard 2.75 Ottawa sand mortar, wet curing, etc.) is

$$f_1 = -2029 + 18.19C_3S + 224.7SO_3 + 349.1K_2O - 168.8Loss - 228.2Insol + 0.5509S_s - 16.82Air \tag{3.2a}$$

$$f_3 = -2950 + 41.51C_3S + 22.05C_3A + 432.5\ SO_3 + 327.3K_2O - 249.9Loss + 0.7573S_s - 49.55Air \tag{3.2b}$$

$$f_7 = -4131 + 56.16C_3S + 90.45C_3A + 378.1SO_3 + 592.8K_{20} - 39.24MgO - 68.66Loss + 1.07S_s - 59.99Air \tag{3.2c}$$

$$f_{28} = 1075 + 42.08C_3S + 53.03C_3A + 23.60C_4AF + 0.5729S_s - 95.61 Air \tag{3.2d}$$

$$f_{365} = 6518 - 103.4C_3A - 687.7Na_2O + 0.4345S_s - 100.2Air \tag{3.2e}$$

$$f_{1825} = 5331 + 16.25C_2S - 85.22C_3A - 1091Na_2O - 107.9MgO + 0.5375S_s - 507.8Insol - 106.5Air \tag{3.2f}$$

$$f_{3650} = 7833 + 18.77C_2S - 161.5C_3A - 71.0C_4AF - 157Na_2O - 723Insol + 0.2496S_s - 122.2Air \tag{3.2g}$$

where

f_t = standard copressive strength of mortar at the age of t days, psi

$C_3S, C_2S, C_3A,$ and C_4AF = calculated (Bogue) amounts of the four cement compounds, mass %

$SO_3, K_2O, Na_2O,$ and MgO = sulfate, potassium, sodium, and magnesium contents, respectively, mass %

$Loss$ = loss on ignition, mass %

$Insol$ = insoluble portion of the cement, mass %

S_s = fineness of cement by the air permeability (Blaine) method, cm^2/g

Air = air content of the fresh mortar, volume %

Blaine and his co-workers (1968a) also presented formulas for relative compressive strengths, that is, strength-gain ratios for various ages. Two of their pertinent formulas can be written with a little license in the following form:

$$\frac{f_7}{f_{28}} = \frac{1}{2.825 - 0.014C_3S - 0.0297C_3A - 0.000137S_s} \tag{3.3}$$

$$\frac{f_{365}}{f_{28}} = 2.30 - 0.0103C_3S - 0.0356C_3A - \tag{3.4}$$

$$0.0136C_4AF - 0.000034S_s$$

where the symbols and the limits of validity are the same as those for Eqs. 3.2a through 3.2g. Ratios calculated by Eq. 3.3 for Gonnerman's cements with $S_s = 2700$ cm^2/g provide good agreement with the experimentally obtained values, but Eq. 3.4 underestimates the late strength.

As mentioned before, Eqs. 3.2a through 3.4 represent strictly the formulas that best fit the strengths obtained experimentally. In several cases the structures of the formulas contradict our present knowledge concerning the hydration of portland cement. Blaine and his co-workers offer no explanation for these contradictions except that:

1. These formulas are useful for practical purposes because they indicate, at least qualitatively, what factors affect the strength development. For instance, they demonstrate that the alkalies increase the compressive strength up to the age of 7 days but "reduce" it at later ages.
2. Perhaps the formulas reflect the effects of the various independent variables on *differences* in the compressive strength values attained by various portland cements rather than the actual contributions of the variables to the strength. In other words, these formulas might help explain why certain cements attain greater strengths than others.

Using again cement 24 from Gonnerman's investigation with the estimated values of Air $= 1\%$, SO$_3$ $= 1.8\%$, K$_2$O $= 0.4\%$, Na$_2$O $= 0.6\%$, MgO $= 2.8\%$, Loss $= 0\%$, Insol $= 0.2\%$, and $S_s = 2700$ cm^2/g, the strength values calculated by Eqs. 3.2a through 3.2f are presented in Figure 3.4. Von Euw and Gourdin (1970) also used computerized regression analysis for determination of the coefficients for 13 independent variables. Their model is similar to the one applied by Blaine et al. with comparable results.

In another effort, Alexander presented a simplified two-component linear model (Alexander, 1972) where the silicates and aluminates are each represented by only a single variable. In this way the associations between the quantities of C$_3$S and C$_2$S as well as between C$_3$A and C$_4$AF are eliminated from the formulas.

The following formulas of this type were offered for the cube strengths of 2.75 standard Ottawa sand mortars under wet curing:

$$f_3 = -1477 + 24.64C_3S + 40.43C_3A + 0.484S_s \qquad (3.5a)$$

$$f_7 = -1245 + 41.16C_3S + 78.84C_3A + 0.344S_s \qquad (3.5b)$$

$$f_{28} = 286 + 27.26C_3S + 146.96C_3A + 0.384S_s \qquad (3.5c)$$

where the symbols are as in Eqs. 3.2a through 3.2f.

The conversion factor from psi to MPa is again 0.0069. Strength values calculated by Eqs. 3.5a through 3.5c for cement 24 are presented in Figure

3.4. Other linear formulas are discussed in a RILEM report (RILEM, 1991). Equations 3.2a through 3.2g provide a fair fit to experimental data (Fig. 3.4), perhaps because they include more strength-affecting factors than Gonnerman's model. Another advantage is that the amounts of most of these factors are independent of each other. However, the rest of the objections raised against Gonnerman's model remain valid. Also, the equations for the different ages do not form a homogeneous system, and their forms are inconsistent. For instance, the C_3A term appears in all of their equations except in the one for the 1-day strength, where one would especially expect it; the C_4AF appears in the 28-day and 10-year strength formulas, but nowhere else.

Essentially, the same objections are valid for the formulas by Alexander, although his equations are more consistent and more attractive, due to their conscious simplicity.

Since neither the inclusion of many additional variables nor the reduction of the number of independent variables to two have improved the linear model substantially, the conclusion may be drawn that the linear form is not suitable to express the strength development adequately in terms of the cement composition at various ages. Therefore, several nonlinear models are discussed below. Additional nonlinear cement models are discussed in a RILEM report (RILEM, 1991).

3.3 EXPONENTIAL MODEL FOR RELATIVE STRENGTH

The exponential model is discussed in a comprehensive manner because it appears to depict the hardening process of portland cement better and within wider limits than other models (S. Popovics, 1998).

The original form of the model expressed the concrete (and mortar) strength in relative terms, that is, in percent. Also, it covered only the first stage of strength development (Section 3.1), the stage where the hydration process is controlled solely by the chemical reactions between cement and water. Subsequent work extended the original form to express the strength in stress units (Section 3.4), included the zero and second stages of strength development (Section 3.5), and took the curing temperature into account (Section 3.7).

Note, however, that the exponential formula is useful even in its original, relative form, such as Eq. 3.7, because it is applicable for the analysis of some of the factors that influence the hardening process, as illustrated in this section.

Original Model

The original form of the exponential cement model represents a hypothetical cement that has the following properties (S. Popovics, 1974, 1976a):

1. The model cement consists of two hardening components. One component is the C_3S, the other is the quasihomogeneous mixture of the other cement ingredients, mostly C_2S.
2. The final strength developed by C_3S is the same as that by the second component.
3. The contribution of C_3A to the strength development is twofold: it is part of the second component, but its main contribution is through an extra, indirect mechanism.
4. The process of hydration is controlled only by the rates of reactions of the two components at all ages.
5. The two components hydrate as first-order reactions, independently of each other; therefore, the strength of the cement model is the sum of the strengths developed by the two components.
6. The strength development starts at age $t = 0$.
7. It expresses the strength in relative terms, as a percentage of the strength at a specified age.

A numerical form of this cement model is the following (S. Popovics, 1967c, 1968b):

$$f_{rel} = 100 \frac{f}{f_{28}} = 100 \frac{C_3(1 - e^{-a_1 t}) + (100 - C_3)(1 - e^{-a_2 t})}{C_3(1 - e^{-28a_1}) + (100 - C_3)(1 - e^{-28a_2})} \qquad (3.6)$$

$$= 100 \frac{100 - C_3 e^{-a_1 t} - (100 - C_3)e^{-a_2 t}}{100 - C_3 e^{-28a_1} - (100 - C_3)e^{-28a_2}} \qquad (3.7)$$

where f_{rel} = relative strength of portland cement concrete (paste, mortar), % of the 28-day strength

t = age of the specimen at testing, days

C_3 = computed C_3S content of the portland cement, mass %

a_1 and a_2 = rate parameters of the first and second hardening components, respectively, which are independent of the strength, age, and C_3S content but may be a function of the temperature, C_3A content, and any other factor that influences the course of hydration (fineness, gypsum content, admixtures, water-cement ratio, curing and testing method, etc.), 1/day

There is an extra benefit resulting from the mathematical structure of Eqs. 3.6 and 3.7. This is that the interfering effects of the conditions of clinkering of the cement, cooling, and so on, on the strength development are reduced considerably by the ratio structure of the model. Note that f_{rel} can be called the *age factor,* that is, its product with the actual 28 day strength provides the strength in a stress unit at any t age.

Each of the two hardening processes has its own *specific rate of hardening,* that is, (rate of hardening)/ (remaining strength development) at a given age for the two components, represented by the two *a* parameters. It is significant that both of these parameters may increase linearly with an increase in the C_3A content of the cement indicating the catalytic effects on the calcium silicates in the model cement. The *decelerations* of the hardening of both the C_3S and the second component are also proportional at any given age to the remaining strength development at that time and the two proportionality factors are the *squares* of the same a_1 and a_2 parameters (S. Popovics, 1968c).

Calibration of the Model

Numerically, the a_1 and a_2 parameters can be obtained from the strength results of at least two pertinent trial mixtures made with two portland cements of different compound composition. The higher the number of different trial mixtures, the more reliable the *a* values. Gonnerman published pertinent strengths results produced by 71 different portland cements of identical fineness of griding (Gonnerman, 1934). The range of composition of these cements was purposely expanded beyond that of normal portland cements. The compressive strengths of the cements were determined from standard 1:2.75 Ottawa sand mortar cubes moist-cured at room temperature. The strength tests were performed at ages 1, 3, 7, and 28 days, 3 months, and 1 and 2 years. A statistical analysis of these results produced the following *a* values (S. Popovics, 1968c):

- For the compressive strength of standard 1:2.75 Ottawa sand mortars with wet curing:

$$a_1 = 0.0067C_3A + 0.10 \tag{3.8}$$

$$a_2 = 0.0018C_3A + 0.005 \tag{3.9}$$

where C_3A represents the potential tricalcium aluminate in the portland cement computed according to the Bogue method, expressed in mass percent.
- For the tensile strength of standard 1:3 Ottawa sand mortars with wet curing:

$$a_1 = 0.004C_3A + 0.65 \tag{3.10}$$

$$a_2 = 0.007C_3A + 0.04 \tag{3.11}$$

The overall goodness of fit of Eq. 3.7 with Eqs. 3.8 and 3.9 to the pertinent test results by Gonnerman is illustrated in Figure 3.5, where 396 computed and observed values are compared. The overall difference between these computed and observed values is less than 10%. Although this goodness of fit is

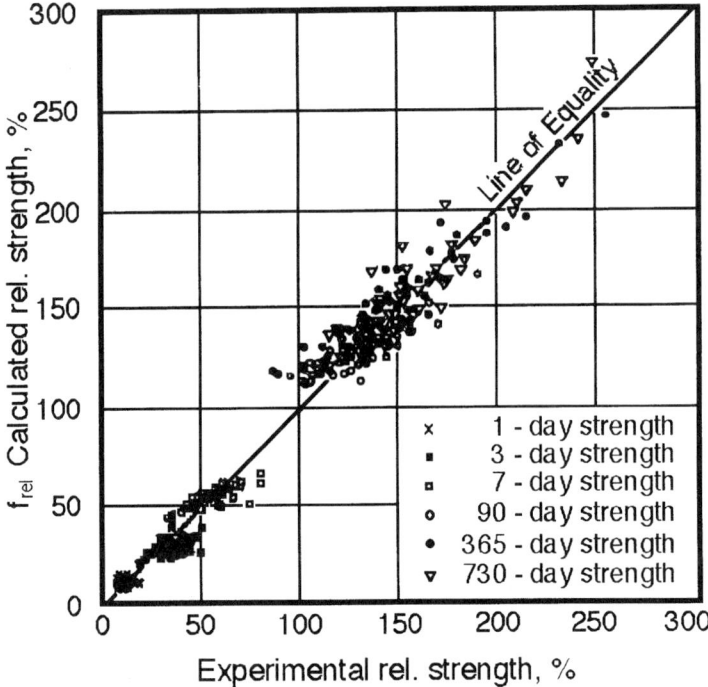

Figure 3.5 Comparison of 396 of Gonnerman's experimental values with computed values of relative compressive strength of 1:2.75 Ottawa sand mortars computed by Eq. 3.7 with Eqs. 3.8 and 3.9. (From S. Popovics, 1967c. Copyright TRB. Reprinted with permission.)

not better than what Eq. 3.1 provides, the significant distinction is that the exponential model covers a much wider compound composition with only two parameters, namely a_1 and a_2, in contrast with the 28 parameters of Eq. 3.1 in Table 3.2.

Experiments performed by Klieger in the long-time studies (LTS) of the Portland Cement Association (PCA) on concrete and standard mortars (Klieger and Isberner, 1957) with 29 portland cements of different compositions at various ages also support the applicability of Eq. 3.7 (S. Popovics, 1967c). Additional details have been reported elsewhere (S. Popovics, 1967b, 1968b, 1969c). Air entrainment does not appear to affect the validity of the exponential model.

The a values obtained for Klieger's Ottawa sand *mortar* strengths are presented in the upper part of Table 3.3 where the C_3A is expressed again in mass %. Note that these specific rates are not quite identical with the corresponding a values of Eqs. 3.8 through 3.11. This is probably due to a lower curing temperature used by Gonnerman. Thus the values in Table 3.3 are valid again only under the circumstances that were used by Klieger (limits of C_3A content, fineness, curing and testing conditions, etc.). The a values re-

TABLE 3.3 Parameters of Eq. 3.7 for selected mortars and concretes

Type of Test	Approx. w/c by Weight	a_1 (1/day)	a_2 (1/day)
Standard Mortars			
Tensile strength of mortar			
(ASTM C190-49)		0.80	$0.02C_3A$
Compressive strength of mortar			
(ASTM C109-49)		0.20	$0.005C_3A$
Flexural strength of mortar		0.45	$0.01C_3A$
Concretes			
Compressive strength, 6 bags/cu yd[a]			
(modified cube test)	0.43	0.40	$0.002C_3A + 0.01$
Flexural strength, 6 bags/cu yd[a]			
(third-point load)	0.43	0.70	$0.001C_3A + 0.02$
Compressive strength, $4\frac{1}{2}$ bags/cu yd[a]	0.54	0.30	$0.005C_3A$
Flexural strength, $4\frac{1}{2}$ bags/cu yd[a]	0.54	0.50	$0.005C_3A$
Compressive strength, 3 bags/cu yd[a]	0.80	0.15	$0.003C_3A$
Flexural strength, 3 bags/cu yd[a]	0.80	0.25	$0.004C_3A$

Source: S. Popovics (1967c). (Copyright TRB. Reprinted with permission.)
[a] 1 bag/cu yd = 94 lb/cu yd = 56 kg/m³; C_3A is in mass%.

lated to his *concrete* strengths are presented in the lower part of Table 3.3. The application of the formulas are illustrated below.

Example 3.2   The compositions of three of Gonnerman's cements, including cement 24, are shown in Figure 3.6. The computed C_3S contents of all three cements are practically the same, but the C_3A contents differ. Illustrate the effect of C_3A content on the relative strength development at various ages by using Eq. 3.7 along with Eqs. 3.8 and 3.9.

The *a* parameters of cement 24 are calculated by Eqs 3.8 and 3.9: $a_1 = 0.14$ 1/day and $a_2 = 0.018$ 1/day. By substituting these values into Eq. 3.7, the following is obtained:

$$f_{\text{rel},24} = 100 \left(\frac{100 - 41e^{-0.14t} - 59e^{-0.018t}}{100 - 41e^{-3.92} - 59e^{-0.50}} \right)$$
$$= 100(1.58 - 0.65e^{-0.14t} - 0.93e^{-0.018t})$$

Similarly, the equations of the curves for relative compressive strength versus age for the cements 1 and 7, respectively, are

$$f_{\text{rel},1} = 100(1.25 - 0.54e^{-0.22t} - 0.71e^{-0.037t})$$
$$f_{\text{rel},7} = 100(2.12 - 0.89e^{-0.10t} - 1.23e^{-0.005t})$$

Values calculated by these formulas are presented in Figure 3.6 by lines along with observed results of points from Gonnerman's experiments. The significant effect of C_3A on the strength development is clearly reflected by the model. Note how little the *relative* compressive strength values of the three cements differ at ages younger than 28 days. The major effect of C_3A appears only at later ages.

Example 3.3 The relative strength development f_{rel}^I of the first component (C_3S) of the exponential cement model and that of the second component f_{ret}^{II} can be illustrated for the three portland cements of Example 3.2, as:

From Eq. 3.6,

$$f_{rel}^I = 100(1 - e^{-a_1 t})$$

and

$$f_{rel}^{II} = 100(1 - e^{-a_2 t})$$

The development of f^I and f^{II} is controlled solely by the rates of reactions a_1 and a_2 at all ages without any consideration of the dormant period or the

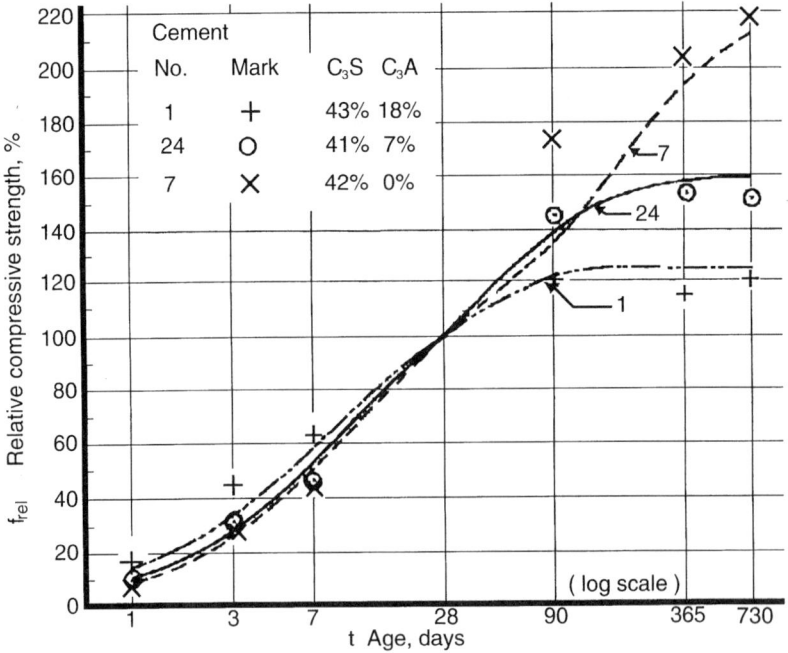

Figure 3.6 Effect of C_3A content on the kinetics of hardening of portland cement in 1:2.75 Ottawa sand mortars. Lines represent Eq. 3.7 with Eqs. 3.8 and 3.9; points represent Gonnerman's experimental values. (S. Popovics, 1967c, 1992. Copyright TRB. Reprinted with permission.)

period of diffusion control. These a parameters calculated from Eqs. 3.8 and 3.9 are the same as in Example 3.2. Therefore, the calculated relative compressive strength developments of the two components of cement 24 are

$$f^{I}_{rel,24} = 100(1 - e^{-0.14t})$$

and

$$f^{II}_{rel,24} = 100(1 - e^{-0.018t})$$

Similar formulas can be obtained for the two components of cements 1 and 7. The values calculated by these formulas are presented in Figure 3.7. The significant effect of C_3A on the strength development is obvious.

Further Comparison with Experimental Data

1. The effect of the strength type on the strength development is illustrated in Figure 3.8. Here experimentally obtained values with standard Ottawa sand mortars (points) and calculated values (lines) of the relative tensile, flexural, and compressive strengths are presented. The experimental values were published by Klieger (1957), and the calculated values were obtained by Eq. 3.7 with the appropriate values of a_1 and a_2 from the upper part of Table 3.3.

2. Figure 3.9 shows points representing experimental compressive strength values by Gonnerman for his cements, having approximately 6.5 mass % C_3A content, and lines representing corresponding values calculated by Eq. 3.7

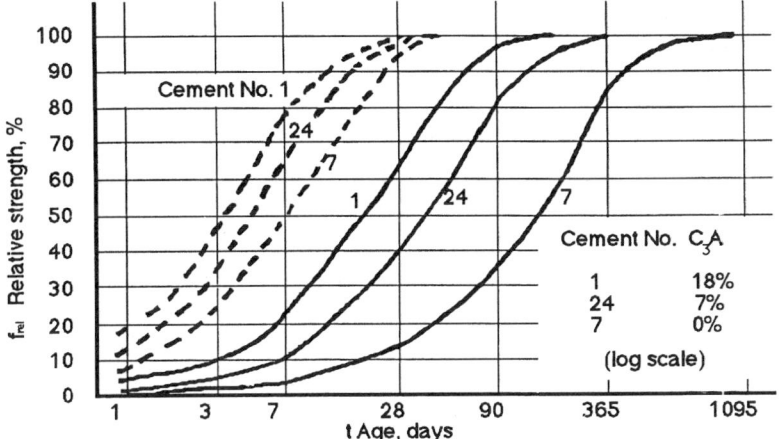

Figure 3.7 Relative strength development of the first component f^I of the exponential model (dashed lines) and that of the second component f^{II} (continuous lnes) as a function of age for three portland cements of Figure 3.6. The hardening is assumed to be controlled solely by the rates of reactions at all ages.

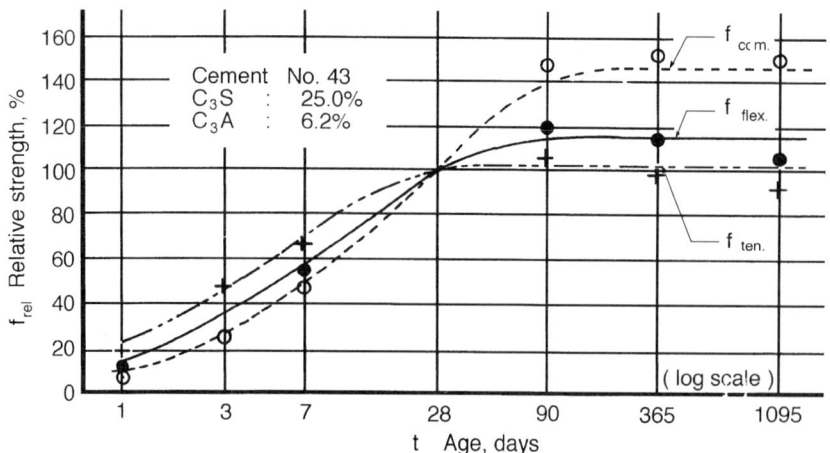

Figure 3.8 Effect of strength type on the kinetics of hardening of portland cement in Ottawa sand mortars. Points represent experimental values; lines represent Eq. 3.7 with appropriate *a* values. (From S. Popovics, 1967c, 1992a. Copyright TRB. Reprinted with permission.)

with Eqs. 3.8 and 3.9 for various ages. It can be seen that the exponential model reflects quite well the effect of the C_3S content on the strength development.

3. Figure 3.10 shows the relationship between the 7-day *compressive* strengths of Ottawa sand *mortars* and the related compressive strengths of 252 kg/m³ ($4\frac{1}{2}$ kg/cu yd = 423 lb/cu yd) *concretes* made with the same cements. Points again represent the experimental values by Klieger and lines represent values that were calculated by Eq. 3.7 with the appropriate values of a_1 and a_2 from Table 3.3. Figure 3.11 shows the relationship between the 7-day *flexural* strength of mortars and those of the 252-kg/m³ ($4\frac{1}{2}$ bag/cu yd = 423 lb/cu yd) concretes made with the same cements. These figures not only show that the relationship between corresponding flexural strengths of mortars and those of concretes is dependent on the C_3A content of the cement, while the similar relationship for compressive strength is not, but also that the exponential model is sensitive enough to reflect this.

4. Equation 3.7 can be used for estimation of the *ultimate strength*, that is, the strength obtained after very long wet curing, expressed as a percentage of the 28-day strength. This equation takes on the following form for large *t* values (S. Popovics, 1972a):

$$f_{rel}^0 = 100 \left[\frac{100}{100 - C_3 e^{-28a_1} - (100 - C_3)e^{-28a_2}} \right] \tag{3.12}$$

$$\approx 100 \left[\frac{100}{98 - (100 - C_3)e^{-28a_2}} \right] \tag{3.13}$$

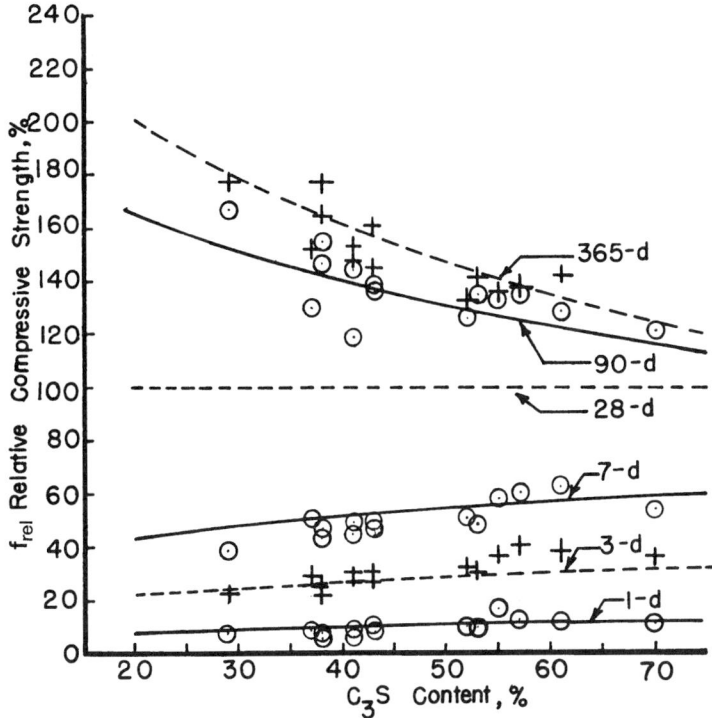

Figure 3.9 Relative compressive strengths of 1:2.75 Ottawa sand mortars as a function of the C_3S content for portland cements with computed C_3A content between 5 and 8%. Lines represent Eq. 3.7 with Eqs. 3.8 and 3.9. Points represent Gonnerman's experimental values. (From S. Popovics, 1967c. Copyright TRB. Reprinted with permission.)

where f_{rel}^0 = relative ultimate compressive strength, % of the 28-day strength. The other symbols are the same as the symbols of Eq. 3.7.

Figure 3.12 presents the relationship between the compound composition of portland cement and the relative ultimate compressive strength. Points represent experimental values reported by Gonnerman (1934) for cements of approximately 40 and 60 % C_3S contents, respectively, with constant (1.8%) SO_3 content, and the same fineness (144 m²/kg by air permeability). The lines designate values that were calculated from Eq. 3.13 for 40 and 60% C_3S contents, respectively, with the a_2 value of Eq. 3.9. Considering the inherent fluctuation of the results in such kind of experiments, the goodness of fit appears reasonably good in Figure 3.12.

Properties of the Exponential Model

Experimental results discussed above support the validity of the exponential model within wide limits, even though both the dormant period and the dif-

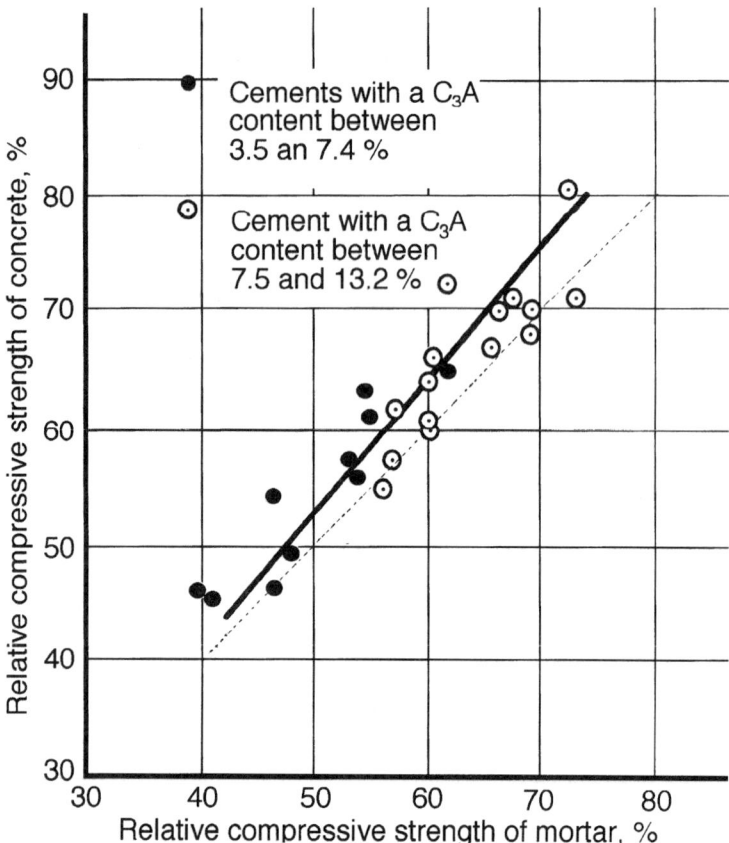

Figure 3.10 Relationship between the 7-day relative compressive strengths of Ottawa sand mortars and $4\frac{1}{2}$-bag/cu yd (423 lb/cu yd = 252 kg/m³) concretes made with the same cements. Experimental values are represented by points, computed values by the solid line; the dotted line represents equality. (From S. Popovics, 1967c. Copyright TRB. Reprinted with permission.)

fusion period are disregarded. This support may mean that perhaps there is a similarity between the hardening process represented by the model and the actual process. Therefore, it may be instructive to see what Eq. 3.7 has to say about the hardening process of the model portland cement.

1. The more intensive the early hardening, the greater the value of a_1; and the longer and larger the hardening after 28 days, the smaller becomes the value of a_2 up to a certain limit. In Eqs. 3.8 through 3.11 the values of a_1 is about seven to 10 times higher than a_2 for these mortars within the usual range of C_3A content; therefore, the model cement of Eq. 3.7 hardens as if the C_3S developed the full value of its compressive

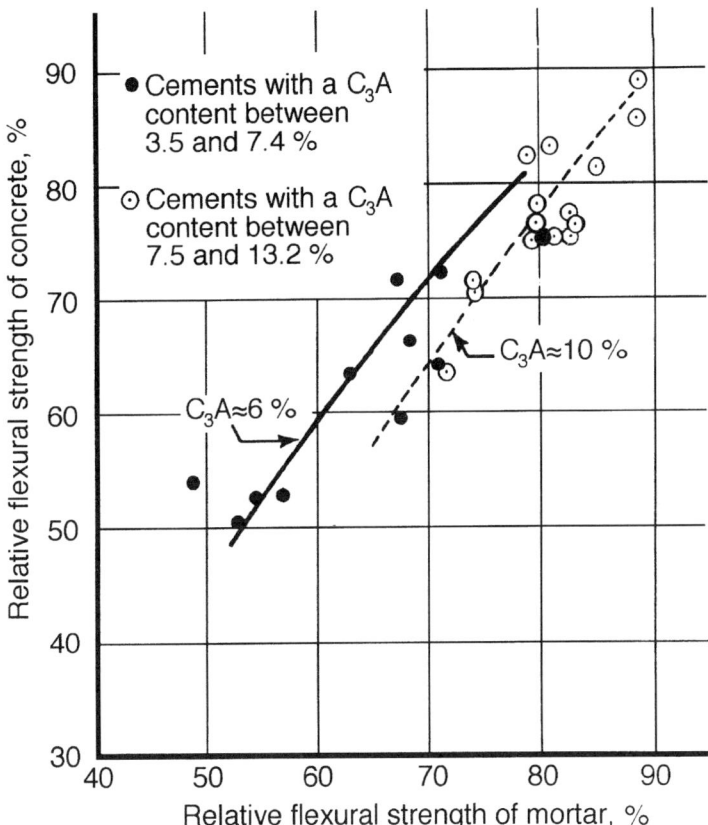

Figure 3.11 Relationship between the 7-day relative flexural strengths of Ottawa sand mortars and $4\frac{1}{2}$-bag/cu yd (423 lb/cu yd = 252 kg/m³) concretes made with the same cements. Experimental values are represented by points, computed values by lines. (From S. Popovics, 1967c. Copyright TRB. Reprinted with permission.)

strength by the age of about 28 days (Fig. 3.7). After that any further strength increase seems to be due only to the hardening of the second component. It may also be noted that the suitable a_1 and a_2 values are much higher for the tensile strength than the corresponding values for the compressive strength resulting in an earlier stop of tensile strength development (Figure 3.8).

2. The *form* of the equations for a, such as those in Table 3.3, is just as important as the numerical values of the coefficients because the form reflects the effect of C_3A with respect to the kinetics of the hardening. Namely, the linear relationships between the a rates and the C_3A content indicate that the C_3A acts as a *catalyst* on the two hardening components

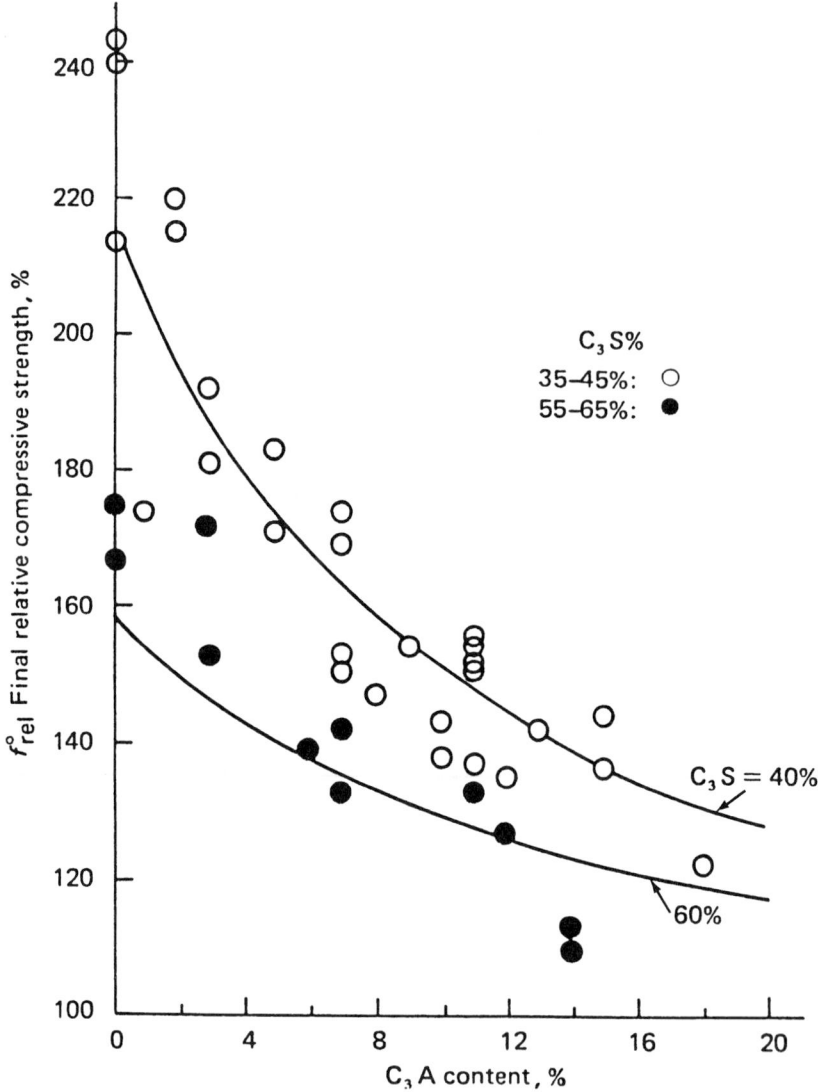

Figure 3.12   Final relative compressive strengths of portland cements in mortars as a function of the compound composition. Lines represent Eq. 3.13 with Eq. 3.9; points represent Gonnerman's experimental values. (From S. Popovics, 1972a.)

of the model during hydration (S. Popovics, 1974, 1976a). Whether this calaytic effect does exist in actual portland cements, or C_3A has some physical effect on the developing structure of the CSH gel, or C_3A has another complex role in the hydraiton, as advanced by Hansen (1970), is still an open question.

3. The a_1 and a_2 values represent the specific *rate of hardening* of the two components, that is, (rate of hardening)/(remaining strength development at a given age). The squares of a's represent the specific *deceleration* of hardening, that is, (deceleration of hardening)/(remaining strength development at a given age) (S. Popovics, 1968b). Consequently, the *specific rate* of hardening is a linear function, and the specific deceleration of hardening is a quadratic function of the C_3A content with a reasonable degree of approximation. In other words, an increase in the C_3A content increases the early strength through an increased specific rate of hardening, but, simultaneously, it increases the deceleration of the hardening quadratically, thus the late strength development will be less.

It is important to point out again that not only the C_3A but also any other factors that increase the early strength (higher curing temperature, etc.) would produce lower late-strength development because of the restrictive effect of the intensified deceleration.

4. Example 3.2 and Figure 3.6 illustrate several additional features of the strength development as follows:
   - The effect of C_3A content on the development of relative strength is highly significant through its effect on the rate and deceleration of hardening, respectively.
   - A straight line approximates the compressive strength versus age relationship in a semilogarithmic system only within the limits of 3 and 90 days. Beyond these age limits this approximation is no longer valid.

5. A comparison of the a values for the concretes of two different cement contents in Table 3.3 reveals that the development of relative strength is quicker and the deceleration is stronger for higher cement contents and for lower water-cement ratios, other factors being equal.

PROBLEM 3.1   Show that Eq. 3.7 can be derived from the premise that the decelerations of the hardening of the two components at a given age are proportional to the remaining strength development at that time.

Solution   Mathematically for the first hardening component

$$-\frac{d^2f_1}{dt^2} = a_1^2(f_o - f_t)$$ (3.13a)

where $-d^2f_1/dt^2$ = deceleration of strength development of the first component (C_3S) of the model,
   t = age of the concrete at testing,
   f_1 = strength of the first component at the given t age,

f_o = strength of the first component after very long wet curing,

a_1 = experimental parameter that is independent of the strength and age, but may be a function of the fineness and composition of the cement, concrete composition, curing and testing methods, etc.

The integration of Eq. 3.13a with the boundary conditions $f = 0$ when $t = 0$ and $f = f_o$ when $t = \infty$ produces the strength developed by the first component. The strength developed by the second component can be obtained similarly with $a_2 \neq a_1$. The sum of the two strengths, followed by the elimination of the usually unknown term f_o by expressing the strength in relative terms, results in Eqs. 3.6 and 3.7.

Extension of the Model to Properties Other Than Strength

Certain properties of the hardening cement paste can also be modeled in terms of the compound composition. For instance, Eq. 3.7 was extended for the description of the development of nonevaporable water content, and water sorption of hardening pastes. Experimentally obtained nonevaporable water contents supported the exponential model concept but only with a modified form of Eq. 3.7 (S. Popovics, 1969). Less successful was the attempt to extend Eq. 3.7 to the kinetics of vapor sorption, that is, the specific surface of the calcium silicate hydrates, perhaps because of the inherent uncertainty of the sorption measurement.

The relationship between the heat of hydration and compound composition of portland cement is illustrated in Figures 3.13 and 3.14 (S. Popovics, 1992). The equations are supported quite well by experimental data published by Verbeck (1950). Note that there seems to be no interaction between the hydration of calcium silicates and calcium aluminates in the heat of hydration model, unlike in Eq. 3.7. This seems to contradict the hypothesis that the C_3A acts as a catalyst on the CSH gel. In any case, one cannot expect single linear relationships between the strength and the corresponding other properties of the hydrating cement paste.

Advantages and Disadvantages

The advantages of the exponential model for relative strengths (i.e., Eq. 3.7) are

1. It works; that is, it is supported by experimental results within wide limits.

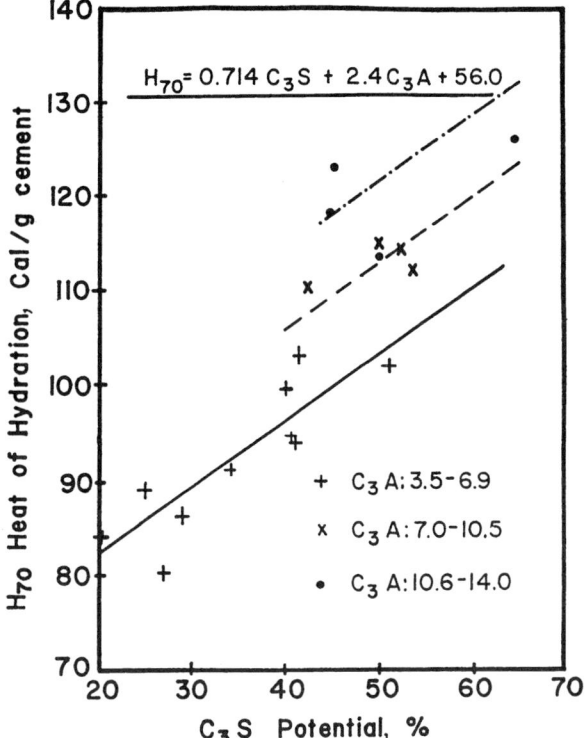

Figure 3.13 Heat of hydration as a function of clinker composition. The paste specimens were cured at 70°F (21°C) for 6.5 years. (From S. Popovics, 1992. Copyright Noyes. Reprinted with permission.)

2. The model is conceptually simple. It has only two independent variables, namely the C_3S and C_3A contents. More importantly, it needs only two parameters (a_1 and a_2) instead of the many of the other models, for instance, those in Table 3.2.

3. Its interpretation from the standpoint of cement chemistry may be meaningful.

Specifically, it indicates that

- The hydration of the two components of the model fall in the category of *first-order reactions*.

- The C_3A acts as a catalyst on the hydration of the calcium silicates in the model cement. Since this is strongly supported by test results, it is possible that the same catalytic action takes place in the hydration of actual portland cements.

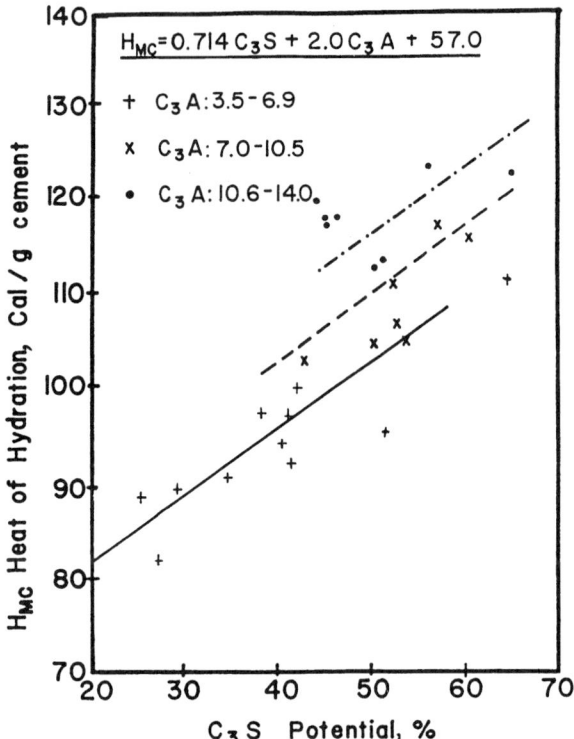

Figure 3.14 Heat of hydration as a function of clinker composition. The paste specimens were cured under simulated mass storage for 6.5 years. (From S. Popovics, 1992. Copyright Noyes. Reprinted with permission.)

4. The model may also be used conveniently for quantitative investigations of other effects on the strength development, such as the effect of admixtures or cement fineness.

5. The model is applicable for various strengths of mortars and concretes.

6. It may also be applicable to other characteristics of the hydration process, such as the amount of chemically bound water, specific surface of the hydration products, heat of hydration, etc.

7. The structure of the formula is statistically sound.

The disadvantages of Eq. 3.7 are as follows:

1. It uses only one hardening mechanism; this is the one that is controlled by the rates of reactions (first stage of strength development). In actual portland cements, however, there are two more stages, namely, the dormant period at very early ages (zero stage) and the diffusion-controlled hydration (second stage). The consequence of this simplification is that

Eq. 3.7 overestimates a little, although regularly, the compressive strengths at each age when the actual strengths are low, and underestimates them when the actual strengths are high (Fig. 3.5). It will be shown in Section 3.5 that this regular error is reduced by the inclusion of all three stages of strength development into the exponential cement model.

2. Equation 3.7 does not include the SO_3 content, the minor constituents and admixtures as independent variables, although the effect of temperature might be included (Section 3.7). It is likely, however, that further investigation would result in the inclusion of these factors into the model.

3. Equation 3.7 provides only the relative strengths. However, this weakness will be eliminated in the next section.

3.4 EXPONENTIAL MODEL FOR STRENGTH IN STRESS UNITS

There are several ways for an exponential model to be used to estimate the strength expressed in a stress unit instead of as relative strength. One possibility is to establish experimentally the actual strength of a mixture at age t and substitute it into Eq. 3.7. This is illustrated in Eq. 3.14. Another method is to combine Eq. 3.7 with another equation that predicts the strength, such as Eqs. 3.5a through 3.5c or the Abrams' formula. This is illustrated in Eq. 3.20.

Direct Exponential Model for Standard Mortars (S. Popovics, 1981b)

A simple approach is the transformation of Eq. 3.7 into the following formula for the compressive and tensile strengths:

$$f_t = f_{90} \frac{100 - C_3 e^{-b_1 t} - (100 - C_3)e^{-b_2 t}}{100 - C_3 e^{-90 b_1} - (100 - C_3)e^{-90 b_2}} \tag{3.14}$$

where f_t = estimated mortar or concrete strength at the age of t days, psi or MPa or kg/cm^2

f_{90} = strength of mortar or concrete at the age of 90 days, in the same stress unit as f_t

For the standard *compressive* strength of any of the five types of portland cements obtained with nonair-entrained *mortars* of ASTM C109 type:

$$f_{90} = \frac{5500 S_s}{S_o} \quad \text{(psi)} \tag{3.15}$$

For the *compressive* strength of the same cements and mortars as above but when the curing takes place in moist air instead of water:

$$f_{90} = \frac{5900S_s}{S_o} \quad \text{(psi)} \tag{3.16}$$

The modified rate parameter of the first hardening component,

$$b_1 = \frac{S_s}{S_o} a_1 \quad \text{(1/day)} \tag{3.17}$$

The modified rate parameter of the second hardening component,

$$b_2 = \frac{S_s}{S_o} a_2 \quad \text{(1/day)} \tag{3.18}$$

where the *a* parameters are as defined in Section 3.3, and S_s and S_o = specific surface of the cement in question and that of a typical Type I cement, respectively. If no actual surface values are available, S_s can be taken for other than Type III as 180 m²/kg (turbidimeter). S_o can be taken as 180 m²/kg (turbidimeter ASTM C115) that is approximately 300 m²/kg by air permeability (ASTM C204). The other symbols are identical with the symbols of Eq. 3.7. The conversion factor from psi to MPa is 0.0069.

Equations 3.15 and 3.16 are approximations implying that the strength at the age of 90 days is independent of the compound composition of the cement. Their acceptability is supported not only by the mortar strength results by Klieger and by Gonnerman mentioned above but also by other experiments (Gonnerman and Lerch, 1952; Neville, 1973; U.S. Bureau of Reclamation, 1966; Price, 1951; Orchard, 1962).

For the *tensile* strength of a standard briquette of ASTM C190 type but with moist-air curing;

$$f_{90} = 580 \quad \text{(psi)} \tag{3.19}$$

Note that the modification of the rate factors *b* in Eqs. 3.17 and 3.18 by the cement fineness *S* is based only on a few experimental results. It is possible that this will change when further data become available. f_{90} for air-entraining portland cements is approximately 15% less than indicated by Eqs. 3.15 and 3.16 but not enough pertinent data are available for these cements to make a more definitive statement.

An example of satisfactory goodness of fit of Eq. 3.14 to experimental compressive strengths is given in Figure 3.4. A more general illustration of this is presented in Figure 3.15. Five of Klieger's LTS cements, one of each standard type, were used. The C₃S and C₃A contents of these cements are given in Table 3.4. It can be seen that Eq. 3.14 with f_{90} = 37.3 MPa (5400

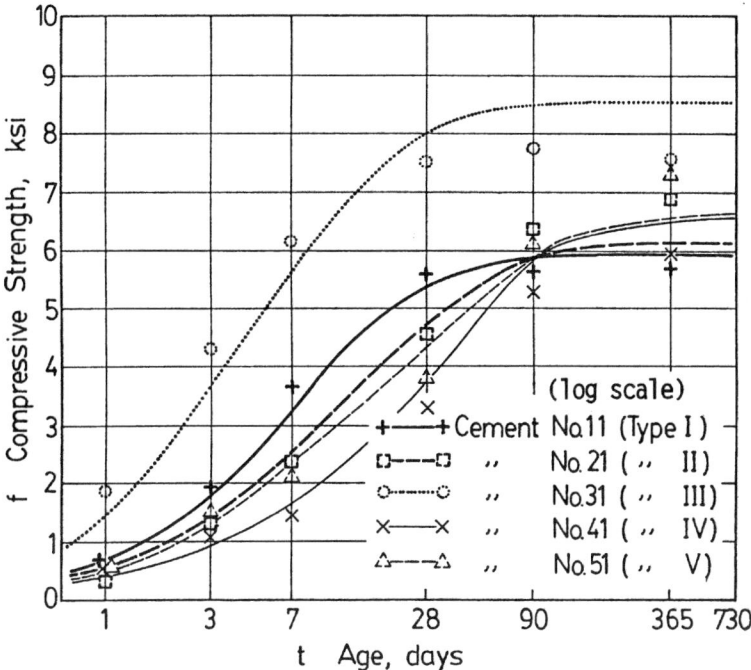

Figure 3.15 Compressive strengths of five standard portland cement types in 1:2.75 Ottawa sand mortars with moist-air curing. Lines represent Eq. 3.14 with Eqs. 3.16, 3.17, and 3.18. 1 ksi = 6.90 MPa. (From S. Popovics, 1981b. Copyright Thomas Telford Publishing. Reprinted with permission.)

TABLE 3.4 Calculated compound compositions of five portland cements from the LTS series

| Cement | | Quantity (mass %) | |
No.	Type	C_3S	C_3A
11	I	50.0	12.1
21	II	40.0	6.4
31	III	56.0	10.8
41	IV	20.0	4.5
51	V	41.0	3.7

Source: Lerch and Ford (1948). (Copyright ACI. Reprinted with permission.)

psi) fits the experimentally obtained compressive strengths of the majority of these cements reasonably well. More specifically, the average error in the strength estimates for the mortar-cube strengths of the five LTS cements and five ages (from 1 through 90 days) is 0.20 MPa (285 psi). This average error is defined as the average of the absolute values of differences between the measured mortar compressive strengths and those calculated by Eqs. 3.14 (S. Popovics, 1967d). The average errors in the estimates of the 7- and 28-day compressive strengths are 0.26 and 0.21 MPa (367 and 301 psi), respectively, which represent approximately a 14.4% and a 7.1% error, respectively.

Equation 3.14, with the appropriate b parameters, also shows good fit to standard mortar *tensile strengths* obtained experimentally with the same five cements. The average errors for the tensile strength of standard mortars with the same five LTS cements are approximately of the same relative magnitude as for the compressive strengths. Further details can be found in the literature (S. Popovics, 1981b).

Direct Exponential Models for Concrete (S. Popovics, 1981a)

The direct exponential model can also be used for the estimation of *concrete* strength. One such use is from the strength of the same concrete obtained experimentally at a *different age,* as discussed above with mortar. The practically important case, of course, is when the 28-day strength is estimated from the result of an early strength determination. The method is illustrated in Figure 1.11. This compares the measured values and corresponding values calculated from the measured 1-day strength with Eq. 1.11 for the standard compressive strength of 335 kg/m^3 (565-lb/cu yd) concretes. The same five portland cements were selected for the comparison. Their compositions are presented in Table 3.4. It can be seen that the goodness of fit between the experimental and calculated strengths is good except for the Type II cement. The indication is that the low estimates for this cement are due to an error in the experimentally obtained 1-day strength. This is a warning that only highly reliable test results may be used for such strength estimation.

Another possibility is the combination of Eq. 3.14 with a *strength versus water-cement ratio formula* (Section 5.5). By using the most popular formula, Abrams' formula, the following model is offered for the *compressive strength* of concrete (S. Popovics, 1981a):

$$f_{co} = \frac{A}{B^{w/c}} \sqrt{\frac{S_s}{S_o}} \frac{100 - C_3 e^{-b_1 t} - (100 - C_3) e^{-b_2 t}}{100 - C_3 e^{-90 b_1} - (100 - C_3) e^{-90 b_2}} \tag{3.20}$$

where f_{co} = estimated compressive strength of concrete, psi, at the age of t days

A = 107 MPa (15,500 psi) when determined on 150 × 300 mm
(6 × 12 in.) cylinders made, cured, and tested according to
ASTM C192 and C39

B = 6.4 for the same conditions as above

w/c = water-cement ratio by mass.

The values for b can be calculated by multiplying the values of a in Table
3.3 by S_s/S_o (Eqs. 3.17 and 3.18). The other symbols are the same as the
symbols of Eqs. 3.15 through 3.19.

The values of 107 MPa (15,500 psi) for A and 6.4 for B in Eq. 3.20 were
established from the analysis of several independent test series (S. Popovics,
1967b). They may change if there is a change in the units; type of aggregate;
methods of making, curing, and testing the specimen; type and/or the quantity
of admixture; or type of strength. For instance, for air-entrained concretes,
with cement and air contents used typically for structural purposes, the value
of A is approximately 97 MPa (14,000 psi) in Eq. 3.20, but otherwise the
formula remains unchanged.

Example 3.4 Figure 3.16 shows calculated as well as experimental results
of a series of compressive strength results performed with five portland cements representing the ASTM five standard types. The specimens were 150
by 300-mm (6 by 12-in.) cylinders, cured and tested according to ASTM
C192 and C39. Data concerning the reported compositions of the concretes
are as follows:

- Nominal cement content: approximately 280 kg/m³ (470 lb/cu yd)
- Water-cement ratio: w/c = 0.49 by mass
- For the Type III cement: S_s = 273 m²/kg (turbidimeter)
- For the other cements: S_s = S_o = 171 m²/kg (turbidimeter)

Points in the figure represent experimental values reported by Gonnerman and
Lerch (1952); lines represent Eq. 3.20 with the appropriate b values from
Table 3.5, and Eqs. 3.17 and 3.18.

**TABLE 3.5 Characteristics of the kinetics of strength development in Figure
3.18.**

Cement Type	a_1 (1/day)	a_2 (1/day)	t_s (days)	t_d (days)	f_{ctd} (MPa)
I	0.48	0.11	0.10	43.8	42.56
II	0.27	0.052	0.0177	50.6	37.73
III	0.80	0.072	0.0153	35.6	39.66

For the *flexural strength* of normal-weight, nonair-entrained structural concretes, the following formula is valid:

$$f_{fl} = \frac{1300}{2.55^{w/c}} \frac{100 - C_3 e^{-b_1 t} - (100 - C_3)e^{-b_2 t}}{100 - C_3 e^{-90 b_1} - (100 - C_3)e^{-90 b_2}} \tag{3.21}$$

where f_{fl} is the estimated flexural strength of concrete in psi at the age of t days when determined on beams 150×150 mm (6×6 in.) in cross section, and when made cured and tested with third-point loading according to ASTM C192 and C78. The other symbols are identical with the symbols of Eq. 3.20. Values for the flexural b rate parameters can be obtained again by multiplying the pertinent values of a in Table 3.5 by S_s/S_o. The A and B values for Eq. 3.21 were established earlier from the analysis of several independent test series (S. Popovics, 1967b).

The major difference between Eqs. 3.20 and 3.21 is that the latter does not directly contain the specific surface term of the cement. This reflects the experimentally demonstrated fact that the cement fineness has no sizable effect on the flexural strength of concrete at later ages. The limits of validity of Eqs. 3.20 and 3.21 are those of Abrams' formula, as well as the age from 1 day through 1 year and compound composition and fineness as they occur in the five ASTM standard portland cement types as defined in ASTM C150. Figures 3.16 and 3.17 illustrate the satisfactory goodness of fit of Eqs. 3.20 and 3.21, respectively. Further illustrations can be found in the literature (S. Popovics, 1981a).

Advantages and Disadvantages of the Direct Model

The exponential model for strengths in stress units (i.e., Eq. 13.14) has essentially the same advantages and disadvantages that Eq. 3.7 has. Yet they show improvements in two important directions. First, the model provides the estimates of strengths in stress units instead of the relative strengths in percent. Second, it includes the effect of cement fineness on the strength development, although in a preliminary manner. The formula appears to be applicable for all five standard cement types and for a time period spanning from 1 day to 90 days, or even to 1 year. It is also reassuring that the structure of the model is sound and simple. It indicates, for instance, that the specific surface of the cement influences the rates of hardening in a linear manner in accordance with the expectation. It may be that the goodness of fit of these equations is somewhat less than that of Eq. 3.7, due to the utilization of additional assumptions. Nevertheless, Eq. 3.14 with Eqs. 3.15 through 3.19 as well as Eqs. 3.20 and 3.21 with the appropriate b parameters appear still better than any other formula offered for the same purpose.

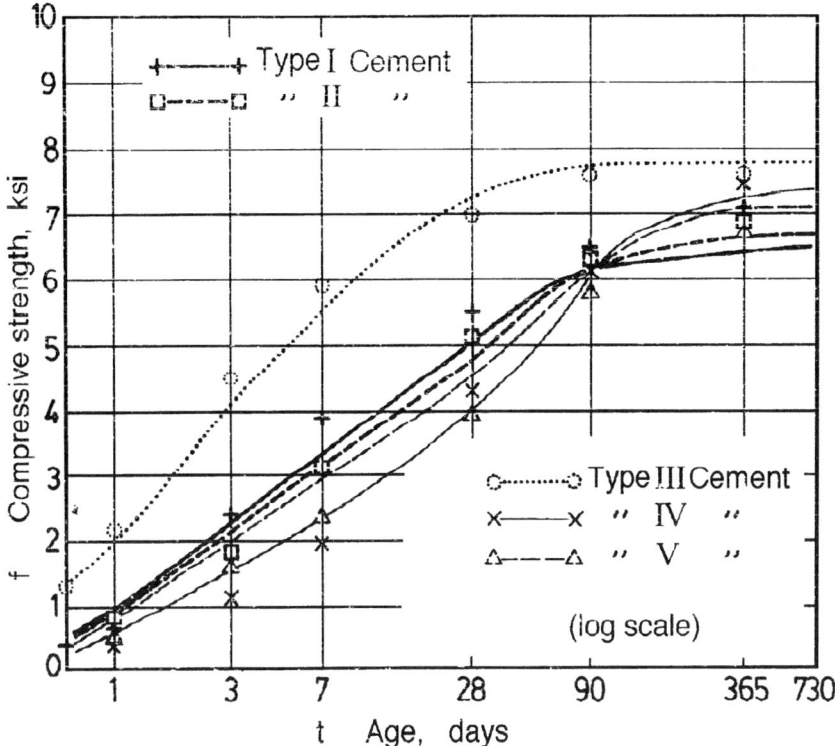

Figure 3.16 Compressive strengths of concretes as a function of cement type and age at testing. Points represent Gonnerman's experimental values; lines represent values calculated from Eq. 3.20 with the appropriate b values. 1 ksi = 6.90 MPa. (From S. Popovics, 1981a. Copyright ACI. Reprinted with permission.)

Complete Quadratic Model

If one expands the exponential terms in Eq. 3.14 with $S_s = S_o$ into the well-known series of the exponential function, and omits terms of second and higher degrees, the following formula is obtained:

$$f_t = g + hC_3S + iC_3S \times C_3A + jC_3A \tag{3.22}$$

where g, h, i, and j are experimental parameters that are independent of the C_3S, C_2S, and C_3A contents of the cement but may be affected by any other factor that influences the strength development. Best-fit values for these parameters are not available at this time.

The effect of the cement fineness on the strength may be expressed numerically by adding a kS_s term to Eq. 3.22, where S_s is the specific surface of the cement and k is an experimental parameter that depends on the age of

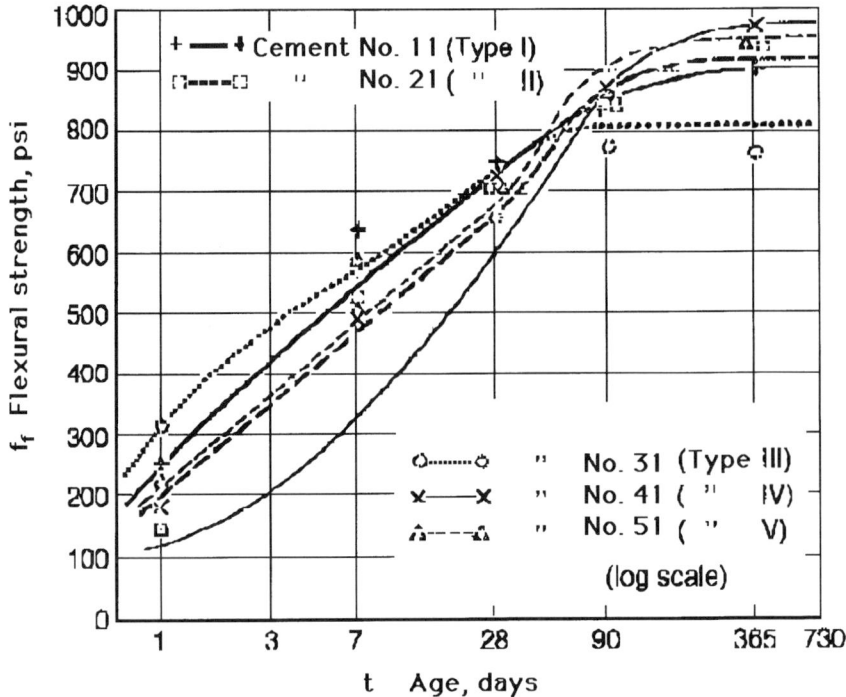

Figure 3.17   Flexural strength of concretes as a function of cement type and age at testing. Points represent Klieger's experimental values; lines represent values calculated from Eq. 3.21 with the appropriate *b* values. 1 ksi = 6.90 MPa. (From S. Popovics, 1981a. Copyright ACI. Reprinted with permission.)

the strength specimens, cement content, and on most of the other factors mentioned above in connection with the other pertinent experimental parameters. Limited experimental data support this linear approximation (S. Popovics, 1992).

Truncated Quadratic Model

To reconcile the hardening model with the experimental observation that the strength developed by the calcium silicates depends on the proportion of C_3A in the cement, Alexander deviated from the linearity of the early models. As the result of a regression analysis, he offered the following formulas of second degree (Alexander, 1972):

$$f_3 = -1248 + 22.32C_3S(1 + 0.030C_3A) + 0.468S_s \qquad (3.23a)$$

$$f_7 = -772 + 34.20C_3S(1 + 0.043C_3A) + 0.329S_s \qquad (3.23b)$$

$$f_{28} = 1129 + 15.36C_3S(1 + 0.181C_3A) + 0.325S_s \qquad (3.23c)$$

where the symbols and the limits of validity are the same as those for Eqs. 3.5a through 3.5c. The conversion factor from psi to MPa is 0.0069.

Again using cement 24 from Gonnerman's investigation, the strength values calculated by Eqs. 3.23a through 3.23c are practically the same for this cement as those by Eqs. 3.22a through 3.22c shown in Figure 3.4. Alexander believes these quadratic equations to be more realistic than any of the linear models because the nonlinear model provides improved agreement between the conclusions from regression analysis and those from cement chemistry. He finds no consistent evidence that the sensitivity of the hydration of C_3S to C_3A varied with gypsum content. He also points out, however, that Eqs. 3.23a through 3.23c are not supported better by the corresponding experimental strengths than his Eqs. 3.5a through 3.5c. This can also be seen from Figure 3.4. Note that Eqs. 3.23a through 3.23c can also be obtained from Eq. 3.22 with $j = 0$. This again supports the validity of the exponential model. It also explains the term *truncated* in the name of Alexander's cement model.

3.5 EXTENSION OF THE EXPONENTIAL MODEL TO THE DORMANT AND DIFFUSION PERIODS (S. Popovics, 1987c)

As mentioned earlier, the exponential model represents certain chemical reactions of the hydration during the hardening process. Therefore, strictly speaking, this model covers only the *first stage* of the strength development that is controlled by chemical reactions. Although it has been shown that it fits experimental strengths after standard curing reasonably well, it can be refined still further by the extension of the model to the other two stages. This would extend its limits of validity, for instance, to curing temperatures other than 22.8°C (Section 3.7).

Dormant Period

Preceding the first stage of strength development, the *zero stage* covers the period before, during, and immediately after the setting process (Section 3.1). This is the $t < t_s$ period in Figure 3.18 the length of which is several hours at the standard 22.8°C (73°F) curing temperature but longer at lower temperatures and shorter at higher ones. The chemistry and physical chemistry of the early hydration differ from those in the subsequent first stage; therefore, the strength development is also different for the description of which a formula other than the exponential model is needed. However, the strength developed during the zero stage is so small that its contribution to the overall strength of concrete is negligible, so it does not seem worthwhile to complicate a formula for strength development by the addition of an extra term. It is more practical to assume a "dormant period" (i.e., zero strength for a short time) up to t_s and start the strength development of the exponential model from that point. The position of t_s is determined statistically from the con-

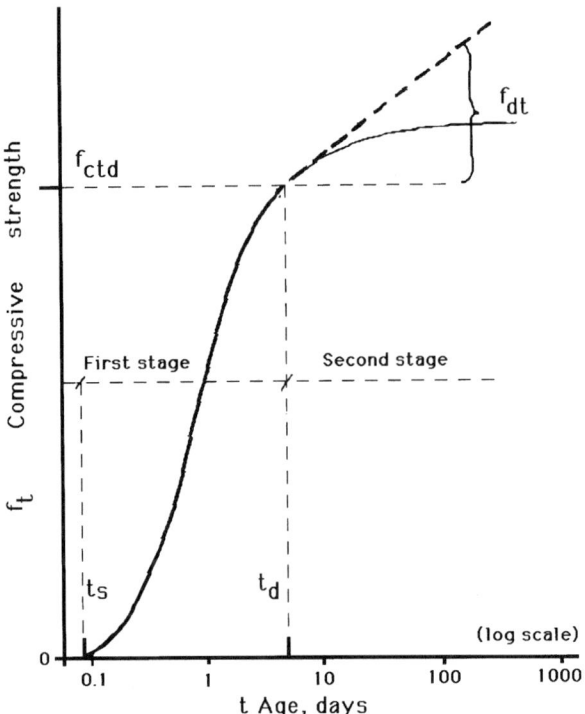

Figure 3.18 Schematic representation of strength development of the model cement as a function of age. Solid line, strength values representing Eq. 3.24; dashed line, strength values representing Eqs. 3.25 and 3.26. The thick lines represent calculated f_t. (From S. Popovics, 1987c. Copyright Cement and Concrete Research. Reprinted with permission.)

dition that the values calculated from the exponential model fit best the experimental strength values of 1 day and older. This approximation means that the exponential model does not necessarily cover strengths for ages of less than 1 day. t_s is naturally related to the setting time of an actual portland cement, although not identical with it.

First Stage of Strength Development

Using the symbols in Figure 3.18, the numerical form of the strength development f_{ct} in the extended cement model for the *first stage* (i.e., the refined form of Eq. 13.20 for $t_s < t < t_d$), can be written as follows:

$$f_{ct} = f_0 f_{28} \frac{100 - C_3 e^{-a_1(t-t_s)(S/300))} - (100 - C_3)e^{-a_2(t-t_s)(S/300)}}{100 - C_3 e^{-a_1(28-t_s)(S/300)} - (100 - C_3)e^{-a_2(28-t_s)(S/300)}} \quad (3.24)$$

where f_{ct} = strength of portland cement mortar or concrete in the same stress unit as f_{28}, when the age t is between t_s and t_d

f_{28} = strength of the same mixture at 28 days but with Type I portland cement and cured for 28 days in the standard manner at 22.8°C (73°F), in the same unit as f_{ct}

f_o = hypothetical strength that the concrete would achieve if the control of the strength development did not stop at t_d but continued indefinitely, %/100 of the 28-day strength

t_s = hypothetical age when the hardening process is assumed to start in the cement model, before which the strength is assumed to be zero, days

t_d = hypothetical age when the first stage of strength development is assumed to end and the second stage to start, days

S = specific surface of the cement, m²/kg; S_o = 300 m²/kg by the air permeability method (ASTM C204)

The other symbols are identical with the symbols of Eq. 3.20. Equation 3.24 is represented by continuous thick line in Figure 3.18.

Diffusion Period

The *second stage* of strength development may be long (months, years) and controls the late-strength development, which can be sizable. Thus, the addition of a new term to the exponential model is justified for the representation of the second stage.

The minimum information that a cement model needs to cover the second stage is:

- The beginning of this stage, that is, the division point between control by chemical reactions and diffusion control; this is t_d in Figure 3.18.
- The type of the function that approximates the strength development during the second stage.

The t_d point was determined again by stepwise approximation to fit the cement model best to experimental strengths. For the description of the strength development in the diffusion stage, (i.e., when $t > t_d$) a hyperbolic function would probably have the widest limits of validity. For the sake of simplicity, however, a logarithmic function was selected, as follows:

$$f_{dt} = \omega \log \frac{t}{t_d} \qquad (3.25)$$

where f_{dt} = strength developed after the age of t_d (i.e., during the second stage), in the same stress units as f_{28} (Fig. 3.18)

ω = experimental parameter that may be a function of the cement composition but seems independent of the curing temperature

This equation is represented by the dashed line in Figure 3.18 and seems to be valid at least up to 1 year.

Therefore, the total strength f_t of a mortar or concrete during the second stage of strength development is the sum of Eqs. 3.24 and 3.25:

$$f_t = f_{ctd} + f_{dt} \qquad (3.26)$$

The kinetic characteristics in Eq. 3.26 for a 22.8°C (73°F) constant curing temperature are listed in Table 3.5. This includes the strength f_{ctd}, which is the strength related to t_d. The *rate* of strength development of portland cements, as calculated for the three stages of the exponential cement model, is discussed in Example 4.1.

The extended cement model is also suitable for the quantitative illustration of the effect of curing temperature on the *kinetics of strength development.* This and several other factors that influence concrete strength are discussed in Sections 3.7 and 3.8.

3.6   EFFECT OF THE CURING TEMPERATURE ON CONCRETE STRENGTH

Qualitative Assessment

It has been recognized from the beginning of concrete production that the ambient temperature, especially the curing temperature, has a major effect on the properties of concrete, including strength (ACI, 1971b). The effects of extreme weather conditions are mostly undesirable because they may delay construction and/or their mitigation may cost extra money. All these have practical significance during and/or after construction in cold weather, in a topical climate, and in steam curing of concrete. Thus it is not surprising that a large amount of literature is available concerning the effect of temperature on concrete, going back at least to 1915 (Samarai et al., 1983a).

All three stages of the *strength development* are affected by the concrete temperature, although to different degrees. The *zero stage*, which includes the dormant period and a part of the setting process (Section 3.1), is probably affected most, and the second stage, least. The temperature affects the *duration* of the zero stage and this can affect strength development in the first stage. The higher the temperature, the shorter the length of the zero stage. Since the length of the first stage is related to the setting times, Figure 3.19

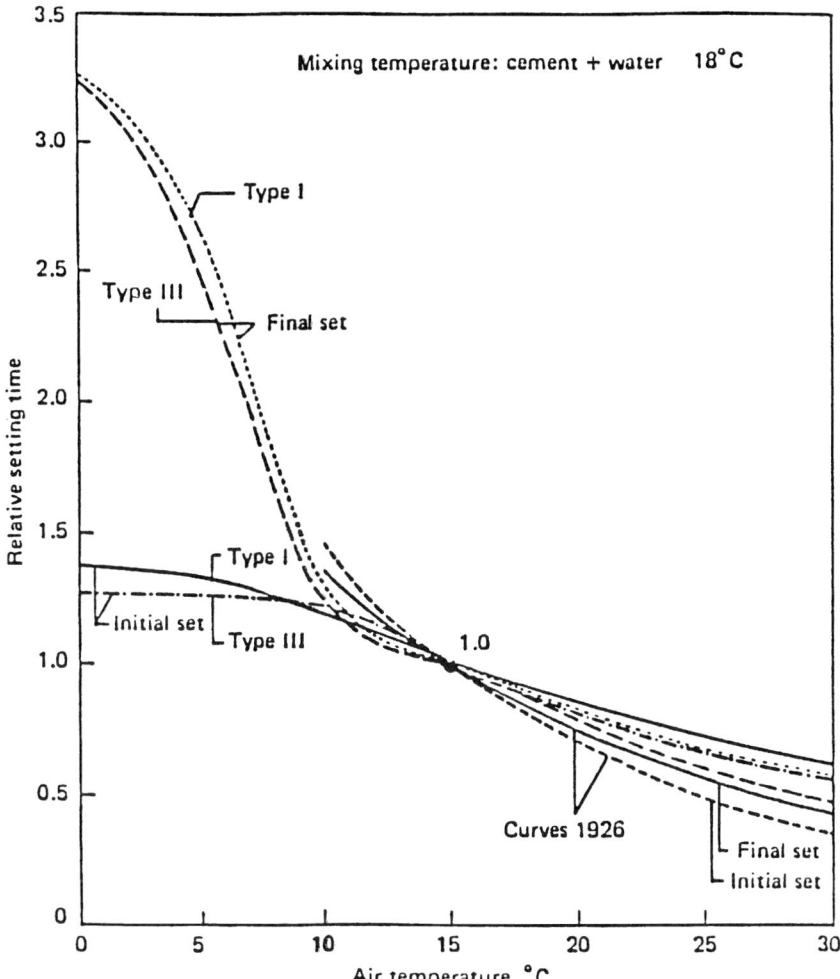

Figure 3.19 Influence of the curing temperature on the settings of Types I and III portland cement pastes. (From Voellmy, 1956.)

provides an approximate picture of the effect of temperature on the length of the zero stage.

The higher the ambient temperature, the higher the *early* strengths in the *first stage* of strength development. This is due partly to the earlier start of the first stage under these circumstances, and partly to the fact that higher temperatures speed up the chemical reactions in the hydration. As illustrated in Figures 3.20 and 3.21, this higher strength period can last from a few days to a few weeks, depending on the circumstances. These figures also demonstrate that the more intensive the early strength development, the lower the final strength of the concrete, even under otherwise favorable curing condi-

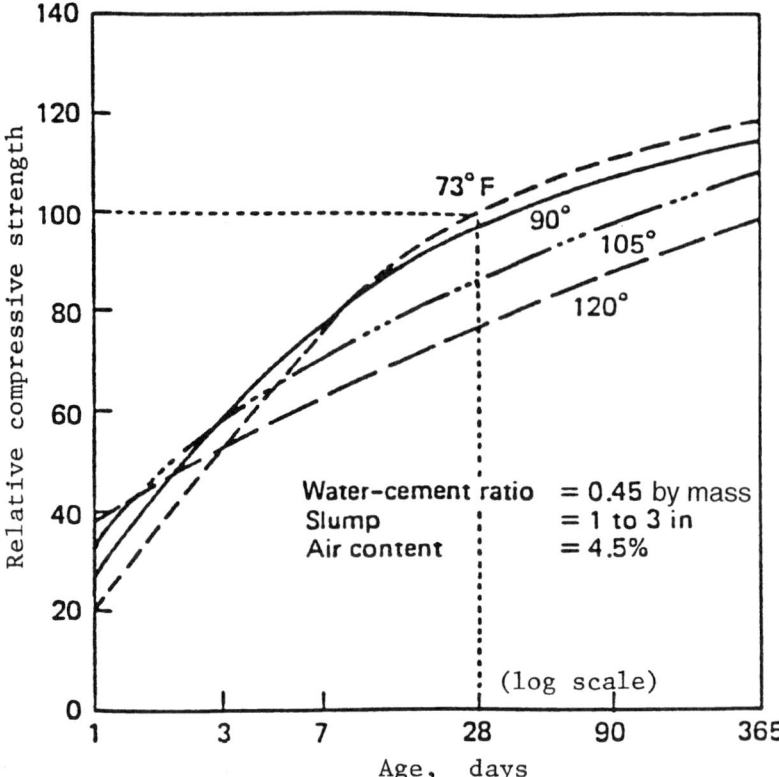

Figure 3.20 Effect of elevated temperatures of casting and moist curing on compressive strength of concrete. The strength is expressed as a percent of 28-day, 73°F-cured concrete. Specimens cast and moist cured at temperature indicated for first 28 days. All moist cured at 73°F thereafter. (From Portland Cement Association, 1965. Copyright PCA. Reprinted with permission.)

tions. Even when the concrete is cast and kept at a given temperature only for a few hours and then cured at standard temperature until break, the higher the initial temperature, the lower the 28-day strength (Price, 1951). This late-strength reduction is mitigated by the addition of calcium chloride or other accelerators (S. Popovics, 1987d, 1988). The effect of cement type on strength development can also be seen under various constant temperatures.

This also means that there is a temperature during the early life of the concrete that may be considered optimum with regard to strength at later ages, or more strictly, at comparable degrees of hydration. This temperature is somewhat influenced by cement type (Klieger, 1958). According to Figure 3.22, this curing temperature is 13°C (55°F) for Type I cement and 4.5°C (40°F) for Type III for strengths after 28 days for Klieger's concretes. In general, the best curing temperature for most concretes is the "normal" temperature, between about 15 and 40°C (60 through 105°F).

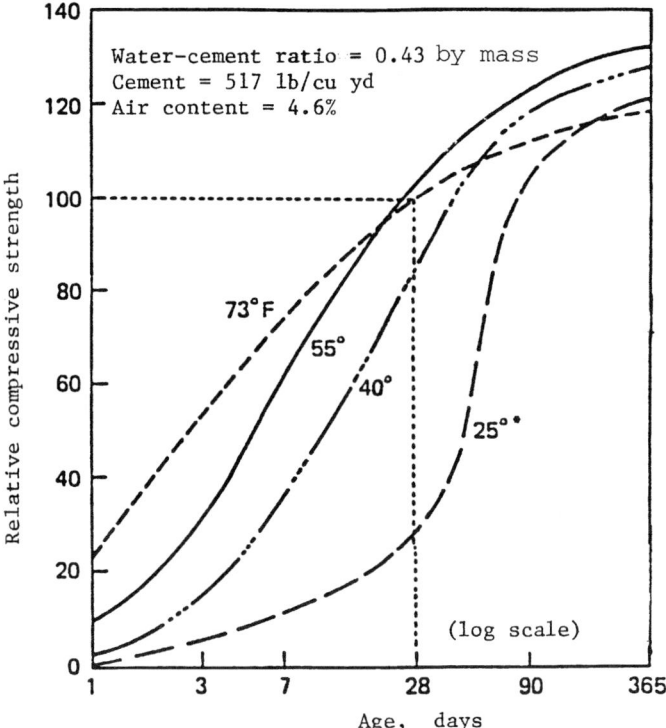

Figure 3.21 Effect of low temperatures of casting and moist curing on compressive strength of concrete. The strength is expressed as a percent of 28-day, 73°F-cured concrete. Specimens cast and moist cured at temperature indicated for first 28 days. All moist cured at 73°F therafter. *Cast at 40°F, stored at 25°F for first 28 days. (From Portland Cement Association, 1965. Copyright PCA. Reprinted with permission.)

A possible explanation for the inverse correlation between the early strength and late strength is that certain strength-affecting properties of the hydration products are modified by a change in the rate of hardening, and the cause of the rate change is secondary at most. For instance, CSH gels under the effects of intensifying physical or chemical factors produce a somewhat higher specific surface at early ages (Kantro et al., 1961). This means finer texture and, presumably, lower porosity at early ages, producing higher strengths at early ages but preventing hydration later.

Problems with Concreting Under Cold Temperatures

A curing temperature *below the freezing point* at early ages is a problem, as it reduces the concrete strength. In one case (Bernhardt et al., 1956) the decrease in the 28-day compressive strength was about 30 to 40% when the concrete was frozen *immediately* after being placed in the molds. When the concrete was frozen after 8 hr, the decrease was much smaller, and when the

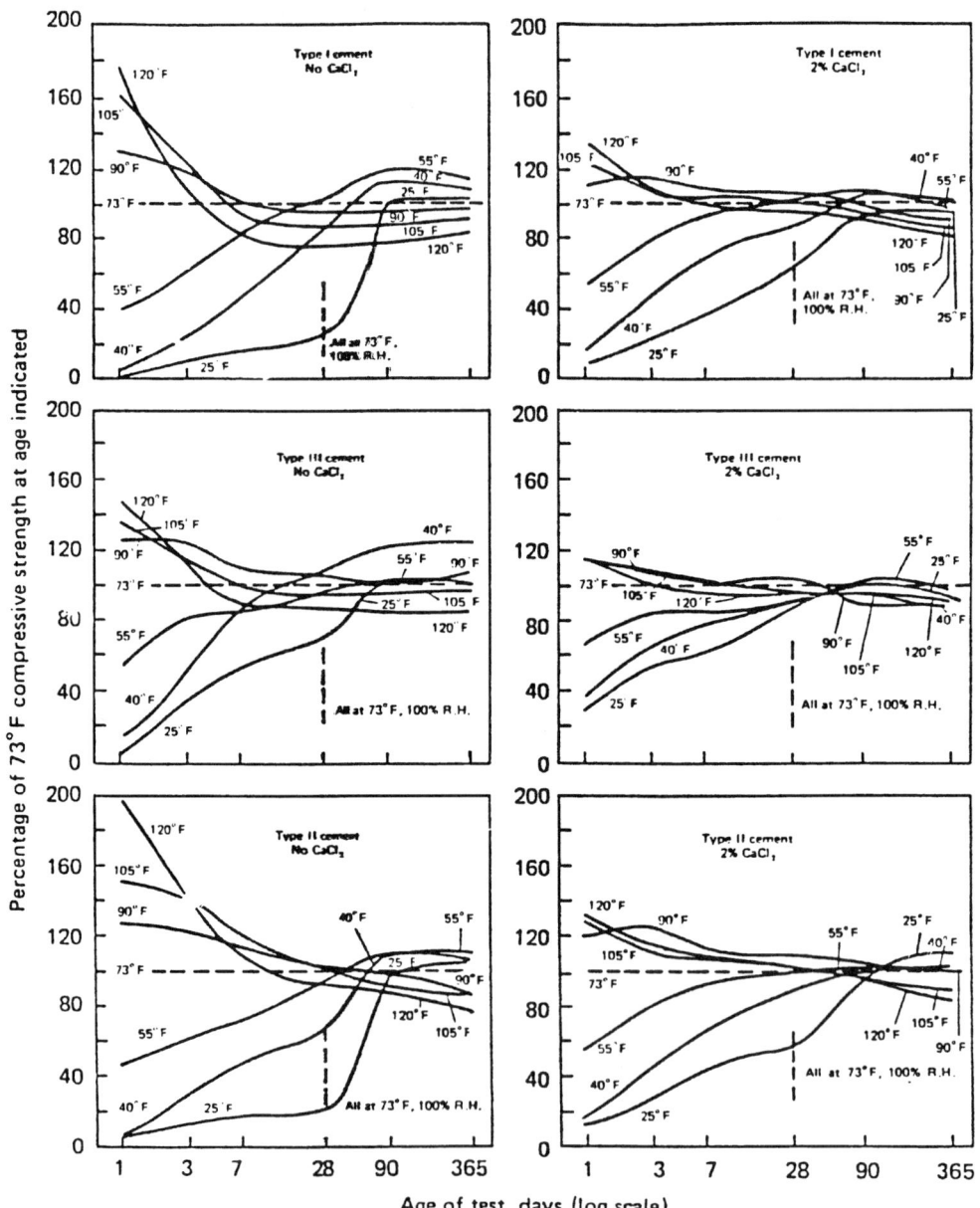

Figure 3.22 Effect of temperature on relative compressive strength of concretes made with different types of cement with and without an accelerator. Air content of all concretes, 4.5 ± 0.5% (neutralized Vinsol resin solution added at mixer); cement content of all concretes, 5.5 sacks/cu yd. Within the individual boxes, the net water-cement ratios of the concretes are approximately equal. (From Klieger, 1958. Copyright ACI. Reprinted with permission.)

freezing took place after 24 hr, the strength reduction was negligible. A rule of thumb is that after the concrete has reached about 3.5 MPa (500 psi) compressive strength, freezing the concrete does not reduce the strength significantly (ACI Committee 306, 1988b). Neither the length of the freezing time (between 2 and 28 days) nor the freezing temperature seems to affect the strength reduction.

Concretes damaged by freezing in this manner do not recover strength equivalent to unfrozen concrete. The freezing damage can be attributed to the ice from the mixing water observed in the freshly frozen concretes in the form of crystals mostly on the surface of the aggregate particles. These crystals force the mortar to render space, thereby causing debonding which is not reversible, at least not without revibration. When the concrete thaws, the ice melts away but leaves behind water-filled cavities. These will not be filled by the continuing hydration during later curing at temperatures above the freezing point. If the freezing takes place a few hours after placement, hydration will have used up some of the free water, fewer ice crystals can be formed, thus less damage will occur. After a while the freezing will cause no damage because not enough free water will be available for ice formation. This happened to the concrete in Bernhardt's experiments around the age of 12 hr: no ice crystals—no strength reduction. Such undamaged concrete can develop considerable strength later on even at very low curing temperatures.

Air entrainment appears to help the frost resistance of the fresh concrete (Nerenst and Plum, 1956).

The freezing point in fresh concrete is below 0°C (32°F) primarily because the liquid phase contains a large amount of salts in solution. Thus, concrete can gain strength even when exposed to below-freezing temperatures after an initial curing at temperature above freezing point, although the strength development is slow. However, the strength development increases markedly when later the curing temperature exceeds 10°C (50°F). This is demonstrated experimentally in Figure 3.21 and by Malhotra (1971b).

Early low strengths caused by low curing temperature delay construction. Under such circumstances the concrete rquires longer time to develop the strength for save removal of forms and for safe loading. This problem can be reduced by certain protection techniques. Guidelines for this have been prepared (ACl Committee 306, 1986,1988; Carino, 1988). The two main goals are (1) protection of fresh concrete from freezing at early ages, and (2) assurance for adequate strength gain.

To prevent early-age freezing and assure adequate strength gain, arrangements for covering, insulating, housing, or heating newly placed concrete should be made before placement. Specifically, the following are recommended:

- Insulation of the concrete to reduce heat loss
- Placement of the concrete at a temperature of 10°C (50°F) or higher

- Provision of an additional heat source to maintain the concrete at a temperature that will give the needed strength development
- Use of higher cement content than the design specifies for strength, use of Type III cement, and/or use of accelerating admixture

Problems with Concreting Under Tropical Temperatures

The primary difficulty with *fresh concrete* temperatures *above "normal"* is that the zero stage is too short, that is, the concrete stiffens before it can be compacted and finished (Samarai et al., 1983b). The specific harmful consequences of this on the *fresh* concrete are (ACI Committee 305, 1982):

1. Increased water demand
2. Increased rate of slump loss and corresponding tendency to add water at job site
3. Increased rate of setting, resulting in greater difficulty with handling, finishing, and curing, thus increasing the possibility of cold joints
4. Increased tendency for plastic cracking
5. Increased difficulty in controlling entrained air content

Additional undesirable hot-weather effects on *hardening concrete* may include (Samarai et al., 1983c):

1. Increased tendency for drying shrinkage and differential thermocracking
2. Decreased durability
3. Decreased uniformity of the surface appearance
4. Increased creep

Again there are guidelines for the reduction of these difficulties to make construction in tropical climates practical (ACI Committee 305, 1982). One of the important points of protection is to keep the concrete temperature low during placement. Control of concrete temperature through the temperature of ingredients can only be done at the point of batching and mixing. A rule of thumb is that for concrete of conventional proportions, a reduction in concrete temperature of 0.5°C (1°F) requires a reduction, in the cement temperature of about 4°C (8°F), in the water temperature of about 2°C (4°F), or about 1°C (2°F) in the aggregate temperature. Since the greatest portion of concrete is aggregate, reduction of aggregate temperature brings about the greatest reduction in concrete temperature. Thus all practical means should be employed to keep the aggregate as cool as possible. This can be done by shading the supplies, sprinkling or fog spraying the coarse aggregate, and by other means. Ice is a far more effective cooler than cold water. Use of liquid nitrogen for cooling is relatively expensive but may be justified under special circumstances.

Another important point is the proper wet curing of the fresh concrete in hot weather for at least several days. This protects the concrete from sudden moisture loss and eliminates plastic shrinkage (ACI Committee, 264 1992). The effects of curing temperature on the *flexural strength* of concrete are similar to those on the compressive strength, although the extent is usually not as great.

Steam Curing

A practical use of methods for curing concrete at elevated temperatures is the accelerated curing of concrete discussed in Section 1.3. Another, more important application is *steam curing at atmospheric pressure*. This is the most widely used method at the present time to obtain high early strengths in the production of precast concrete structural members, pipes, masonry units, and prestressed products. Steam curing of all these units follows the same basic rules, but the details of the curing procedures may be different for each (ACI Committee 517, 1980). The basic procedure is that the fresh concrete is placed in the mold, compacted and finished, and left at room temperature for awhile. This is called the *presteaming period*. Then the concrete is placed in a steam chamber and the steam is heated up gradually. This period is called the *temperature-rise period*. The concrete is kept at the maximum steam temperature specified, called the *curing temperature*, for several hours. This is called the *steaming period*. After this, the chamber and the concrete are allowed to cool down gradually. This is called the *cooling period*. After cooling, the concrete is taken out of the chamber, the mold removed, and the concrete element stored. The four periods form the *steam-curing cycle*. The length of each period depends on the composition of the concrete as well as on the type of concrete unit. For instance, the recommended presteaming period for a lightweight masonry unit is 1 hour, for a normal-weight masonry unit it is 2 hours, and for concrete pipes, it is 2 to 3 hours. The recommended maximum curing temperatures for these units are 82 to 88°C (180 to 190°F), 66 to 74°C (150 to 165°F), and around 66°C (150°F), respectively. The steaming period is typically 12 hours or more, controlled primarily by the fact that casting beds and forms are generally reused on a 24-hour cycle.

All portions of the curing cycle are interrelated. For instance, lack of an adequate presteaming period could be partially corrected by a slower rate of temperature rise. Conversely, a long presteaming period could be followed advantageously by a relatively rapid rate of temperature rise (ACI Committee 517, 1980).

A different kind of steam curing is that carried out under a *pressure higher than the atmospheric pressure* (ACI, 1972). This curing takes place at about 120 to 180°C (250 to 350°F), usually in an autoclave. The duration of the steaming period is usually about 8 to 12 hours, including about a 2-hour presteaming period, about a 2-hour temperature-rise period, and a cooling period of not less than 15 minutes. Fly ash or some other pozzolanic material is used frequently in such concretes to achieve higher final concrete strengths.

The main advantage of steam curing under high pressure is the high early strength produced. For instance, the compressive strength at the age of 1 day can be as high as, or higher than, the 28-day strength under standard wet curing. Also, the sulfate resistance of such concretes is increased, and the shrinkage is reduced. On the other hand, the bond between reinforcement and concrete is reduced.

3.7 CALCULATION OF THE EFFECT OF TEMPERATURE ON CONCRETE STRENGTH

Exponential Model

Knowledge of the effects of temperature on concrete strengths in qualitative terms has served the construction industry to develop techniques for economical concreting in cold as well as tropical climates. It would still be useful both for cement manufacturers and cement users to have a quantitative description of these effects.

An approach to obtaining such a description is the use of the exponential cement model. With reference to Figure 3.18, one can examine separately the effects of temperature on the zero, first, and second stages of strength development. Since experimental data have supported this three-stage model for the standard curing temperature, it is a reasonable assumption that strengths of concretes cured at constant temperatures other than 23°C (73°F) will also be reproducible by generalization of the exponential model for various curing temperatures.

This generalization was based on the recognition that the strength development of a portland cement is affected by lowering (raising) the curing temperature in four different, somewhat interrelated ways:

1. Delays (speeds up) the beginning of the hardening t_s
2. Slows down (increases) the rate of chemical reactions between cement and water, and even more so the deceleration of these reactions, that is, reduces (increases) a_1 and a_2
3. Delays (speeds up) the time of transition t_d from the first stage of hydration (control by the rate of chemical reactions) to the second stage
4. Increases (decreases) the final strength f_o

The proper combination of these effects produces the overall strength of portland cement mortars and concretes at various ages and curing temperatures. The separation of the two control stages in the strength development (i.e., t_d) is important because the hydration during the first stage is influenced intensively by the curing temperature, whereas this influence during the second stage is not as great.

Calculated Values Versus Strengths Obtained Experimentally

The new characteristics of the kinetics of hardening, [i.e., the values of t_s, t_d, f_{ctd}, and f_{dt} (Fig. 3.18)] can be determined from experimental data by stepwise approximation in the same way as was done with the a_1 and a_2 parameters. The experimental data were taken from a paper by Klieger (1958). Accordingly, ASTM Types I, II, and III portland cements were used to make concrete for the determination of their compressive and flexural strengths. Two cements were used for each type. The nominal cement content of the concretes was 310 kg/m^3 (520 lb/cu yd), the air content was between 4 and 5%, including entrained air, and the water content was adjusted for each mixture to produce a slump of 50 to 100 mm (2 to 4 in.). The concretes were mixed and placed at constant temperatures of 4.44, 12.78, 22.78, 32.22, 40.5, and 48.86°C (40, 55, 73, 90, 105 and 120°F), respectively. This means that the strength results were obtained under isothermal conditions. Specimens were prepared for strength testing at ages of 1, 3, 7, and 28 days, 3 months, and 1 year.

For specimens that were broken at the age of 28 days or earlier, all concretes were cured continuously moist at the mixing and placing temperature until testing the strength. After the initial 28 days, the specimens remaining for tests at 3 months and 1 year were cured further continuously moist but at 23°C (73°F) until breaking. The compressive strength tests were performed on beam ends remaining from flexural tests as 150-mm (6-in.) modified cube tests. Further details concerning the concrete and tests can be found in Klieger's original paper (Klieger, 1958).

The kinetic characteristics obtained from Klieger's data were used to calculate compressive strength values at ages from 1 day through 1 year for concretes cured at six different temperatures from Eqs. 3.24, 3.25, and 3.26. The results are presented in Table 3.6 for Type I portland cement, along with the corresponding experimental data obtained by Klieger. It can be seen that the goodness of fit is quite good between the calculated and experimental strengths. The calculated and experimental compressive strengths developed by Types II and III cements show similar good fit (S. Popovics, 1987c). Further comparisons are shown in Figure 3.23.

Formulas were also obtained for several characteristics of the kinetics of strength development of the model cement as a function of the curing temperature and cement type. These are presented in Table 3.7. Table 3.6 also contains illustrative values of these characteristics of Type I cement for several temperatures.

Note that Klieger's experiments used here were performed at constant curing temperatures, the results presented above are valid only for such curing conditions. The applicability of the method for curing under variable temperature conditions is the subject of further research.

Interpretation of the Kinetics

The characteristics for the kinetics of three types of model cements in Table 3.7 indicate the following:

TABLE 3.6 Calculated and experimental strengths obtained at various curing temperatures, Type I cement[a]

Curing Temp (°C)	Source of Strength	a_1 (1/day)	a_2 (1/day)	t_d (days)	f_{ctd} (MPa)	f_o (%/100)	Compressive Strength (MPa)					
							1 day	3 days	7 days	28 days	90 days	365 days
4.44	Calc.	0.08	0.014	98.6	55.03	1.49	0.20	5.61	14.36	36.73	53.88	58.85
	Exptl.						0.56	5.11	14.98	36.26	49.84	54.11
12.78	Calc.	0.20	0.060	68.1	48.00	1.14	4.12	13.71	25.77	43.83	48.81	52.90
	Exptl.						4.06	13.51	26.25	43.75	52.25	55.93
22.78	Calc.	0.48	0.11	43.8	42.56	1.00	9.42	21.52	31.59	41.85	44.67	48.75
	Exptl.						9.87	21.84	31.43	42.35	44.52	50.33
22.78	Calc.	0.43	0.12	43.8	38.32	1.00	7.67	18.43	28.25	37.68	40.40	44.51
	Exptl.						8.05	18.20	28.42	38.08	42.00	45.15
32.22	Calc.	0.70	0.12	28.8	37.43	0.99	10.65	20.96	28.67	37.37	40.75	44.84
	Exptl.						10.43	21.63	29.12	36.68	39.90	43.68
40.56	Calc.	0.95	0.13	19.9	34.72	0.95	12.27	21.38	27.98	35.71	39.12	43.20
	Exptl.						12.95	22.40	26.95	33.04	36.75	41.02
48.86	Calc.	3.0	0.20	13.8	27.31	0.74	14.53	19.50	24.44	29.38	32.79	36.87
	Exptl.						14.28	19.88	24.08	28.77	33.11	37.24

Source: S. Popovics (1987c). (Copyright Cement and Concrete Research. Reprinted with permission.)
[a]1 MPa = 145 psi.

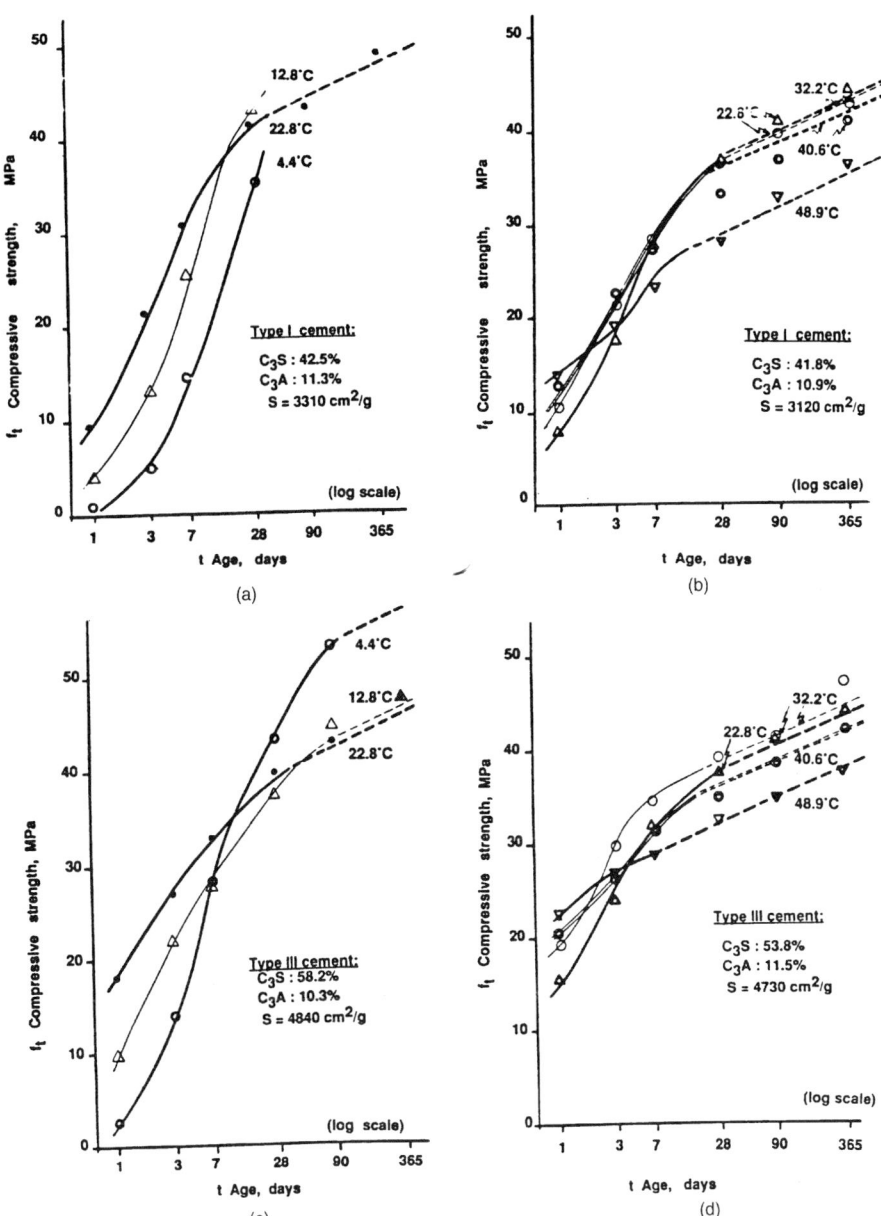

Figure 3.23   Comparison of concrete strengths calculated from the cement model (lines) to values obtained experimentally (points) for (a) and (b) Type I cements and (c) and (d) Type III cements. Solid lines represent strength development controlled by the rate of reactions, dashed lines represent diffusion control. (From S. Popovics, 1987c. Copyright Cement and Concrete Research. Reprinted with permission.)

TABLE 3.7 Predictive equations for the kinetics of strength development of portland cements[a]

Hydration Characteristics		Equations		
Name	Symbol	For Type I Cement	For Type II Cement	For Type III Cement
Beginning of hardening (days)	t_s	$\dfrac{215}{(t^0+10)^2} - 0.10$	$\dfrac{190}{(t^0+10)^2} - 0.12$	$\dfrac{165}{(t^0+10)^2} - 0.10$
Beginning of the diffusion control (days)	t_d	$120 \times 10^{-0.0192 t^0}$	$115 \times 10^{-0.0156 t^0}$	$122 \times 10^{-0.0235 t^0}$
Compressive strength developed during the first stage (MPa)	f_{ct}	$f_0 f_{28} \dfrac{100 - C_3 e^{-a_1(t-t_s)S/300} - (100 - C_3)\, e^{-a_2(t-t_s)S/300}}{100 - C_3 e^{-a_1(28-t_s)S/300} - (100 - C_3)e^{-a_2(28-t_s)S/300}}$		
Compressive strength developed during the diffusion control (MPa)	f_{dt}	$6.62 \log \dfrac{t}{t_d}$	$7.10 \log \dfrac{t}{t_d}$	$5.93 \log \dfrac{t}{t_d}$
Relative final compressive strength (%/100)	f_0	$1.0 - 33 \times 10^{-6}(t^0 - 29)^3$	$1.1 - 30.2 \times 10^{-6}(t^0 - 29)^3$	$1.0 - 30.2 \times 10^{-6}(t^0 - 29)^3$

Source: S. Popovics (1987c). (Copyright Cement and Concrete Research. Reprinted with permission.)

[a] t^0 = constant curing temperature, °C. The log is of 10 base. The other symbols are as in Eqs. 3.24 and 3.25. 1 MPa = 145 psi.

1. As expected, the rate of hardening of the C_3S (i.e., the a_1 parameter) is the largest for Types I and III intermittently and smallest for Type II. Note also that these rates increase rapidly with increasing curing temperature. This increase is more than an order of magnitude for each cement type as the curing temperature increases from 4 to 48°C (40 to 120°F).

2. The starting time of hardening in the cement model, t_s, is naturally related to the setting time of an actual portland cement (Fig. 3.18). For instance, t_s is influenced by the same factors, and in similar ways, as the setting time, including the rapid reduction of both with increasing curing temperature. Nevertheless, t_s is not the standard setting time of a portland cement at various curing temperatures. It is regularly less than the time of initial setting of an actual portland cement as measured by the standard Vicat method. At higher temperatures, it can be even negative. Thus one may consider t_s as a fictive time for the model, a relative indicator of the beginning of hardening at a given temperature as compared to that at standard curing temperature. Note that when the curing temperature is -10°C (12°F) or less, the model cement does not start hardening. The value of t_s affects primarily the very early strengths.

3. The time of transition in the cement model, t_d, from the first stage of hardening to the second stage (Fig. 3.18) shortens exponentially with increasing curing temperatures, as shown by the equations in Table 3.7. This reduction is the largest in the case of Type III cement and the smallest for Type II. Consequently, the hardening is rate controlled in Type II cement longer than in Type I, which, however, is still longer than in Type III for every temperature. This was recognized earlier but only in qualitative terms.

   t_d is also controlled, negatively, with the speed of hardening, that is, C_3S and C_3A contents, fineness, and so on, of the cement. Similar tendencies appear with the strength development of the various cements during the second stage; that is, the higher the rate of hardening, the smaller the strength increase in the second stage. Note that the strength development in the second stage, f_{dt}, is less sensitive to the temperature of curing than is the chemical reaction–controlled strength development. The t_d and f_{dt} values affect the strength at later ages.

4. Table 3.6 demonstrates that the concrete strength at the age of t_d (i.e., f_{ctd}) is not a constant value, as one might expect. It decreases with increasing curing temperature, although much less than t_d. Apart from the two extreme temperatures, the value of f_{ctd} is between 35 and 45 MPa (5000 and 6500 psi) for each temperature regardless of the cement type for the tested concretes. Note, however, that the ratio f_{ctd}/f_o seems to be constant for a wide range of temperatures for a given cement type. For instance, for Type I cements in Table 3.6, the average of these ratios is 38.7 with very little fluctuation. For Type II cements (with the ex-

ception of curing at 4.44°C) from the same test series, the average ratio is 34.9, and for Type III cements it is 37.5. This constancy seems to suggest that the switch from control by chemical reactions to diffusion control of the strength development of a given cement type takes place when the concrete strength reaches a definite percentage of the final strength f_o.

5. According to Table 3.7, the factor ω of Eq. 3.25 for f_{dt} may be a function of the cement type (or rate of hardening) but seems independent of the curing temperature.

6. The final strength f_o decreases with increasing curing temperatures approximately with identical rates for each cement type, as the pertinent equations in Table 3.7 as well as the numerical values in Table 3.6 indicate. Numerically, f_o is, again, a function of the speed of hardening. The faster the hardening, the lower the final strength. For instance, f_o is the highest for Type II cement and lowest for Type III cement, indicating the inverse effect of C_3S and C_3A contents on the final strength. This has also been demonstrated experimentally (S. Popovics, 1992). When the curing temperature is close to 29°C (85°F), the final strength developed by Type II cement is about 10% higher than that of Types I and III. The f_o equations in Table 3.7 also show that the final strengths change little as long as the curing temperature is not too far from 29 °C (85°F). However, the changes increase rapidly (with the third power) with further increases or decreases in the curing temperature. Carino and Lew (1983) also reported that the value of f_o decreases with increasing curing temperatures, but they recommended a linear relationship for this function the limits of validity of which are narower than those of the cubed form presented in Table 3.7.

The equations show that a low value of f_o reproduces numerically the strength-reducing effect of the higher (early) curing temperature on the f_{ct} portion of the concrete strength at later ages. The value of f_o primarily affects the strength at later ages.

The equations in Table 3.7 are approximations with restricted limits of validity. The restrictions originate from the intentional simplification of the formulas. For instance, the actual f_o value does not increase indefinitely as the cubed function indicates but rather levels off around 0°C as well as above 60°C, respectively. Such a curve, however, would require a much more complicated mathematical form. Also, the zero-stage mechanism was disregarded. Nevertheless, all these formulas seem to be useful tools between 4 and 50°C (40 and 122°F) for the estimation of several characteristics of the kinetics of hardening numerically that we could do only in qualitative terms up to now.

Formulas similar to the equations in Table 3.7 have not been found in the literature for the kinetics of hardening. They are novel; thus it is not surprising that they still have rough edges. For instance, there are a few deviations in the proposed a_1 and a_2 values from a smooth function of curing temperature. There are several reasons for such irregularities. One is the simplification of using one t_s value in the cement model instead of two: one for the first hardening component, C_3S, and another for the second component. The same is valid for some of the other kinetic characteristics. Because of this, the primary importance of the formulas presented is not so much the actual numbers they provide. It seems more important that they show the general approach for the establishment of similar but improved formulas for the determination of how the cement characteristics influence the t_d, f_o, and other values under various temperature conditions.

Carino–Tank Model (Carino and Tank, 1992)

The purpose of this model, called the *rate constant model,* is the same as that of the exponential model represented by the equations in Table 3.7, namely the description of the strength development under various constant curing temperatures. However, the rate constant model is simpler and less general. The mathematical form of this model is, as follows:

$$f_t = \frac{f_o k_T (t - t_s)}{1 + k_T (t - t_s)} \tag{3.27}$$

where k_T is an experimental parameter, the *rate constant* for strength development, at the curing temperature T, 1/day.

The differences between Eq. 3.27 and the extended exponential model are that Eq. 3.27:

- Uses a hyperbola rather than an exponential function for the description of strength development
- Treats the portland cement as a single, quasihomogeneous hardening material instead of a blend of two hardening components
- Does not distinguish between the hardening controlled by chemical reactions and hardening controlled by diffusion
- Does not take the compound composition of the cement into consideration

These simplifications restrict the limits of validity of Eq. 3.27. Nevertheless, Carino and Tank report fairly good agreement between strengths calculated from Eq. 3.27 and a group of experimental strengths for several different mortars and concretes. Here again, the results are valid only for such curing conditions. The applicability of the method for curing under variable temperature conditions is the subject to further research.

Original Maturity Concept

What Is Maturity?   Maturity is a mathematical model, a quantitative tool for the estimation of the strength of a concrete in terms of its temperature history. Although the exponential model has the potential to reproduce the strength development of concretes cured under various constant temperatures, the maturity method is for the estimation of the concrete strength even when the curing temperature is not constant. The maturity method is simpler and more practical than the exponential or hyperbolic models, although its reliability is limited. Also, this method does not reveal anything about the kinetics of the strength development.

A major application of the maturity method is the in-situ estimation of the compressive strength of concrete at *early ages*. Another application is the estimation of the strength of a concrete from the measured strength of the same concrete obtained under different temperature conditions. Examples are Eqs. 1.9 and 1.10.

Mathematical Form of Maturity   The concept of maturity is a combination of the age and temperature history of concrete for the quantification of this combined effect on the strength development of concrete. It was originally defined as the age of the concrete multiplied by the average curing temperature that it has maintained above freezing (Nurse, 1949; Saul, 1951). This product is called *maturity*.

The mathematical form of this definition of maturity is the maturity function:

$$M = \sum (T - T_0) \, \Delta t \qquad (3.28)$$

where M = maturity at age t
    T = average temperature of the concrete during time interval Δt
    T_o = constant datum temperature; temperatures below T_o are not considered in the calculation of the maturity.

When the equation of the temperature as a function of age $T = f(t)$ is known, Eq. 3.28 can be written as

$$M = \int_{t_s}^{t} [(f(t) - T_o] \, dt \qquad (3.29)$$

For instance, when the temperature is constant, the temperature function $f(t)$ becomes constant, k_T. Thus Eq. 3.29 becomes

$$M = k_T(t - t_s) \qquad (3.30)$$

where k_T = rate constant for the given curing temperature T
t = age
t_s = age when the hardening is assumed to start, that is, when M = 0

The graphical equivalent of Eq. 3.28 or 3.29 is the shaded area A shown in Figure 3.24, where concrete temperature is plotted against age. This is the area between the horizontal line representing the datum temperature and the temperature curve between t_s and t age limits. Saul recommended T_o = −10°C, Plowman T_o = −12°C for the datum temperature.

A more general way to express the maturity for varying temperature is the following:

$$M = \int_{t_s}^{t} k(T)\, dt \tag{3.31}$$

where $k(T)$ is the *rate constant* as a function of temperature.

Determination of the Maturity   The degree of maturity is normally determined by using *thermocouples* cast into the concrete pour and reading the temperature at intervals, or by *maturity meters* (Carino, 1991a) installed during placement of the concrete. In the first case, the thermocouples can be connected to a computer which integrates temperature with time and provides the maturity directly through digital display. The maturity meter consists of

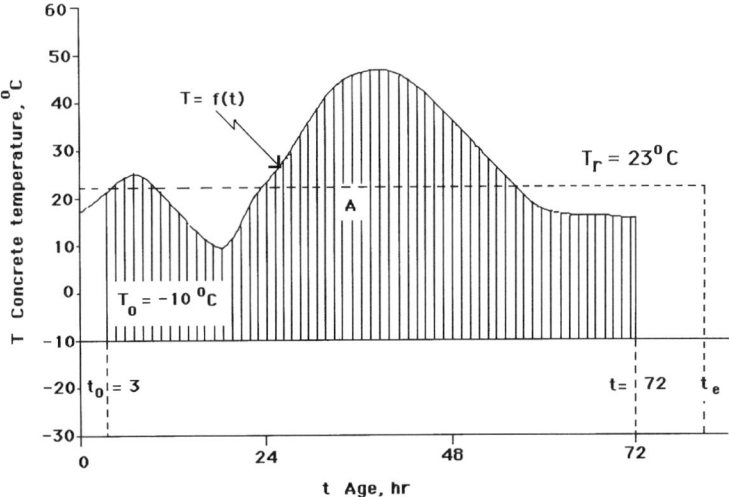

Figure 3.24 Shaded area A represents the maturity M according to Eq. 3.29, which is the Saul–Nurse definition. t_e, equivalent age; T_r, reference temperature.

a bridge circuit incorporating a thermistor that varies with temperature in a predetermined way. The circuit is arranged so that for any temperature above the datum temperature of, say, $-10°C$ ($14°F$), the bridge is unbalanced by the action of the thermistor. The integrator circuits contain a capacitor that electronically creates a pulse current which drives the display counter. Hence, if the concrete is at a high temperature, a large number of pulses are fed to the counter in a given time and the maturity will be greater than if the temperature is low and the pulses are fewer. Once the calibration curve or a strength formula, (i.e., a correlation between maturity and corresponding strength) has been established for a particular concrete mixture, the strength of that concrete can be estimated from the maturity of the concrete alone (Mukherjee, 1975; Hulshizer et al., 1984).

Strength Formulas   Saul (1951) formulated the relationship between strength and maturity in a qualitative manner called the *maturity rule:* "Concrete of the same mix at the same maturity (reckoned in temperature–time) has approximately the same strength whatever combination of temperature and time go to make up that maturity." For practical use, the maturity rule should be in quantitative form: that is, a formula or calibration curve is needed for the relationship between concrete strength and maturity. Any such formula can be based on the following assumptions, implied in the maturity rule:

- There is a unique *strength versus maturity* relationship for each mixture, similar to the formulas for constant curing temperature.
- The ultimate strength f_o is independent of the history of curing temperature.

On this basis, several different empirical formulas (exponential, logarithmic, etc.) have been offered for the estimation of the strength of a concrete under a variable curing temperature from the value of maturity. Two of these have been presented as Eqs. 1.9 and 1.10. A hyperbolic function for the strength versus maturity relationship (Bernhardt, 1956a; Chin, 1971) has also been recommended. A modified form of this is the following:

$$f = \frac{M - M_o}{(1/A) + [(M - M_o)/f_o]} \tag{3.32}$$

where f = concrete strength
M = maturity
M_o = maturity at which the strength starts developing
A = initial slope of the strength versus maturity curve
f_o = final or ultimate strength that is obtained after very long wet curing under the given temperature conditions

Later investigations found that when the maturity is calculated from Eq.

3.29, the validity of Eq. 3.32 is quite limited. The primary source of the error is the assumption of the independence of f_o from the temperature history. As the pertinent data in Table 3.6 as well as the f_o equations in Table 3.7 demonstrate, this assumption is an unacceptable oversimplification. Other experiments also show that for equal maturities, specimens exposed to low early-age temperature were weaker at early maturities and stronger at later maturities than specimens exposed to a higher early age temperature (Carino, 1991a; RILEM, 1956). Malhotra summarized the limitations of the original maturity method, as follows (Malhotra, 1980a):

• The age limits for Eq. 3.32 are about 3 to 28 days at normal temperature.
• The initial temperature of the concrete should be between 15 and 30°C.

Within these limits, experience with this method has been satisfactory (Bickley, 1975; Sadgrove, 1975; Lew and Reichard, 1978; Malhotra, 1973b). Nevertheless, several modifications have been recommended (Malhotra, 1971), none of which has been completely successful.

Another formula for the compressive strength versus maturity relationship was obtained from the combination of Eq. 3.7 and Eq. 3.29 (S. Popovics, 1969c):

$$f_{rel}^a = 100[100 - C_3e^{-3.6M/(273+T)} - (100 - C_3)e^{-0.36M/(273+T)}] \quad (3.32a)$$

where f_{rel}^a = relative compressive strength as percentage of the strength after long wet curing, %
       T = temperature of concrete, °C
      C_3 = computed C_3S content % by mass
      M = maturity, °C × day.

Strengths calculated with Eq. 3.32a for a given maturity are lower when the curing temperature is higher. Experimental results show similar tendency. This equation also indicates that the strength versus maturity relationship is affected by the cement type which is also in accordance with experience (S. Popovics, 1969c).

Concrete Properties Other Than Compressive Strength   The maturity concept is also valid for the estimation of some other *individual* concrete properties with the same limitations as for the compressive strength. Such properties are flexural strength, splitting and bond strengths (Villareal, 1975; Lew and Reichard, 1978), dynamic and static moduli of elasticity, and ultrasonic pulse velocity (Elvery and Evans, 1964). However, *relationships* between properties, such as compressive strength versus static modulus of elasticity, or dynamic modulus versus pulse velocity, are less dependent on the maturity. Consequently, where such relationships are involved with maturity, knowledge of details of the curing history is essential.

Modification of the Maturity

Selection of the Rate Constant   One way to improve the strength estimate from maturity is to use Eq. 3.31 for maturity calculations. For this one has to select a suitable function for the rate constant $k(T)$, and use this maturity in Eq. 3.32 for the strength calculation. For instance, when the temperature is constant, $k(T) = k_T = $ constant, the maturity is given by Eq. 3.30, substitution of which into Eq. 3.32 provides Eq. 3.27.

The next simplest case is when a linear function is assumed for the rate constant; that is,

$$k(T) = K(T - T_o) \tag{3.33}$$

where K = experimental parameter
   T_o = temperature corresponding to $k(T) = 0$

If the maturity is calculated with this equation, the resulting strength equation will be Eq. 3.32. In other words, use of Eq. 3.33 leaves all the restrictions of Eq. 3.32 intact. Improvement was reported, however, by use of the Arrhenius equation for the calculation of $k(T)$ (Copeland et al., 1960). The value of the activation energy for the concrete in question can be determined experimentally, for instance from compressive strength tests. Further improvement may be achieved by the consideration that the portland cement consists of two hardening components, as in the exponential model, and by determination of the activation energy for each.

Another recommendation is the following exponential formula:

$$k(t) = Ae^{BT} \tag{3.34}$$

where A and B are experimental parameters in $1/\text{day}$ and $1/°C$, respectively. It was reported that Eq. 3.34 produced as good agreement with a group of experimental data as the Arrhenius equation, considering portland cement as a quasihomogeneous material.

Equivalent Age and Relative Strength   Another direction for the improvement of strength estimation is the application of the concept of *equivalent age* of curing. This is defined as the duration of the curing period at the constant reference temperature T_r that would result in the same value of maturity as the actual temperature history of the curing. Mathematically, we have

$$t_e = \frac{\sum (T - T_o)}{T_r - T_o} \Delta t \tag{3.35}$$

where t_e = equivalent age at the reference temperature
   T_r = constant reference temperature

The equivalent age can also be expressed with the original maturity function Eq. 3.28 as

$$t_e = \frac{m}{T_r - T_o} \qquad (3.36)$$

Graphical illustration of Eqs. 3.35 and 3.36 is given in Figure 3.24.

Relative strength f_{rel} here is the ratio of the concrete strength f at a given age to the final strength f_o, or

$$f_{rel} = \frac{f}{f_o} \qquad (3.37)$$

Note that the relative strength concept was discussed and used in earlier sections.

It has been shown that the combination of the equivalent age and the relative strength together provide better correlation with concrete strength than does M calculated from Eq. 3.28. This is so because this approach does not use the assumption that f_o is independent of the temperature history. This recognition led to the following rewording of the maturity rule (Carino, 1991a): Samples of a given concrete mixture that have the same equivalent age and have had a sufficient supply of moisture for hydration will have developed equal fractions of their ultimate strength f_o irrespective of their actual temperature histories. This statement is in accordance with item 4 on p. 227.

Several empirical formulas have been published for the calculation of t_e from the history of the curing temperature. Another possible utilization of t_e for calculation of the concrete strength is to substitute t_e into one of the strength formulas for constant curing temperature, such as Eqs. 3.24, 3.25, and 3.26.

One More Modification (S. Popovics, 1990b) As discussed earlier, after a certain limit age, any further strength development is more or less independent of the curing temperature. That is, estimation of strength after lengthy curing, say, 90 days, requires calculation of the maturity only up to a certain age rather than of the full 90 days. A rough support for this hypothesis comes from the following experiment. Concrete was prepared with Type III cement, aggregate–cement ratio = 3 by mass, and water-cement ratio = 0.57 by mass. Some of the specimens were cured under standard conditions at 23°C for 16 days, then broken for compressive strength; the others were cured under standard conditions for 15 days, then put into a water bath, boiled for 20 hr, cooled down, and broken. Despite the more than 20% difference between the calculated maturity (312 versus 380°C-day) of the two series, the compressive strengths were practically the same, approximately 40 MPa (6000 psi).

The limit age for the calculation of the maturity may be the time when the control of the hydration by chemical reactions ends and diffusion control begins. This age depends on a number of factors (Table 3.7). Nevertheless, a suitable mathematical model (such as the exponential model) may provide approximate values of the the the beginning of the diffusion control.

3.8 OTHER MAJOR FACTORS AFFECTING THE STRENGTH OF CONCRETE

General

In addition to the obvious influence of the quality and quantity of the concrete ingredients and the curing temperature, there are many factors that can, and do, influence the strength of concrete to a smaller or larger extent. Major influencing factors are:

1. The concrete-making procedures, such as batching, mixing, delivering, placing, and consolidating the fresh concrete
2. Age at testing
3. Wetness of curing
4. Air content and porosity
5. Testing procedures, including the shape and size of the specimen, capping, type of the testing machine, and speed of loading

Concrete-making procedures and their effects on concrete strength are not discussed in this book. Some of the other effects are the topic of this section.

Concrete Strength Versus Age

It is anticipated, and experienced, that the strength of a concrete increases with the progress of the hydration, that is, with age. When the hydration process stops, the strength development also stops. As mentioned earlier, the hydration stops when any of the following three conditions occurs:

1. No more unhydrated cement is available in the concrete.
2. There is not enough free water.
3. Diffusion can no longer take place.

Regardless of the hydration process, no retrogression in strength is permitted at any age. Such retrogression is always the warning sign of deterioration of the concrete due to some attack from outside, faulty cement, reactive aggregate, or other cause.

Quite a few formulas have been offered for the prediction of concrete strength as a function of its age at testing but only for the cases when the hydration can progress freely. Such a formula is of particular value in the scheduling of form stripping, removal of shores, handling or erecting precast structural elements, and transferring prestressing forces.

Time–strength prediction formulas for isothermal curing (i.e., when the concrete is cured at *constant temperature*) can be divided into three groups. The strength is estimated:

1. Only from age
2. From age and some other characteristic(s) of the concrete composition
3. From age and an earlier strength

Strength Prediction from Age   An example is the following frequently used logarithmic formula:

$$f = a_l \log t + b_l \qquad (3.38)$$

where f = strength of concrete at age t
$\quad\quad\quad t$ = age
$\quad a_l$ and b_l = empirical parameters for the given concrete

The limits of validity of this formula are quite narrow. It has been shown, for instance by Figure 3.4 or 3.6, that any of the f versus log t curves can be approximated adequately by a straight line only in a very short time period in the semilog system of coordinates. Equation 3.38 also predicts, falsely, that the strength of a concrete approximates infinity with increasing age. In reality the strength can level off quite early. The age when this happens depends on the compound composition of the cement and as Figure 3.8 demonstrates, on the type of strength, among other factors. The gain of tensile strength stops earlier than the gain of compressive strength. The a_l and b_l parameters in Eq. 3.38 are to be determined by trial mixes for a given concrete and curing conditions.

The limits of validity are wider for the following hyperbolic function (Goral, 1956):

$$f = \frac{t}{a_h t + b_h} \qquad (3.39)$$

where the symbols are the same as in Eq. 3.38. When the age approaches infinity, the concrete strength calculated by Eq. 3.39 approaches the ultimate strength f_o as

$$f_o = \frac{1}{a_h} \qquad (3.40)$$

Two more comments:

1. Equation 3.39 appears to be applicable also to flexural strength, and
2. Experimental data support Eq. 3.39 better than Eq. 3.38 but not for concretes younger than 3 days.

The a_h and b_h parameters should be determined by trial mixes here, too.

Prediction from the Age and Other Characteristics   A generalization of Eq. 3.39 can be obtained from Eqs. 5.36 through 5.39, as follows (S. Popovics, 1990a). For Type I cements,

$$f_c = \frac{A_{I,c}}{(46/t + 2.5)^{w/c}} \qquad (3.41)$$

and $\qquad (3.42)$

$$f_{fl} = \frac{A_{I,fl}}{(15/t + 2.5)^{w/c}}$$

For Type III cements,

$$f_c = \frac{A_{III,c}}{(27/t + 6)^{w/c}} \qquad (3.43)$$

and $\qquad (3.44)$

$$f_{fl} = \frac{A_{III,fl}}{(7.5/t + 2.5)^{w/c}}$$

where f_c and f_{fl} = compressive and flexural strengths of concrete, respectively, in a stress unit

A = experimental parameter of Abrams' formula (Chapter 5) in the same units as f,

t = age of concrete, days

w/c = water-cement ratio by mass

Equations 3.41 through 3.44 are not valid for concretes younger than 3 days. Considerations discussed in Section 3.5 suggest that the lower limit of validity could be extended to ages younger than 3 days for these equations by using the expression $t - t_s$ instead of t, where t_s is the age when the strength development begins (Fig. 3.18). Equations 3.41 and 3.42 as well as Eqs. 3.43 and 3.44 can also be used for the estimation of the f_{fl}/f_c ratio in terms of age, water-cement ratio, and cement type (Eqs. 2.6 and 2.7, and Fig. 2.3). The application of Eqs. 3.41 through 3.44 for concrete proportioning is demonstrated in the User's Manual on the accompanying disk.

Strength Prediction from Age and from Measured Strength(s)   The exponential model (Sections 3.4 and 3.5) is a good tool to estimate the strength of a concrete for a given age from the strength measured at another age. Equations 3.38 and 3.39 can also be modified to include the result f_1 of a strength test obtained at a t_1 age. For instance, from Eq. 3.38,

$$f = a_L \log \frac{t}{t_1} + f_1 \tag{3.45}$$

A modification of Eq. 3.39 is the following (Branson and Christianson, 1971):

$$f_t = f_{28} \frac{t}{a_H t + b_H} \tag{3.46}$$

where f_t and f_{28} = compressive strengths at the age of t and 28 days, respectively

t = age, days

a_H and b_H = empirical parameters

The parameters a_H and b_H are functions of both the type of cement used and the type of curing employed. The use of normal-weight sand and light-weight coarse aggregate or all-lightweight aggregate does not appear to affect these parameters significantly (ACI, 1982a). The ranges of a_H and b_H, using both standard moist curing and steam curing and Types I and III cement are a_H = 0.67 through 0.98 and b_H = 0.05 through 9.25. Typical values for these parameters are also presented in the literature (Branson and Christianson, 1971; ACI, 1982a).

Another formula, the CEB-FIP formula quoted by Mehta and Monteiro (1993), can be written in the form:

$$f_t = f_{28} \exp \left\{ s \left[1 - \left(\frac{28}{t} \right)^{0.5} \right] \right\} \tag{3.47}$$

where s is a parameter that is a function of the cement type: s = 0.25 for Type I cement and 0.2 for Type III cement. The other symbols are as in Eq. 3.46.

It is likely that the strength at a given age can be predicted better from *two* earlier strength test results than from a single test. Creskoff (1945) was probably the first to recommend this in the linear form

$$f_{28} = \frac{a_c(f_3 + 2f_7)}{3 + b_c} \tag{3.48}$$

where f_3, f_7, and f_{28} = 3-, 7- and 28-day strengths, respectively
a_c and b_c = experimental parameters

A more general logarithmic form of the same concept is

$$f = \frac{f_2 - f_1}{\log(t_2/t_1)} \log \frac{t}{t_1} + f_1 \tag{3.49}$$

Effects of the Moisture Content (S. Popovics, 1986b)

The presence of water has considerable influence on the behavior of hardened concrete. For instance:

· Keeping the concrete wet by curing at early ages prevents cracking.
· The presence of water helps continue the hydration (Section 3.1).
· Water freezing inside the concrete can seriously hurt it.
· Free water can affect some other engineering properties of concrete.

The last item includes the *strength* of concrete. As illustrated in Figure 3.3 for several cases, the rule is that the duration and magnitude of strength development are longer and larger, respectively, in a wet-cured concrete than in a concrete that is dry cured. The strength finally stops increasing after awhile, as discussed in Section 3.1 (Gilkey, 1937; U.S. Deptartment of the Interior, 1981).

There is, however, an exception to this rule; namely, that the compressive strength of a portland cement concrete is higher when the specimen is dry, or a surface layer is dry, when tested than that of the same specimen in a fully or surface-saturated condition (Munday and Dhir, 1984; Pihlajavaara, 1965; Bloem, 1968). Strength reduction caused by the presence of moisture has also been observed in rocks (Jumikis, 1983) and porous glass (Pihlaja-vaara, 1964). Bilgeri et al. (1991) report that such strength reduction is small, however, even negligible when German blast-furnace slag cements are used.

A hypothetical explanation for the compressive strength of dry concrete being higher than that of a comparable wet concrete is based on the observation that drying decreases the volume of the hardened cement paste. The volume reduction is caused by surface tension that increases in water-filled small pores during drying, reducing the average distance between surfaces in the hardened cement gel (Wittmann, 1968). This, in turn, increases the secondary bonds between these surfaces, thereby increasing the strength, provided that the paste specimen is dried in such a careful way as to avoid excessive stresses during drying (Powers, 1962). Since rewetting the hardened paste causes volume increase, and therefore an increase in the average distance between surfaces of the cement gel, this would explain the lower strengths of wet cement paste and concrete specimens.

Another hypothetical explanation is based on internal pore pressure which develops in a wet concrete during the compressive strength test, as follows: The increasing external load on the concrete specimen during strength testing develops an increasing internal pressure not only on the solid components of the concrete but also on the liquid in the pores, trying to squeeze the liquid out of the specimen. Since, however, the migration of the liquid is not free due to the smallness of the capillary pore sizes, the hindrance produces a pressure on the contacting pore walls which increases as the external load on the specimen is increasing. This pore pressure intensifies the crack propagation in the concrete, which then reduces the magnitude of the external load the specimen can carry; that is, the measured compressive strength will be less in the presence of moisture in the specimen. This explanation is valid, strictly speaking, for the case when the specimen is fully saturated since even a small percentage of air in the pores diminishes the changes in pore pressure caused by loading. Note, however, that if the concrete is only partially saturated, the air squeezes out from the pores readily and the specimen may become saturated during the test. The higher the degree of saturation, the sooner full saturation is obtained, therefore the lower the measured compressive strength.

Effects of the Moisture Gradient

Neither of the explanations above takes the *distribution* of the free moisture in the concrete into consideration. It is thereby implied that the *average* of the moisture content in the concrete is the strength-controlling factor. The first indication that this is not true, that is, that the moisture *distribution* in the specimen does also have an effect on concrete strength has come from *flexural* tests. Even a short, say, 30-minute pretest drying produces a sizable reduction in the flexural strength (Walker and Boloem, 1957a). The explanation for this is that drying immediately before testing produces a moisture gradient in the specimen; this creates tensile stresses in the surface layer, which reduces the measured flexural strength of the concrete. This explanation is supported by the observation that sudden cooling of a beam, which produces similar tensile stresses in the surface layer, also creates a reduction in the flexural strength (S. Popovics, 1967a).

The effect of the distribution of moisture on the *compressive strength* has received less attention. This effect can be quite sizable, as has been demonstrated experimentally (S. Popovics, 1986b). In these experiments a Type I portland cement, natural sand, and crushed stone of 12.5-mm (0.5-in.) maximum particle size were used. Some series contained a chloride-free accelerator (S. Popovics, 1987d, 1990c). Concretes that had an epoxy as admixture were also tested.

The aggregate–cement ratio was kept constant as 4.75 by mass in all mixtures, but the amount of mixing water was adjusted so as to produce practically the same consistency: medium plastic for all mixtures. Immediately after mixing, 50-mm (2-in.) cubes were prepared essentially according to ASTM C109 for testing the compressive strengths. Cubes were selected for the investigation because the relatively small size of the specimen accentuates the effect of curing on the compressive strength.

All specimens were kept in the molds for 24 hours but after that the specimens from each batch were divided into eight groups. Each group was cured under different moisture conditions. The curing temperature for all the specimens was the constant standard 23 ± 1.5°C (73 ± 3°F).

The applied curing methods are presented in Figure 3.25. In this figure M stands for moist (fog-room) curing, A for air curing, and W for curing in lime water. The lengths of curing periods of M, A, and W are represented by shaded, blank, and crosshatched areas, respectively. Accordingly, CM and CA represent continuous moist and continuous air curings, respectively, until testing. $M + 3$ stands for an M continuous moist curing during the first 25 days followed by a 3-day finishing moist curing. $A + M$ represents alternating air and moist curings in the periods shown, in ending with a 4-day finishing moist curing. Conversely, $M + A$ represents alternating moist and air curings ending with a 4-day finishing air curing. $A + W$ and $W + A$ are similar to $A + M$ and $M + A$, respectively, except that moist (fog-room) curing was substituted for by curing in lime water.

Compressive strengths were determined at the ages of 3, 7, and 28 days from each of the first four groups and at 28 days from each of the last four

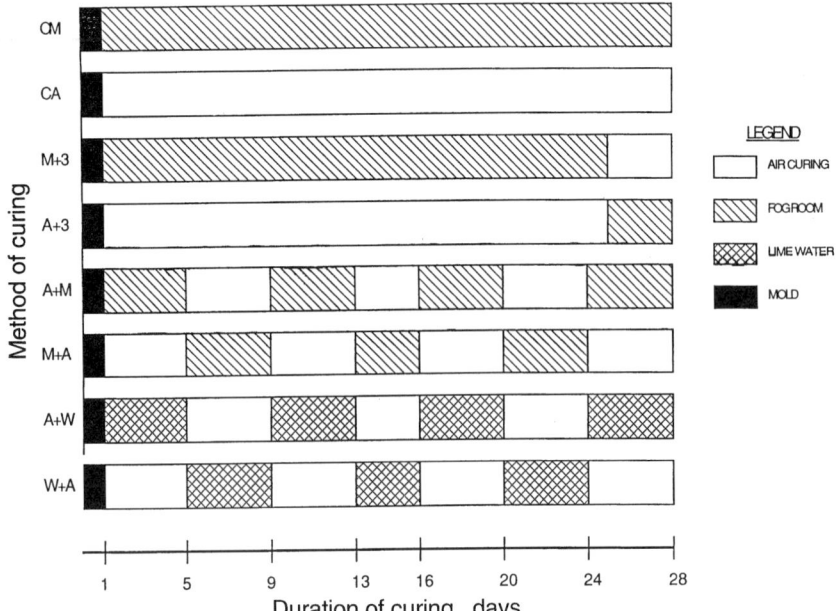

Figure 3.25 Eight methods of curing used in the experiments in Figure 3.26. (From S. Popovics, 1986b. Copyright ACI. Reprinted with permission.)

groups. Three specimens were tested at each age and the average of the three results was used as "compressive strength." Specimen weights were also determined before the strength tests.

Typical strength results are presented in Figure 3.26. Additional results can be found in the literature (S. Popovics, 1986b). These serve as a basis for the analysis presented below.

1. The highest compressive strengths of the concrete specimens were produced at any age by continuous moist curing followed by 3-day air curing before the break, that is, by M + 3 curing. These strengths were regularly higher at any age than the concrete strengths obtained after CM curing. At the same time, the unit weight of the concrete after M + 3 drying was about 2% lower. Continuous dry curing (CA) reduced the strengths but A + 3 curing provided even lower strengths, although the unit weight of the wetted concrete was about 1% higher. In the latter case the 28-day control strengths were lower then the 7-day strengths (Fig. 3.26).

2. The strength increases of concrete specimens due to 3-day finishing air drying are regularly greater than the strength-reducing effect of 3-day finishing rewetting. For instance, the differences between the strength after M + 3 curing and the comparable strength of the same mixture after CM curing at the same age are more than 7 MPa (1000 psi) for most concretes at both 7 and 28 days. The strength reductions produced by the A + 3 curing compared to the CA curing are smaller but still significant.

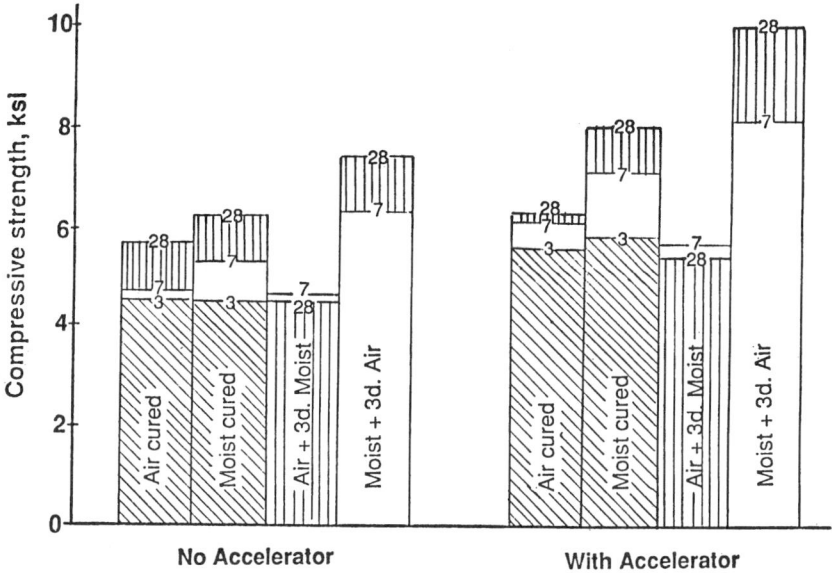

Figure 3.26 Effect of curing method on the compressive strength of concrete. 1 ksi = 6.90 MPa. (From S. Popovics, 1986. Copyright ACI. Reprinted with permission.)

3. When the periods of air curing and wet curing alternate in every 3 or 4 days, the effects of the curing method applied during the last few days before testing the compressive strength appear less than those when the change in curing method occurs only once, namely a few days before the strength test.

4. No significant difference was found between the strengths of specimens cured in fog room and those cured under lime-water.

5. The use of the chloride-free accelerator produced approximately 25% strength increase, regardless of the curing method.

6. Four days of air curing, after the first 24 hours in the mold, produced a weight loss of approximately 2.5%, whereas the same air curing after 24 days of alternating wet and dry curings produced about a 2% weight loss. Four days of wet curing after the first 24 hours in the mold produced about a 1% weight gain; whereas the same wet curing after 24 days of alternating wet and dry curings produced about a 2% weight gain. These data confirm the conventional wisdom that there is more free moisture in concrete at early ages than later. No significant difference was found between the weights of concrete specimens cured in a fog room and those cured under water.

The test results presented, along with many others, show unequivocally that the concrete compressive strength is influenced not only by the moisture content but also, significantly, by the moisture distribution in the concrete specimen. A measure of the difference in moisture content from one point to another is the *moisture gradient*. If this is defined as positive in the direction of increasing moisture content, it appears that negative gradients in the direction from outside to inside of the concrete decrease the compressive strength, whereas positive ones increase it.

Neither of the two hypothetical mechanisms mentioned earlier concerning the change of concrete strength by curing can explain the large effect of moisture gradient on the compressive strength. A more suitable possible mechanism is described below.

In the case when the moisture gradient is positive from outside to inside in the concrete, the outside layer wants to shrink because it is dryer than the inside portion of the specimen. This shrinkage is restricted, which develops a lateral biaxial compression on the restricting inside part of the specimen. This, then, increases the uniaxial compressive strength in the third direction. The opposite happens when the moisture gradient is negative. Namely, the outside layer, having a higher moisture content, tries to expand more than the inside, which creates a lateral biaxial tension in the specimen. This then reduces the compressive strength in the third direction.

One could also hypothesize, as follows: Since the permeability of a cement paste is greater in a dryer state than in a saturated state, the pore liquid can move more easily in the pores of a drying concrete under pressure toward the external surface, creating less biaxial tension. This then shows as an increased compressive load-carrying capacity of the concrete in the third direction. The problem with this hypothetical mechanism is that it is difficult to apply it to explain the strength-reducing effect of finishing rewetting. A quantitative determination of the moisture distribution in the concrete specimens (Nilsson, 1980) would probably indicate which mechanism is correct; such measurements have not, however, been made with these specimens.

The practical significance of these results is that it contributes to a better interpretation of the results of compressive strength tests. Also, it indicates that the fluctuation of strength results can be reduced by tightening the curing specifications. From a theoretical standpoint, the demonstrated large effects of drying and rewetting on the compressive strength of concrete point to the significant role of secondary bonds in the strength of hardened portland cement pastes.

Effects of the Specimen Shape and Size on the Measured Strength

One of the factors that make the strength of concrete so elusive is the effect of the shape and size of the specimen. Specimens made of the same concrete but having *shapes* of different geometries, such as cube versus cylinder, or even different sizes of the same geometry, such as small cube versus large cube, produce more or less different strengths. Specifically, the compressive strength measured according to European standards (on cube) is higher than the strength of the same concrete measured according to American standards (on cylinder). Thus conversion of strengths from one specimen shape to another, and/or from one size to another, is needed, if for nothing else, for better international cooperation. What makes this situation confusing is that there is no reliable correlation between these measured strengths, even when it comes to limited cases, such as cubes and cylinders, as Tables 1.1 and 1.2 demonstrate for compressive strength.

It is shown later that statistics can explain, at least qualitatively, some of the effects of specimen size on various concrete strengths. Also, it was shown in Section 1.2 that in the case of compressive strength, the friction between the steel compression plates of the testing machine and the concrete specimen also has a major role in explaining the effect of specimen shape (RILEM, 1997). Note that a relatively large number of studies have been published on this topic. A limited research of the literature revealed more than 50 papers on the topic, starting in 1903.

Empirical Approach   The relationships presented in Tables 1.1 and 1.2 for strength conversions were obtained from results of comparative experiments. Another, more general relationship is that of Neville, offering the following empirical relationship for compressive strengths (Neville, 1966):

$$f = f_{150} \left[0.56 + 0.7 \frac{d}{(V/6h) + h} \right] \qquad (3.50)$$

where f = compressive strength obtained with a cylindrical, cubic, or prismatic specimen, psi

f_{150} = compressive strength of the same concrete obtained with a 6-in. (150-mm) cube, psi

V = volume of the specimen, in^3

h and d = height and maximum lateral dimension of the specimen, respectively, in.

This formula is not valid for specimens such as very slender cylinders where the lack of stability is the cause of failure, or for cases when the maximum particle size of the aggregate is larger than one-third of the smallest dimension of the specimen. An important fault of Eq. 3.50 becomes obvious when it is used to estimate the strength for a 150-mm (6-in.) cube: The formula provides $f_{150} = 0.8878f_{150}$, which is an obvious contradiction. The reason for this is not clear, especially since experimental results for other specimen sizes support Eq. 3.50 quite well for all three types of specimens.

Equation 3.50 can be further simplified for specimens having cross-sectional areas close to circle or square (S. Popovics, 1967d). One of these formulas is:

$$f = f_{150}\left(0.56 + \frac{0.7}{0.15d + h/d}\right) \tag{3.51}$$

Further simplification, for cases where the effect of slenderness can be disregarded, results in

$$f = \frac{f_{150}}{d^{0.095}} \approx 0.85f_{150}\left(\frac{6}{d}\right)^{0.095} \tag{3.52}$$

where the symbols are the same as in Eq. 3.50. It follows that

$$f_{rel} = \frac{f}{f_{150}} = \left(\frac{6}{d}\right)^{0.1} \tag{3.53}$$

represents a sort of upper limit for this relationship, as shown in Figure 3.27. This formula has the advantage that it provides the value of 1.0 for f_{rel}, as it is supposed to when $d = 6$ in. (150 mm).

When the geometry of the specimen remains the same, only the *slenderness* is variable, such as cylinders where the h/d ratio is the only variable, the rule is that the more slender the specimen, the lower the measured compressive strength. This is true, however, up to the point where the stability starts controlling the failure. Factors for correcting compressive strength of one slenderness to another of geometrically similar specimens are available in the literature, such as in ASTM C42–94 for cylinders. According to this standard,

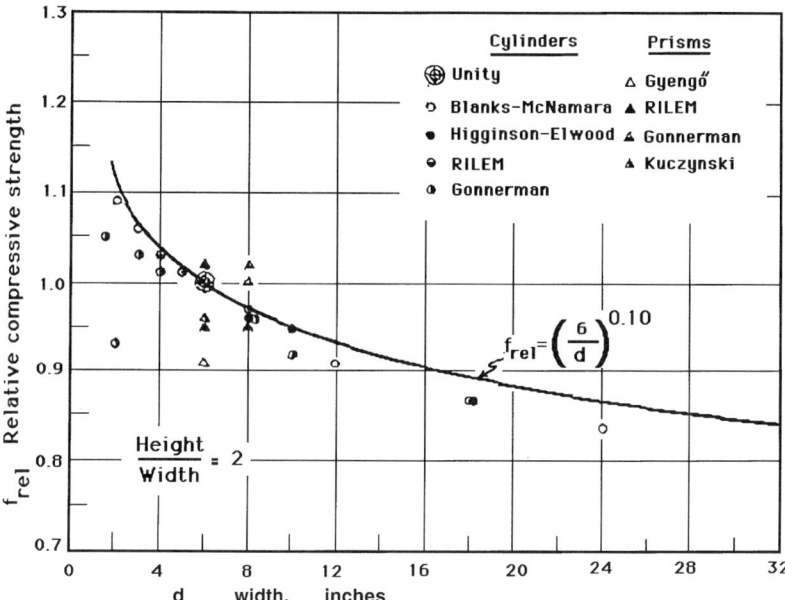

Figure 3.27 Influence of the size of prism and cylinder, respectively, on the measured compressive strength of concrete. 1 in. = 25.4 mm.

if the h/d ratio of the specimen is less than 1.94, the following correction factors should be used:

h/d	1.75	1.50	1.25	1.00
Strength correction factor	0.98	0.96	0.93	0.87

Experience has also shown that under usual circumstances the effect of slenderness becomes slight when this ratio is larger than 2.

When the *size* of the specimen is the variable, such as in the case of cubes of different sizes or cylinders of different sizes but with the sam h/d ratio, the rule is that the larger the specimen, the smaller the measured strength f. An empirical equation for this relationship for cubes is (Voellmy, 1957–58)

$$f_d = \frac{9 + 0.75d}{4 + d} f_{20} \tag{3.54}$$

where f_d = compressive strength expected from a cube of d cm, in a stress unit

f_{20} = compressive strength obtained from a 20-cm (8-in., 200-mm) cube, in the same stress units as f_d

d = cube size, cm

Another illustration, for cylinders of $h/d = 2$, is the following (U.S. Department of the Interior, 1981):

d [mm (in.)]	50 (2)	75 (3)	150 (6)	200 (8)	300 (12)	450 (18)	600 (24)	900 (36)
Rel. f (%)	109	106	100	96	91	86	84	82

All these data of the size effect on compressive strength are based on specimens made in molds. When cube specimens were *cut out* from a larger mass of concrete, the strengths obtained were found independent of the cube size (Voellmy, 1957–1958). The explanation offered is that the size effect discussed above is due to the differing hardening conditions in smaller and in larger molded specimens (temperature, shrinkage, etc.). Indeed, the differences in the hardening conditions may be a factor in the case of cube specimens. However, this is not the full answer for the size effect. Compressive strengths obtained with stone cubes also show size effect (Bonzel, 1959). Furthermore, statistical considerations, presented below, do indicate further effects of the size of the specimens. Another effect of the specimen geometry, the friction between the concrete specimen and the steel plattens of the testing machine, has also been demonstrated experimentally (Section 1.2). So the problem of size effect on the compressive strength of concrete can still use further research.

Flexural and *other strengths* are also affected by the size and shape of the specimen. These effects have also been investigated, although to a much smaller extent than the compressive strength. For instance, presently the commonly used specimen for flexural strength determinations is a beam of 150 ×150 mm (6 × 6 in.) in cross section and 525 mm (21 in.) long that is tested on a span of 450 mm (18 in.) (Section 2.1). An increase in the depth of the beam from 100 to 200 mm (4 to 8 in.), while keeping the length–depth ratio at 6 causes a reduction in the apparent strength calculated by Eq. 2.1 of about 20% (Meyer, 1963b; Wright, 1952). Also, as the span length of a 200 × 200 mm (8 × 8 in.) beam is increased from 500 to 1000 mm (20 to 40 in), the flexural strength calculated by Eq. 2.1 decreased by about 10% (Voellmy, 1957–1958).

A somewhat different approach used by Reagel and Willis in investigating the same problem resulted in smaller effects (Reagel and Willis, 1931). Their experiment consisted of making and testing 768 carefully prepared concrete beams by four laboratories where only the width, height, and length of the beams were variables. The maximum particle size of the limestone aggregate was 1 in. (25.4 mm). The cement was a standard portland cement, equivalent to Type I, the water-cement ratio was 0.703 by mass. The specimens were wet-cured and tested in three-point loading according to the pertinent ASTM standards at the age of 28 days. The strengths were calculated by Eq. 2.1. The analysis of the results reveals linear relationships between the flexural strengths and the geometric characteristics of the beams. The slope of each of these relationships is then an indicator of an individual size effect on the flexural strength. When the overall average flexural strengths are plotted against the height of the beam, width of beam and length of the span, respectively, the following approximate slope values were obtained:

In U.S. customary units,

$$s(f_{fl} \text{ vs. } h) = -16.5 \text{ psi/in.} \tag{3.55}$$

$$s(f_{fl} \text{ vs. } w) = 2.33 \text{ psi/in.} \tag{3.56}$$

$$s(f_{fl} \text{ vs. } l) = -0.583 \text{ psi/in.} \tag{3.57}$$

in SI units,

$$s'(f_{fl} \text{ vs. } h') = -4.48 \text{ MPa/m} \tag{3.55a}$$

$$s'(f_{fl} \text{ vs. } w') = 0.633 \text{ MPa/m} \tag{3.56a}$$

$$s'(f_{fl} \text{ vs. } l') = -0.158 \text{ MPa/m} \tag{3.57a}$$

and in metric units,

$$s''(f_{fl} \text{ vs. } h'') = -4.57 \text{ kg/cm}^2/\text{cm} \tag{3.55b}$$
$$s''(f_{fl} \text{ vs. } w'') = 0.645 \text{ kg/cm}^2/\text{cm} \tag{3.56b}$$
$$s''(f_{fl} \text{ vs. } l'') = -0.161 \text{ kg/cm}^2/\text{cm} \tag{3.57b}$$

where $s(\cdot)$, $s'(\cdot)$, and $s''(\cdot)$ = slope of the straight line representing the effect of the geometric characteristic on the flexural strength, psi/in., MPa/m, and kg/cm²/cm, respectively
   f_{fl} = flexural strength of the concrete, psi, MPa, or kg/cm², respectively
   h, w, and l = height, width, and length of span of the beam, in., m, or cm

The primary reason for presenting the data and formulas above is not to provide exact numbers but rather to indicate trends. So the trend is that there is:

1. A definite decrease in the measured flexural strength with an increase in the depth of the beam.
2. A smaller increase with an increase in the width.
3. An even smaller, perhaps negligible strength decrease with an increase in the length of span.

When flexural strengths with beams of low length/depth ratio are tested and the strengths are calculated in the usual manner by Eq. 2.1, the results are further distorted by the prevailing high shear stresses (Mullin and Knoell, 1970).

Statistical Approach As mentioned in Section 2.4, the weakest-link principle has been used as one approach for estimation of the effect of the size of specimen on the measured strength of concrete. The Griffith criterion implies that the measured strength should decrease as the area under maximum stress increases, because as this area increases, so does the number of flaws and consequently, the chance of finding a more damaging flaw. The numerical form of this problem is identical to the statistical study of the distribution of the smallest values in samples of size n drawn from a population having some cumulative distribution function $F(x)$ (Orchard, 1974). Peirce (1926) was probably the first to recognize this statistical identity in his study of the strength of cotton threads. Since then, quite a few numerical solutions have been presented for this problem based on different distribution functions. A good evaluation of some of these methods was published by Epstein (1948). The most popular method was developed by Weibull (Weibull, 1938, 1939; Hudson and Fairhurst, 1971), who was the first to apply the weakest-link principle to the strength of brittle materials.

For the sake of simplification of the calculations, Weibull assumed that $F_o(f)$, the probability of a breakage of a "link" as a function at stress f, is given by $F_o(f) = 1 - \exp[-(f/f_o)\gamma]$, where f_o and γ are constant characteristics of the material. On this basis he calculated that two specimens of a material having different volumes V_1 and V_2 have equal risks of rupture for homogeneous states of stresses f_1 and f_2 when the ratio of these stresses at failure is

$$\frac{f_1}{f_2} = \left(\frac{V_2}{V_1}\right)^{1/\gamma} \tag{3.58}$$

where the factor γ, called the uniformity factor, increases with the number of flaws in the material.

Equation 3.58 shows qualitatively that the lower the number of flaws per unit volume in the specimen (i.e., the higher the concrete strength) the greater the effect of the specimen size. However, the quantitative comparison of this equation to experimental data would be possible in complete generality only if the measured strengths were independent of the shape of the specimen, which they are not. If, however, Eq. 3.58 is restricted to geometrically similar specimens (i.e., where the height/diameter ratio and the shape of the cross section are constant) the equation can be written in the form

$$\frac{f_1}{f_2} = \left(\frac{a_2}{a_1}\right)^{3/\gamma} \tag{3.59}$$

where a_1 and a_2 are values characteristic of the diameter of specimens V_1 and V_2, respectively. This equation with $\gamma = 31.6$ is identical with the empirical Eq. 3.53, which supports the validity of these equations.

Similarly, the s_1/s_2 ratio of standard deviations of strengths obtained on two sizes of geometrically similar specimens under uniform stresses is (McClintok 1966)

$$\frac{s_1}{s_2} = \left(\frac{a_2}{a_1}\right)^{1/\gamma} \tag{3.60}$$

where the symbols are as in Eq. 3.59.

Equation 3.60 indicates that the scattering of strength results about the average should decrease with a decrease in the average strength and with an increase in the volume of the specimen. A consequence of this is that more specimens should be tested for statistical evaluation of strength results, for equal reliability, when the specimens are small.

These equations can also be generalized for nonhomogeneous states of stresses, as Frankel showed for flexural strengths of cement mortars (Frankel, 1948). Analysis of pertinent data reveals that most of these statistical predictions, concerning size effect, scatter of strength, and so on, are only qualitatively supported by experiments in full generality. Quantitatively, there are discrepancies between calculated and measured values. One reason for this is, of course, that the measured strength is not independent of the shape of the specimen. Others are: the restricted reliability of the present form of the Griffith criterion; the assumption that the hardening conditions are identical in specimens of differing sizes and shapes; the use of certain assumptions, such as the weakest link concept; and the supposition that the cracks do not interfere with each other. All these represent oversimplifications as far as the failure of concretelike materials are concerned. Thus these statistical theories can be considered only marginally successful.

Approach Based on Fracture Mechanics Bazant and co-workers approached the size effect on various strengths from the viewpoint of fracture mechanics (Bazant 1976; Bazant and Sener, 1988; Bazant et al., 1991). Experimental strength data appear to support his size-effect formula within practical limits; that is, the approach is a rational foundation for the quantitative estimation of this effect. The extrapolation of his formula to infinite specimen size also provides a material parameter in addition to fracture toughness (Section 2.4).

Effects of Test Conditions on the Measured Strength

There are also other factors that can influence the *results* of strength tests even when the actual strength of the concrete remains unchanged. Most of these factors produce an apparent reduction in strength. Three examples are presented below. Additional cases are presented elsewhere (Richardson, 1991).

Deviations from the Test Conditions Specified   (Gonnerman, 1924) Any irregularity in the testing machine (poor calibration, uneven bearing faces, etc.) can cause major errors in the result of a strength test. Not centering the specimen in the testing machine is another irregularity, although small eccentricities do not reduce the measured *compressive* strength seriously. Neither were major strength reductions observed when the loaded ends of the specimen were not quite parallel, or if they were parallel but the axis was inclined somewhat.

More serious error source is the roughness or unevenness of the loaded surfaces of the compression specimen. These produce stress concentrations which cause reduction in the measured strength. For instance, a *convexity* of 0.01 in. (0.25 mm) produced apparent strength reduction of about 35% in two test series and a convexity of 0.05 in. (1.25 mm) of about 60%. Such strength reduction increases with increasing concrete strength. The effect of *concave* ends is smaller, although it can still be significant. Thus plane end surfaces are essential to eliminate improper reductions in the strength results (ASTM C617–1994). If full contact between *flexural* specimens and the load-applying blocks and supports is not obtained, capping, grinding, or shimming with leather strips is required at these points by ASTM C78–94.

Rate of Loading   Strengths of most materials are affected by the rate of loading or rate of strain. Concrete is no exception. The higher the rate of loading during the testing, the higher the measured concrete strength. This observation is valid not only for compressive strength (Bischoff, 1991) but also for other strengths (Wright, 1952, Komlos, 1969, Raphael, 1984), although not necessarily to the same extent. Within the range of customary testing procedures, however, this effect is not large (Halasz, 1975). As compared with a normal rate of loading, which is about 5 kPa/s (35 psi/sec), loading at 0.15 kPa/s (1 psi/s) reduces the indicated strength approximately 12%, and loading at 145 kPa/s (1000 psi/sec) increases the indicated strength approximately 12% (Troxel et al., 1968). Beyond a critical strain rate, however, large strength increases occur. This critical strain rate is approximately 5/s for tension, and 60/s for compression (Ross, 1995).

Several formulas have been recommended for the reproduction of the effect of rate of loading on concrete strength (Soroushian, 1986). One of these is the following (Jones and Richard, 1936):

$$f = f_1(1 + k \log R) \tag{3.61}$$

where f = measured strength at the given rate of loading
   R = the given rate of loading
   f_1 = strength at a rate of 0.15 kPa/s (1 psi/sec)
   k = experimental parameter

The limit of validity of this equation, according to one test series (McHenry and Shideler, 1956), is approximately between 0.05 and 10,000 MPa/min (7 and 1.5×10^6 psi/min). It appears that Eq. 3.61 is also applicable to flexural strength (McNeely and Lash 1963).

Note that Poisson's ratio of concrete at a certain level decreases with the rate of compressive loading but the inverse seems true in tensile loading (Takeda, 1972).

Several explanations have been offered for the effect of rate of loading. One is that the higher the rate of loading, the less likely it is that the cracks can find the weakest paths to propagate (Section 2.4). A further source of rate sensitivity is creep (or stress relaxation). The creep effect is negligible at very fast loading rates where inertia (or wave propagation) has a major contribution. The creep effect becomes important only at sufficiently slow loading rates where the inertia effects vanish (Bazant, 1995). The presence of free water in the concrete can also increase the rate effect (Rossi, 1991, 1992; Reinhardt, 1990; Ross, 1995).

Effects of the Testing Machine   The two most popular types of machines for testing the compressive strength are the hydraulic and the screw-gear machines. For hydraulic machines it is customary to adjust the pumping rate to obtain the desired rate of load increase. This cannot be done with screw-gear machines. A fixed setting of a screw-gear machine results in a variable increase in load per unit time. The rate increase depends on the characteristics of the driving motor, the rigidity of the testing machine, the size of the specimen, and the modulus of elasticity of the concrete (Troxell et al., 1968). The differences between the two types of testing machines and variations of these characteristics can influence the results of the strength test. Also, the movement of the head of a rigid machine cannot follow the compression of the concrete specimen under load, thus a higher strength is recorded. On the other hand, a less rigid machine records a lower load at failure, often accompanied by explosion (L'Hermite, 1955).

4

STRUCTURE OF THE HARDENED CEMENT PASTE AND CONCRETE

The internal structure of concrete, like many others composites, is quite complex. The length scale of this structure ranges from the nanometer scale of the hydrated calcium silicates to the centimeter scale of the concrete. Since the internal structure defines most of the concrete properties, several aspects, mostly those related to concrete strength, are discussed below.

4.1 STRUCTURE OF THE FRESH CEMENT PASTE

The structure of the fresh cement paste has major effects on the structure and properties of the hardened paste. As Powers put it, "whatever the structure of a given specimen of concrete may be, it is one that grew out of the structure of the original mixture and one that incorporates the gross flaws and anisotropisms that developed before the mixture became too firm to settle" (Powers, 1968). Concrete thus behaves like a living entity—plant, animal, or human—in that many of the basic characteristics of the entity are established by the genes from the minute of conception.

The initial structure of a fresh cement paste depends on the volume fraction, particle size distribution, chemical composition of the cement particles, and the presence of admixtures (Cement and Concrete Association, 1976). This structure is such that it makes the fresh paste plastic, that is, makes it capable of being molded without losing continuity and retaining a shape. This is so because the cement particles and air bubbles are dispersed in an aqueous solution, and especially because the interparticle forces tend to hold particles

254

together while preventing actual point-to-point contact. This also causes the *flocculent state* in fresh cement paste. More specifically, the plastic state is due to the coexistence of forces of attraction and repulsion between cement particles. *Attraction* is due to relatively long range intermolecular forces known as *van der Waals forces*. The forces of *repulsion* are due to

- Electrostatic repulsion that is caused by negative ions being held near each cement particle by positive ions selectively adsorbed from the surrounding aqueous solution (Papadakis, 1963)
- Adsorbed water molecules covering the surface of the cement particles

As a result of the opposing interparticle forces, a pair of cement particles has a minimum of potential energy with respect to those forces when the particles are separated from each other by a certain small distance, amounting to perhaps 10 or so water molecule diameters. Hence cement particles tend to assume positions with respect to each other corresponding to balance of internal forces. Particles in this balanced position are said to be in potential troughs. Any mechanical displacement of particles with respect to each other requires a certain amount of work to "lift" the particles out of their potential troughs. Thus interparticle forces give the freshly mixed paste a structure that has firmness or shearing strength, although only a low degree.

A freshly mixed paste can be softened by diminishing the depth of the potential troughs. This can be done by using more water or an appropriate surface-active agent (admixture) that is able to increase repulsion (Fig, 4.1). Although the cement particles in a fresh paste are dispersed individually throughout the volume of the mixing water, they cannot be dispersed uniformly because the interparticle forces tend to hold the particles very close to each other. As a result, the cement particles form clusters, or floccules, which in turn merge into a three-dimensional irregular network that appears as a continuous mass. This gives the entire system the aspect of a single large flock, but with floccule interstices (Powers, 1945; Bombled and Kalvenes, 1967; Papadakis, 1955; Greenberg and Meyer, 1963). The flocculation in fresh cement pastes was observed directly by Uchikawa by SEM with a sample-freezing, back-scattered electron image method (Uchikawa et al., 1987). An indirect proof of the flocculation is that the longitudinal ultrasonic pulse velocity is very low in fresh cement pastes and concretes (S. Popovics et al., 1994). Thus the fresh cement paste can be considered either as a concentrated suspension of cement particles dispersed in an aqueous solution or as a weak porous permeable solid containing continuous capillaries filled with the aqueous solution (Powers, 1964). The major portion of the spaces between cement particles is filled with liquid and the rest with air. The nature of this flocculent structure is that an increase in the water content of the paste decreased not only the packing denseness of the floccules but also the packing denseness of the particles within the floccules.

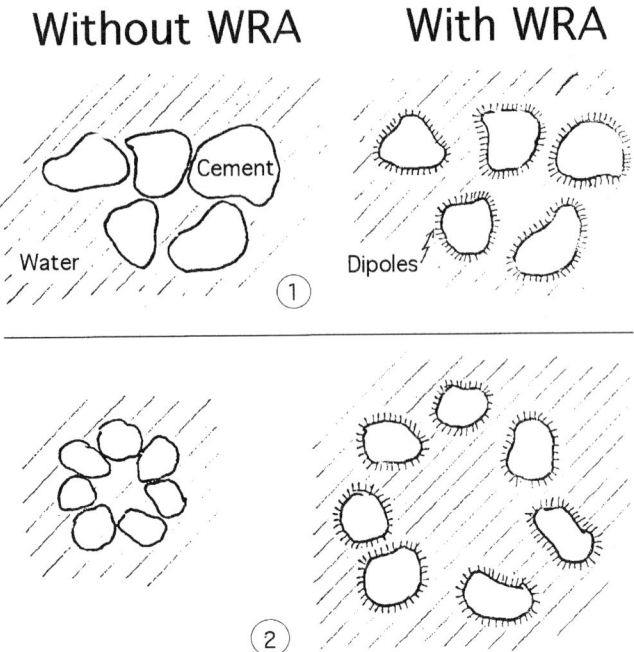

Figure 4.1 Water-reducing admixtures (WRA) disperse the floccules that the cement particles form in water and release a portion of the retained water.

4.2   CEMENT GEL AND ITS SPECIFIC SURFACE

A brief summary of the structure of cement paste on the molecular level (nanostructure) is given below. Further details can be found in the literature (Popovics 1992a; Cement and Concrete Associations 1976; Double and Hellawell, 1977; Hansen, T. C. 1986).

Cement Gel

The solid material in a hardened cement paste consists of several phases: the gel phase, which is the colloid fraction of the hardened cement paste on the nanometer length scale, crystalline materials (mainly calcium hydroxide), unhydrated cement particles, and pores of different types, sizes, and shapes filled with liquid and air on the micrometer length scale. The spatial distribution of these phases determines the internal distribution of stresses created by internal and external forces on concrete. Also, the different solid phases are not equally important in their contribution to the ultimate load-carrying capacity of the cement paste (Grudemo, 1974).

When all the cement particles have hydrated completely, about half of the cement gel occupies the sites that were originally occupied by cement particles, and the other half occupies space outside the original boundaries of particles. Since the external volume does remain relatively constant, the doubling of volume of the cement grains cannot take place by a symmetrical growth of cement gel. That is, the process by which unhydrated cement becomes cement gel must be such that the cement gel is produced only where there is sufficient space to accommodate it. That is, the cement gel evidently conforms to the shape of whatever space is available to it.

Studies of hardened portland cement paste with electron microscope (Copeland and Schulz, 1962; Copeland et al., 1967; Grudemo, 1964; Mills, 1968; Richartz, 1968; Hadley, 1972; Krokosky, 1970; Midgley, 1971; Hale, 1971; Diamond, 1972) as well as adsorption studies have revealed that the cement gel consists of porous, exceedingly ill-formed crystalline, almost colloidal products, that is, gel particles of no definite number and of irregular configuration. Liquid can be present in the pores in various forms. The colloidal material first appears as needlelike or lathlike bodies that bridge the pores between cement grains forming bonds and start reducing the paste porosity. In the course of hydration they grow laterally and become irregular, still very small, thin sheets which frequently agglomerate in conical forms, figures, and rosettes. In this gel are embedded several more-or-less well-crystallized materials, mainly hexagonal calcium hydroxide, a number of crystals of aluminates and sulfoaluminates, and unhydrated cement particles (Copeland and Schulz, 1962; Copeland and Verbeck, 1974). Crystalline calcium hydroxide usually constitutes 20 to 30% of the weight of dry cement gel. The morphology of a young, hardening portland cement paste is shown in Figure 4.2. The needlelike crystals appearing in large number in the photo are *ettringite*.

Specific Surface

Various measurements have indicated that the cement gel consists of very small particles; therefore, their specific surface is very large. This is perhaps the most significant characteristic of the gel, because the size of the specific surface influences many technically important properties of the cement pastes (Powers and Brownyard, 1947). About 75% of the cement gel is calcium silicate hydrate (CSH) or *tobermorite* gel, and at least 80% of the specific surface of the hydrated paste is CSH gel surface.

Increases in the fineness of the cement and curing temperature, or reduction of the water-cement ratio of the paste, may increase the specific surface of the gel. Several other factors, such as age at testing, cement composition, and admixtures, may also have similar effect. The size of these changes is usually small numerically; nevertheless, they may have noticeable consequences concerning the properties of the hardened paste. A notable exception is the case of high-pressure steam curing, which causes a profound alteration in the structure, producing a much coarser structure in the cement gel. The reactions are also fundamentally different from those occurring during hydration at temp-

Figure 4.2 SEM picture of portland cement Type III paste of water-cement ratio 0.35 by mass at 1 day. Magnification: 2000×. (From S. Popovics et al., 1987. Copyright ACI. Reprinted with permission.)

eratures under 100°C; in contrast, they have much in common with those taking place in the production of the autoclaved lime–silica products (Kalousek, 1968; Taylor et al., 1972).

The most commonly used methods for measuring the specific surface of a CSH gel are based on the theory of physical adsorption. The essence of this is that a sample of dry, hardened cement paste, when exposed to a suitable gas (vapor) at a given temperature, would take up (absorb) gas in definite amounts. These amounts are a function of the vapor pressure and the magnitude of the *internal surface* exposed to air. Thus, in sorption studies, the relationship between the amount of gas adsorbed and its relative vapor pressure is determined at constant temperature. The resulting curve is known as an *adsorption isotherm*. From this one can calculate the magnitude of this internal surface area, usually expressed as specific surface (Soroka, 1979). A pertinent method of calculation developed by Brunauer, Emmett, and Teller (1938), called the *BET method,* can be written for water vapor as

$$SS = 3800\frac{V_m}{c_h} \tag{4.1}$$

where SS = specific surface of the hydrated cement gel, m^2/g
              V_m = quantity of water required to cover 1 cm^2 of surface with a layer of water molecule deep, g
              c_h = quantity of hydrated cement, g

The BET method is applicable in the relative vapor pressure range 0.05 to 0.35, provided that the adsorption isotherm is reversible in this range.

Many gases (vapors) are suitable for such adsorption measurements. Theoretically, and also in many practical cases, the specific surface determined from adsorption is independent of the adsorbed material, that is, of the adsorbate applied. Unfortunately, the hardened cement gel is an exception: The specfic surface determined by *water vapor* as the adsorbate is usually over 200 m^2/g or over 550 m^2/cm^3 of cement gel, which is approximately three to five times as large as the specific surface determined by *nitrogen* adsorption. Thus the question is: Which value represents the true specific surface of the hydrated cement gel? Powers and Brunauer, among others, accept the value obtained with water vapor (Brunauer et al., 1970; Powers, 1960). On the other hand, Canadian researchers (Feldman, Sereda, etc.) prefer the specific surface value measured by nitrogen adsorption (Feldman, 1968, 1969, 1971a).

There is no definite proof at present concerning which of the adsorption methods provides the true value for the specific surface of the cement gel. This means, in more general terms, that there has not been any hard evidence of any specific structure (random, layered, etc.) of the hardened cement gel at the molecular (nanometer) level. Light microscopy and SEM can only resolve at the micrometer level. Thus only hypothetical paste structures, called *structural cement models,* can be established by indirect tests, such as adsorption measurements or porosimetry. The indirect tests give no information as to how the pores are arranged in space; thus conclusions about the pore and molecular structures in hardening cement paste have to be drawn by inference. Nevertheless, such models are capable of simulating, at least qualitatively, the observed strength, deformability, volume change, permeability, and durability characteristics of hardened pastes and concretes. A few such models are discussed below.

Powers–Brunauer Model for the Structure of Cement Gel

This model of the cement gel is historically the first such model and represents a dramatic breakthrough in the science of cement hydration. It offers a large (water vapor) specific surface which is approximately 200 m^2/g and consequently, accepts the overwhelming role of the secondary bonds in the strength of the paste (Powers and Brownyard, 1947). Although these bonds are weak, they can provide high compressive strength for the paste when they combine with a large specific surface. It is also assumed, however, that a small fraction of the boundary of a gel particle is chemically bonded (cross-linked) to neighboring particles. Only very small particles can produce a specific surface which is approximately 200 m²/g of cement gel. This is illustrated in Problem 4.1.

PROBLEM 4.1   Calculate the diameter of spheres of equal size that produce a 200-m²/g specific surface. Use 2.5 g/cm³ for the density of the material in question (cement gel).

Solution   If the number of spheres in 1 g of material is N, then

$$N\frac{d^3\pi}{6} = \frac{1}{2.5} \quad cm^3/g$$

and

$$Nd^2\pi = 2 \times 10^6 \quad cm^2/g$$

From this, $d = 1.2 \times 10^{-6}$ cm = 120Å.

Although the particles in the cement gel are not spherical, this value provides an idea about the average size of the particles in the cement gel. X-ray diffractometric measurements by Grudemo (1979) show similar sizes.

The model also contain a considerable number of pores of different sizes filled with air or liquid. Porosity of the cement paste is defined by Powers as the fraction of the volume of a saturated specimen occupied by evaporable water. The various pores in this cement paste model are classified by Powers (1958) into three systems:

1. The system formed by the smallest pores is called the *gel pores*. These pores are the spaces left between the gel particles of the developing hydration products; that is, this system is the intrinsic porosity of the hardened gel. Simple calculations based on the combination of the results of adsorption and density measurements with the model concept of Powers and Brunauer provided 15 Å for the average width of the gel pores and 28% of the gel volume for their quantity (Fig. 4.3). Since then, higher estimates for these values have also been suggested.

2. Another void system is made up of that part of the originally water-filled space that has not become filled with the porous cement gel. The sizes of these pores are up to 500 Å, that is, they are larger at least by an order of magnitude than the gel pores and are called *capillary pores*. Capillary pores can be seen in a scanning electron microscope; gel pores cannot.

3. A third pore system is formed by *air voids* resulting either from incomplete compaction (entrapped air), from intentionally entrained air, or both. These voids are substantially larger than the capillary pores, starting around 0.05 mm in diameter (entrained air), and can be as large as several millimeters or even larger (entrapped air). Bubbles of entrained air can be seen through

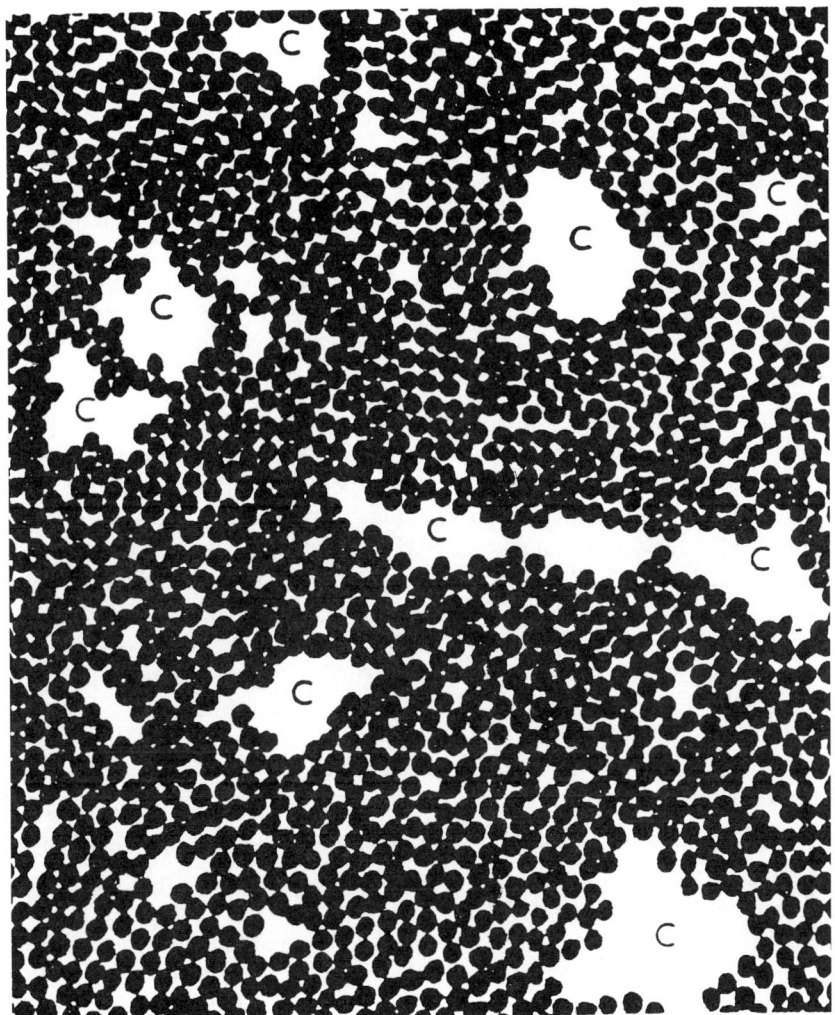

Figure 4.3 Simplified form of the Powers–Brunauer model for the structure of cement gel. Solid dots represent gel particles; interstitial spaces are gel pores; spaces such as those marked C are capillary pores, (Not to scale.) (From Powers, 1958. Copyright Portland Cement Association. Reprinted with permission.)

an optical microscope or a magnifying lens. Air voids formed by entrapped air are usually visible to the unaided eye. Assumptions concerning the shapes of the gel particles or pores are not an essential part of this model.

   The water, or more precisely, the water solution of alkali and other salts can be present in three different forms in the hardened paste according to this model:

1. Water combined chemically in the hydration product, and as such, is a part of the solid. This form of the water is called *chemically bound* or *combined* or *nonevaporable water.* The amount of the nonevaporable water is given by the mass loss of the dried cement paste on heating at 1000°C (1832°F).

2. Water adsorbed, that is, water held by short-range forces on the surface of the gel particles, including in the gel pores. This is called the *gel water.*

3. Water in a free state, that is, when the water is in the capillary pores and air voids of the hardened paste but beyond the active range of the surface forces of the solids. This is called the *free* or *capillary* water.

The gel and capillary water together are called *evaporable water.* The amount of the evaporable water is given by the mass loss of the paste during a not-too-intensive drying process, such as keeping it at 105°C (221°F). Note, however, that there are no definite dividing lines between the various forms of water in the hardened cement paste. For instance, it has been observed that on heating at 105°C (221°F), not only gel and capillary water are lost but also some of the nonevaporable water. If this extra water loss is significant, as some critics of the Powers–Brunauer model believe, the pore content of the hardened paste is not equal to the volume of the measured evaporable water but is less, and calculations based on this equality are incorrect.

Feldman–Sereda Model

This model adopts the gel nature of the hydrated silicates, the determination of the specific surface by adsorption measurements, and the porous character of the cement gel from the Powers–Brunauer model. Otherwise, however, several important modifications are introduced. One of these is the assumption of a layered structure of the cement gel having pores as spaces between these layers (Fig. 4.4). Another is the assumption concerning a special form of water, the *interlayer* (*zeolitic*) *water* present in the hardened paste. The third is the use of nitrogen for the adsorption measurements, which provide a much smaller specific surface for the hardened cement gel than the water vapor adsorption measurements. Note that the specific surface of the Feldman –Sereda model is still quite large (around 50 m²/g), so the proposition is still acceptable that mostly physical (i.e., weak) bonds between gel particles are responsible for the strength of the paste (Feldman 1968; Soroka 1968, 1969). It is also assumed generally that the major portion of volume changes of concrete, such as shrinkage and creep, is the result of the loss of interlayer water.

Feldman and Sereda introduced their model to reduce certain weaknesses that they perceived in the Powers–Brunauer model. Their main objection was the interpretation of the water vapor adsorption isotherms by Powers and

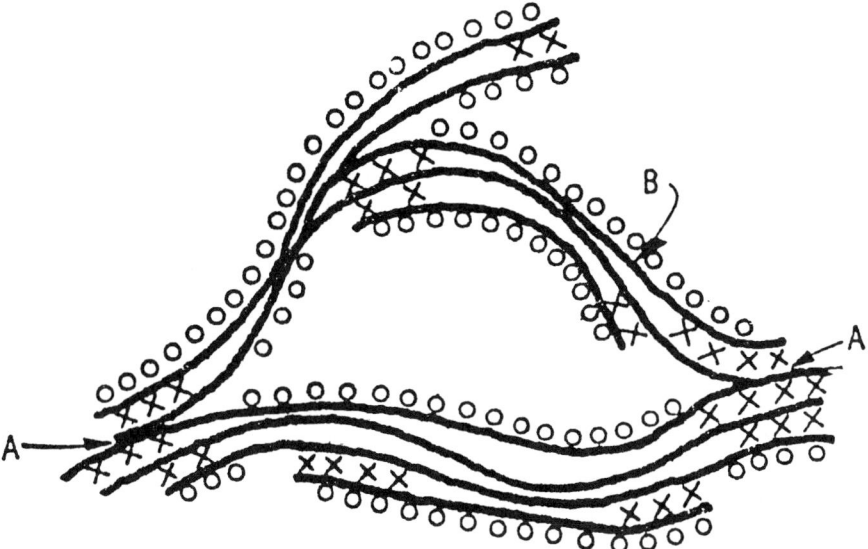

Figure 4.4 Simplified form of the Feldman–Sereda model for the structure of cement gel. A, interparticle bonds; x, interlayer water; B, tobermorite sheets; o, physically adsorbed water. (From Feldman and Sereda, 1968. Copyright RILEM. Reprinted with permission.)

Brunaner, which, they felt, resulted in specific surfaces for the cement gel much higher than the true value. They explained this opinion by pointing out the fact that the water vapor adsorption isotherms are irreversible for hardened cement paste over any pressure range (Feldman, 1989). In other words, the isotherm that is obtained on wetting is different from that obtained on drying; thus the two iostherms form a hysteresis. For such isotherms the BET method provides incorrect results. Feldman and Sereda hypothesized that the hysteresis of the isotherm over the low-pressure range is due to the presence, or more specifically, to the removal of interlayer water. This in turn assumes a layered structure for the cement gel. Therefore, they broke down the amount of evaporable water into two portions:

- The actually (physically) adsorbed water
- The interlayer (zeolitic) water, a part of which behaves as a solid but whose other part is evaporable at drying at 105°C (221°F)

The amount of interlayer water is increased by the water vapor used for the adsorption measurement. The isotherm of the water actually adsorbed is reversible in the low-pressure range, but the isotherm of the interlayer water is not. Since the specific surface depends only on the amount of the water actually adsorbed, the use of water vapor isotherms, representing the sum of the adsorbed water and interlayer, provides erroneously high specific surfaces.

The amount of the interlayer water can be eliminated from calculation of the specific surface of the cement gel by using a suitable adsorbate, such as nitrogen, instead of water vapor. Naturally, the nitrogen-provided specific surface is considerably less than the water vapor specific surface. Also, if water vapor adsorption includes interlayer water, the volume of this interlayer water should equal the difference between the total volume of evaporable water and the volume of adsorbed nitrogen (Soroka, 1979). Experimental evidence supports this expectation.

Comparison of the Two Models

Both the Powers–Brunauer and Feldman–Sereda models are based on adsorption tests. Powers and Brunauer attempted to explain, without the interlayer water, the sizable difference between the water vapor specific surface and the nitrogen specific surface. They attributed this difference to the fact that the diameter of nitrogen molecule is larger (4.05 Å) than that of the water (3.25 Å). This larger diameter, they feel, restricts the penetration of nitrogen into the smaller pores, thus cannot reach some of the surface that is accessible to the water molecules. It has been questioned by others, however, whether such a small difference in molecule diameter could produce four or five times larger specific surfaces. Powers and Brunauer's answer is that it can, especially if the gel pores have a narrow-necked (ink bottle) shape (see Fig. 4.12 c, d) with necks 3.3 to 4.0 Å in diameter.

Another, perhaps more satisfactory explanation for this difference is based on the observation that specific interactions do occur between adsorbent and polar adsorbates. Accordingly, Copeland and Verbeck (1974) would expect that the interaction energy between water molecules and cement gel surface would be much larger than that between methyl alcohol, or nitrogen especially, and cement gel. Such specificity should enable water to penetrate spaces that other molecules cannot enter. One may also mention that:

- Specific surface values of dried cement pastes calculated from *low-angle x-ray scattering* measurements are abut the same as obtained by water absorption (Winslow and Diamond, 1973, 1974).
- Specific surfaces calculated from *water permeability* of hardened cement pastes are even higher (Powers, 1978).

Note, however, that specific surfaces obtained by the two latter methods are also questionable because of the simplifying assumptions used in the calculations.

Thus presently, the debate is not over which model represents the actual structure of hardened portland cement paste better. Nevertheless, it appears that a larger portion of the available indirect experimental evidence supports the Feldman–Sereda model. This includes helium flow measurements (Feldman, 1971b, 1972; Sereda et al., 1980) as well as the behavior of hardened cement paste, especially its time-dependent deformations.

Other Models for the Cement Gel

1. A model developed by Tamas (Tamas and Fabry, 1973; Tamas and Varady, 1975) is also based on adsorption. It provides the smallest specific surface for the gel, determined by *argon adsorption,* and suggests that the primary source of the strength of the cement paste is chemical, that is, primary or strong, bonds resulting from limited polymerization of the calcium silicates, although van der Waals forces and hydrogen bonds also contribute to a certain extent. The role of the *monomer* is played by the SiO_4^{4-} tetrahedron of the portland cement clinker. The type of reaction between the calcium silicates and water is assumed to be condensation polymerization where short-chain silicate anion polymers, called *oligomers,* are formed through binding the SiO_4^{4-} tetrahedrons together by bridging oxygen ions; the by-product is CaO, or more precisely, its hydrated $[Ca(OH)_2]$ and/or carbonated $[CaCO_3]$ forms. In other words, the hardened CSH gel in this model is a conglomeration of silicate anion oligomers of differing degrees of polymerization which form an irregular, three-dimensional net and in which there are also cross bonds between oligomers. The exact size distribution of the oligomers is not known yet, but it appears that the average oligomer contains 8 to 10 Si atoms. This would correspond to a 0.4 to 0.5 average degree of polymerization. The existence of silicate anion oligomers in the hardened portland cement paste has been proved by a chemical method called *silylation* (Tamas et al., 1976).

2. The Taylor model is based on the Ca/Si ratio, conditions of formation, silicate anion type and its changes with time, densities and H_2O/Ca ratios for varying drying conditions, thermogravimetric curve, and x-ray and selected-area electron diffraction patterns (Taylor, 1993). It does not use adsorption data. The CSH gel formed in pastes of portland cement under ordinary conditions is assumed to have a disordered layer structure in which most of the layers are structurally imperfect ones of jennite (approximately $C_9S_6H_{11}$), and others are similarly related to 1.4-nm tobermorite (approximately $C_5S_6H_9$). Nevertheless, there are differences between the cement gel and these two crystalline substances. One is that the silicate anions in both 1.4-nm tobermorite and jennite are long chains, but those in CSH are short ones, chains having lengths of 2, 5, 8, . . . , $(3n - 1)$ tetrahedra. In other words, the structure of the gel is assumed to be a modified form of the crystalline materials, modified by the omission of many of the silicate tetrahedra. This means that the chains in CHS are kinked in such a way as to repeat at intervals of three tetrahedra. These short chains are then bridge together by a third tetrahedron. The bridging tetrahedra are the ones that can be absent, the $(3n - 1)$ chain lengths thus being accounted for.

3. The Richardson–Groves model is similar to the Taylor model. For instance, it is based on x-ray diffraction measurements and retained the highly disordered layer structure of CHS of the Taylor model. The main difference is that the Richardson–Grove model is concerned primarily with the nature and relative proportions of the ions and molecules present but not with how they are arranged. The gel is thus treated as a solid solution of a calcium silicate component and $Ca(OH)_2$, together with water molecules. That is, the Richardson–Grove model is essentially a constitutional rather than a structural model. According to Taylor, such a model is convenient for a discussion of such properties as mean Ca/Si ratios or mean chain length. However, many other properties of the hardened cement paste, of which the range of local compositions is one, require the use of a structural model. The two approaches complement each other.

4. Xi and Jennings (1993) present a good description of several additional cement models. The *Munich model* of Wittmann was developed from adsorption measurements. It describes te CSH as a three-dimensional network of separate colloidal particles without any particular internal structure. The chemical bonds and the van der Waals attraction are equally important in bonding of the gel particles. The *Tokyo model* is also based on adsorption results and is, in a sense, a combination of the Powers–Brunauer and Feldman–Sereda models. The gel particles are composed of porous layers. The pores in the layers are classified as intercrystallite type and intracrystallite type. The intercrystallite pores are similar to the micropores in the Powers–Brunauer model, whereas the intracrystallite pores correspond to the Feldman–Sereda model. The pores between gel particles are capillary pores. The *Jennings–Tennis model* is made up of spheres 5 nm in diameter with a specific surface of 400 m^2/g. The internal structure of the spheres is layered similarly to the Taylor model. They are assembled into one or two types of structure, each with its own structure of specific porosity. Certain material properties can be predicted by varying the proportion of each of the three structures formed in the cement paste, which depend on material parameters, such as the water-cement ratio. CSH is not considered a homogeneous material; the structures of inner and outer hydration products (Section 3.1) are different.

5. A different kind of model has been offered by Garboczi (1993). This is a computer-generated *micro*structure model for simulation of the development of the internal structure of cement paste during hydration. The molecular arrangement, that is, the nanometer structure, of the cement gel is ignored. The cement particles are considered to be a single phase, similar to C_3S. Having the cement particles made up of individual pixels allows random shapes to be represented and allows for material redistribution to simulate microstructure development during hydration. Each cycle of the model has three steps. In the first step, all cement particles in contact with water are identified. Some of these are dissolved at random and are then placed into

the capillary pore system. In the second step, all those pixels dissolved in the pore space that are destined to become either CSH or CH undergo random walks through the pore space. Reaction and product formation occur in the third step. When all dissolved species have been consumed, new cement surfaces in contact with water are identified, and a new cycle of dissolution, diffusion, and reaction begins until all cement is hydrated or until there are no more cement surfaces in contact with water. This model does not allow diffusion of liquid through the layers of CSH that cover unreacted cement. Mineral admixtures such as fly ash or silica fume can also be incorporated into the model and their effects simulated. A combination of this model with concepts of percolation theory (Bentz et al., 1993) has been helpful in understanding the relationship between the random microstructure of hardening cement pastes and their transport properties, even when the microstructure is changing with time. Changes in transport properties over time are important for predictions of durability and service life under various environmental conditions. Examples are penetration of salt solution through bridge decks (Bentz et al., 1996) or attack of sulfate solutions on concrete.

Rate of Hardening Versus Paste Structure

It has been discussed that there is a correlation between high early strength and subsequent lower strength gains, resulting in relatively low final strengths, especially flexural strength. A more specific form of this statement is provided by the exponential model (Section 3.3). This is that if a factor increases that specific *rate* of hardening linearly, the same factor intensifies simultaneously the *deceleration* quadratically. So the higher the rate of the early strength development, the sooner the strength increase will stop. The lower final strength has been attributed to the differential thermal expansion of the constituents of concrete causing internal cracking (Mironov, 1966; Budnikov and Erschler, 1966; Butt et al., 1969; Venuat, 1974).

More evidence is available for another possible physical explanation for this inverse correlation. Certain strength-affecting properties of the hydration products are modified by a change in the rate of hardening (S. Popovics, 1967b). It has been shown that cementlike CSH gels under the effect of intensifying physical or chemical factors (temperature, etc.) produce somewhat higher specific surface at early ages (Kantro et al., 1961, 1962). This means finer texture and, presumably, lower porosity at early ages, a trend that may reverse itself at later ages. So a high rate of early hardening appears to lower the strengths at later ages through its influence on the structure of cement gel. Specific cases are presented below.

1. Verbeck and Helmuth (1969) state that the higher the curing temperature, the higher the rate of early hardening and the higher the concentration of the cement gel should become in the zone immediately surrounding the hydrating cement grain. Experimental evidence is available to demonstrate

that this relatively impermeable rim around the cement grain not only retards subsequent hydration but may also significantly reduce the ultimate degree of hydration. This, in turn, is detrimental to the final strength.

2. The microporosity of a steam-cured cement paste is usually lower than that of a paste hardened at conventional temperatures for 28 days (Butt et al., 1969).

3. It will be shown in Chapter 5 that the specific rate of hardening increases with a decrease in the water-cement ratio (Verbeck and Helmuth, 1969). This means that a decrease in the water-cement ratio increases the early strength relatively more than the strength at 28 days or later. The assumed mechanism for this is that the lower the water-cement ratio, the smaller the initial capillary pore content in the paste, therefore the more intensively the early strengths can develop. On the other hand, this higher rate of hardening results in a denser cement gel, which in turn is increasingly detrimental to further hydration at later ages (i.e., the strength deceleration is intensified). This hypothetical mechanism gains direct support from pore size versus water-cement ratio measurements by Verbeck and Helmuth shown in Fig. 4.13.

4. The effect of the fineness of grinding on the kinetics of strength development shows the same trend. Experimental data show that a portland cement of one-size grading develops lower strengths at later ages than cement of the same clinker with a grading off identical specific surface which contains finer and coarser grains (Kuhl, 1961), Here the beneficial effect of the larger cement grains on the final strength may again result from their slower hydration. Another factor may be the lower voids contents of cements with mixed particle sizes. These result in smaller original capillary porosity in the paste, which in turn is beneficial for rapid strength development.

Example 4.1 Using the notations in Figure 3.18, the derivatives of Eqs. 3.24 and 3.25 provide an informative, although not necessarily exact picture of the kinetics of strength development of portland cements in terms of their compound composition, fineness, and age of the concrete at testing. v_i is the rate of strength development in stress unit/day.

1. Zero stage or dormant period, $0 < t < t_s$:

$$v_O = 0 \text{ by definition}$$

2. First stage of the hardening, $t_s \leq t \leq t_d$:

$$v_1 = \frac{f_o f_{28} S}{300} \frac{a_1 C_3 e^{-a1(t - t_s)S/300} + a_2(100 - C_3)e^{-a2(t - t_s)S/300}}{100 - C_3 e^{-a1(28 - t_s)S/300} - (100 - C_3)e^{-a2(28 - t_s)S/300}}$$

(4.2)

When $t = t_s$, that is, when the rate of strength development is the *maximum,* then

$$(v_1)_{max} = \frac{f_o f_{28} S}{300} \; \frac{a_1 C_3 + a_2(100 - C_3)}{100 - C_3 e^{-a_1(28 \, - \, t_s)S/300} - (100 - C_3)e^{-a_2(28 \, - \, t_s)S/300}} \tag{4.3}$$

3. Second stage of the hardening or diffusion period, $t > t_d$:

$$v_2 = 0.4343\omega \frac{t_d}{t} \tag{4.4}$$

When $t = t_d$, then

$$(v_2)_{td} = 0.4343\omega \tag{4.5}$$

Similar relationships can be derived for the *deceleration* of strength development.

4.3 INTERPARTICLE BONDS

Physical Bonds

The forces bonding together the particles of the hydration products are very important because they control many essential properties of the hardened, air-pore-free cement paste or concrete. The most conspicuous of these properties is strength. Although the always present pores reduce this hypthetical pore-free or intrinsic strength drastically, the higher the number of these bonds and the stronger they are, the stronger the cement paste. Therefore, the importance of the actual magnitude of the specific surface of the CSH gel is more than academic since the answer can provide information concerning the nature of bonds in the cement gel: If the actual specific surface of the gel is large, say 200 m^2/g, the hypothesis may be acceptable that the strength of a hardened cement paste originates from weak physical (or secondary, or van der Waals) bonds, called *adhesion,* acting between the gel particles. If, however, the actual specific surface is much less, say, one-tenth, the presence of much stronger, *chemical* (primary), mostly covalent bonds should be assumed as the main source of strength.

Brunauer illustrated the origin of strength as follows (Brunauer and Copeland, 1964): The hydrous amorphous silica has greater specific surface than the cement gel, yet the latter is a better cementing material because it possess more than double the surface energy (around 390 ergs/cm^2). On the other hand, $Ca(OH)_2$ exerts three times as much surface energy as the cement gel, yet it is a poorer cementing material because the specific surface of the most finely divided calcium hydroxide has only one-tenth of the specific surface of the cement gel. For this reason and because of their chemical vulnerability, the calcium hydroxide crystals are weak spots in the hardened cement paste. But their contribution to the strength development can be increased considerably, especially at later ages, by combining them chemically into forming a calcium silicate hydrate, which again is a gel-like, strong material. Most commonly, pozzolanic materials are used for this purpose at regular temperatures, and ground quartz in the case of autoclave curing.

According to one school of thought, the interparticle forces in a hardened cement paste are probably of purely physical origin. That this is possible was demonstrated by L'Hermite in 1936 by mixtures of pulverized basalt and 8% water compacted with high pressure. These mixtures provided compressive strengths up to 450 kg/cm² (about 6500 psi) immediately after form removal, that is, without any measurable contribution from chemical reactions (L'Hermite, 1936).

A similar, more recent demonstration comes from Czernin (1962):

- The compacted mixture of 100 g of coarse quartz sand (the specific surface is around 20 cm²/g) and 20 g of water is without adhesion, therefore without strength.
- The compacted mixture of quartz sand of cement fineness (the specific surface is around 2000 cm²/g) and 20 g of water acquires some adhesion.
- The compacted mixture of 100 g of pulverized sand with a specific surface of around 20,000 cm²/g and 20 g of water has a strength of measurable order.

The appearance of strength in this series of inert powder–water mixtures is due to the increase in secondary bonds between solid particles (Philleo, 1966). This increase is the result of more numerous contacts between the particles, that is, to the increased area of contact, which, in turn, is the result of the increase in the specific surface. Therefore, it is not surprising that cement pastes develop such high strengths from a similar source if the specific surface of the cement gel is in the neighborhood of 2,000,000 cm²/g.

One can speculate that in addition to the adhesion, there is a limited amount of points of chemical (i.e., primary) bonding between particles, or between gel particles and other bodies. After all, a hardened cement paste retains its strength when immersed in water, whereas dry clay containing only secondary bonds does not. Recent experimental evidence described below seems to indicate, however, that chemical bonds, defined in the conventional sense, could be essentially absent from the hardened cement paste. Instead, the interparticle bonds could be solid-to-solid contacts where a varying proportion of long and short-range forces are involved.

Cement Compacts

This hypothetical mechanism of the strength of hardened cement paste is not merely an extrapolation of the L'Hermite–Czernin demonstrations. There are experiments that support this mechanism directly. For instance, the strength of pressed ceramics (Spriggs and Vasilos, 1963) is influenced by the grain size in the same way as shown by quartz powder. The most indicative are, however, comparative experiments with *cement compacts* (Sereda et al., 1966; Feldman and Sereda, 1963). The cement compacts are usually made of powdered, fully hydrated cement passing sieve 100 (0.15 mm) obtained by hydrating it in rotating bottles with a large amount of water. This powder is consolidated to the required denseness (i.e., porosity) by a corresponding high pressure.

In a pertinent experiment (Feldman and Sereda, 1968; Soroka, 1968), disks 0.05 in. (1.27 mm) in thickness made of three cement systems were tested for modulus of elasticity and hardness. These properties are obviously correlated to strength. The three cement systems involved are (Fig. 4.5):

I. Conventional cement pastes in which porosity varied from 40 to 70%, water-cement ratio from 0.4 to 1.20 by mass, and degree of hydration from 73 to 98%.

II. Pastes similar to those in series I but compacted *after* hardening, in which porosity varied from 25 to 60%; and

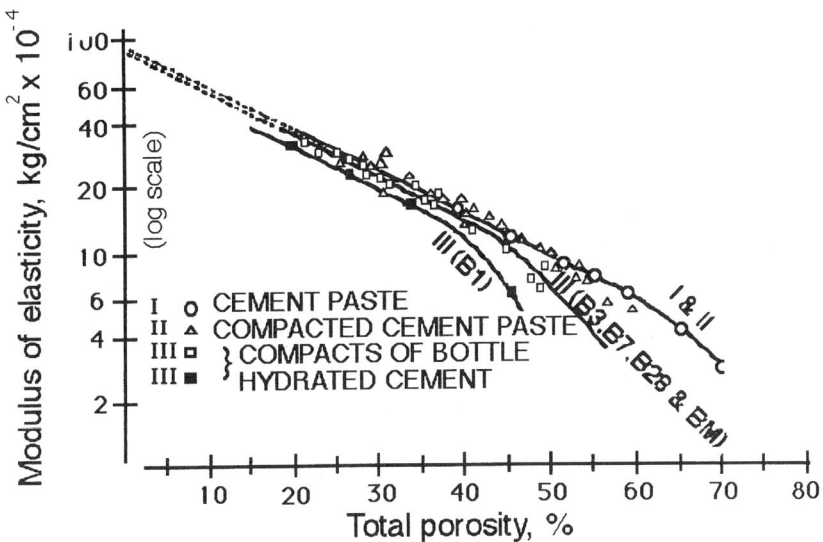

Figure 4.5 Modulus of elasticity versus porosity for cement systems. 1 kg/cm^2 = 14.2 psi = 0.098 MPa. (From Soroka, 1968. Copyright RILEM. Reprinted with permission.)

III. Cement compacts containing fully hydrated and unhydrated cement grains in which the porosity varied from 20 to 55% and the degree of hydration from 0 to 90%.

The most significant result of this experiment is that the modulus of elasticity versus porosity relation is the same for the compacted cement pastes in series II as for the cement pastes in series I, despite the fact that in II the paste samples having an original porosity in the range 40 to 70% were compacted *after hardening* to porosities in the range 25 to 60% (Fig. 4.5). This is strong evidence for the absence of chemical bonds because it is unlikely that these would not be broken (if they were present) when the pore volume is reduced by half during compaction of the hardened specimens. The fact that the values for the modulus of the compacts of bottle hydrated cement in series III fit so closely to that of the conventional paste and compacted paste lends support to the idea that this system has no or very few chemical bonds.

The findings above are in striking contrast with the results obtained with corresponding gypsum systems (Soroka, 1968), where chemical bonds can be identified. In this case the modulus for the in-situ hydrated plaster of paris ($CaSO_4 \cdot \frac{1}{2}H_2O$, that is, gypsum) was an order of magnitude greater than for the compacts at the same porosity.

The results from modulus of elasticity and hardness measurements show that the degree of hydration is a significant factor only when its value is below 50% as represented by system III (B1) in Figure 4.5. This point is further demonstrated by the data related to systems I and II in the same figure. Here, although the degree of hydration varies in the various series, it exceeds 50% in all cases and the results from both systems can be represented by a single line.

These results suggest that when cement is hydrated in excess of 50%, it is possible to obtain bonds between the hydrated and unhydrated particles in the paste and in the compacts that are as strong and as many as between hydrated particles themselves. Thus presumably, the nature of bonds in these three different mixtures is also similar; namely, the result from the solid-to-solid contacts made by bringing the surfaces together with the hydration products originating from the chemical reaction, or with appropriate pressure resulting from the externally applied load. As pointed out earlier, this process also reduces the paste porosity.

The interparticle bonds resulting from solid-to solid contact in hardened cement pastes allows for the possibility of one particle breaking its bond with one neighboring particle and remaking of a similar bond with another neighbor causing only permanent deformation (creep, shrinkage), but no reduction in strength, when the system is subjected to sustained load (Soroka, 1968, 1979).

Another hypothesis for the applicability of this concept of interparticle bonds is as follows: When a speciman of mature hardened cement paste is dried in such a way as to avoid excessive stresses during drying, the specimen becomes stronger as its water content is reduced, even if some of the chemically combined water is removed. In terms of interparticle forces, as considered by one school of thought (Powers 1960), this gain of strength could be accounted for the reduction in average distance between surfaces in the cement gel (Section 3.8).

Primary Bonds

There are also publications indicating that the hydrated system consists mainly of primary (i.e., strong) bonds (Krokosky, 1970; Tamas and Varady, 1975; Tamas and Fabry, 1973). The conflict between the weak- and primary-bond hypotheses has not yet been resolved.

Bonds Versus Porosity

In brief, the strength of a hardened cement paste is determined primarily by two factors: the number and intensity of the interparticle bonds in the hydration products as the source of the strength, and the porosity of the paste as a strength-reducing factor. However, these two factors are not independent of each other, which causes an ambiguity. Therefore, before we leave the interparticle bonds in the hardened cement paste and turn to the role of porosity, this ambiguity should be discussed.

Note that porosity reduction cannot explain the strength increases in the sand samples of the Czernin demonstration earlier in this section. This is so because it is unlikely that the powdered sand sample had a lower total porosity than the coarse sand sample, let alone so much lower, which could explain the observed strength increase of several orders of magnitude. For the qualitative justification of this opinion, it is enough to recall that the total voids contents, that is porosity, of geometrically similar particle size distributions are independent of the maximum particle size (S. Popovics 1992; Powers, 1968).

Incidentally, this is a more general form of the better known statement that the volume of voids between spheres of equal diameter in dense, regular arrangements is independent of the sphere size. The specific surface and the number of particles increase with a decrease in the sphere size; therefore, the number of contacts between particles in a unit volume increases as the size of the pores decreases, but the total volume of porosity remains the same. Thus one should conclude that the development of new bonds has the dominant role in the strength increase in the sand samples in the Czernin demonstration, not a reduction in porosity. Strength results reported by Jambor lead to a similar conclusion (Jambor, 1979).

In the hardening cement paste, the new bonds are produced by the developing hydration products, which, however, simultaneously reduce the capillary porosity by filling these pores. At present we do not know what percent of an observed strength increase in a cement paste is due to the increased number of bonds and what percent to the increased volume of load-carrying solids, or reduction in porosity. Such a distinction is difficult even conceptually because these factors are interrelated. Besides, it is easier to measure porosity than the bonds from solid-to-solid contact. For these practical reasons it has become customary to attribute such a strength increase, or practically any strength increase, solely to a reduction in the paste porosity.

4.4 POROSITY IN CEMENT PASTE

The type and size of the paste porosity has very important effects on the properties of concrete, especially on the strength and durability of the hardened concrete (Modry, 1973). Also, the unit weight of a mixture is affected inversely by the porosity. Chapter 5 is devoted to the effects on strength. The durability aspects are not discussed in this book.

Description of Porosity

From the standpoint of chemistry, the term *porosity,* specifically in the hardened cement paste, is defined by Powers as the fraction of the volume of a saturated specimen occupied by evaporable water (Powers, 1958). From the standpoint of strength, porosity can be defined as a dispersed phase in the cement paste portion of concrete whose strength is negligible compared to the strength of the matrix. In other words, porosity means voids filled with air, liquid, or other strengthless material. Assuming that the porosity in the aggregate particles does not influence the concrete strength, which simplification is quite reasonable except for special lightweight aggregates, the effective porosity means voids in the hardened cement paste, which is the matrix portion of the concrete.

Porosity in a cement paste includes smaller and larger pores, such as gel pores, capillary pores, and air voids. Another classification is macro- (capillary and air) and micro- (gel) pores. Macropores are spaces that are large enough for the water to form menisci as the pores are filled or emptied; thus the water in it behaves as bulk water. Micropores are so narrow that water cannot form menisci; therefore, it has a different behavior from bulk water. Water in micropores acts to keep the solid layers apart by exerting a disjoining pressure (Mindess and Young, 1981).

The gel pores are an intrinsic part of the hydrated cement that one hardly can influence. Therefore, it may be omitted from further discussion. The capillary pores are filled originally with liquid (an aqueous solution) and the air voids with air in the fresh paste. Later, air may replace part of the liquid in the capillary pores and liquid can penetrate the air voids; that is, any pore can be filled up with air or liquid or some other weak material, or any combination of these (Czernin, 1962; Powers, 1958b). For the sake of simplicity, only air- and liquid-filled pores are considered here.

Capillary pores have tortuous, tubelike shapes, and air voids are shorter with much larger diameters. Pores formed by entrained air are close to spherical shapes. In addition capillary pores and air voids differ from each other in two other aspects:

1. The origin of the two types of porosity is different. The air voids in the cement paste are the results of incomplete consolidation, or entrained air, or both. The capillary pores in a hardened cement paste are that portion of the originally water-filled spaces in the fresh cement paste which has not become filled with the cement gel as the result of hydration; and

2. The volume of the original air voids V_a (Fig. 4.6) remains essentially constant during the life of the cement paste or concrete (Mielenz, 1969). In other words, the number of large voids in a hardened concrete (paste) is practically the same at any age as the air content of the concrete in

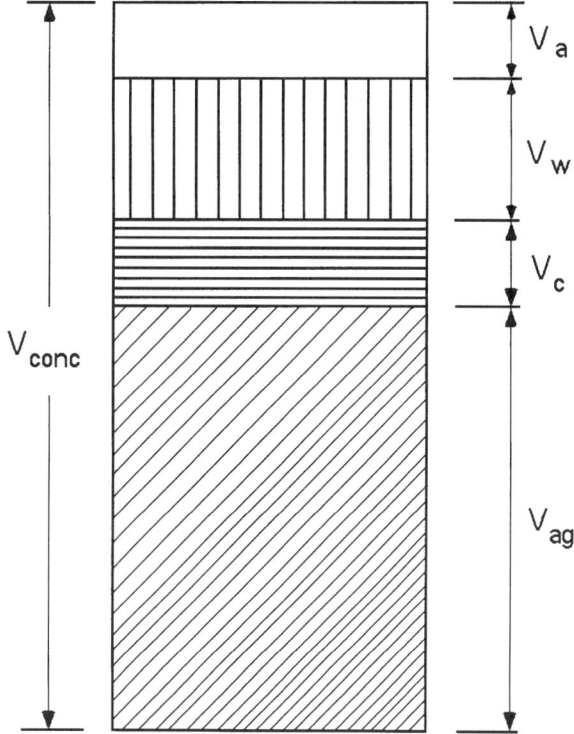

Figure 4.6 Schematic representation of the composition of fresh concrete. Although the ingredients in a well-made concrete are distributed uniformly, it is easier to illustrate the concrete composition if they re imagined as melted and separated into voidless layers and air. $V =$ volume; $W =$ weight; $U_a = W_{conc}/V_{conc}$; $U_o = W_{conc}/(V_w + V_c + V_{ag})$; $V_a = V_{conc} - (V_w + V_c + V_{ag})$; $a = 100V_a/V_{conc} = 100(U_o - U_a)/U_o$.

the fresh state. In contradistinction, the volume of capillary porosity, V_w, decreases with age of concrete under normal conditions because the solid hydration products fill up these pores gradually.

Pores and voids are distributed randomly but statistically uniformly in a properly made concrete. Concrete containing voids too large for the size of the concrete specimen are not considered here.

Fresh Paste

The voids between cement particles in a fresh cement paste can be filled with air and/or liquid. Their total volume can be expressed in several different ways. These expressions are based on one of the two axiomatic equations of concrete technology. One of these is

$$V_c + V_{ag} + V_w + V_a = V_{conc} \tag{4.6}$$

where V_c and V_{ag} = absolute volume of the cement and the aggregate, respectively, in the concrete
V_w and V_a = volume of the water and air, respectively, in the concrete
V_{conc} = volume of the concrete sample

The other axiomatic equation is

$$W_c + W_{ag} + W_w = W_{conc} \tag{4.7}$$

where the W values represent the weight of the cement, aggregate, and water, respectively, in the concrete sample. The terms V_{ag} and W_{ag} are missing from these equations when they are applied to cement paste. Figure 4.6 is the graphical representation of Eqs. 4.6 and 4.7. For the sake of simplicity, the designations $W_c = c$ and $W_w = w$ will be used in the material that follows.

The porosity can be expressed as fraction of percentage of the total volume of the paste specimen:

$$p_o = \frac{V_a + V_w}{V} = p'_o + p''_o \tag{4.8}$$

where p_o = initial total porosity, %/100
V = volume of the fresh paste specimen, including air pores
p'_o and p''_o = relative quantities of the air-filled and water-filled pores, respectively, %/100

The other symbols are as in Eq. 4.6. V_a is dependent on the degree of consolidation (i.e., compaction) and the amount of entrained air. V_w is the volume of the mixing water.

The air-filled porosity is called the *air content* and is usually expressed in relative terms as

$$a = 100p'_o = 100\left(1 - \frac{U_a}{U_o}\right) \tag{4.9}$$

$$= 100\left(1 - U_a\frac{w/c + 1/G_c}{w/c + 1}\right) \tag{4.10}$$

where a = air content, % by volume

U_a = unit weight of the cement paste containing air/pores, g/cm^3

U_o = unit weight of the cement paste computed on an air-free basis, g/cm^3

w = mass of the liquid in the fresh specimen

c = mass of the cement in the same units as that of the water

w/c = water-cement ratio, by mass

G_c = specific gravity of the cement

Example 4.2   If $a = 10\%$, this means that, say, 1 cu ft (or 1 m^3) of paste or concrete contains 0.90 cu ft (or 0.90 m^3) of air-free paste or concrete and 0.10 cu ft (or 0.10 m^3) of total air voids.

When $V_a = 0$, the specimen is said to be completely consolidated because it is air-free (i.e., all the pores are filled up with liquid in the paste); therefore, its volume cannot be reduced by additional consolidation. V_w the water-filled volume, depends on the water content of the paste. The relationship between the water-filled porosity p''_o and the water-cement ratio is, from Eq. 4.8,

$$p''_o = \frac{V_w}{V} = \frac{c(w/c)}{c(w/c) + c/G_c} = \frac{w/c}{w/c + 1/G_c} \tag{4.11}$$

$$= \frac{(w/c)_a}{(w/c)_a + 1} = \frac{1}{1 + (c/w)_a} \tag{4.12}$$

where $(w/c)_a$ is the water-cement ratio by absolute volume. The other symbols are as in Eqs. 4.8 through 4.10. Equation 4.11 is illustrated in Fig. 4.7 for the case when the specific gravity of the cement is 3.125.

The total initial porosity p'_o in a fresh cement paste is the sum of p'_o and p''_o (i.e., Eq. 4.8). This can also be written in the form

$$p_o = \frac{(w/c)(1 - 0.01a)}{w/c + 1/G_c} + 0.01a \tag{4.13}$$

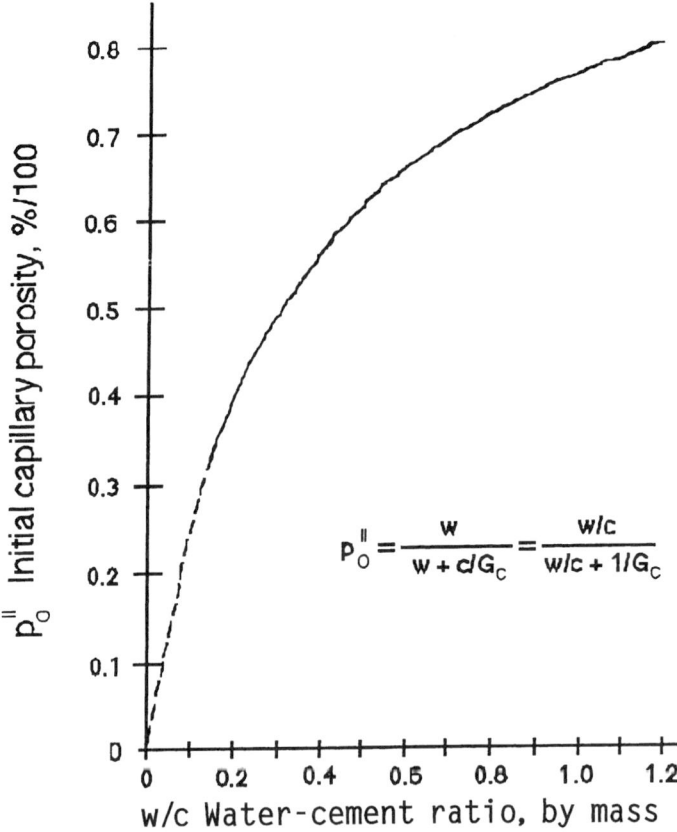

Figure 4.7   Initial capillary porosity of a fully consolidated fresh cement paste as a function of the water-cement ratio for $G_c = 3.125$.

The first term of this equation shows that the capillary porosity in the total volume here is less than it is in the air-free paste (Eqs. 4.11 and 4.12) proportionate to the air content.

PROBLEM 4.2   Derive Eq. 4.13 from fundamentals. Use the symbols of Eqs. 4.6 and 4.7. Also, W_p is the mass of the paste specimen and U_p is its unit weight, as in Figure 4.6.

Solution   By definition, for a fresh cement paste,

$$p_o = p'_o + p''_o = \frac{V_w}{V_w + V_c + V_a} + \frac{V_a}{V_w + V_c + V_a}$$

on a volume basis. The same concept on a mass basis is

$$p_o = \frac{w}{w + c/G_c + 0.01a(W_p/U_p)} + 0.01a$$

$$= \frac{w/c}{\dfrac{w}{c} + \dfrac{1}{G_c} + \dfrac{0.01a}{1 - 0.01a}\left(\dfrac{w}{c} + \dfrac{1}{G_c}\right)} + 0.01a$$

$$= \frac{w/c}{\left(\dfrac{w}{c} + \dfrac{1}{G_c}\right)\left(1 + \dfrac{0.01a}{1 - 0.01a}\right)} + 0.01a$$

$$= \frac{(w/c)(1 - 0.01a)}{w/c + 1/G_c} + 0.01a$$

Hardened Paste

The porosity described by Eq. 4.8 or 4.11 does not remain long in the paste. The developing hydration products start filling the capillary pores gradually since the completion of mixing, although essentially no bulk volume change occurs during this process (Figs. 4.8 and 4.9). The reduction of capillary porosity proceeds until one of the following conditions occurs: (1) no more

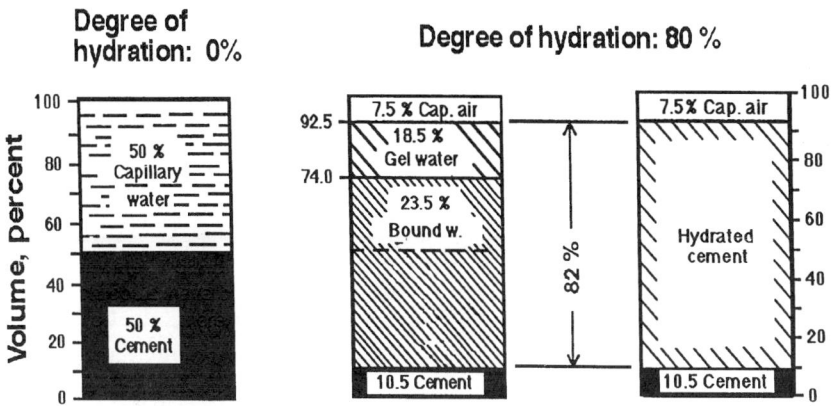

Figure 4.8 Schematic presentation of two stages of hydration of a fully compacted cement paste with a water-cement ratio of 0.32 by mass. When curing conditions are applied where neither access to water nor evaporation are possible, the reactions stop at about 80% of the complete hydration because all the available capillary water has been used up at this point. If cured under water, 7.5% of additional water would have been available by filling up the emptied pores, which could have hydrated another 6% of the cement. Further hydration, however, is impossible because there are no more spaces available for new products. (From Czernin, 1962.)

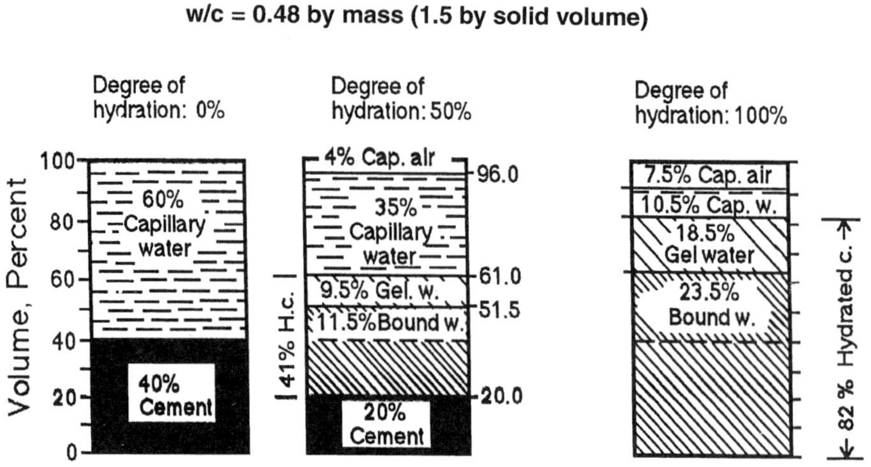

w/c = 0.48 by mass (1.5 by solid volume)

Figure 4.9 Scehmatic presentation of three stages of hydration of a fully compacted cement paste with a water-cement ratio of 0.48 by mass. Here the water and space conditions permit complete hydration even when curing is applied such that neither access to water nor evaporation are possible. In this case 18% capillary pores remain in the specimen, out of which 10.5% is filled with water. If cured under water, all the 18% pores would have been filled up with water; the other volumes would remain unchanged. (From Czernin, 1962.)

unhydrated cement is available for reactions, (2) not enough free water is available, or (3) all te capillary pores are filled up with solid hydration products. The last condition can occur when the water-cement ratio is less than approximately 0.35 by mass.

The remaining porosity, p, in a hardening paste can also be calculated at any stage of the hydration since it is the initial porosity p_o plus the space created by the reduction in the quantity of the unhydrated cement minus the volume of the hydration products. That is,

$$p = \frac{V_w + V_a + V_p - V_h}{V} = p_o - \frac{V_h - V_p}{V} \qquad (4.14)$$

where V_p = volume of portland cement used up by hydration

V_h = volume of the solid hydration products

The other symbols are in Eq. 4.6.

The point should be made here that of the existing several solid hydration products, essentially only the cement gel contributes to the strength development of the paste (Grudemo, 1974), Therefore, not V_h but only the volume of cement gel, V_g, should be used in the calculations when the porosity is correlated with strength:

$$p = \frac{V_w + V_a + V_p - V_g}{V} \tag{4.15}$$

Another way to characterize the porosity in the hardened paste is to show numerically to what extent the space available is filled up by the cement gel. Such a characteristic is the V_{nw}/V_w, ratio where V_{nw} designates the quantity of nonevaporable water and V_w is the mixing water. This ratio represents the fraction of the space available *initially* that is occupied by solid hydration products, provided that the air content of the fresh paste was negligible. A somewhat different characteristic is the gel-space ratio introduced (Powers and Brownyard, 1947) and modified by Powers (Powers, 1949, 1958). The latter is similar in concept to the Feret cement-space ratio, leading to a modified gel-space ratio X_F, defined as the ratio of the volume of the cement gel to the actual space available:

$$X_F = \frac{V_g}{V_w + V_a + V_p} \tag{4.16}$$

where the symbols are as in Eq. 4.14.

The value of X_F can vary between 0 and 1. $X_F = 1$ represents the theoretical case when all the capillary pores and air voids are filled up completely by cement gel. This does not happen in actuality. What does happen is that unless the original water-cement ratio exceeds approximately 0.7 by mass, the cement gel eventually may destroy the continuity of the capillaries (Powers, 1959). Destruction of capillary continuity does not, however, destroy the continuity of the pore system as a whole, because water flow can occur through the gel pores. Just how far the hydration process has developed can be characterized by the *degree of hydration* α. $\alpha = 0$ means that no reactions have occurred between cement and water, and $\alpha = 1$ means that 100% of the cement has hydrated. Intermediate values can be defined in several different ways, which provide different values for α (RILEM Tentative Recommendation, 1981). For instance, a logical definition is the following:

$$\alpha = \frac{\text{quantity of cement gel formed}}{\text{quantity of cement gel formed at complete hydration}} \tag{4.17}$$

Unfortunately, no satisfactory method is available for direct determination of the quantity of cement gel in a specimen, but x-ray analysis can measure the quantity of unhydrated cement. Therefore, the following definition is more suitable for practical purposes:

$$\alpha = 1 - \frac{\text{quantity of unhydrated cement}}{\text{original quantity of cement}} \tag{4.18}$$

which represents the fraction of cement that has become hydrated.

The measurement of porosity of hardened pastes depends on knowledge of the density of either the evaporable water or the solid phase of the paste. It is difficult or even impossible at present to measure either in an unambiguous manner. A good estimate is that the density of the absorbed water is 0.94 g/cm³, in which case the density of the cement gel is 2.54 g/cm³.

PROBLEM 4.3   Develop numerical forms for Eqs. 4.15 and 4.16. The following data given after Powers (1958a): the specific gravity of the cement, $G_c = 3.125$. The hydration of 1 cm³ of portland cement produces 2.06 cm³ of cement gel (containing gel pores) plus the nongel constituents. Assume that no bulk volume changes occur during hydration.

Solution
  1. For Eq. 4.15:

$$V_w = c\frac{w}{c} = w$$

$$V_a = 0.01aV = A \tag{4.19}$$

$$V_p = \frac{c}{3.125}\alpha = 0.32c\alpha$$

where α is the degree of hydration according to Eq. 4.18.

$$V_g = 2.06\frac{c}{3.125}\alpha = 0.66c\alpha$$

$$V = c\frac{w + A}{c} + \frac{c}{3.125} = w + A + 0.32c$$

Thus, after substitution into Eq. 4.15, we obtain

$$p = \frac{(w + A)/c - 0.34\alpha}{(w + A)/c + 0.32} \tag{4.20}$$

  2. For Eq. 4.16, after substitution of the pertinent terms in the equations, the result is

$$X_F = \frac{0.66\alpha}{(w + A)/c + 0.32\alpha} \tag{4.21}$$

Unit Weight

Unit weight or bulk density U provides global information about the internal structure of concrete because it is inversely proportional to the porosity. Unit weight is defined by ACl 116R as "the mass of a material (including solid particles and any contained water) per unit volume including voids." Mathematically, for concrete we have

$$U = c_1 + w_1 + ag_1 \qquad (4.22)$$

where the lowercase letters represent the cement, water, and aggregate contents, respectively, in 1 unit volume of concrete. The usual way to determine the unit weight experimentally is by measuring the weight W and volume V of the sample and calculate the unit weight as

$$U = \frac{W}{V} \qquad (4.23)$$

Details of such a test method are standardized in ASTM C138-92. Unit weight can also be calculated from the cement, water, and air contents of the concrete without a trial mix, as shown below.

PROBLEM 4.4   Develop a numerical form for Eq. 4.23 for the calculation of the fresh unit weight from the concrete composition.

Solution (S. Popovics, 1964c, 1974b)   Simple arithmetic manipulation of Eqs. 4.6 and 4.7 provide the following theoretically sound formula in terms of cement, water, and air contents:

$$U_{1.18} = 0.037c\left[1 + G\left(16.85\frac{100 - a}{c} - \frac{w}{c} - 0.32\right) + \frac{w}{c}\right] \qquad (4.24)$$

A somewhat different case results when the *consistency* (slump) is considered instead of the water content. The following semiempirical formulas can be obtained: For structural concretes of 4-in. (100-mm) slump,

$$(U_{1.18}) = 0.043c + 10 + G[54 - 0.8(m - 6)^2 - 0.015c - 0.6a] \qquad (4.25)$$

For structural concretes of 1-in. (25-mm) slump,

$$(U_{1.18}) = 0.043c + 8 + G[55.6 - 0.8(m - 6)^2 - 0.015c - 0.6a] \qquad (4.26)$$

Similar formulas can be obtained when the aggregate–cement ratio is given.
   Experimental data show that the cement and water contents, as well as the water-cement ratio (i.e., water-filled porosity), have hardly any effect on the unit weight. Thus

$$U_{1.18} = (U_{1.18}) \approx G(45 - 0.6a) + 33 \qquad (4.27)$$

where $U_{1.18} = (U_{1.18})$ = unit weight of fresh concrete in which the maximum particle size D of the aggregate is 1.18 in. (30 mm), calculated from the cement, water, and air contents, and from the cement and air contents and slump, respectively, lb/cu ft

c and w = cement content and water content, respectively, lb/cu yd

G = average specific gravity of the aggregate

a = air content, %

Experimental data support Eq. 4.27 quite well (Fig. 4.10).

Empirical generalization of these equations for any maximum particle size D'', within practical limits, is the following:

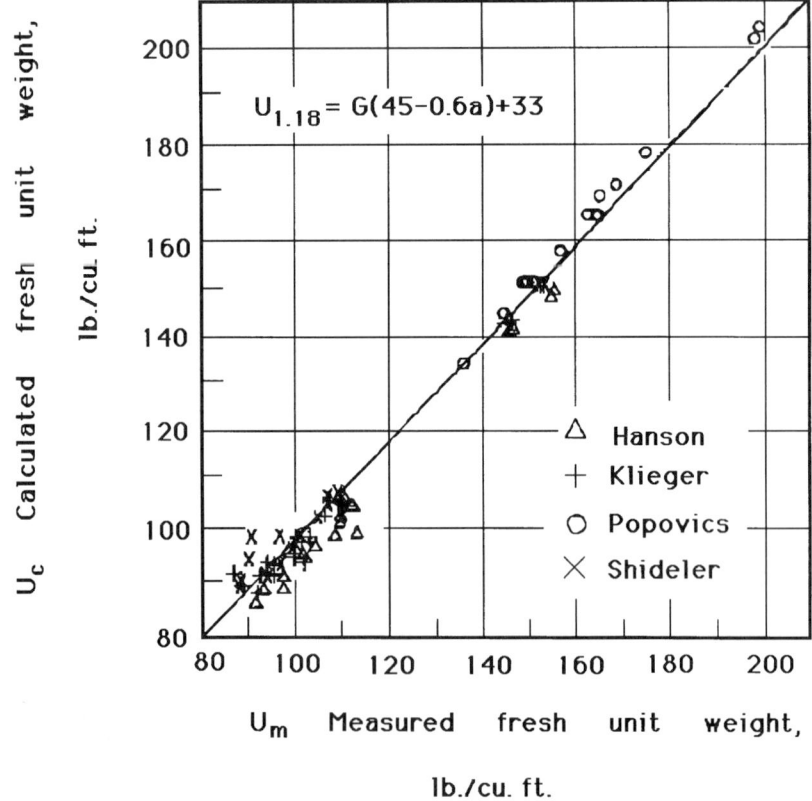

$$U_{1.18} = G(45 - 0.6a) + 33$$

Figure 4.10 Comparison of experimental values by various investigators with computed values of the unit weight of structural concrete mixtures using Eq. 4.27. 1 lb/cu ft = 0.0624 kg /m³. (From S. Popovics 1964c. Copyright Thomas Telford Publishing. Reprinted with permission.)

$$U_{D''} = U_{1.18} + 13.2 \log \frac{D''}{1.18} \qquad (4.28)$$

The SI equivalents of Eqs. 4.24 through 4.28 are

$$U_{30} = c\left[1 + G\left(\frac{1000 - 10a}{c'} - \frac{w'}{c'} - 0.32\right) + \frac{w'}{c'}\right] \qquad (4.24a)$$

and for structural concretes of 100-mm (4-in.) slump,

$$(U_{30}) = 1.2c' + 160 + G[865 - 12(m - 6)^2 - 0.5c' - 10a] \qquad (4.25a)$$

and for structural concretes of 25-mm (1-in.) slump;

$$(U_{30}) = 1.2c' + 130 + G[880 - 12(m - 6)^2 - 0.5c' - 10a] \qquad (4.26a)$$

After simplification,

$$U_{30} = (U_{30}) \approx G(720 - 10a) + 530 \qquad (4.27a)$$

Furthermore,

$$U_D = U_{30} + 210 \log \frac{D}{30} \qquad (4.28a)$$

where U_D and U_{30} = unit weight of the fresh concrete in which the maximum particle size D of the aggregate is D mm or 30 mm, respectively, kg/m^3

c' and w' = cement and water contents, respectively, kg/m^3

The other symbols are as in Eqs. 4.24 and 4.24a.

Incidentally, simplified equations are available for the relation between the fresh unit weight and the final (dry) unit weight of the same structural concrete, as follows (S. Popovics, 1964):

$$U_{final} = U_{fresh} - 8 \text{ lb/cu ft for}$$

$$\text{concretes having a 1.5-in slump} \qquad (4.29)$$

and

$$U_{final} = U_{fresh} - 10 \text{ lb/cu ft for}$$

$$\text{concretes having a 4-in. slump} \qquad (4.30)$$

or in the SI system,

$$U'_{final} = U'_{fresh} - 130 \text{ kg/m}^3 \text{ for}$$

$$\text{concretes having a 38-mm slump} \qquad (4.31)$$

and

$$U'_{final} = U'_{fresh} - 160 \text{ kg/m}^3 \text{ for}$$ $$(4.32)$$

$$\text{concretes having a 100-mm slump}$$

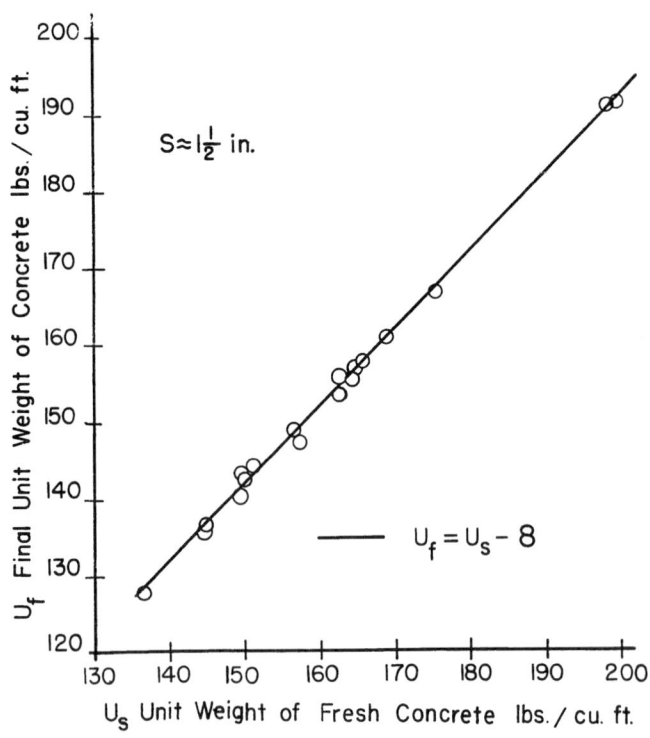

Figure 4.11 Relation between the fresh unit weight and final unit weight of the same concrete when the consistency is stiff. 1 in. = 25.4 mm.) (From S. Popovics, 1964c. Copyright Thomas Telford Publishing. Reprinted with permission.)

Equation 4.29 is illustrated in Figure 4.11.

Poor Size Distribution and Its Measurement

In addition to the magnitude of total porosity, the form in which the porosity is present (i.e., the pore size distribution) is also a meaningful characteristic in the hardened cement paste.

Although visual inspection, including optical and electron microscopes, can provide useful information about larger pores (Diamond, 1993), the two methods used most frequently for direct measurement of the distribution of pore sizes in hardened cement pastes are the mercury *intrusion porosimetry* and the capillary *condensation* (Modry, 1973). Unfortunately, there are discrepancies between the results of these two types of measurement: The total pore volume as measured by mercury intrusion is more nearly equal to the volume of evaporable water in saturated pastes than is the pore volume as determined from sorption isotherms (Copeland and Verbeck, 1974). In addition, both methods have weaknesses in common, which introduce uncertainties for the results. First, the cement paste must be completely dry in order to perform the tests, and drying may cause significant changes in the pore structure. Second, the conversion from measured data to pore sizes includes assumptions, such as the shape of the pores, which may or may not be valid. Third, corrections applied to eliminate some of the disturbing factors, such as the expansion of the intrusion apparatus or the compression of the mercury due to the pressure, may not produce the right results. It has been shown that such assumptions and corrections can have a significant effect on porosimetry results (Cook, 1993). Thus the calculated pore size distributions are only approximations.

Mercury porosimetry is based on the fact that mercury behaves as a non-wetting liquid towards most substances. Consequently, it does not penetrate into the openings and cracks of these substances spontaneously, but rather, one must apply pressure to make it do so. If a sample of a porous solid is sealed into a vessel tapered into a tube, evacuated, filled with mercury, and put under a given pressure, the mercury fills up the pores in a solid material up to a certain minimum size. If smaller pores are to be filled, the pressure should be increased. The pertinent relationship between pressure and pore size is described by the Washburn equation (Van Brakel et al., 1981):

$$r = -2\gamma\frac{\cos\phi}{P} \tag{4.33}$$

where r = radius of the pore
γ = surface tension of mercury
ϕ = contact angle between mercury and the wall of the pore
P = pressure acting on the pore

For the typical values of $\gamma = 480$ dyn/cm and $\phi = 140°$, Eq. 4.33 takes the form

$$r = \frac{75,000}{P} \qquad (4.34)$$

where r = radius of the pore, Å
P = pressure, kg/cm^2

Porosimeters fall into two groups: low- and high-pressure instruments. Low-pressure porosimeters operate from zero to atmospheric pressure, high-pressure units above. Most commercially available high-pressure units have a pressure range reaching 200 MPa, although units up to 400 MPa are also available. The minimum measurable pore size is around 20 Å, the maximum can be extended up to 1.0 mm with special procedures. Mercury porosimetry is most reliable for measurements of pore sizes between about 40 Å and 15 μm, as demonstrated, for instance, by scanning electron microscopy (Diamond, 1971).

The volume of the mercury that penetrates the pores is measured and can be recorded automatically as a function of the pressure. From this the quantities of various pore sizes can be calculated. Certain simplifying assumptions are used in these calculations, such as that the value of surface energy and contact angle are constant.

It follows from the procedure that the pore size distribution obtained can be the correct one only if the larger pores are perfectly interconnected by wide-enough channels or the pore channels are of grandually decreasing sizes. This way the first pores to be filled by mercury are the largest, and subsequent pores are progressively smaller. Figure 4.12a and b show systems where mercury porosimetry would produce reliable results. Unfortunately, it appears that cement pastes do not contain many such types of pores. More common are those shown in Figure 4.12c, d, and e, where the porosimetry produces more or less erroneous results. This error manifests itself in two ways:

1. Pore sizes obtained by mercury intrusion are the pore entry sizes rather than the true sizes. This was referred to in Section 4.2 as the "ink bottle" phenomenon.
2. There are quite large air voids that are inaccessible to mercury intrusion because they are completely surrounded by pores too fine to be recorded by the mercury porosity meter (Fig. 4.12c, d, and e). For instance, in the case of Fig. 4.12c, the mercury porosimetry would record the first large void. The second large void would be interpreted as the diameter of the connecting channel (i.e., the pore entry). In the case of Figure 4.12d, the large void is connected by pore channels that are too fine to admit mercury under the pressure available; therefore, this is missed

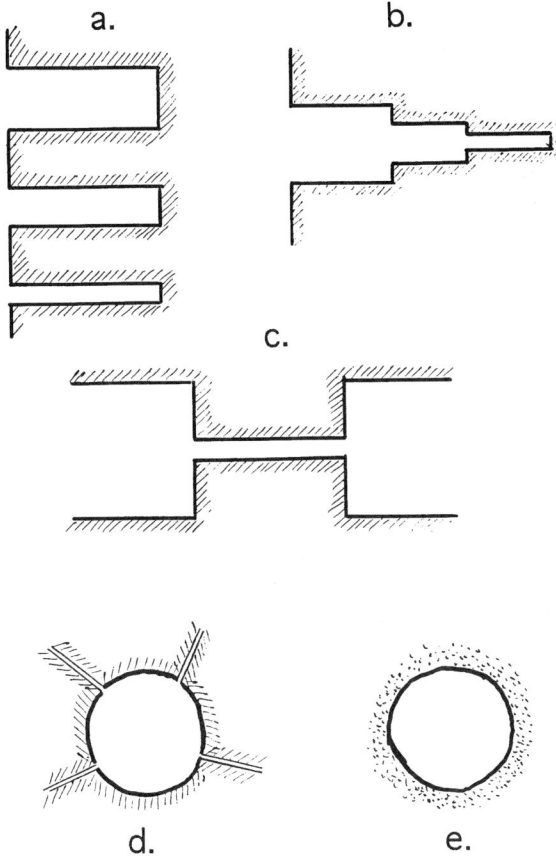

Figure 4.12 Schematic representation of various pore types in hardened cement paste. (From Alford, 1981.)

entirely from the total porosity. Figure 4.12*e* is a special case of Figure 4.12*d* in that the large void is surrounded by hydration products that are too fine structured to be intruded by mercury.

So the question is: Just what kind of, and how much, porosity remain unrevealed by mercury porosimetry? The answer is that both ultrafine pores and large voids are lost in a surprisingly large quantity (Diamond, 1993). For instance, Winslow and Diamond report a case (Winslow and Diamond, 1970) where 35% of the total porosity remained unintruded (i.e., undetected by mercury porosimetry) in a 28-day-old cement paste of 0.4 water-cement ratio by mass. First this "lost porosity" was considered to consist completely of pores too fine to be intruded by mercury, but later the inaccessibility to en-capsulated large voids was suggested as an additional contributor (Diamond, 1971). In another investigation, Alford and Rahman (1981) supplemented mercury intrusion proosimetry with absorption measurements and optical microscopy. They found that in the case of a hardened cement paste of 0.8 water-cement ratio by mass, only 1.43% of the total pore volume has been intruded by mercury into pores greater than 15 μm in diameter, although the total lost porosity was 18.27%. It was deduced that cement pastes of high water-cement ratios have interconnected pore structures and the predominant pore types are those shown in Figure 4.12c and d. At a water-cement ratio of 0.3 by mass, the pore system was far less interconnected, the predominant pore types were most probably those shown in Figures 4.12d and e, and the lost porosity was 37.03%. They concluded that the overwhelming majority of the porosity "lost" by mercury intrusion porosimetry was in pore sizes above 15 μm and not, as was suspected earlier, in the ultrafine pores. Despite all the uncertainties, mercury porosimetry is the most popular among methods providing similar information, because of its comparative simplicity and speed of determination.

Verbeck and Helmuth (1969) reported data concerning pore size distributions over 11 years in moist-cured cement pastes at three water-cement ratios (Fig. 4.13). The data were obtained with a high-pressure mercury porosimeter. The pore size distribution curves show that two pore systems can be separated somewhat arbitrarily considering that the curves display two clearly defined maxima, corresponding, in a sense, to the gel pore and capillary pore concepts, respectively. The curves also show that the sizes of the gel pores, and particularly the sizes of the capillary pores, increase with an increase in the water-cement ratio. It should be noted, however, that more recent measurements indicate neither capillary nor gel spaces but merely spaces between fine individual particles of cement hydration products (Winslow and Diamond, 1970). The reason for this contradiction is unclear. In any case, the size

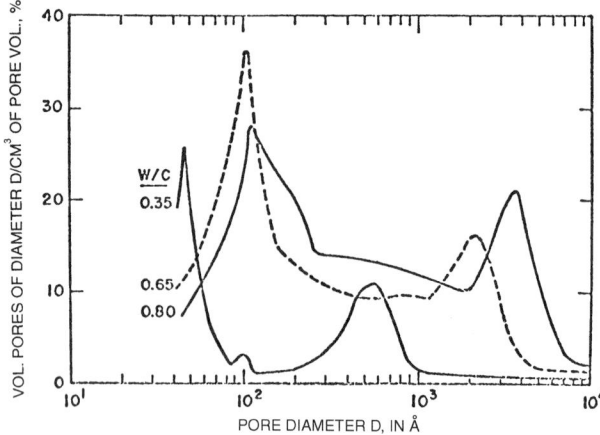

Figure 4.13 Pore size distribution in cement paste moist cured for 11 years. (After Verbeck and Helmuth, 1969.)

distribution of the capillary pores may also be affected by the particle size distribution of the cement used, curing temperature, admixtures, and so on (Jambor, 1974; Barovsky et al., 1979).

Adsorption measurements can also be used for determination of porosity, especially of small pores less than 300 Å in size, such as gel pores (Sneck and Oinonen, 1970). For example, the number of gel pores can be determined by adsorption measurements performed below 0.45 relative water pressure, and the sum of gel and capillary pores by adsorption measurements at 1.0 relative water pressure, that is, at saturation. This is so because the smaller the diameter of the pore, the lower the water pressure. Consequently, in the low-pressure range, the smaller pores become filled with water, whereas the larger ones remain empty (Soroka, 1979). Note that the measurements of very small pores is rarely needed since the contribution at these pore sizes is very small. Optical (Walker, 1974, 1979a, 1979b; Mullen and Bodvasrron, 1978; Bonzel and Siebel, 1982; Gouda, 1982; Anon, 1982), gas chromatography, and absorption (Sereda et al., 1980; Harris et al., 1974) methods are also available for determination of the quantity of air voids.

The pore diameter range that can be measured both by mercury penetration and by nitrogen adsorption is between 180 and 300 Å, and by water vapor adsorption between 180 and 1000 Å. In these ranges mercury penetration provides results that can be greater by several factors than the results obtained by adsorption (Sneck and Oinonen, 1970). Note also that there may be significant differences in pore size distributions measured by nitrogen adsorption compared to those measured by water vapor adsorption (Section 4.1). Thus it is not meaningful to compare pore size distributions obtained by different test methods.

In brief, it appears that at present, the pore size distribution in hardened cement pastes might be obtained best by applying a combination of test methods:

- The total volume of porosity should be determined (e.g., by absorption).
- Pores larger than, say, 15 μm in diameter should be point-counted using optical microscopy.
- Mercury porosimetry can be used to determine the pores of diameters between 15 μm and 40 Å.
- Adsorption methods might be used for pores of diameter below 40 Å.

4.5 INTERNAL STRUCTURE OF CONCRETE (S. Popovics, 1992)

Porosity of the Cement Paste in Concrete

The cement paste with all the pores acts as a matrix in concrete, separating as well as holding the aggregate particles together. That is, the *matrix* portion of a concrete can be defined as the volume in the concrete other than aggregate, except for reinforcement or other inert embedded pieces. As shown later in this section, the porosity in the matrix is not distributed in a randomly uniform manner; it has higher capillary porosity where it is in contact with the surface of an aggregate particle than farther away. This interfacial zone, which can make up one-third or more of the matrix, is weaker and more permeable than the bulk of the matrix.

The matrix can have additional large voids besides the gel and capillary pores or air voids discussed earlier. These additional voids can be caused by inadequate quantity of cement paste or mortar in the mixture resulting from poor proportioning of concrete; or excessive settlement of the mortar from bleeding (S. Popovics, 1973, 1982b). Incidentally, the pores resulting from bleeding are hybrid in nature because, on the one hand, they can be considered as pores from incomplete consolidation since they can be eliminated by re-vibration, and their effect is similar to that of air-filled pores. On the other hand, they are filled with liquid, which makes them difficult to detect. Such voids from settlement are omitted from further discussion.

There are several standard test methods for determination of the air content of fresh concrete and voids in hardened concrete, such as those specified in ASTM C457-90 or ASTM C642-90.

Aggregate Grading and the Internal Structure of Concrete
(S. Popovics, 1992)

The particle size distribution (i.e., the *grading* of a concrete aggregate) is considered good if it can produce, with minimum water content, a concrete that is satisfactory both in fresh and hardened states. The amount of a given fine aggregate and that of a coarse aggregate are well balanced in such a good grading. Quantitatively, this balance depends on numerous factors, such as the maximum particle size, particle shape, cement content, method of compaction, and fineness of the sand. Nevertheless, the behavior of the concrete in the fresh state, as well as visual inspection of the internal macrostructure of concrete, can reveal if the grading used is close enough to the desired optimum (S. Popovics, 1973).

An example for this is given in Figure 4.14, which shows the internal structure of an air-entrained gravel concrete through a cut surface. The fineness modulus of the continuous grading is 5.7. The maximum particle size is $1\frac{1}{2}$ in. (38.1 mm). The cement content of the concrete is 570 lb/cu yd (340 kg/cm^3) of portland cement plus 170 lb/cu yd (100 kg/m^3) of fly ash. The water-cement ratio is 0.50 by mass; the air content is about 3%. The unit weight of the fresh concrete is 146 lb/cu ft (2340 kg/m^3), and the slump is about 1 in. (25 mm). This concrete has a compressive strength of 5960 psi (41.1 MPa) and a flexural strength of 525 psi (3.6 MPa) at the age of 28 days.

The high strength values indicate, indirectly, that this concrete, and thus the grading, is good. It is a good grading indeed, but not quite optimum. The

Figure 4.14 Concrete made with traditional sand and gravel of continuous grading. Fineness modulus = 5.7. (1 in. = 25.4 mm.) (From S. Popovics, 1973d. Copyright TRB. Reprinted with permission.)

behavior of this fresh concrete in the laboratory, especially during compaction by rodding performed according to ASTM C192, indicated that there was a slight excess in the amount of mortar. This observation is supplemented by the fact that the optimum fineness modulus of the aggregate recommended for maximum strength of the particular mixture is approximately 6.2.

Figure 4.14 therefore, shows, a concrete with a well-graded aggregate: The matrix is a dense mortar in which a number of coarse particles are embedded in a random manner, although perhaps a few more gravel pieces could have been placed in it without overcrowding the internal structure. This evaluation is, of course, highly subjective and only qualitative. To make this approach numerical, a simple analysis of the internal macrostructure of hardened concrete, the intercepts of mortar layers between coarse aggregate particles in a cut, plain concrete surface, can be utilized. The coarser the overall grading, the smaller these intercepts become, along with the actual thicknesses of the coatings of mortar surrounding the coarse aggregate particles. The mortar intercepts can be measured with the linear traverse method, which is similar to the procedure used in petrography or as described in ASTM C457 for air-content determination in the hardened concrete, except that there is no need for a microscope in this case. A random cut is taken through the concrete; a regular grid is placed randomly thereon; and linear intercepts are measured along the lines of the grid with the lengths as intercepts in mortar among coarse aggregate particles. These measured intercepts can be summed and averaged, the result of which is a number called the *average mortar intercept.*

Application of the method of mortar intercept to the concrete section in Figure 4.14 and the sections of other concretes produced the following conclusions:

- An average mortar intercept of 0.14 in. (3.5 mm) seems to be the necessary minimum in traditional concretes [i.e., for concretes with traditional gradings of the fine and coarse aggregates for continuous grading with 1 in. (25.4 mm) maximum particle size and with medium cement content] for adequate workability and good strength.
- This intercept criterion is valid not only for gravel concretes but also for crushed coarse particles. In other words, this criterion for good grading seems to be *independent of particle shape.*

Thus one define the coarsest permissible gradings for any particle shape and for traditional concretes as those that provide a dense mortar and an average mortar-layer intercept of 0.14 in. (3.5 mm) in the finished concrete. Any amount of additional mortar would be technically and/or economically undesirable. Less than this minimum mortar in a continuously graded concrete would not provide enough lubrication for the coarse aggregate particles for a workability that is adequate for hand compaction. This does not mean, however, that smaller mortar layer intercepts always result in inadequate workability. For one thing, a reduction in workability caused by a moderate reduction in mortar quantity can be overcome by intensive mechanical compaction, such as vibration. Second, less mortar can also do the job when its lubricating ability is improved. This can be done to a certain extent either by using certain admixtures, increasing the cement content, or using a finer sand.

This latter statement is illustrated in Figure 4.15. This shows a concrete that contains 62%, by absolute volume, of reef shell of $1\frac{1}{2}$-in. (37.5-mm) maximum particle size and of angularity number[†] AN = 30, and 32% beach sand practically all the particles of which passed sieve No. 30 (600 μm) and were retained on sieve No. 100 (150 μm). The fineness modulus of this sand is approximately 1.5. Trial mixes demonstrated that the shell used is the maximum amount that still provides a reasonable workability for this concrete (S. Popovics, 1976b). The average intercept among the shell particles is 0.120 in. (3.04 mm), the low value of which is attributed to the too-fine grading of the applied (beach) sand. Note, however, that when crushed stone (AN = 10)

[†]Angularity number AN = $\nu - 33$, where ν is the percent of voids in a coarse aggregate sample when the sample of a narrow size range is compacted in a prescribed manner in a specific container. The value of AN is 0 for perfectly spherical aggregate particles. AN = 30 represents an extremely irregular shape (S. Popovics, 1992).

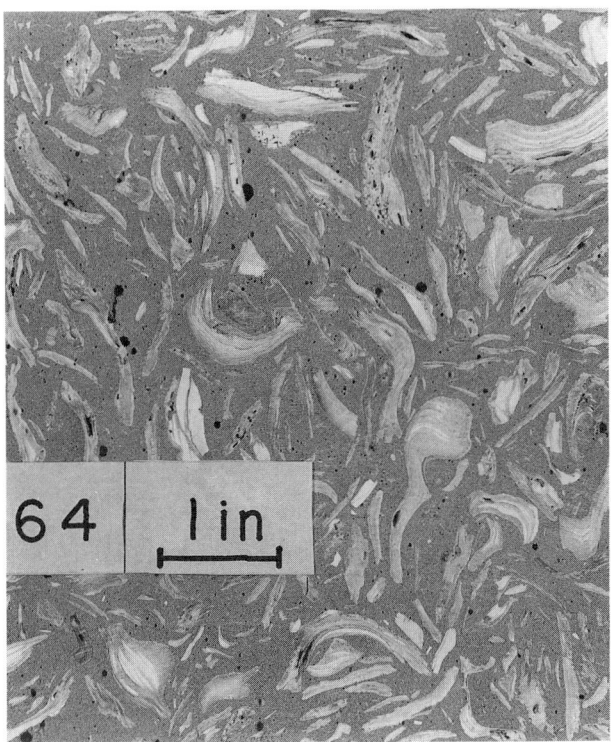

Figure 4.15 Concrete made with 40% by mass beach sand and 60% crushed reef shell. Angularity number = 30. Fineness modulus = 4.95. 1 in. = 25.4 mm. (From S. Popovics, 1973d, 1976b. Copyright TRB. Reprinted with permission.)

or gravel (AN = 5) is used to make concrete with a composition similar to the reef shell concrete above (Fig. 4.16), the average mortar intercept values are 0.148 in. (3.75 mm) and 0.159 in. (4.03 mm), respectively, despite the identical gradings. This means that more coarse aggregate particles can be packed without interference into a concrete when the particle shape is spherical than into a identical concrete when the shape is unfavorable.

The internal structure of a concrete depends not only on the coarseness of the grading and the particle shape of the aggregate but also on the details of the particle size distribution employed. This is demonstrated in Figure 4.17, which is a portion of a larger investigation (S. Popovics, 1964). Here three nonair-entrained concretes of differing but identically coarse gradings and of otherwise identical compositions are presented. The cement contents are 520 lb/cu yd (310 kg/m^3), and the water-cement ratios are 0.62 by mass. Concrete 10 was made with a continuous grading (C$_1$), concrete 15 with a one-gap grading (O$_1$), and concrete 18 with a two-gap grading (T$_1$). Details of the particle size distributions are shown in Figure 4.18. These three gradings were

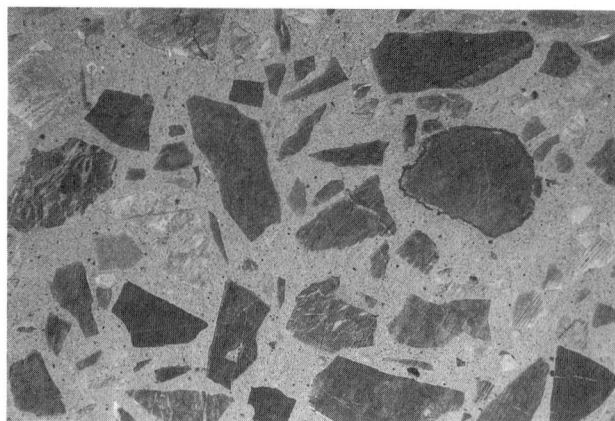

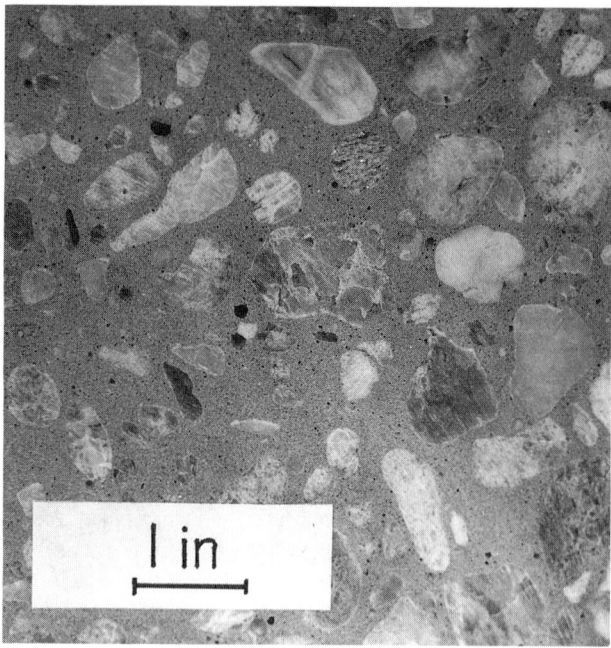

Figure 4.16 Crushed stone and gravel concretes. The gradings and concrete compositions are similar to those of the reef shell concrete in Fig. 4.15. Angularity numbers are 10 and 5, respectively. 1 in. = 25.4 mm. (From S. Popovics, 1973d. Copyright TRB. Reprinted with permission.)

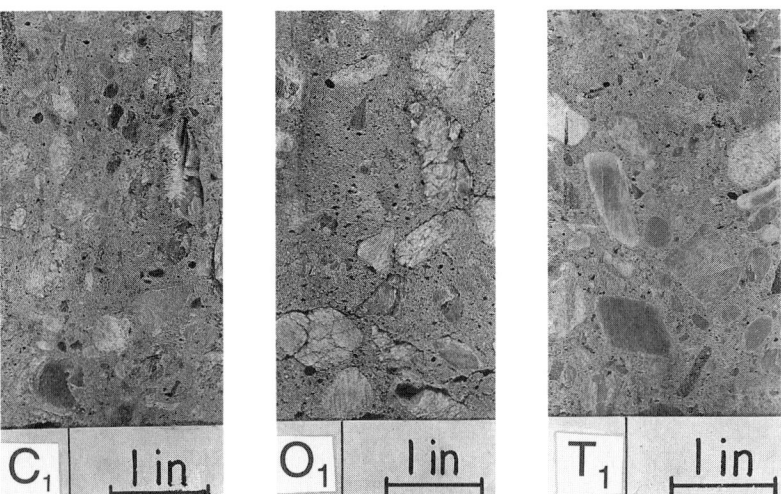

Figure 4.17 Concretes made with continuous (C_1), one-gap (O_1), and two-gap (T_1) gradings of Fig. 4.18. 1 in. = 25.4 mm. (From S. Popovics, 1961b.)

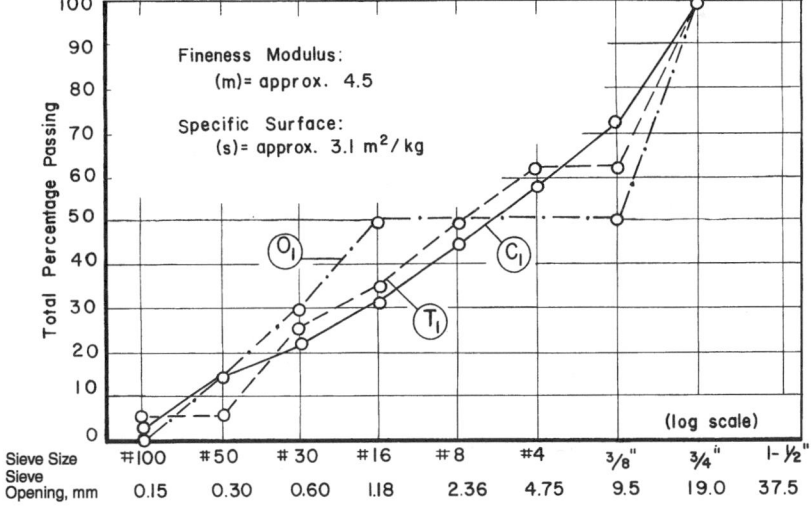

Figure 4.18 Gradings having the same *D, m,* and *s* values, that is, the same coarseness. (From S. Popovics, 1961b.)

set up so that the three important characteristics of coarseness—the maximum particle size, the fineness modulus, and the calculated specific surface of the aggregate—were kept practically constant despite the obvious differences in the particle size distributions.

The differences in the internal structures in Figure 4.17 are quite obvious. The concrete made with the grading of one large gap (O_1), especially, appears different; the grading seems coarser (which is actually not true), and excessive mortar seems to be present (which is true). The latter condition again supports the statement that a thinner average mortar layer is acceptable when a finer sand is used. The appearance of the three graded aggregates is shown in Figure 4.19.

The linear traverse method can provide the proportion of cement paste or mortar in the concrete with fair accuracy (Kelly et al., 1958; Axon, 1962). It is also conceivable to obtain the equivalent of an aggregate sieve analysis (Terrier and Hornain, 1967; Stroeven, 1973) or the specific surface (DeHoff and Rhines 1968; Holliday, 1966) of the aggregate from linear traverse measurements, similar to the methods used for the determination of the parameters of the air-void system in hardened concrete (ASTM C457). But the uncertainties are formidable even for spherical particles. If the aggregate is assumed cuboidal, the situation becomes more complex because lines passing through a cube can be longer or shorter than the cube side. Irregular shapes and mixed shapes present even more difficult problems. Thus the reliability of such a grading analysis in highly questionable (Figg and Bowden, 1971).

Paste–Aggregate Interface

It is truism that the concrete strength may be influenced by three factors: (1) the strength of the matrix, mostly hardened cement paste; (2) the strength of

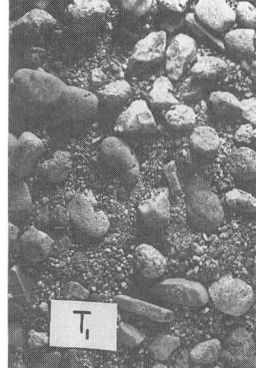

Figure 4.19 Appearance of the gradings presented in Fig. 4.18: C_1, continuous; O_1, one-gap; T_1, two-gap. 1 in. = 25.4 mm. (From S. Popovics, 1961b.)

the aggregate; and (3) the strength of the bond (i.e., the strength of the material in the ring, called the *interfacial zone*, between the bulk matrix and aggregate surface). The individual share of each of these factors in the concrete strength may be different in different concretes, but in the overwhelming majority of structural concretes the bond is the weakest link. It is on the aggregate–matrix interface where crack propagation usually starts under load, leading to gradual failure of the concrete (Section 2.4). There are many other composite materials in which this interface is the weakest link, for various reasons.

The thickness of the interfacial zone can vary typically within 10 and 50 μm, and the water-cement ratio is higher here. Therefore, it contains higher porosity and a higher volume fraction of calcium hydroxide than does the cement paste farther away from the aggregate (Scrivener, 1988). Consequently, the material in the ring is weaker and more permeable than the bulk of the matrix. Although the ring (i.e., the interfacial zone) is thin, it still consumes a significant portion—one-forth to one-third of the cement paste volume— in the mortar or concrete (Garboczi et al., 1995). It is not surprising, then, that the material in the interface zone has a major influence on many properties of the hardened concrete, including strength, elastic properties, electrical conductivity, fluid permeability, and diffusivity. An illustrative example is the observation that the fluid permeability of concrete is usually higher by one or two orders of magnitude than that of its matrix. This is attributed to the high permeability of the interfacial zone. Note that in special cases, such as lightweight aggregates, the interfacial zone can be denser than the bulk cement paste.

In the case of concrete, a hypothetical explanation for the more porous structure of the ring is that at the time of mixing, a traditional aggregate (i.e., not lightweight) particle becomes surrounded by a water film several microns thick. When hydration starts, the hydration products combine with this water film; thus their water-cement ratio becomes higher than the original ratio (Maso, 1980). Another possible mechanism for the production of higher porosity in the interfacial zone comes from the combination of two factors: (1) the particle packing effect, which arises from the fact that particles cannot pack together as well near a flat edge as in free space and (2) the one-sided growth effect that arises from the fact that the hydration products come only from the cement side but not from the aggregate side. The existence of the ring has been proven directly by optical and electron microscopy. It has also been shown that the ring frequently contains microcracks (Section 2.4) even before loading, and voids from bleeding (S. Popovics, 1973), further weakening the interface.

Lowering the water-cement ratio increases the concrete strength not only because it makes the matrix stronger but also because the material in the transitional ring becomes denser and stronger. After all, the bond strength follows Abrams' rule just as does any other concrete strength (Section 5.11). A similar role is attributed to silica fume–superplasticizer combinations (Diamond, 1993). Not only would the dispersed silica fume improve the packing of the particles around the aggregate, but it could also consume some of the extra CH (Bentz and Garboczi, 1991). Also, concretes made with crushed aggregate typically have higher flexural strengths than those of comparable concretes with the same water-cement ratios but with round coarse aggregate particles. The reason for this is the stronger bond of the cement paste to the crushed stone particles, presumably because of the rough aggregate surface. On the other hand, concrete strength, especially tensile and flexural strength, will be less if the aggregate used is coated with certain materials, such as oil or dust, because such coating weakens the bond between the matrix and aggregate. So the surface condition of the aggregate has a major role in the bond strength.

The last observation raises a theoretically and practically important question: Is there any coating that can *increase* the bond strength? To find the answer, several test series were performed (S. Popovics, 1987b). The basic procedure was that smooth quartz gravel particles were coated with an adhesive promotor (silane) or a thin layer of self-hardening adhesive material (epoxy) in which fine particles were embedded before hardening to make the aggregate surface chemically more reactive, or rougher, or both. In one test series the fine particles were covered with a portland cement powder. After epoxy hardening, concrete was made with the coated aggregates and its strength tested. The test results showed that contrary to expectations, making the gravel surface rougher did not increase the concrete strength. Even the strength increase produced by the portland cement coating was negligible. Does this mean that the coatings did not improve the bond? Not necessarily. The coating might have increased the bond between the aggregate coating and the transition ring, but this is not enough. Coating does not strengthen the weak transition ring inside where the failure is initiated; thus the overall concrete strength is not changed by the coating. So, the higher flexural strength of concrete made with crushed stone cannot be attributed solely to the rougher surface texture of the aggregate.

Finally, a paradox should be mentioned here. It would follow from the discussion above that the smaller the surface area of the aggregate in a unit volume of concrete, the higher the concrete strength since the smaller the total amount of the interface ring. For instance, comparable concretes with high cement paste content, that is, low aggregate content, would have higher strengths than lean concretes with identical water-cement ratios. Yet it is shown in Section 5.9 that the opposite is true. The reason for this paradox is not clear at this time.

5

RELATIONSHIP BETWEEN COMPOSITION AND STRENGTH OF CONCRETE

5.1 NATURE OF THE RELATIONSHIPS

As pointed out in Chapter 4, the strength of a hardened *cement paste* is determined partly by the quality and number of bonds between particles of the solid hydration products, and partly by the porosity. The strength of a hardened *concrete* may depend not only on these factors but also, in a secondary manner, on certain properties of the aggregate. Such properties are aggregate strength, deformability, chemical and mineralogical composition, absorption, and perhaps most important, the properties that control the bond between the cement paste and the aggregate surface. Therefore, it would be natural to use these factors in the fracture mechanism used in deriving quantitative relationships for concrete strength as a function of its composition.

Unfortunately, as shown earlier, the science of concrete is not developed enough at present to try such a fundamental numerical approach to concrete strength. We do not know enough quantitatively about the origin of strength in cement pastes, the nature of the bond between paste and aggregate, or the effects of aggregate properties on concrete strength. Neither can fracture mechanics offer quantitative relationships for concrete strength. This again leaves the porosity in the paste portion of the hardened concrete as the only basic factor that we can correlate quantitatively with the strength of concrete, with only a vague reference to the Griffith fracture criterion (Section 2.4). Thus the only practical possibility at present is to attribute almost any strength increase in a hardening concrete to a decrease in the porosity of its matrix. This is the sense in which in future discussions we will state that concrete strength is controlled by porosity.

In brief, it is a good working hypothesis within practical limits that composition affects concrete strength as it affects capillary porosity and the air content of the matrix. One may say that comparable concretes provide (1) lower strengths with higher matrix porosity, (2) higher strengths with lower matrix porosity, and (3) the same strength with identical porosity in most practical cases regardless of the details of composition.

Note that the porosity in the aggregate particles may influence several concrete properties, primarily the durability. However, its effect on concrete strength is only secondary, through its effects on the strength, density, absorption, deformability, and surface properties of the aggregate particles. It would follow logically from this that the magnitude of porosity could be expressed more meaningfully in terms of the volume of the matrix rather than that of the concrete. Nevertheless, only the latter is used in practice, probably because it is simpler and still provides an acceptable correlation with concrete strength.

One can include the effects of factors other than porosity into strength formulas by empirical coefficients covering certain limited values of these factors. For instance, a given set of such empirical coefficients is usually valid for a given cement type, a certain group of aggregates, a narrow range of curing conditions, and a given age. This, of course, restricts the limits of validity of these formulas. Nevertheless, even with these severe theoretical and practical limitations, it is possible to produce formulas for the relationships between concrete strength and composition which are satisfactory for many technical purposes. The reason for this is that the effect of matrix porosity is overwhelming and predictable in most practical cases.

In this sense, the matrix portion of a concrete is the combination of all materials outside the aggregate particles and/or other embedded solid pieces (reinforcement, etc.). The major portion of the matrix is cement paste; the rest is pores filled with air and/or liquid.

5.2 EFFECT OF QUALITY AND QUANTITY OF CEMENT AND MATRIX ON CONCRETE STRENGTH

Effect of Cement Quality

As expected, the strength-developing capability of a cement has a considerable influence on the concrete strength through the quality of the bonds between particles of the solid hydration products (Chapter 3). The best engineering information about the strength-developing ability of a cement can be obtained by making concrete of appropriate composition with the cement in question and testing the *concrete strength* under strictly controlled conditions. There are certain foreign standards, the British standard for instance, that specify both mortar and concrete as alternatives for the strength test of cement. Yet expediency usually dictates the less-time-consuming *testing of mortar* for quality control of cement in laboratories. Such mortar tests are prescribed by the ASTM for compressive (C109), tensile (C190 withdrawn), and flexural (C348 and C349) strength determinations. Strength tests on *cement pastes* are rarely used because they do not adequately evaluate the concrete-making properties of the cement under test, mostly because the applicable water-cement ratio is too low.

Some of the pertinent standard test methods, including their advantages and disadvantages for the concrete technologist, are discussed in the literature (S. Popovics and Pfeifer, 1992).

A more fundamental question is: Which of the mortar methods best shows the concrete-making quality of the cement under test? Most concrete technologists would agree that the results of the briquette test is the least informative (Brickett, 1928). Compressive strength data obtained from standard mortars show better correlation with concrete strengths than tensile data. If a portland cement produces higher compressive strength than another in mortars, the first cement will be likely to produce a concrete of higher strength as well. This is, however, less likely with blended cements. Quantitative relationships between mortar and concrete strengths can also be satisfactory, particularly when the water-cement ratio is practically the same in the mortar and concrete (Walker and Bloem, 1958; Hummel, 1959; S. Popovics and Ujhelyi, 1953; Bonzel and Dahms, 1967).

When the water-cement ratios differ in the two kinds of mixture, the concrete strength–mortar strength ratio also depends on the testing age (Fig. 5.1). The reason for this is that the kinetics of strength development is influenced by the water-cement ratio, as will be shown later. The same statement is valid for the flexural strength (S. Popovics, 1967a, 1967b). Results of comparative strength tests performed according to the cement standards of various countries are available in the literature (Meyer, 1966; Walz, 1963; Foster and Blaine, 1968).

Effects of Cement and Water Contents

The first responses of the student of concrete technology very likely are that (1) the higher the cement content, the higher the strength of comparable concretes; and (2) changes in the water content would not affect the concrete strength within wide practical limits. It is a paradox that experimental data contradict the seemingly logical first response often, and the second response almost always. Experimental data do show, however, that the effect of changes in cement content, or changes in water content, or both, on the concrete strength can be expressed with the ratio of these two quantities (*water-cement ratio*) in a practical, although approximate manner. The general rule is that comparable concretes provide (1) lower strengths with higher water-cement ratios, (2) higher strengths with lower water-cement ratios, and (3) the same strength with the same water-cement ratios. This is independent of the cement and water contents separately within practical limits, at least as a first but

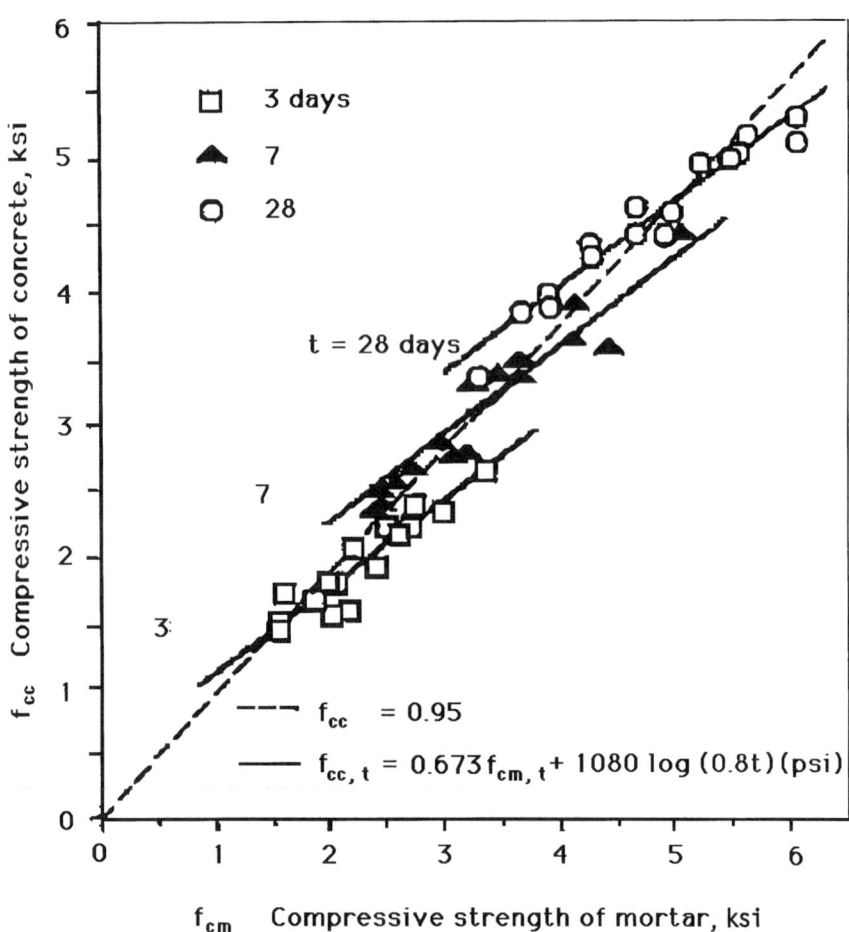

Figure 5.1 Observed and calculated values for the relationship between the compressive strengths of constant-slump concrete and corresponding ASTM C109 mortar at various ages. (From S. Popovics, 1967b. Copyright ASTM. Reprinted with permission.) $f_{cc,t}$ and $f_{cm,t}$ in psi, t is in days. 1 ksi = 6.90 MPa.

fairly good approximation. The strength change caused by a change in the water-cement ratio can be drastic, especially in high-strength concretes. The ratio of water to cement is used in the form of water-cement ratio, or more precisely, the *effective* water-cement ratio.

Water-Cement Ratio and Its Role

The *effective water-cement ratio* is defined as the ratio of the quantity of free water to the quantity of cement in the fresh concrete (Newman, 1959). *Free water* is defined here as the water quantity that is in the matrix immediately after final consolidation (Section 4.4). Water evaporated earlier or lost from the matrix in other ways before or during consolidation (e.g, by aggregate absorption, bleeding, etc.) is excluded from the effective water-cement ratio. On the other hand, water lost from the paste after the consolidation remains part of the effective water-cement ratio. Such water loss does reduce the water-cement ratio in the paste, but this by itself does not make the concrete stronger. Water, absorbed after final compaction, leaves pores behind filled with air in the paste. Thus the strength remains the same as it would be without absorption. Only if such pores are eliminated, (for instance, by re-vibration) will the concrete become stronger corresponding to the reduction in the water-cement ratio. Experimental results also seem to support the idea that water-absorption by the aggregate after compaction does not cause increase in the concrete strength. For instance, experiments by Klieger (1957b) and by Newman (1959) showed that high-strength concretes made with saturated aggregate produce higher strengths than comparable concretes made with air-dry aggregate.

In subsequent discussions, the term water-cement ratio always means the effective water-cement ratio.

As expected from the earlier discussions, the basis of the concrete strength versus water-cement ratio relation is that an increase in the water-cement ratio produces more capillary pores in the matrix portion of concrete. This makes the matrix weaker in comparable cases regardless of changes in the cement and/or water contents (Newman, 1965). A colloquial way to express this is the following: The higher the water-cement ratio, the more diluted the cement paste becomes, therefore the weaker it will be at any stage of hydration. That is, as a first approximation, changes in the cement and/or water contents affect the concrete strength within wide practical limits as they change the water-cement ratio.

The water-cement ratio is a convenient concept because one can use one variable, the water-cement ratio, instead of two from the concrete composition, the cement and water contents, from the viewpoint of many concrete properties, including strength. The relationship between concrete strength and water-cement ratio is only approximate because it may be affected by secondary factors. Fortunately, the approximation is acceptable in most practical cases (Section 5.9).

The quantity of water and the quantity of cement are often expressed in identical units, such as pound or kilogram, which provides the *water-cement ratio by mass*. It also used to be popular in the United States to express the amount of water in gallons and the amount of cement in number of bags in 1 cu yd of concrete; then the unit for the water-cement ratio is gallons per bag. The conversion factor between these two types of water-cement ratios is 11.3 gal/bag for an American gallon and 94 lb of cement per bag.

Another possibility is to express the water and cement quantities in *absolute volumes*. The water-cement ratio calculated in this way is related to the water-cement ratio by mass, as follows:

$$(w/c)_a = G_c w/c \qquad\qquad (5.1)$$

where $(w/c)_a$ and w/c = water-cement ratio by absolute volume and by
    mass, respectively
    G_c = specific gravity of the cement

The water-cement ratio, in any case, provides the quantity of water for 1 unit quantity of cement in the mixture.

The theoretical importance of the water-cement ratio is that it determines the magnitude of capillary porosity in the fresh cement paste, as shown, for instance, by Eq. 4.11. Consequently, it controls many technically important properties of the hardened concrete to a great extent, including strengths, as discussed in Sections 5.3 and 5.4.

Porosity in the Matrix of Concrete

As shown in Section 4.4, there are several different types of pores in the hardened cement paste. In the matrix portion of hardened concrete additional large air-filled voids can be present. These can result from not enough cement paste or mortar in the mixture, and large water-filled voids from bleeding (S. Popovics 1982b). Each pore type may have different effects on the strength of concrete. Since, not enough experience has been gained for the more definite establishment of these differences, however, here again the simplifying assumption is used that all the different *large voids* in the matrix influence the concrete strength to the same extent. However, this extent is not the same as that of the *capillary pores*. The large voids are called *air voids* regardless of whether they are filled with water or air. In practical terms, it is the porosity of the matrix existing at the time of the strength tests that usually controls the concrete strength. Unfortunately, the direct measurements, and even more so the prediction of the capillary porosity in a cement paste at any age except $t = 0$, are quite uncertain, therefore inaccurate (Copeland and Verbeck, 1974).

Therefore, a more practical approach is used determining the total porosity, (i.e., capillary porosity + air voids) in the matrix of the *freshly compacted* concrete. The effect of age on the porosity is considered later. The advantage of this approach is that the water-cement ratio can be used as a substitute for the initial capillary porosity p_o and the air content $100p_o = a$ for the amount of matrix air voids in the fresh concrete (Section 4.4). Both values can be obtained easily and with reasonable accuracy.

The basis of this approach is a special form of Powers' view that whatever the structure of a hardened cement paste may be, it is one that grew out of the structure of the fresh cement paste (Powers, 1968). Therefore, the initial porosity of the fresh paste has a decisive influence on the properties of the paste later in the hardened state.

There is no evidence for any interaction between the effect of capillary pores and the effect of air voids on concrete strength. Thus, a change, say, in the capillary pores does not influence directly the effect of air voids on strength, and vice versa. This makes it possible to discuss the effect of capillary porosity and that of air content on concrete strength separately. The effects of the cement, water, and paste contents, those of aggregate as well as admixtures (i.e., factors other than porosity) will be discussed later. Finally, it will be shown how all these relationships are affected by air content.

5.3 EFFECT OF CAPILLARY POROSITY ON PASTE STRENGTH

Role of Porosity in Strength Reduction

The strengths of hardened cement pastes and concretes as well as that of any brittle material decreases rapidly with an increase in porosity (Modry, 1973). The reason for the rapidity of the strength reduction is a combination of three factors: (1) not only do the pores decrease the quantity of solid material but (2) they also reduce the number of bonds, and most important, (3) they act as stress concentrations—the sharper the pore, the greater the stress concentration (Section 2.4). We do not know at present what percentage of a strength reduction is due to loss of bonds and what percentage to stress concentrations. As was shown earlier, such a distinction is difficult even conceptually since these factors are interrelated. Besides, it is easier to measure porosity than bonds. Therefore, the present practice is to attribute strength reductions of concretes solely to capillary pores and air voids in its matrix.

This simplifying assumption may be used to most factors affecting the matrix strength (age, etc.), since they act through changing the porosity (Locher, 1977; Winslow and Diamond, 1970; Sandstedt et al., 1973).

The amount of air voids can be controlled through air entrainment and compaction. The amount of capillary voids in hardening pastes decreases with a decrease in the original water-cement ratio (Fig. 4.7). It also decreases with the progression of hydration, that is, with age. All these reductions in porosity produce more than proportional strength increases.

Relationship Between Porosity and Paste Strength

A considerable amount of experimental evidence demonstrates that a decrease (or increase) in porosity of the hardened cement paste causes an increase (or decrease) in its strength. For instance, the strength increase caused by carbonation, that is, by carbon dioxide taken up by a hardening cement paste, takes place parallel to a reduction in capillary porosity (Pihlajavaara, 1968). Also, Verbeck and Helmuth (1969) proved experimentally that the effect of porosity on the strength of cement paste is essentially the same as that on the strength of many other brittle materials. So it is more a matter of the number of pores, (i.e., more the degree of space filling) that controls the strength reduction rather than the type of the material (Beaudoin and MacInnis, 1971).

Experiments with cement compacts also support this view. Specifically, standard mortar-cube strengths correlate linearly with mechanical properties measured on compacts of bottle hydrated cement as well as on compacts of unhydrated cement cured in water (Chandra et al., 1968).

In addition, experimental data show that the simplification attributing most of the strength changes solely to changes in porosity is acceptable not only in a qualitative sense but also quantitatively. An illustrative example is presented in Figure 5.2. This also shows the importance of bonds. Note that the normal-cured, normal-cured and partially pressed, and steam-cured specimens provide practically the same compressive strengths when the capillary porosities are the same, but the strengths of the hot-pressed specimens are much

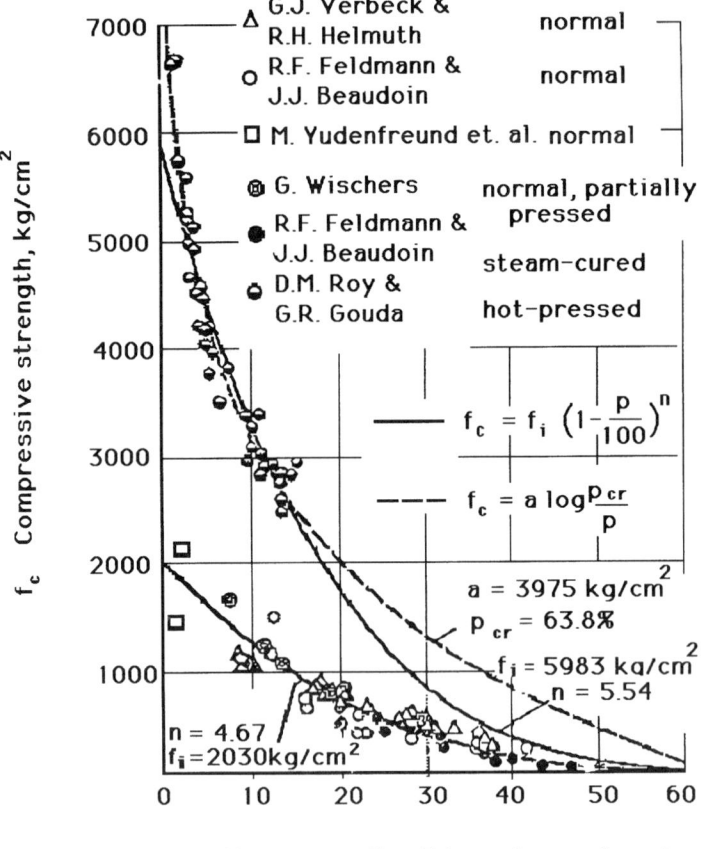

p   Capillary porosity, % by volume of paste

Figure 5.2 Effect of capillary porosity on the compressive strength of cement pastes hardened under different conditions. 1 kg/cm² = 14.2 psi = 0.098 MPa. (From Locher, 1977. Copyright Beton Verlag. Reprinted with permission.)

higher. This shows that the applied hot pressure created stronger bonds than did the other three curing methods.

The upper limits for the paste strengths can also be seen in Figure 5.2. This is the extrapolated, hypothetical strength called the *intrinsic strength* which would be obtained if the capillary porosity were zero in the hardened paste. Note that the power function in this figure provides a finite intrinsic strength f_i, whereas the intrinsic strength of the log formula is infinitely large. On the other hand, the strength predicted by the power function is zero only when the porosity is 100% whereas the log formula provides zero strength at a value of $p_{cr} < 100\%$, which is more realistic. The porosity p_{cr} is called the *critical porosity.* The numerical value of p_{cr} may be less for compressive strength than for flexural strength (S. Popovics, 1989b).

These equations also imply that all the various types of relatively large air-filled voids in the paste (Section 4.4) influence the concrete strength to the same extent. This assumption is probably not true, strictly speaking (S. Popovics, 1989).

Effect of Hydration

In Figure 5.3 compression test results on cement pastes of six different water-cement ratios are plotted against the amount of hydration products, as reported by Taplin (1959). The specimens were tested at ages varying from a few hours to several thousand hours. The strength results for each water-cement ratio show good correlation with the amount of hydration products, regardless of the testing age. That is, the strength increases take place through the reduction of the capillary porosity by the developing hydration products, regardless of the age. This shows again that the porosity in the paste existing at the time of the strength test controls the strength. Findings by others support this view (Hummel, 1959; S. Popovics, 1974; Powers, 1958; Chomahidze and Chikobani, 1982).

As expected, the porosity versus strength relationship is nonlinear for which there are only empirical formulas available at this time (Roy and Gondas 1974; Skalny and Bajza, 1970). An exponential formula seems to be valid within wide limits (S. Popovics, 1969a; Kondo et al., 1972; Manning and Hope, 1971), as discussed in Section 5.12. The *MacKenzie model* has been mentioned as another possible approach (Lawrence et al., 1971). The well-fitting lines in Figure 5.2 show that power functions or logarithmic functions may also be used within certain limits for pastes. In Figure 5.3 the compressive strengths of pastes are again expressed as a power function of the amount of hydration products. The power (exponent) in this test series is around 2.2 for water-cement ratios from 0.25 by weight to 0.50, or even from 0.157 to 0.65 (S. Popovics, 1974a). A similar formula with an exponent of 1.5 can be used for Taplin's flexural strength results. That is, within the given limits of validity, the compressive strength as well as the flexural strength can be ex-

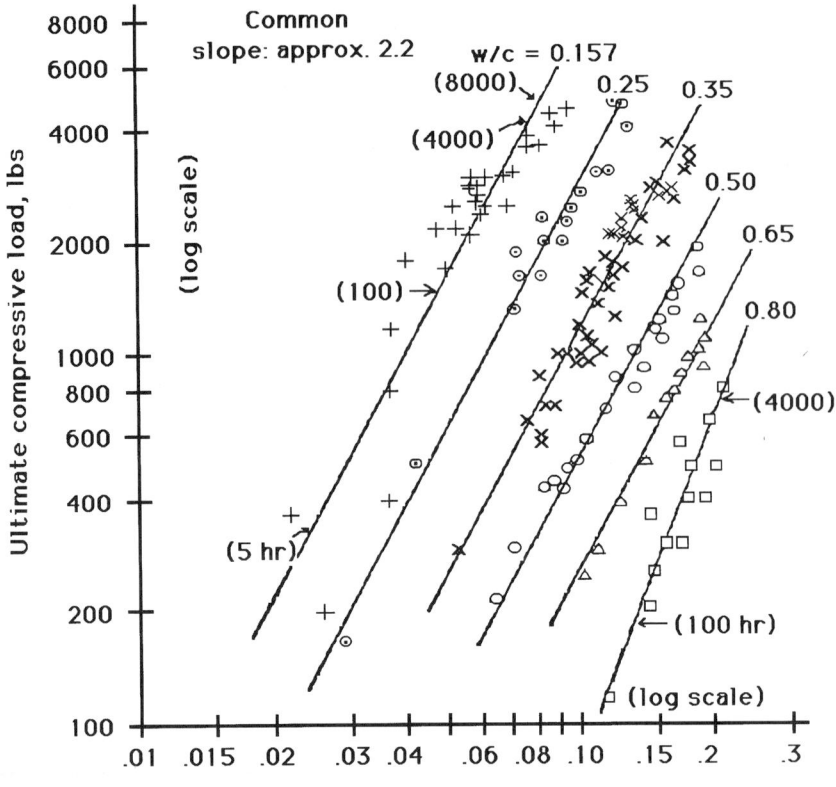

Amount of hydration products per g of cement

Figure 5.3 Effects of water-cement ratio, by mass, and amount of hydration products on the compressive strength of pastes. The numbers within parentheses show the approximate age at testing. 1 lb = 0.454 kg. Data from Taplin (1959). (From S. Popovics, 1974a.)

pressed in terms of the total amount of hydration products per gram of cement in the cement paste at the time of testing:

$$f_c = K_c V^{2.2} \tag{5.2}$$

and

$$f_f = K_f V^{1.5} \tag{5.3}$$

where f_c and f_f = compressive strength and flexural strength of the pastes, respectively

V = total amount of hydration products

K_c and K_f = experimental coefficients that are independent of the strength, amounts of hydration products and age of the specimen; they may depend, however, on the applied methods of curing and testing, units, water-cement ratio, etc.

The use of constant exponents in Eqs. 5.2 and 5.3 is a simplification because higher water-cement ratios should give larger exponents (S. Popovics, 1969a).

Not only Eqs. 5.2 and 5.3 but also independent analyses by other researchers have demonstrated that the hydration (i.e., a change in porosity) affects the compressive strength c of a paste or concrete more than does its flexural or tensile strength (Guttmann, 1935; Werner, 1931; S. Popovics, 1973; Johnston, 1970). This, in turn, explains the experimentally demonstrated facts that (1) a change in the water-cement ratio affects the flexural and tensile strengths relatively less than the corresponding compressive strength (Section 5.11), and (2) the period of increase of the flexural and tensile strengths with the age is shorter than that of the compressive strength (Fig. 3.8). Also, the f_c/f_f ratio calculated from Eqs. 5.2 and 5.3 increases proportionally with approximately the -1.4 power of the porosity.

Another power function, recommended by Powers for the compressive strength versus modified gel–space ratio relationship, is (Powers, 1958a)

$$f_c = 29{,}000X_F^3 \qquad (5.4)$$

which can be approximated as

$$f_c \approx 35{,}000X_F - 14{,}000 \qquad (5.5)$$

where f_c = compressive strength of cement paste specimens, psi
        X_F = modified gel–space ratio, as defined by Eqs. 4.16 and 4.21.

A similar relationship was established by Fulton (1963).

Figure 5.4 reveals that Eq. 5.5 provides a better fit than Eq. 5.4 within 0.5 and 0.9 of X_F. This may mean that an exponent smaller than 3 ($X_F^{2.63}$) fits the experimental data better. On the other hand, Eq. 5.4 implies that the compressive strength of the completely dense cement paste ($X_F = 1$) is approximately 29,000 psi (200 MPa $\simeq$ 2000 kg/cm²). This is in close agreement with the pertinent value in Figure 5.2, but there the exponent is 4.67.

These contradictions are not surprising because the equations were obtained by curve fitting. They are valid only for results obtained under the conditions used by the investigators. Powers' experimental results represent paste specimens from three different mixtures that were tested at six different ages, ranging from 7 days to 2 years. Thus, significantly such a relationship is independent, within wide limits, of the original water-cement ratio, air content, or the age of testing. Consequently, in comparable hardening pastes the specimen is expected to have the highest strength that has the highest denseness at the age of testing. This statement is independent of the age or the water-cement ratio separately if the degree of hydration exceeds 50%. Equations 5.4 and 5.5 also imply through the definition of X_F by Eq. 4.21 that a change in the water content has the same effect on the compressive

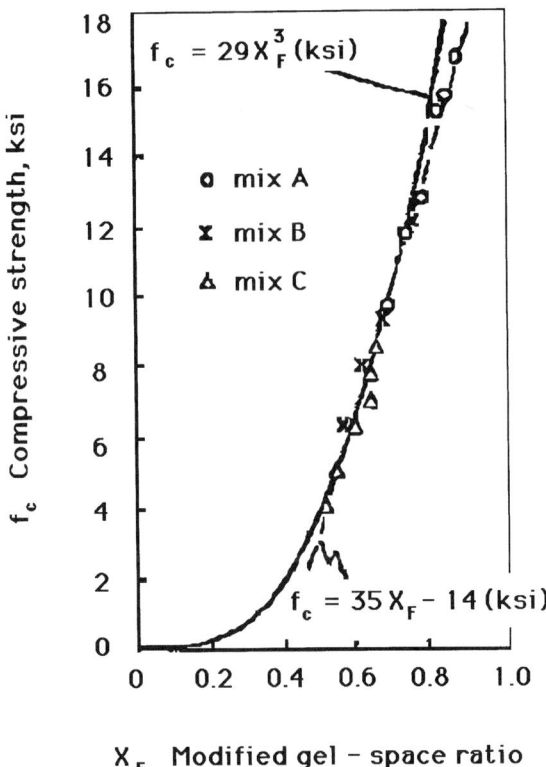

Figure 5.4 Relationship between compressive strengths and modified gel-space ratio of cement pastes. 1 ksi = 6.90 MPa. (From Powers, 1958a. Copyright Portland Cement Association. Reprinted with permission.)

strength as the same change in the volume of the air voids. In other words, the implication is that the strength-reducing effect of capillary pores is the same as that of air voids. This assumption has not been proven for cement pastes and has been shown false for concretes (Section 5.12).

Note also that a higher degree of hydration in a specimen does not automatically mean a higher strength. The original volume of capillary pores may have been so high, due to a high water-cement ratio, that even the larger amount of cement gel produced could not reduce it adequately. Another consideration is that the hydration in pastes of water-cement ratios less than 0.35 by weight may stop at a low degree of hydration because of the lack of further capillary spaces to accept more gel; nevertheless, the strength will be high due to a high denseness of the paste. Good correlation between strength development and a hydration characteristic (such as the development of the amount of hydration products, or that of nonevaporable water content, or that of heat of hydration, or surface area of the gel) by itself can be obtained only with pastes of roughly identical original pore content (Powers and Brownyard,

1947). This explains the paradox shown by Figure 5.3 that the degree of hydration in cement pastes at a given age increases with the increase in the original water-cement ratio (Taplin, 1959; Hunt, 1969; Copeland and Kantro, 1964, 1968) while the strength decreases.

The importance of the discussion above lies in the demonstration that strength changes in hardened cement pastes can be correlated numerically with changes in the porosity of the paste. Considering that the initial capillary porosity is a unique function of the water-cement ratio (Eq. 4.11), the strengths of comparable cement pastes can also be correlated with the water-cement ratio. Equation 4.11 also shows that when the capillary porosity is low, the porosity increases more rapidly with an increase in the water-cement ratio than at higher porosities. Thus a given increase in the water-cement ratio should produce a higher strength loss at lower water-cement ratios than at higher ones. It is shown in the next section that this expectation is fulfilled.

The *porosity of the aggregate* influences the unit weight, deformability, and durability of the hardened concrete significantly, but besides the case where very light aggregate is used, not its strength. Thus the concrete strength can be controlled to a large extent by controlling the water-cement ratio and the air content in the matrix portion of the concrete. This is discussed in the next section.

5.4 CONCRETE STRENGTH VERSUS WATER-CEMENT RATIO

Relationship in General: Abrams' Rule

As discussed in Section 5.2, the first approximation of the concrete strength versus water-cement ratio† relationship is that *comparable* concretes provide: (1) lower strengths with higher water-cement ratios, (2) higher strengths with lower water-cement ratios, and (3) the same strength with identical water-cement ratios. This relationship is called in *Abrams' law* or, more accurately, *Abrams' rule* after his famous paper on this subject published in 1918 (Abrams, 1918). It should be noted for the sake of historical accuracy, however, that European investigators had contributed to this topic long before that. Graf, (1960) says, for instance, that Zielinszki was probably the first to examine systematically the effect of water-cement ratio on the strength of mortars (Zielenszki and Zhuk, 1901; Zielenszki, 1909). Feret in his early work to relate strength to the denseness of mortars also contributed to this topic (Feret, 1892).

Abrams' rule is usually applied for compressive strengths of portland cement concretes, although it is also valid for other strengths and for many other hydraulic binders (Section 5.11).

---

†With the exception of some of the formulas, the term *water-cement ratio* is also used in this book for *water-cementitious materials ratio,* for the sake of brevity, unless indicated otherwise.

Qualitative Aspects of Abrams' Rule

The basic form of Abrams' rule claims that *the strengths of comparable concretes depend solely on their water-cement ratios regardless of their compositions.* Comparable concretes are made of materials of similar qualities, are prepared, cured, and tested under identical conditions, and are compacted to the same air content. There are many experimental results in the literature showing that the degree of approximation of this relationship is good within practical limits of concrete composition. An example is given in Figure 5.5. Here compressive strength results as well as fitted curves are presented as a function of the water-cement ratio for concretes made with the same structural lightweight aggregate but with different mix proportions (Kaplan, 1960b). Since it was shown in Section 5.3 that the strength of a hardened cement paste is also controlled by the water-cement ratio, one can rephrase the basic form of Abrams' rule. This second form states that the concrete strength is

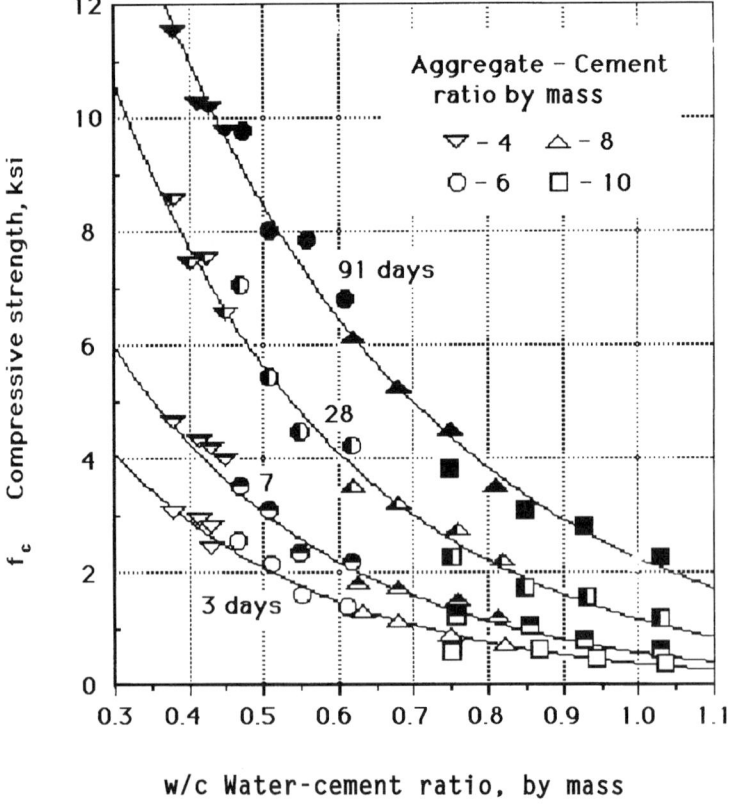

Figure 5.5 Compressive strength of concrete of various ages as a function of water-cement ratio. 1 ksi = 6.90 MPa. (From Kaplan, 1960b. Copyright Thomas Telford Publishing. Reprinted with permission.)

controlled solely by the strength of matrix. A third form of the Abrams' rule is that the addition of aggregate does not influence the matrix strength, thus the concrete strength, regardless of the composition, as long as the concretes remain comparable. The third formulation of the strength versus water-cement ratio relationship shows immediately that the validity of the relationship has definite restrictions: namely, if the quality or quantity of aggregate does affect the concrete strength, the relationship is not valid. For instance, when Gilkey, proving the obvious, substituted charcoal for traditional coarse aggregate in a concrete, he got strength reduction despite the identical water-cement ratio. (Gilkey, 1961). Another counterexample is the case where so little cement paste is used that it is not enough to fill up the voids between aggregate particles. One could say that in both of these cases the concretes were not comparable; thus the counterexamples do not invalidate the Abrams' rule. So the term *comparable* is important; therefore, it needs a sharper definition.

To say that different concretes are comparable as long as their strength is controlled solely by their matrix portion is true, but it is just an argument in a circle, that is, a repetition of the second form of Abrams' rule. It is more meaningful to list the conditions for the validity of the concrete strength versus water-cement ratio relationship. So Abrams' rule is valid for different compositions if *all* the following conditions are fulfilled in the various concretes to be compared:

1. The strength-developing capabilities of the cements used are identical.
2. The quantities and strength-influencing effects of the admixtures used are identical.
3. The concrete specimens are prepared, cured, and tested under the same conditions.
4. The concrete ingredients (cement, water, aggregate particles, admixtures) are distributed uniformly in the concrete.
5. The air contents are the same in the concretes, the air voids are distributed uniformly in the concrete, and none of the voids is too large for the size of the specimens.
6. The aggregate particles are stronger than the matrix; that is, the fracture propagates more in the matrix than in the particles.
7. The bond between the aggregate surfaces and matrix is equally strong in the concretes compared and is strong enough to transfer the major portion of stresses in the matrix to the aggregate before the concrete is crushed by the load.
8. The strength-affecting physical and/or chemical processes in the concretes (drying, aggregate reactivity, etc.), beyond the cement hydration, are not overwhelming (cracking, etc.) and are the same,
9. The nonhomogeneity or composite nature of concrete, the origin of which is in the differing characteristics of matrix and aggregate par-

ticles, affects the strength of the compared concretes to the same extent.

10. The contribution of the aggregate skeleton, resulting from interlocking of the aggregate particles during loading, to the concrete strength is the same in the various concretes (S. Popovics 1971b).

If all these conditions are not fulfilled simultaneously, which is the usual case, concretes of the same water-cement ratio do not produce the same strength. However, the deviations may be small in many practical cases. The effects of some of these factors on the strength versus water-cement ratio relationship is discussed later.

Other Consequences of Changes in Water-Cement Ratio

The water-cement ratio in a concrete can be changed by (1) changing the cement content while keeping the water content constant, (2) changing the water content while keeping the cement content constant, and (3) changing both. Although Abrams' rule does not distinguish between these methods, the way the water-cement ratio is changed can have an effect on some of the concrete properties. When the water-cement ratio is reduced only by the reduction of the water content, the paste content of the concrete will also be reduced. When the water-cement ratio is reduced only by an increase in the cement content, the paste content will be increased. This change in the paste content has a small but definite effect on the concrete strength versus water-cement ratio relationship, as will be shown in Section 5.9. More important, when the cement content is changed and the water content is kept constant, the concrete consistency does not change too much. In other words, an increase in the cement content, up to a certain limit, does produce an increase in the concrete strength, while keeping the consistency unchanged, because this reduces the water-cement ratio.

On the other hand, when the cement content is kept constant and the water content is changed, the concrete consistency can change drastically (S. Popovics, 1982b). When the consistency of a concrete is so stiff that its workability is just barely suitable for the available method of compaction, any reduction in the water-cement ratio caused by a decrease in the water content will lead to an overly dry consistency. The consequence of this is a reduction in concrete strength because the impaired workability will result in incomplete compaction (i.e., a higher air content in the concrete (Fig. 5.6). Of course, this increase in the air content can be eliminated by using a more efficient method of compaction, for instance, vibration rather than hand rodding. In this case the reduced water-cement ratio will provide the higher strength again, as predicted by Abrams' rule.

The water-cement ratio affects not only the concrete strengths at a given age, but also the kinetics of the hardening process (S. Popovics, 1967c, 1969c; Graf et al., 1960; Jevtic, 1959; Ackroyd, 1964; Wischers, 1964–1965; Henk,

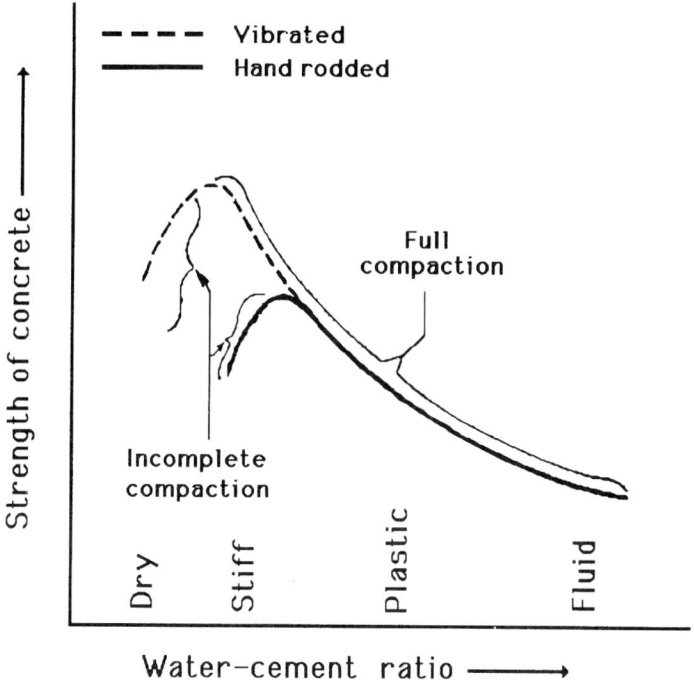

Figure 5.6 Effect of water content, as expressed by the water-cement ratio, on the strength and consistency of concrete.

1966; Basalla, 1965; Meyer, 1963b; Gaynor, 1968). A concrete of low water-cement ratio has not only a higher strength but also a higher initial hardening rate and a greater deceleration of hardening than those of comparable concretes with higher water-cement ratios. This means that concretes with low water-cement ratios have relatively high early strengths but small strength increases later compared to concretes with higher water-cement ratios. The higher capillary porosity in the pastes of higher water-cement ratios results in lower strengths, especially at the beginning of the hardening. However, the availability of this large pore content helps the development of solid hydration products. Therefore, the process of hydration and the strength development are more intensive at later ages in pastes of higher water-cement ratios than in comparable pastes of low water-cement ratios. This phenomenon is illustrated in Figure 5.1, where the water-cement ratio of the mortar is lower than that of the concrete.

Accelerated Strengths and Steam Curing

Abrams' rule is also valid for strength results obtained after curing at elevated temperatures. Such is the case when the concrete specimens are cured by steam, or by any of the standard A, B, and C accelerated test methods of

ASTM C684-89 (Section 1.3). An example of this is presented in Figure 5.7. It can be seen that there is a specific relationship for each concrete mixtures tested after procedure A curing that can be expressed in formulas. Other, somewhat similar relationships have been established for procedures B and C as well as for steam-cured concretes (S. Popovics and Pfeifer, 1992). Compressive strength results obtained by Siviero (1994) after steam curing at the age of 17 hours also show the existence of good relationships between compressive strength and water-cement ratio. The pertinent coefficients for the formulas can be obtained by trial mixtures in the usual manner (Section 5.6). The limits of validity here seem to be the same as those for the strengths obtained traditionally.

5.5 FORMULAS FOR RELATIONSHIP BETWEEN COMPRESSIVE STRENGTH OF CONCRETE AND WATER-CEMENT RATIO

Formulas with One Independent Variable

As has been shown, the validity of Abrams' rule is restricted. Strictly speaking, differing strength versus water-cement ratio, or f versus w/c, curves are valid for different air contents, aggregate types, maximum particle sizes, cement and water contents, admixtures, and for any of the 10 conditions for the validity of Abrams' rule discussed in Section 5.4. However, there are many practical cases where some of these curves can be combined into a single curve with good approximation, as shown in Figure 5.5. Such a curve represents a single formula with one independent variable, the w/c, with acceptable approximation within certain limits.

These formulas may be called *strength formulas* or predictive formulas since they can be used for the prediction of concrete strength from the water-cement ratio. Many such formulas have been recommended. These are all empirical essentially, although some of them have been derived from a group of assumptions (Schiller, 1958; Hobbs, 1972). Use of other assumptions would have resulted in other formulas, as demonstrated in Example 5.1. The main criterion for selection of the basic forms of these formulas was that their graphs have the shapes of the fitted curves of f versus w/c in Figure 5.5. Several possibilities are presented below (S. Popovics, 1990a).

1. *Exponential function* or Abrams' formula (Abrams, 1918):

$$f = \frac{A_A}{B_A^{w/c}} + C_A = A_A \times 10^{-(\log \beta_A)w/c} + C_A \tag{5.6}$$

where f = strength of concrete
  w/c = water-cement ratio

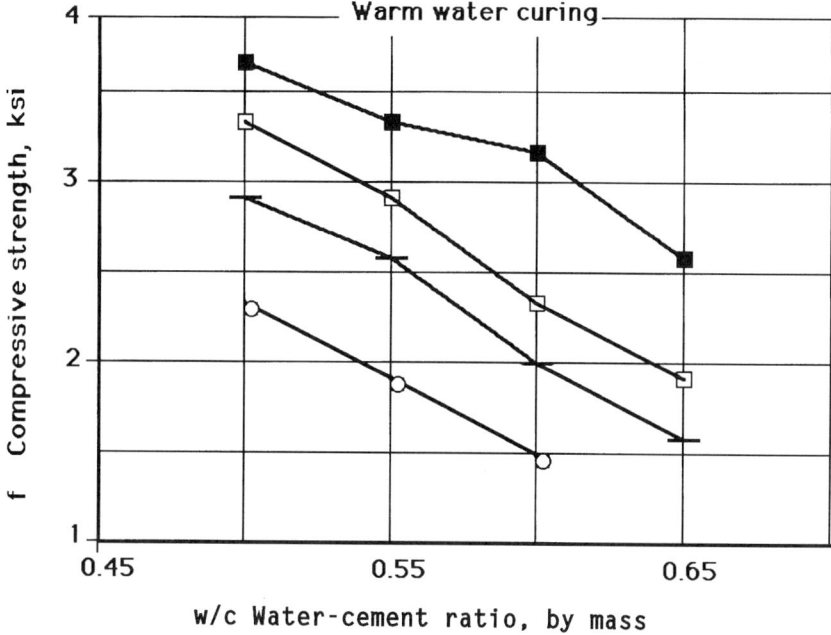

Warm water curing

—	Type I – 500 lb/cy
o	Type II – 500 lb/cy
■	Type III – 500 lb/cy
□	Type III – 750 lb/cy

— Type I – 500 lb/cy   f=2.4567e+4*10^(-1.8292w/c) R^2=0.978

o Type II – 500 lb/cy   f=2.2019e+4*10^(-1.9530w/c) R^2=0.993

■ Type III – 500 lb/cy   f=1.1171e+4*10^(-0.9559w/c) R^2=0.936

□ Type III – 750 lb/cy   f=2.0714e+4*10^(-1.5688w/c) R^2=0.985

Figure 5.7 Strength versus water-cement ratio relationships of three standard cement types for ASTM C684-89 warm-water cured concretes. 1 ksi = 6.90 MPa. (From S. Popovics and Pfeifer, 1992.)

A, B, C = empirical coefficients that are independent of the strength and water-cement ratio; however, they depend on the units, materials, type of strength, test method used, age of testing, and on the conditions of the validity of Abrams' rule.

See also Example 5.1.

2. *Logarithmic function* or Schiller's formula (Schiller, 1958):

$$f = A_S \log(B_S c/w) + C_S \qquad (5.7)$$

$$= A_L - B_L \log(w/c) = A_L - \log[(w/c)^{B_L}] \qquad (5.8)$$

3. *Power function* and *polynomial:*

$$f = A_H(c/w)^{B_H} + C_H \tag{5.9}$$

$$= A_C(w/c)^2 + B_C(w/c) + C_C \tag{5.10}$$

$$= A_I(c/w)^2 + B_I(c/w) + C_I \tag{5.11}$$

$$= A_P(C_P - w/c)^{B_P} = A_R\left(1 - \frac{w/c}{C_R}\right)^{B_R} \tag{5.12}$$

and $$f = A_F(1 + C_F w/c)^{-B_F} \tag{5.13}$$

See also Example 5.1.

4. Zietsman (Fulton, 1964):

$$f = A_z a(c/w) - B_z \sin \pi(\sqrt{2a(c/w) + 1} - 1) \tag{5.14}$$

where the symbols are identical as in Eq. 5.6.

In the United States, Abrams' formula is predominant; in Europe, including the past states of the Soviet Union, it is Eq. 5.9 with the Bolomey formulation. The A_A and B_A coefficients of Eq. 5.6 are discussed further later.

Since all these formulas were developed essentially in an empirical way, no preference can be given to any of them from a theoretical standpoint. It might be that under certain circumstances some of these formulas yield a better approximation to experimental results than the others. Such cases can be when linearity is desirable, or for a certain group of cements, or very low or very high water-cement ratios, or a desirable value of $(w/c)_{cr}$ or f_i.

Strength formulas with variable(s) in addition to the water-cement ratio for the concrete strength versus water-cement ratio relationship are discussed in Section 5.9.

Coefficients of the Formulas

Most of Eqs. 5.6 through 5.14 contain three empirical coefficients that can be obtained by curve-fitting techniques (ACI Committee 228, 1996). Experience has shown, however, that two such coefficients are usually enough to provide a satisfactory fit to experimental data within practical limits of strengths and water-cement ratio. For instance, one can assume that the C coefficients in Eqs. 5.6 through 5.8 are negligible, or that some of the coefficients can be predetermined and remain constant. Examples are for the 28-day compressive strength of normal-weight concretes of constant (negligible) air content, as follows:

1. Eq. 5.6: Abrams (1918), and following him, everybody else, assumed that $C_A = 0$. This simplified formula is illustrated by replotting Figure 5.5 in Figure 5.8 in a semilog system.

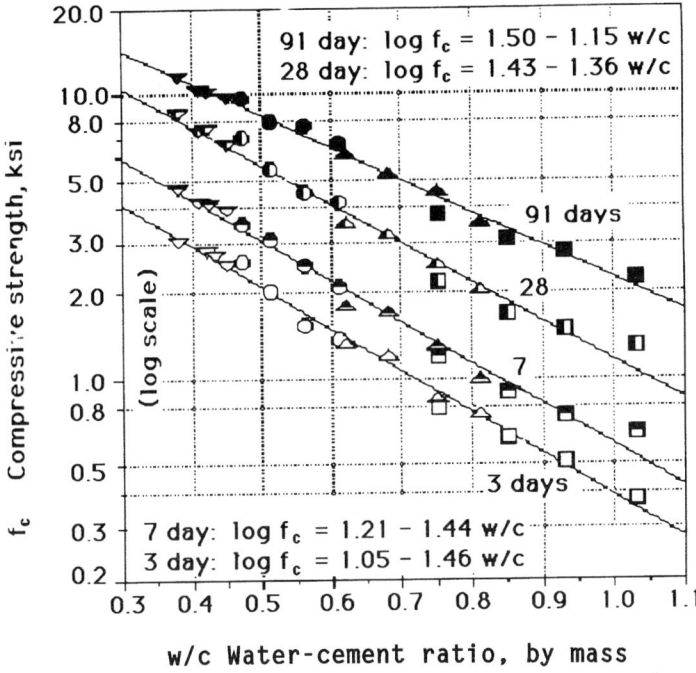

Figure 5.8 Experimental data of Figure 5.5 in the log f_c versus w/c system representing Eq. 5.6. f_c is in ksi. 1 ksi = 6.90 MPa.

According to Lorman (1962), Powers and Lord offered similar exponential formulas.

2. Eq. 5.7: Schiller (1958) recommended $C_S = 0$; S. Popovics found (1969a) $B_S = 1.5$ in one test series (Fig. 5.9) and $B_S = 1.1$ in another (Fig. 5.10) along with $C_S = 0$. This formula is also plotted in Figure 5.2, with the dashed line representing a possible strength versus porosity relationship for cement pastes.

3. Eq. 5.8: In this formula $A_S = B_L$ above.

4. Eq. 5.9: Graf (1960) recommended $B_H = 2$ and $C_H = 0$; Dutron (1955) $B_H = 1.5$ and $C_H = 0$; Riha (1984) $B_H = 1.24$ and $C_H = 0$; Bolomey (1926) $B_H = 1$ and $C_H = -0.5A_H$; Bendel (Lorman, 1962) $B_H = 1$ and $C_H = -0.15A_H$; Lyse (Lorman, 1962) $B_H = 1$; and Hobbs (1985) $A_H = 0.45f_{0.5} + 10.2$, $B_H = 1$, and $C_H = -16.6$, where $f_{0.5}$ is the compressive strength of the concrete with $w/c = 0.5$ by mass.

5. Eq. 5.12: The values of $C_P = C_R$ may vary, at least within 2 and 4, for practical limits of the water-cement ratio without hurting the goodness of fit of the formula significantly. If $C_P = 2$, which may be a little low, $B_P = 4$ with good approximation regardless of the age at least within

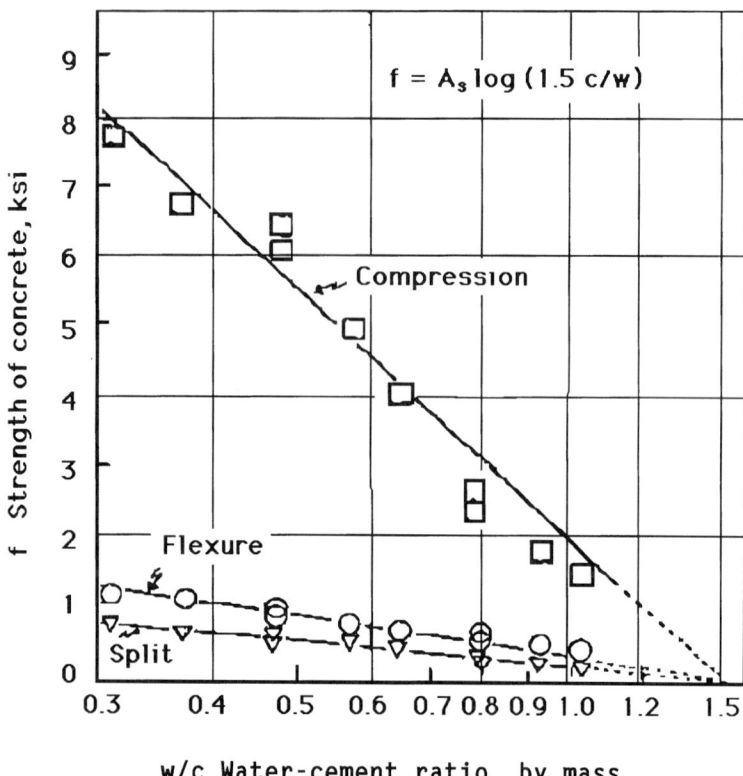

Figure 5.9 Compressive, flexural, and splitting strengths of comparable concretes at the age of 28 days as a function of the water-cement ratio in the f versus log(w/c) system. (From S. Popovics, 1969a. Copyright ASTM. Reprinted with permission.) The lines represent Eq. 5.7. 1 ksi = 6.90 MPa.

3 and 91 days (see: Table 5.1).

Eq. 5.12 is plotted in Figure 5.2 with solid lines representing a possible strength versus porosity relationship for cement pastes.

6. Eq. 5.13: Feret (1892) recommended $C_F = 3.15$ and $B_F = 2$; and Talbot and Richart (1923) $C_F = 3.15$ and $B_F = 2.5$. Slater (Lorman, 1962) offered a similar power function.

7. Eq. 5.14: Faury (1958) also presents a formula attributed to Wieser that is similar to Eq. 5.14.

As Figure 5.2 as well as Figures 5.11 and 5.12 show, the 28-day strength values calculated by Eqs. 5.6 through 5.13 in ksi lie fairly close to each other despite the differences in the formulas. This is so because the range of the practically applicable water-cement ratios is less than 1, by mass, that is, quite narrow. Others also found good agreement among the various formulas for

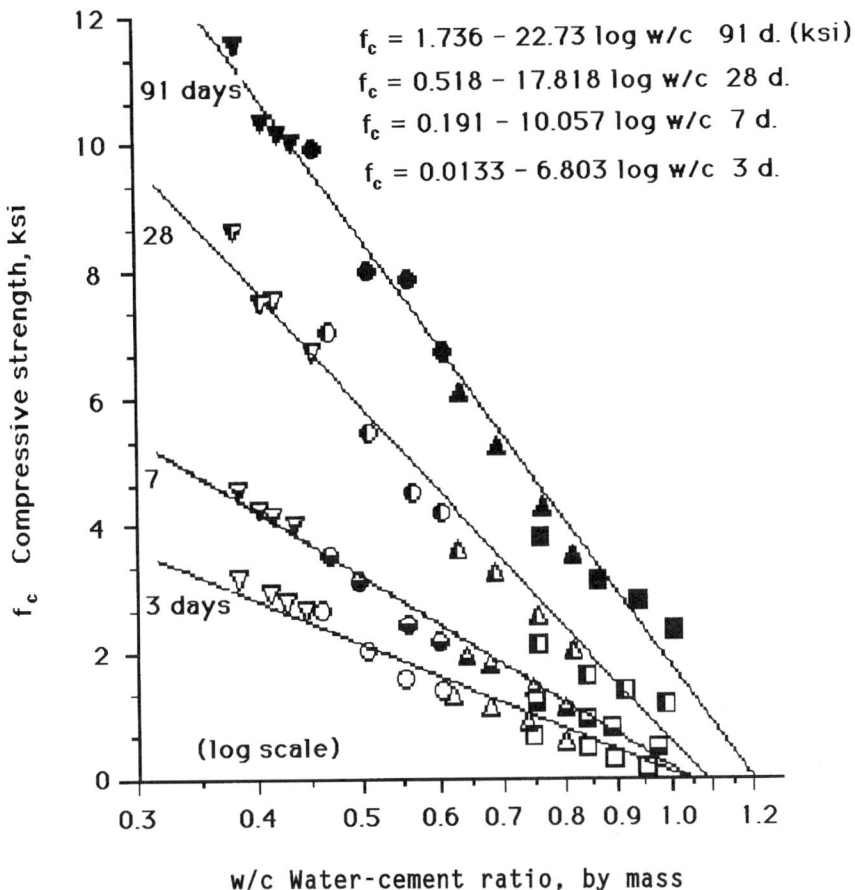

Figure within chart:

$f_c = 1.736 - 22.73 \log w/c$ 91 d. (ksi)

$f_c = 0.518 - 17.818 \log w/c$ 28 d.

$f_c = 0.191 - 10.057 \log w/c$ 7 d.

$f_c = 0.0133 - 6.803 \log w/c$ 3 d.

Figure 5.10   Experimental data of Fig. 5.5 in the f_c versus $\log (w/c)$ system representing Eq. 5.8 (From S. Popovics, 1969a. Copyright ASTM. Reprinted with permission.) The point of convergence of the fitted lines on the w/c axis is approximately $w/c = 1.1$. 1 ksi = 6.90 MPa.

the relationship between strength and water-cement ratio (Ros, 1950; Palotas, 1935; Schiller, 1971). Equations 5.7 through 5.13 can be derived from Eq. 5.6 by expansion into series. This explains mathematically why the curves in Figures 5.11 and 5.12 lie so close to each other.

Interpretation of the Formulas

The strength formulas presented above have mathematical and physical interpretations that are simple if the C coefficients in Eqs. 5.6 through 5.9 are assumed to be 0. In this case Eq. 5.6 provides a straight line in a $y = \log f$ versus $x = w/c$ semilogarithmic system with a slope of $\log B_A$ (Fig. 5.8). This line passes through point $w/c = 0$ and $f_c = A_A$. It is also the mathematical

TABLE 5.1 Parameters of some of the strength-predicting formulas for various ages based on the experimental data in Figure 5.5[a]

Eq. No.	Age (days)	Parameter A (ksi)	B	C	f_i (ksi)	$(w/c)_{cr}$ by Mass	r	Remark
5.6	3	11.252	28.84	0	11.252	inf.	1.00	$f_i = A$
	7	16.073	27.35	0	16.073	inf.	0.99	
	28	27.122	23.07	0	27.122	inf.	0.99	
	91	31.812	14.16	0	31.812	inf.	0.99	
5.8	3	0.133	6.803 ksi	—	inf.	1.046	0.98	$(w/c)_{cr} = 10^{A/B}$; not valid for very high w/c values
	7	0.191	10.057 ksi	—	inf.	1.044	0.98	
	28	0.518	17.818 ksi	—	inf.	1.069	0.98	
	91	1.736	22.730 ksi	—	inf.	1.192	0.99	
5.9	3	0.458	2.135	0	inf.	inf.	0.99	
	7	0.685	2.112	0	inf.	inf.	0.99	
	28	1.357	2.005	0	inf.	inf.	0.99	
	91	2.559	1.676	0	inf.	inf.	0.99	
5.9	3	1.772	1	−1.444 ksi	inf.	1.227	0.99	$(w/c)_{cr} = -A/C$
	7	2.262	1	−2.151 ksi	inf.	1.052	1.00	
	28	4.647	1	−3.642 ksi	inf.	1.276	0.99	
	91	5.854	1	−3.421 ksi	inf.	1.712	0.99	
5.10	3	7.405	−14.547 ksi	7.589 ksi	7.589		0.99	$f_i = C$; not valid for very high w/c values
	7	11.605	−22.383 ksi	11.480 ksi	11.480		1.00	
	28	19.958	−38.852 ksi	20.273 ksi	20.273		0.99	
	91	18.100	−39.768 ksi	23.956 ksi	23.956		1.00	
5.11	3	0.014	1.722 ksi	−1.403 ksi	inf.		0.99	
	7	0.136	2.140 ksi	−1.753 ksi	inf.		1.00	
	28	0.175	4.023 ksi	−3.112 ksi	inf.		0.99	
	91	−1.106	9.806 ksi	−6.660 ksi	inf.		0.99	
5.12	3	0.344	4.474	2	7.644	2	0.99	$f_i = AC^B$; $(w/c)_{cr} = C$; not valid for extremely high w/c values
	7	0.520	4.399	2	10.971	2	0.99	
	28	0.045	4.175	2	18.876	2	0.99	
	91	2.032	3.533	2	23.521	2	0.99	

[a]Graphs calculated with the 28-day parameters above are presented in Figs. 5.11 and 5.12 along with pertinent experimental data from Fig. 5.5. 1 ksi = 6.90 MPa.

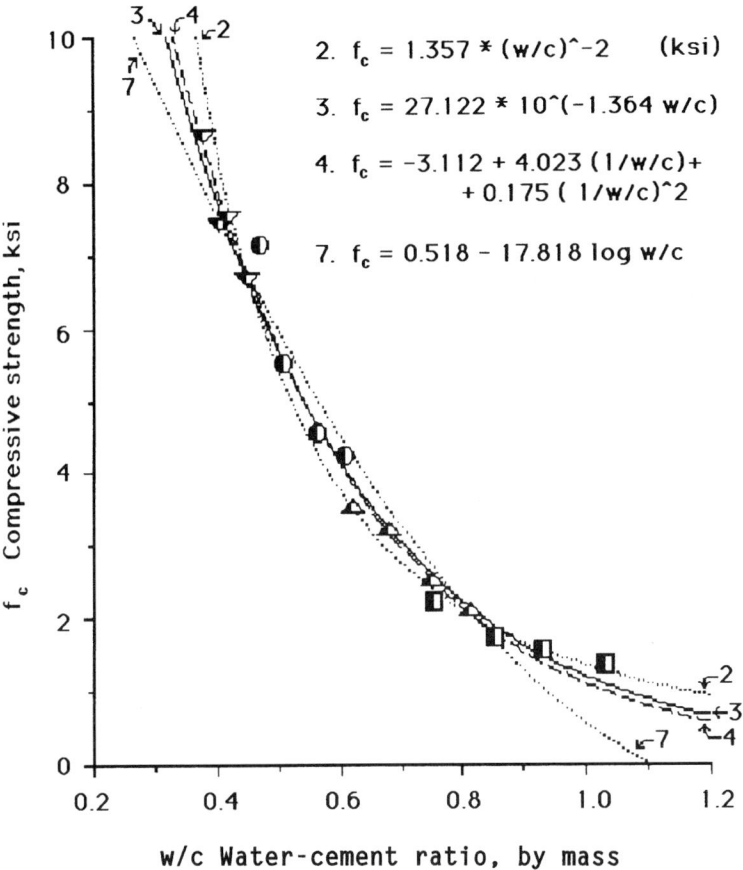

$$2. \ f_c = 1.357 * (w/c)^{-2} \quad (ksi)$$

$$3. \ f_c = 27.122 * 10^{(-1.364 \, w/c)}$$

$$4. \ f_c = -3.112 + 4.023 \, (1/w/c) + \\ + 0.175 \, (1/w/c)^2$$

$$7. \ f_c = 0.518 - 17.818 \log w/c$$

Figure 5.11 Comparison of several formulas for prediction of the 28-day compressive strength from the water-cement ratio. I. The experimental data were taken from Fig. 5.5. 1 ksi = 6.90 MPa.

equivalent of the hypothesis that a *fixed* $\Delta x'$ change in the water-cement ratio always causes the same *fixed* $\Delta f'_{rel}$ *percentage* change in the relative strength of a concrete. That is, this $\Delta f'_{rel}$ value is independent of the age, water-cement ratio, strength, and composition of the concrete as long as the Abrams' rule is valid and coefficient B_A is unchanged. This $\Delta f'_{rel}$ value is defined as:

$$\Delta f'_{rel}(\%) = 100(f_{rel} - 1) \tag{5.15}$$

If, for instance, B_A is 20 within the age limits of 3 and 90 days, then, according to Eq. 5.34, a $w/c - (w/c)_1 = \Delta x' = 0.05$ increase in the water-cement ratio always causes about a 16% decrease in the compressive strength of concrete.

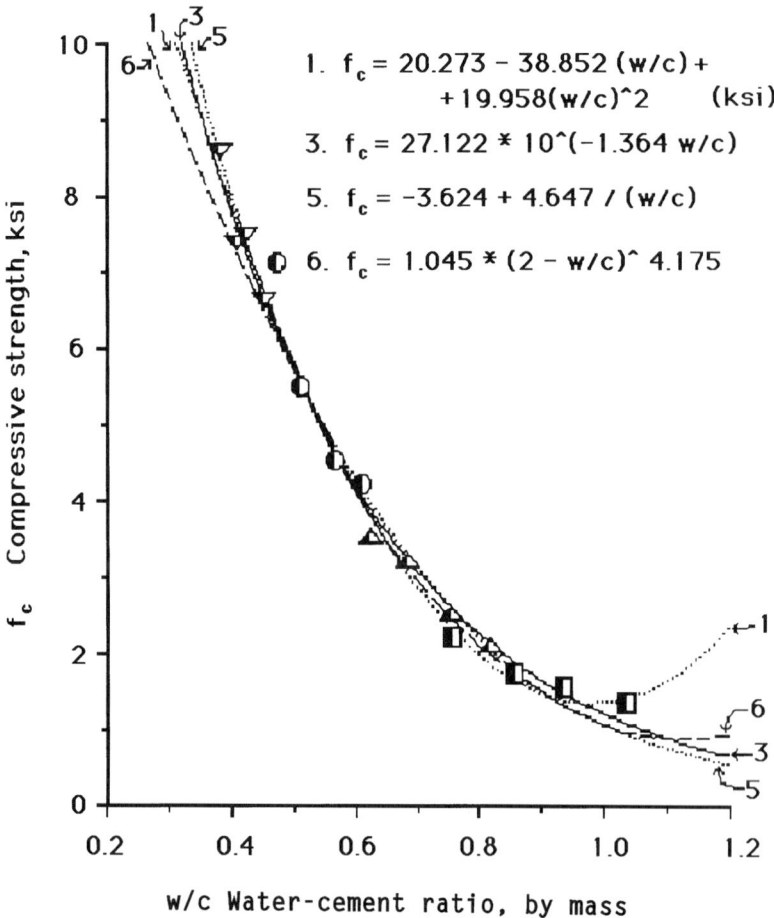

Figure 5.12 intro equations shown within the plot:

1. $f_c = 20.273 - 38.852\,(w/c) + 19.958(w/c)^2$ (ksi)

3. $f_c = 27.122 * 10^{(-1.364\,w/c)}$

5. $f_c = -3.624 + 4.647 / (w/c)$

6. $f_c = 1.045 * (2 - w/c)^{4.175}$

Figure 5.12 Comparison of several formulas for prediction of the 28-day compressive strength from the water-cement ratio. II. The experimental data were taken from Fig. 5.5. 1 ksi = 6.90 MPa.

The intrinsic strength f_i (Section 5.3), that is, the upper limit of the strength of the given concrete predicted by Eq. 5.6, is equal to A_A when $C_A = 0$. This is the extrapolated (i.e., hypothetical) strength the concrete would approach as the water-cement ratio goes to zero. The concrete strength predicted becomes zero only when the water-cement ratio is infinitely large. This is not realistic; nevertheless, only predictions of concrete strengths with very high water-cement ratios are affected by this. Besides, this mostly theoretical objection can be eliminated by selection of the coefficient C_A as follows:

$$C_A = -\frac{A_A}{B_A^{(w/c)_{cr}}} < 0 \qquad (5.16)$$

where $(w/c)_{cr}$ is the (critical) water-cement ratio above which the concrete strength is negligible. $(w/c)_{cr}$ corresponds to the critical porosity p_{cr} concept (Section 5.3) and may be less for compressive strength than for flexural strength. The C_A value for the 28-day standard compressive strength is typically less than 50 psi (0.35 MPa), and for younger concretes even less (i.e., negligible) in most cases. The ratio A_A/B_A represents the concrete strength when $w/c = 1$. Also, it will be shown in the next section that the coefficient B_A is the hypothetical increase in the relative strength f_{rel} when the reduction in the water-cement ratio is unity.

The simplified ($C_S = 0$) Eq. 5.7 provides a straight line in an f versus log w/c semilogarithmic system. As long as B_S is constant, these lines converge in a point $w/c = B_S$ and $f = 0$ for different A_S values (Fig. 5.9). The formula is the mathematical equivalent of the hypothesis that a *fixed percentage* change in the water-cement ratio always causes the same *fixed* change in strength, say 500 psi. The intrinsic strength provided by Eq. 5.7 is infinitely large. Although this is not realistic, only the prediction of strengths with extremely low water-cement ratios are affected by this. Also, this objection can be eliminated by the proper selection of C_S. The critical water-cement ratio $(w/c)_{cr}$ in this formula is B_S, which is usually unrealistically low when $C_S = 0$ (Figs. 5.9 and 5.10). The behavior of Eq. 5.8 is similar (Fig. 5.10). The A_L value represents the concrete strength when $w/c = 1$; that is, numerically this coefficient is comparable to the ratio A_A/B_A of Abrams' formula.

Equation 5.9 is represented by a straight line in an f versus c/w system when $B_H = 1$. It states that a *fixed* change in the cement-water ratio always produces the same *fixed* change in the strength. Another simplified form of Eq. 5.9 ($C_H = 0$) provides a straight line in a log f versus log w/c system. As long as B_H remains constant, these lines are parallel for various concretes (Fig. 5.3). This simplified form is the mathematical equivalent of the hypothesis that the *fixed percentage* change in the water-cement ratio always causes the same *fixed percentage* change in the strength. The intrinsic strength here is infinite. If $B_H = 1$, the critical water-cement ratio $(w/c)_{cr} = -A_H/C_H$, which is, in most cases, unrealistically low. It can be increased, however, by appropriate selection of B_H. $f = A_H + C_H$ when $w/c = 1$. That is, if $C_H = 0$, A_H represents the concrete strength for $w/c = 1$; thus its value is again comparable to the ratio A_A/B_A. Also, it will be shown in the next section that if $B_H = 1$, $A_H = A_B$ represents the hypothetical strength increase when the increase of the cement-water ratio is unity. Thus it is a characteristic of the water sensitivity.

The value of f_i in Eq. 5.10 is C_C, (i.e., comparable to A_A) and in Eq. 5.11 it is infinite. The critical water-cement ratios can be calculated by solving these equations for $f = 0$, although as curves 1 and 6 demonstrate in Figure 5.11, $(w/c)_{cr}$ does not necessarily exist for these formulas. This is, of course, a weakness. In both polynomials $f = A + B + C$ when $w/c = 1$; that is, this sum is comparable to A_A/A_B as well as $A_H + C_H$.

Equation 5.12 also has simple interpretations that the reader can derive. Here the intrinsic strength $f_i = A_R$; that is, the numerical value of this coefficient is comparable to A_A and C_C. The critical water-cement ratio for Eq. 5.12 is $(w/c)_{cr} = C_P = C_R$. This value is comparable to B_S of Eq. 5.7. Note that this is the only one among the strength-predicting formulas discussed that has finite values for both the intrinsic strength and the critical water-cement ratio.

The intrinsic strength for Eq. 5.13 is $f_i = A_f$, which is comparable to A_A, C_C, and A_R; C_F is the specific gravity of the cement, usually taken as 3.15 for portland cement; and the critical water-cement ratio is infinite.

Selection of a Suitable Strength Formula

The mathematical interpretations above help us judge how well experimental strength data support a formula in question and within what limits.

However, the goodness of fit is not the only criterion for the selection of a formula for the strength versus water-cement ratio relationship. For instance, curve 7 in Figure 5.11 provides physically impossible negative strengths for water-cement ratios greater than 1.07 by mass. Another example is Eq. 5.11, which fits the given 28-day strengths very well for low and medium water-cement ratios, yet as curve 1 in Fig. 5.12 shows, it predicts physically impossible strengths for water-cement ratios greater than 0.95 by mass. Curve 6 shows a similar trend.

Sometimes the structure of the strength formula is the decisive factor. For instance, linear formulas may facilitate linear programming. In other cases the values of f_i and/or a_{crit} are important.

Two additional examples concerning the meanings of the coefficients of some of the strength formulas are given below.

PROBLEM 5.1   Develop a formula for the strength change caused by a change in the water-cement ratio based on the Bolomey formulation of Eq. 5.9.

Solution   The Bolomey formula is written traditionally in the form:

$$f = A_B(c/w - C_B) \tag{5.17}$$

From this the strength change could be expressed in a ratio form of f_{rel}, similarly to Eq. 5.26. It is more logical, however, to use a form of difference because it is the mathematical form of the simple physical interpretation of the Bolomey formula, discussed earlier. This difference is the following:

$$\Delta f = f - f_1 = A_B(c/w - c_1/w_1) = A_B\Delta(c/w) \tag{5.18}$$

where $\Delta f = f - f_1$ = strength change caused by the given $\Delta(c/w)$ change

f_1 and f = given concrete strength and any other concrete strength, respectively

$\Delta(c/w) = (c_1/w_1 - c/w) = $ given change in the cement-water ratio
A_B and $C_B = $ coefficients of the Bolomey formula
c_1/w_1 and $c/w = $ cement-water ratio related to strength f_1 and f, respectively

Equation 5.18 shows clearly that A_B (the same as A_H in Eq. 5.9) is a good characteristic of the water sensitivity of concrete strength.

PROBLEM 5.2

a. What is the physical interpretation of the coefficient A_P in Eq. 5.12?
b. Express the water sensitivity as represented by this formula.

Solution

a. A_P is the predicted concrete strength when the water-cement ratio $w/c = C_P - 1$.
b. Since $f_i = A_R$, and $(w/c)_{cr} = C_R$, the larger the abslute value of A_R/C_R under the given set of circumstances, the greater the water sensitivity.

Example 5.1 The development of the concrete strength f versus water-cement ratio w/c relationship is demonstrated for two cases. Any concrete strength versus porosity p formula should produce maximum strength when $p = 0$ and zero strength when the porosity is very high, that is, $p \approx 1$. From this premise, two equations are assumed for the f versus p relationship, from which formulas are derived for the f versus w/c relationship. In both cases the theoretically sound Eq. 4.11 is used, which is repeated here:

$$p = \frac{w/c}{w/c + 1/G_C} \qquad (5.19)$$

where $p = $ capillary porosity, $\%/100$
  $w/c = $ water-cement ratio, by mass
  $G_c = $ specific gravity of the cement

The difference between the two development cases is that the concrete strength versus capillary porosity relationships, which are used as starting points, are different.

1. The simplest form of a *power function* for the f versus p relationship is:

$$f = f_i(1 - p)^n \qquad (5.20)$$

Substituting Eq. 5.19 into 5.20 gives

$$f = f_i \left(1 - \frac{w/c}{w/c + 1/G_c} \right)^n \tag{5.21}$$

$$= f_i \left(\frac{1/G_c}{w/c + 1/G_c} \right)^n \tag{5.22}$$

$$= f_i (1 + G_c w/c)^{-n} \tag{5.23}$$

Changing the symbols so that $f_i = A_F$, $G_c = C_f$, and $n = B_F$, then Eq. 5.23 will become identical with Eq. 5.13.

    2. The simplest form of an *exponential* function for the f versus p relationship is the following:

$$f = f_i b^{-p/(1-p)} \tag{5.24}$$

But from Eq. 5.19,

$$p/(1 - p) = G_c w/c \tag{5.25}$$

Therefore,

$$f = f_i b^{-G_c w/c} \tag{5.26}$$

Changing the symbols so that $f_i = A$ and $b^{G_c} = B$, Eq. 5.26 will become identical with the Abrams' formula, that is, with Eq. 5.6 with $C_A = 0$.

5.6   DETERMINATION OF THE COEFFICIENTS

Rationale

The determination of the coefficients is conceptually simple and practically quick. Linear regression programs are available for the best-fitting coefficients for most of the formulas presented in many statistical software packages, even for pocket calculators (ACI Committee 228, 1996).

    The calculations of the coefficients in the strength formulas are illustrated below. When Eq. 5.6 with $C_A = 0$ is used for the development of the strength versus water-cement ratio relationship, at least two trial mixes are needed, in principle, to determine the coefficients A_A and B_A. The trial mixes should have different water-cement ratios. When an augmented form is used, the number of needed trial mixes is at least three. For traditional concretes these numbers can be reduced, as described below; however, this reduction occurs at the expense of reliability. Conversely, a higher number of trial mixes results in more reliable calculation of the coefficients. All calculations are based on recognition that Abrams' formula provides a straight line in the log f versus *w/c* system of coordinates, which, by the way, does not necessarily provide the best coefficients for the Abrams' formula. Note that the only variable in the composition of concretes involved in the trial mixes can be the water-cement ratio. That is, all specimens should be made from the same materials, prepared, cured, and tested under identical conditions and should have practically the same air content, corresponding to the specific job circumstances.

Illustrative examples for the calculations are presented below.

Use of Multiple Trial Mixtures

Values for the A_A and B_A coefficients of Eq. 5.6 are obtained when trial mixes with *more than two water-cement ratios* are utilized. A set of pertinent formulas comes from the standard linear regression analysis and are:

$$\log B_A = -\frac{n\sum [(w/c) \log f] - \sum \log f \sum (w/c)}{n \sum (w/c)^2 - \left[\sum (w/c)\right]^2} \tag{5.27}$$

and

$$\log A_A = \frac{\sum \log f + \log B_A \sum (w/c)}{n} \tag{5.28}$$

$$= \frac{\sum \log f \sum (w/c)^2 - \sum (w/c) \sum [(w/c) \log f]}{n \sum (w/c)^2 - \left[\sum (w/c)\right]^2} \tag{5.29}$$

where *w/c* = given water-cement ratios of the trial mixes
  f = average strength related to each water-cement ratio
  n = number of different water-cement ratios involved

Note that in case of the Abrams' formula, and also in certain other cases, optimization methods other than the standard linear regression may be more suitable for the determination of the *A* and *B* coefficients (ACI Committee 228, 1996).

Example 5.2 Using the experimentally obtained compressive strengths presented in Fig. 5.5, the coefficients for Eqs. 5.6 through 5.12 were calculated with the standard linear regression for various ages and presented in Table 5.1. The graphical forms of these equations for the 28-day strengths are presented in Figs. 5.11 and 5.12.

PROBLEM 5.3 Calculate the coefficients for Eq. 5.6 for both non–air-entrained and air-entrained concretes from the data recommended by ACI 211 (1991) for the strength versus water-cement ratio at the age of 28 days. These data are reproduced in Table 5.2.

TABLE 5.2 Relationship between water-cement ratio and compressive strength of concrete

Compressive Strength at 28 Days (psi)	Water-Cement Ratio, by Mass	
	Non–Air-Entrained Concrete	Air-Entrained Concrete
6000	0.41	—
5000	0.48	0.40
4000	0.57	0.48
3000	0.68	0.59
2000	0.82	0.74

Source: ACI 211.1-91. (Copyright ACI. Reprinted with permission.)

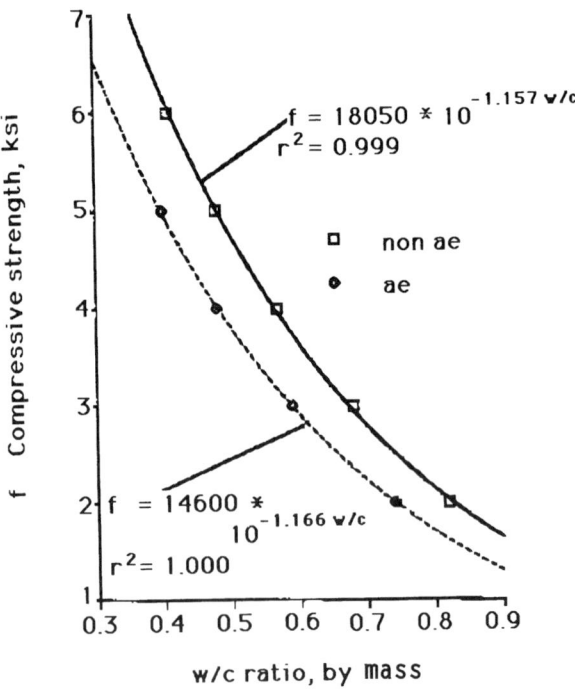

Figure 5.13 Strength formulas in psi for air-entrained (ae) and non–air-entrained (non ae) traditional portland cement concretes representing the recommendations of ACI 211.1–91. 1 ksi = 6.90 MPa.

Solution   Substituting the values in Table 5.2 into Eqs. 5.27 and 5.28 (or 5.29), the results are for non–air-entrained concretes:

$$A_A = 18{,}050 \text{ psi } (124.45 \text{ MPa})$$

$$B_A = 14.36$$

and for air-entrained concretes;

$$A_A = 14{,}600 \text{ psi } (100.67 \text{ MPa})$$

$$B_A = 14.67$$

The graphs and the equations with the coefficients calculated above are shown in Figure 5.13. The values of these coefficients are considerably lower than the comparable values in Table 5.1. This means that the values recommended by ACI are quite conservative, to provide safety. Note also how close the B_A value of the nonair-entrained concrete is to the B_A value of the air-entrained concrete.

Equations 5.27 through 5.29 can be used for calculations of the coefficients from trial mixes in every case, including concretes containing silica fume and/or superplasticizer. This calculation is very simple and fast with the software Prop 21 on the accompanying disk.

Example 5.3   Equations 5.27 through 5.29 can be used not only for proportioning but also for quality control. As the strength results are obtained from the quality control tests of concretes made with the same group of concrete-making materials during construction, these values can be input into Eqs. 5.27 through 5.29. The new coefficient values will probably be close to the original values; if so, they provide more reliable updated values for A and B. If on the other hand, there is a jump in the values of any of the newly calculated coefficients, a warning is given that a change has taken place in the concrete composition. This would justify an adjustment in the water-cement ratio.

Use of Two or Fewer Trial Mixtures

When strength results are available from trial mixtures with *two different water-cement ratios*, Eqs. 5.27 through 5.29 in Example 5.3 reduce to:

$$\log B_A = \frac{\log(f_1/f_2)}{(w/c)_2 - (w/c)_1} \tag{5.30}$$

and

$$\log A_A = \log f_1 + (w/c)_1 \log B_A \qquad (5.31)$$

$$= \log f_1 + \frac{(w/c)_1}{(w/c)_2 - (w/c)_1} \log \frac{f_2}{f_1} \qquad (5.32)$$

where $(w/c)_1$ and $(w/c)_2$ = given water-cement ratios of the trial mixes
f_1 and f_2 = average of the strengths related to $(w/c)_1$ and
$(w/c)_2$, respectively

The reliability of the coefficients calculated depends on the reliability of the strength results f_1 and f_2.

When all the available strength results were obtained from a single trial mixture, or *trial mixtures of the same water-cement ratio*, only one coefficient, either A_A or B_A, can be calculated for the Abrams' formula. Even in such a restricted case, however, approximate values of either A or B may be estimated, and the other value may be calculated in turn. One such case is when the specimens are prepared and cured according to ASTM C39 and C192. Here the values of coefficient A_A for traditional concretes are more or less constant, which can be used in the computer program. For instance, for the standard compressive strength of traditional concretes $A \approx 12{,}000$ psi (≈ 83 MPa), and for flexural strengths $A \approx 1000$ psi (≈ 7 MPa), regardless of whether the cement is of Type I or III, and for all ages of concrete from 3 days on.

Another possibility is to use formulas, such as Eqs. 5.36 through 5.39, again for traditional concretes, for the estimation of values for $B_c = B_A$ that can be included in the calculation. Then A_A can be calculated as follows:

$$A_A = f_a B_c^{w/c} \qquad (5.33)$$

When *no trial mixture results* are available, values presented in Table 5.2 can be used for the 28-day compressive strength versus water-cement ratio relationship but only for standard traditional concretes. A more general approach is for such cases to use the 12,000 psi ($\approx$ 83-MPa) and 1000 psi ($\approx$ 7-MPa) values, respectively, for A, and calculate the values of B from Eqs. 5.36 through 5.39, respectively. The effect of air content can taken into account by Eq. 5.73.

All these calculations are in computerized form in Prop 21 and the User's Manual on the accompanying disk.

Two points should be noted, however:

1. The approximations above are valid only for traditional concretes up to about 5000-psi (about 35-MPa) 28-day strength prepared cured and tested according to the appropriate standards. For higher strengths an augmented formula should be used.

2. Even for such concretes, the reliability of the coefficients obtained is less than those based on two or more trial mixes.

5.7 DISCUSSION OF ABRAMS' FORMULA FOR COMPRESSIVE STRENGTH

Relative Strength

The A_A factor in Eq. 5.6 is the hypothetical upper limit for the strength of the given concrete. Nevertheless, this coefficient is not necessarily a characteristic of the strength-developing capability of the cement, especally if it is a blended cement. Equation 5.6 also provides the possibility of expressing the *relative* change in strength caused by a change in the water-cement ratio within the limits of validity of Abrams' rule. Specifically, if a concrete strength f_1 related to the water-cement ratio $(w/c)_1$ is selected as unity (or 100%), any relative strength f_{rel} related to the water-cement ratio w/c can be obtained from:

$$f_{rel} = \frac{f}{f_1} = B_A^{(w/c)_1 - w/c} = 10^{[(w/c)_1 - w/c]\log B_A} \tag{5.34}$$

That is,

$$\log f_{rel} = [(w/c)_1 - w/c]\log B_A = (w/c)\log B_A \tag{5.35}$$

Equation 5.35 is comparable to Eq. 5.18.

The $\log f_{rel}$ versus w/c relationship is represented by a single straight line with the slope of $\log B_A$ for various concretes as long as Abrams' rule is valid and the coefficient B_A is constant (S. Popovics, 1967b). This is illustrated in Figure 5.14, where the data presented in Figure 5.5 with four curves are replotted as relative strength values in the semilogarithmic system along a single line. The strength values related to the water-cement ratio of 0.60 by mass were chosen arbitrarily as unity.

Equations 5.34 and 5.35 also show that the increase in the calculated relative strength related to a reduction of unity in the water-cement ratio is shown by the numerical value of B_A. Thus the greater the B_A value, the more sensitively the strength of concrete reacts to a change in the water-cement ratio. This is also shown by the observation that the greater the value of B_A, the steeper the slope of the $\log f_{rel}$ versus w/c straight line. That is, the B_A value is an indicator of the *water sensitivity* of concrete strength.

Equations 5.34 and 5.35 provide a powerful tool for the comparison and analysis of concrete strengths under various circumstances as a function of the water-cement ratio (Nagaraj et al., 1990). For instance, a remarkable stability of coefficient B_A is displayed in Figure 5.15. Here the relative standard 28-day compressive strengths of various concretes with Type I portland ce-

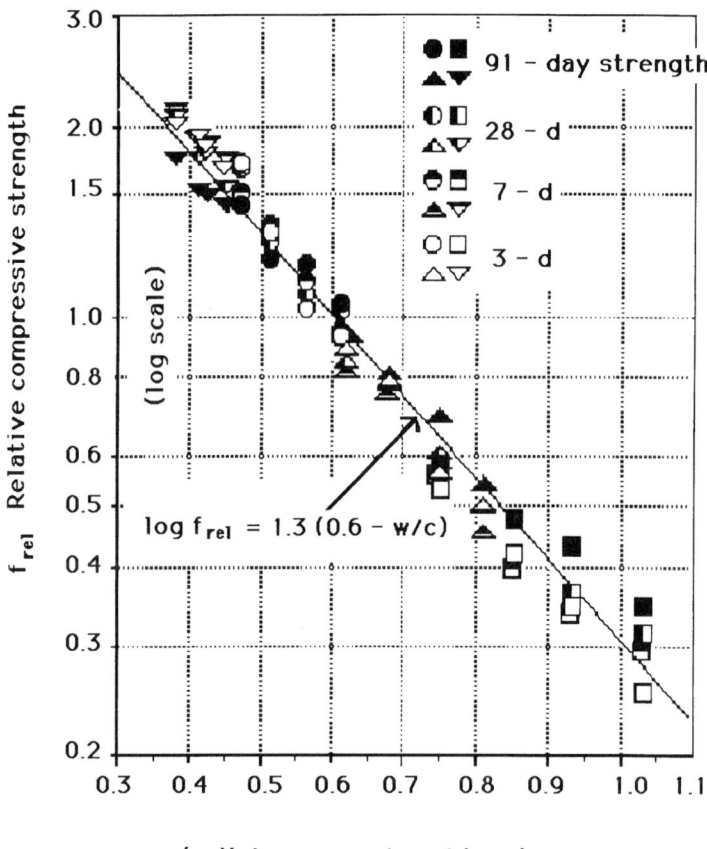

Figure 5.14 Overall relationship between relative compressive strength and water-cement ratio at ages from 3 to 91 days based on data of Fig. 5.5. The strength related to $w/c = 0.60$ is 100%. (From S. Popovics, 1967b. Copyright ASTM. Reprinted with permission.)

ments are plotted. These concretes were produced with different materials in different places at different times. Nevertheless, the strength results do form a single straight line with a good approximation indicating that more-or-less the same coefficient, $B_A = 15$, is applicable to them.

Figure 5.16 also shows the stability of B_A. Here 90-day compressive strengths of five concretes are presented where each concrete was made with a different cement. Yet the strength lines form a family of parallel straight lines with a good approximation, indicating the constancy of the value of B. On the other hand, there are cases where the numerical value of B_A in Eqs. 5.6, 5.34 and 5.35 depends on several factors. Some of these factors are discussed below.

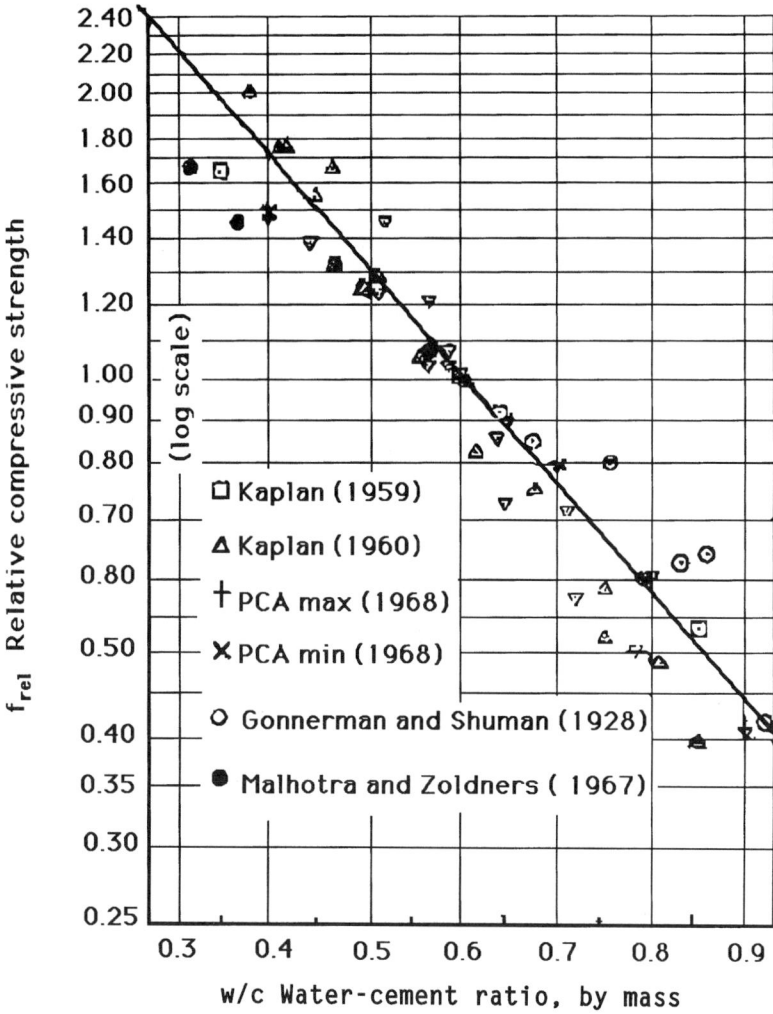

Figure 5.15 Relative compressive strength versus water-cement ratio relationship at the age of 28 days.

Dependence of Coefficient B_A on Age, Type of Strength, and Cement Type

The effect of age of the concrete on B_A is illustrated in Figure 5.17. Here the 3- and 91-day relatively compressive strength values are replotted from Figure 5.14. The 3-day curve is steeper than the 91-day curve; that is, B_3 is larger than B_{91}. It is not difficult to explain the effect of age on the water sensitivity of strength with the kinetics of strength development. As was shown

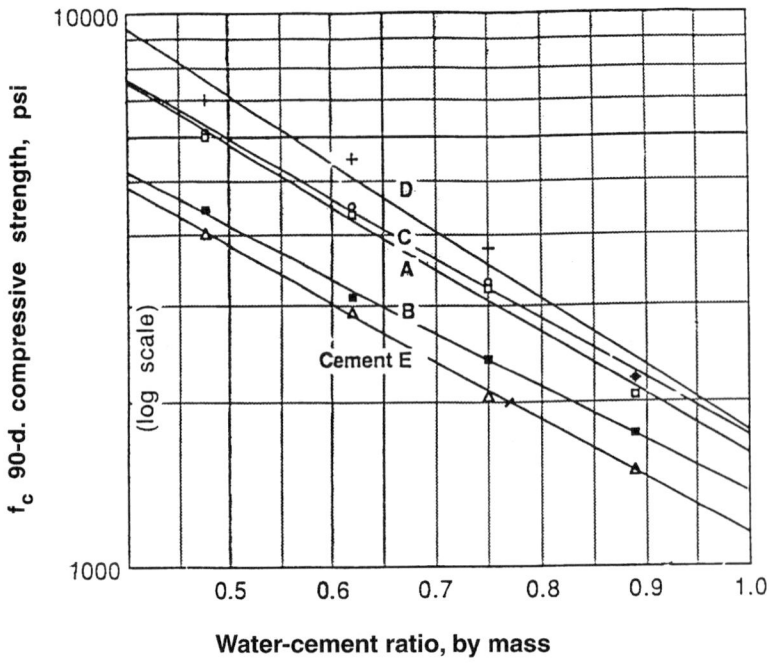

□	**log fc(A)= 4.322-1.1169w/c R^2 = 0.992**
■	**log fc(B)= 4.103-0.9681w/c R^2 = 0.998**
o	**log fc(C)– 4.310-1.0741w/c R^2 = 0.996**
+	**log fc(D)= 4.458-1.2146w/c R^2 = 0.967**
Δ	**log fc(E)= 4.109-1.0525w/c R^2 = 0.999**

Figure 5.16 Compressive strengths of concretes made with various cements, as a function of the water-cement ratio. 1000 psi = 6.90 MPa.

earlier in this chapter, concretes with low water-cement ratios have high early strengths but relatively little strength increase later; whereas concretes with high water-cement ratios have low early strengths but relatively large strength increases at later ages.

Another contributing factor to the difference $B_3 - B_{91}$ may be the composite nature of concrete. Specifically, in normal-weight concretes the matrix is much more deformable at younger ages than the aggregate, which causes extra stress concentrations (Section 2.4). This difference in deformabilities

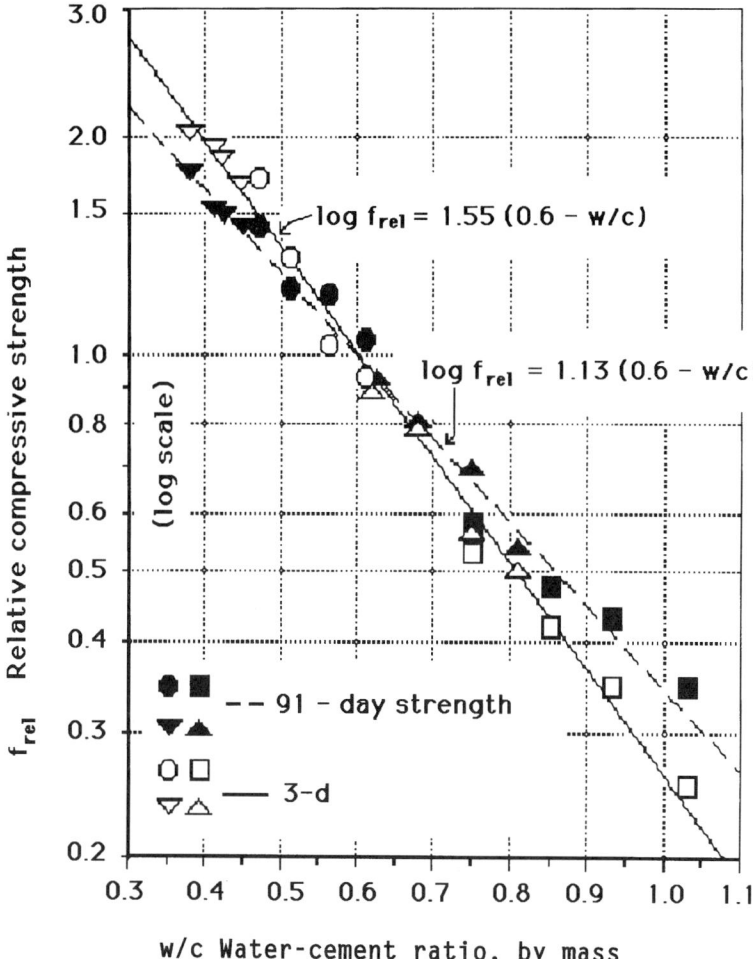

Figure 5.17 Relative compressive strength versus water-cement ratio relationships at the ages of 3 and 91 days, respectively, based on the data of Fig. 5.5. (From S. Popovics, 1967b. Copyright ASTM. Reprinted with permission.)

decreases with age, especially in concretes of high water-cement ratio. Thus the strength concentrations also become relatively less, which, in turn, appears in reduced water sensitivity of the concrete.

The dependence of the parameter B_A on time can also be expressed numerically for certain cases. In Figure 5.18, values of coefficient B_A derived from various test series are plotted as a function of age for Type I portland cements and for compressive and flexural strengths. The experimental values were taken from publications by Malhotra and Zoldners (1967b), Gonnerman and Shuman (1928), the Portland Cement Association (1968), and Kaplan

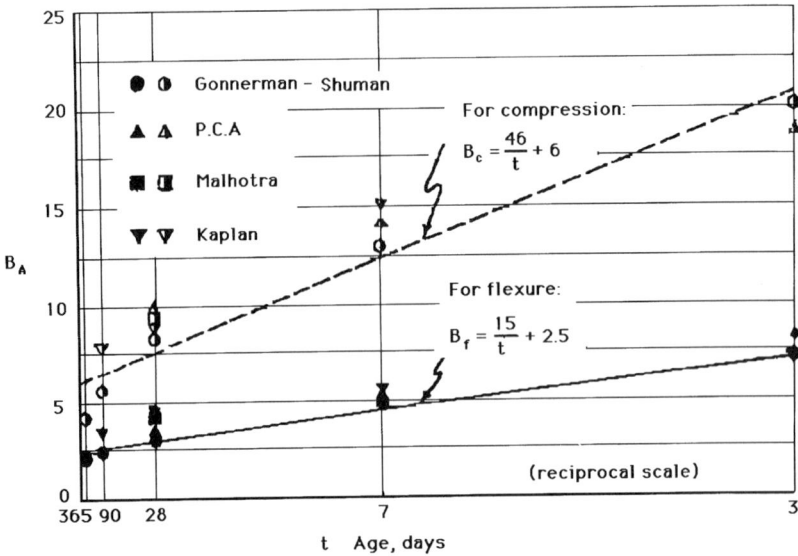

Figure 5.18 Observed and calculated values of the B_A factors for Type I cement as a function of age at testing. (From S. Popovics, 1967b. Copyright ASTM. Reprinted with permission.)

(1959a). It can be seen that a hyperbolic function approximates the experimental values of B quite well within the limits of 3 days and 1 year. These functions, however, provide values too low for the age of 1 day. The true 1-day values of B are around 100 and 40 for compressive and flexural strengths, respectively, of concretes made with typical Type I cements, and 40 and 20, respectively, for Type III cements. One possible reason for the 3-day lower age limit is that the cements used in these experiments had not reached at 1 day the 50% level of hydration. This is the threshold needed for the porosity to take over the control of paste strength (Section 4.3).

Figure 5.19 demonstrates that the sensitivity of the flexural strength to changes in water-cement ratio is considerably less at any age than that of the compressive strength. Also, an analysis of the strength curves recommended by the Portland Cement Association (1968) reveals that Type III cements are less sensitive than Type I cements to changes in water-cement ratio. This is particularly true at early ages. Other comparable strength curves show the same tendency (Basalla, 1965; Meyer, 1963b, 1967; Wischers, 1964; Road Research Laboratory, 1950; McIntosh, 1964). These effects on the coefficient B_A are expressed below as a function of time by empirical formulas for the case when the concrete specimens are made, cured, and tested in a standard manner (ASTM C192 and C39) (S. Popovics, 1967b). For Type I cements,

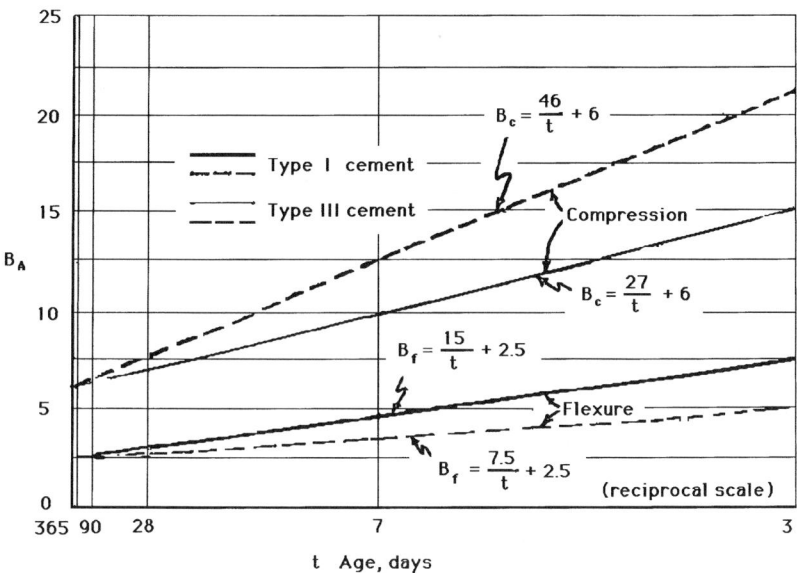

Figure 5.19 Recommended values for the B_A factors for compression and flexure as a function of age at testing and type of cement. (From S. Popovics, 1967b.)

$$B_c = \frac{103}{t + 4} + 5.2 = \frac{46}{t} + 6 \qquad (5.36)$$

$$B_f = \frac{33}{t + 4} + 2.7 = \frac{15}{t} + 2.5 \qquad (5.37)$$

and for Type III cements,

$$B_c = \frac{27}{t} + 6 \qquad (5.38)$$

$$B_f = \frac{7.5}{t} + 2.5 \qquad (5.39)$$

where t = age of concrete, days
   B_c and B_f = factor B_A in Eq. 5.6 for compressive strength and flexural strength, respectively

Equations 5.36 through 5.39 are presented in Figure 5.19. Numerically, Table 5.3 shows the values of B_A calculated by these equations, and the corresponding f_{rel} strength reductions caused by a 0.1 increase in the water-cement ratio by mass. The values of B_A decrease strongly with age. In fact, this decrease is responsible mainly for expressing the strength increase of a concrete with age in Eq. 5.6 because coefficient A_A varies surprisingly little

TABLE 5.3 Values for B and percentage reduction in the compressive and flexural strengths caused by a 0.1 increase in the water-cement ratio (by mass) for various ages and cement types[a]

Type of Cement	Type of Strength	3 days		7 days		28 days		90 days		365 days	
		B	Δf	B	Δf	B	Δf	B	Δf	B	Δf
I	Compressive	21.3	26.4	12.6	22.4	7.64	18.4	6.51	17.1	6.13	16.6
	Flexural	7.50	18.20	4.64	14.30	3.04	10.5	2.67	9.40	2.54	8.90
III	Compressive	15.0	23.7	9.86	20.5	6.97	17.7	6.30	16.8	6.07	16.5
	Flexural	5.00	14.90	3.57	12.0	2.77	9.70	2.58	9.10	2.52	8.80

Source: S. Popovics (1967b). (Copyright ASTM. Reprinted with permission.)
[a]The values were calculated by Eqs. 5.15 and 5.36–5.39.

with age beyond 3 days (Higginson et al., 1963). This seems to indicate that the coefficient B_A is related to the gel–space ratio of the cement paste in the concrete. A finding of Powers (1949) supports this opinion: namely, when the cement in the water-cement ratio is the amount of the *hydrated* cement, the same coefficients can be used in the formula for various ages after a certain age and for a variety of portland cements. Seki et al. (1969) offered a similar relation.

In the special case when the water-cement ratio is fixed, Eqs. 5.36 through 5.39 become the hyperbolic function recommended by Goral (1956) for the concrete strength versus age relationship (Section 3.8). Note that Eqs. 5.36 through 5.39 are valid only for traditional concretes and for portland cements of Types I and III.

Use of the Strength Formulas for Proportioning of Concrete

Equations 5.6 through 5.14 can be used for estimation of the water-cement ratio needed for a specified concrete strength. This concept is regularly used in the various procedures for proportioning of concrete (S. Popovics, 1982b), and discussed in detail in the User's Manual on the accompanying disk, including computerization.

PROBLEM 5.4   A job requires a 27.60-MPa (4000-psi) target compressive strength at the age of 7 days under standard conditions. Type III portland cement is to be used. What is the recommended water-cement ratio for this concrete?

Solution   If Abrams' formula is used for the calculation, the needed water-cement ratio is

$$w/c = \frac{\log(A/f)}{\log B_A} \tag{5.40}$$

where f is the target strength. The other symbols are as in Eq. 5.6. Thus for completion of the calculation, assume a reasonable value for A_A, say, 83 MPa (12,000 psi); calculate the B_A value from Eq. 5.38:

$$B_A = B_c = \frac{27}{7} + 6 = 9.86$$

Estimate the water-cement ratio needed:

$$w/c = \frac{\log(12,000/4000)}{\log 9.86} = \frac{0.4771}{0.9939}$$

$$= 0.48 \text{ by mass}$$

Further examples of proportioning for strength are presented in Prop. 2.1.

The other strength formulas can be used for proportioning in a similar manner.

5.8 EFFECTS OF THE NATURE OF CEMENT

Type of Portland Cement

Equations 5.36 through 5.39 demonstrate that the water sensitivity of concrete strengths with Type III cements is significantly less than that of comparable concretes with Type I cements. Graf also reported that the water sensitivity of the strengths of various portland cements may differ considerably (Graf, 1930). The exact reason for this is not clear at present. The values of a listed in Table 3.3 imply that this has something to do with C_3A content and the fineness of the cement. These, in turn, control the *rate of the strength development*.

The relationship between strength developing capacity and compound composition of cements can be expressed mathematically, as discussed in Sections 3.2 through 3.5. Therefore, it is logical for the simulation of the effects of cement composition on the strength–water-cement ratio relationship to combine a suitable cement model with a strength versus w/c formula. An example is Eq. 3.20 where Eq. 5.6 is substituted for f_{90} in Eq. 3.14 with the appropriate a parameters in Table 3.5

The most significant feature of this approach is that it makes Abrams' formula applicable for all five standard types of portland cement within wide limits of validity with only a few experimental coefficients. For instance, coefficient $B = 6.4$ in Abrams' formula is valid for most portland cements and age. The accuracy of the formula is also reasonable (see: Fig. 3.16).

Blended Cements

Abrams' rule is also valid for blended cements, including when a portion of the portland cement is substituted for by fly ash (Alexander, 1955; S. Popovics, 1982c, 1986; Ghosh, 1976; Slanicka, 1991a). This is also illustrated in Figure 5.20 which was replotted from Ghosh's (1976) data. Thus the Abrams' formula is applicable for prediction of the strength of fly ash concretes from the water-cementitious materials ratio $w/(c + p)$ if the B values are expressed as a suitable function of the fly ash content p. The following formula provides an approximation to the values of B_A derived from Ghosh's 28-day compressive strengths (Fig. 5.20):

$$B_A = 5.5 \times 6.3^{p/c} + 6 \tag{5.41}$$

where p and c are the actual or relative mass of fly ash and that of portland cement, respectively, in the mixture.

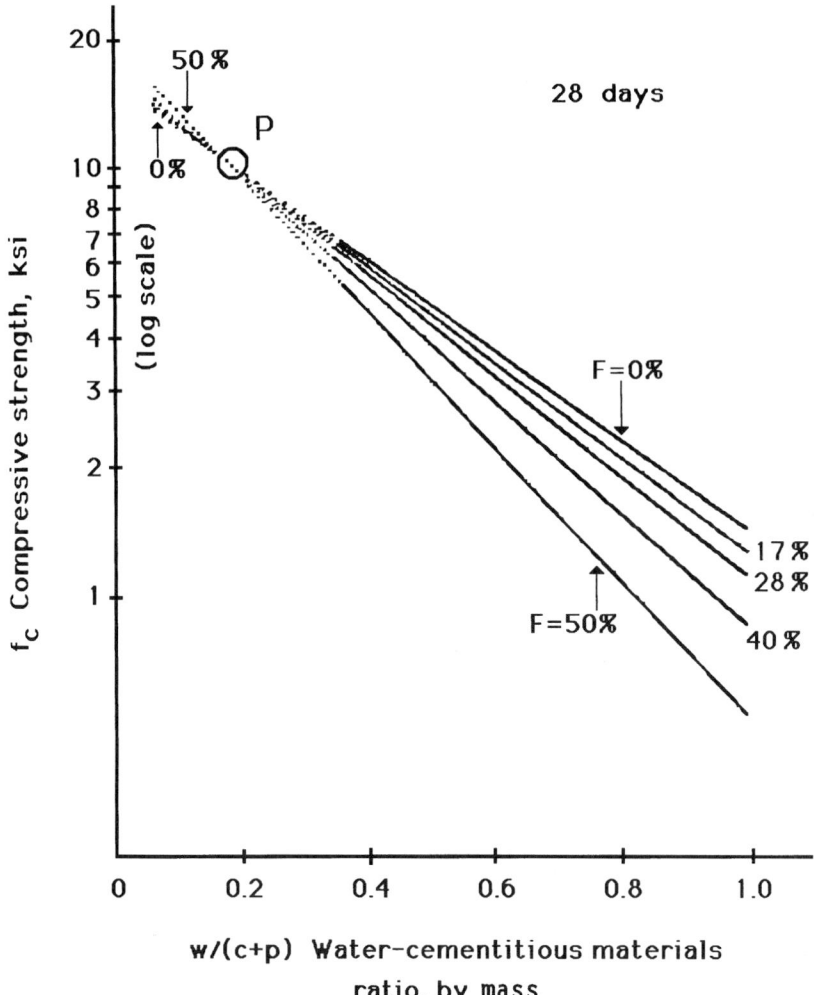

Figure 5.20 Compressive strength of fly ash concrete at the age of 28 days as a function of fly ash content and water-cementitious materials ratio. Data from Ghosh (1976). 1 ksi = 6.90 MPa. (From S. Popovics, 1986c. Copyright ACI. Reprinted with permission.)

Equation 5.41 is applicable within 0 and 1 of p/c, that is, when the fly ash content is within 0 and 50% by mass in the cement-fly ash blend. Equation 5.41 shows that the water sensitivity of the strengths of fly ash concretes increases very rapidly with the increases of fly ash substitution, especially at early ages (Alexander, 1955). This is probably due again to the lower rate of hydration of fly ash cements.

It is likely that Abrams' rule is also valid for concretes containing *silica fume* (Slanicka, 1991b). However, not enough data are available at present to discuss in detail the effect of silica fume on the strength versus water-cement ratio relationship. Approximate data are offered in ACI 211.4R-93 for proportioning purposes. A computerized version of this, called Prop 21, is on the accompanying disk. Nevertheless, it is better practice to establish the pertinent coefficients by trial mixes.

Abrams' rule is also applicable for other *nonportland cements*. Blondiau reported that blast-furnace slag cements showed water sensitivities similar to those of portland cement–fly ash combinations (Blondiau, 1960). Other nonportland cements, such as high-alumina cements (Neville, 1975, 1981), magnesium phosphate–base cements (S. Popovics et al., 1987), and plaster of paris (Schiller, 1971) seem to follow Abrams' rule. The quantity of pertinent experimental data, however, is not enough at present for the development of reliable formulas for such cements.

5.9 EFFECTS OF CEMENT, WATER, AND PASTE CONTENTS

Background

As discussed at the beginning of this chapter, the implication of Abrams' rule is that only the quality of the matrix portion of the concrete, which is hardened cement paste containing pores, controls the strength of comparable concretes; that is, the presence of aggregate has no influence. Therefore, the quantity of the cement, cement paste, or aggregate does not matter from the standpoint of concrete strength within practical limits. The same implication is expressed by the strength formulas discussed so far.

First, the formulas contain only the water-cement ratio as the single independent variable, that is, without any reference to quantities. Second, it is implied by the formulas that the magnitude of the changes in concrete strength with the w/c is independent of whether this ratio is altered by changing the cement content, water content, or both. The disregard of the effects of component quantities on the concrete strength simplifies the formulas, but at the expense of accuracy. For instance, if two comparable concretes have the same water-cement ratio, the strength of the concrete with higher cement content is lower. This difference is small but definite, as has been demonstrated by many investigators (Williams, 1962; Fritsch, 1951; Welch, 1961, Cement and Concrete Association, 1954; McIntosh, 1957; Singh, 1958; Walker and Bloem, 1961; Mather, 1967). Figures 5.21 and 5.22 show typical examples. The phenomenon can also be phrased so that increasing cement content has a smaller effect on concrete strength than is expressed by Abrams' rule based solely on the reduction in w/c.

The extrapolation of this observation to cement paste indicates that a hardened paste would have lower compressive strength than comparable concretes with the same water-cement ratio. Whether this expectation is fulfilled is not certain. Pertinent test results are contradictory. For instance, Stoke et al.

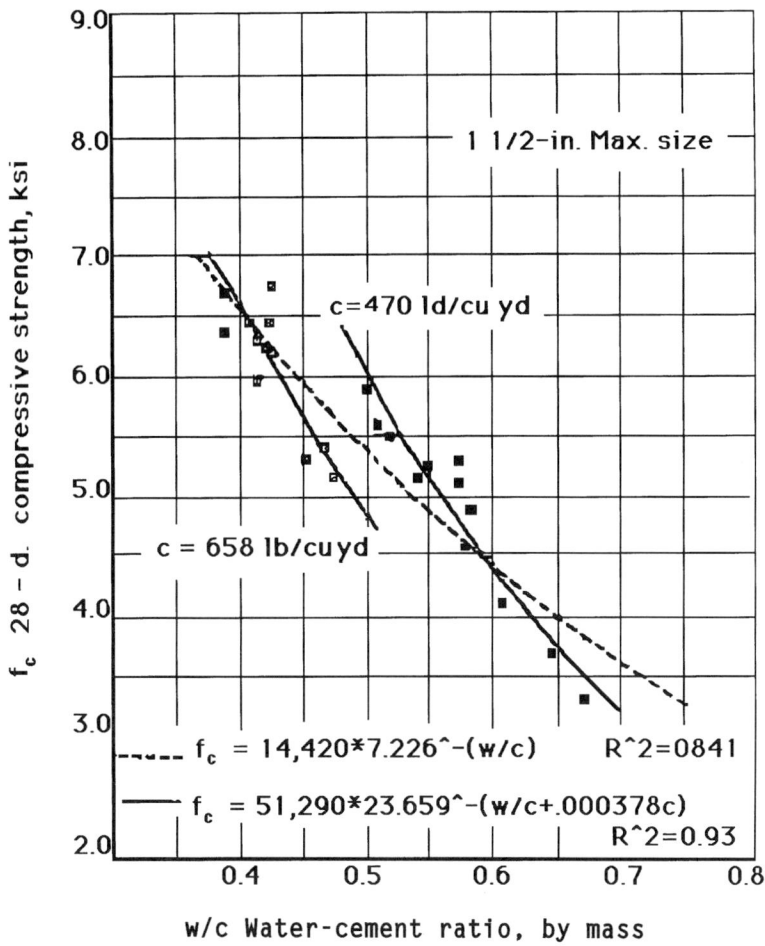

Figure 5.21 Effect of cement content on the concrete strength versus water-cement ratio relationship. Abrams' formula, Eq. 5.6 (model 1), is represented by the dashed line, the formula augmented by the cement, content, Eq. 5.45 (model 2), is represented by the solid lines. 1 ksi = 6.90 MPa. 1 lb/cu yd = 0.59 kg/m³, (From S. Popovics, 1990a. Copyright ACI. Reprinted with permission.)

(1979) obtained higher paste than concrete strengths; whereas Hobbs (1974) reported that in one test series the paste had higher strength than the concrete when the water-cement ratios were as low as 0.35 by mass. With the increase in the water-cement ratio, however, this strength difference reversed itself. The same test series also shows that under triaxial testing the paste is significantly weaker than the comparable concretes.

When the water-cement ratio is kept constant, a higher cement content also means a higher water content, therefore a wetter consistency within practical limits (Figs. 5.23, 5.24), and a higher paste content. Consequently, the aggre-

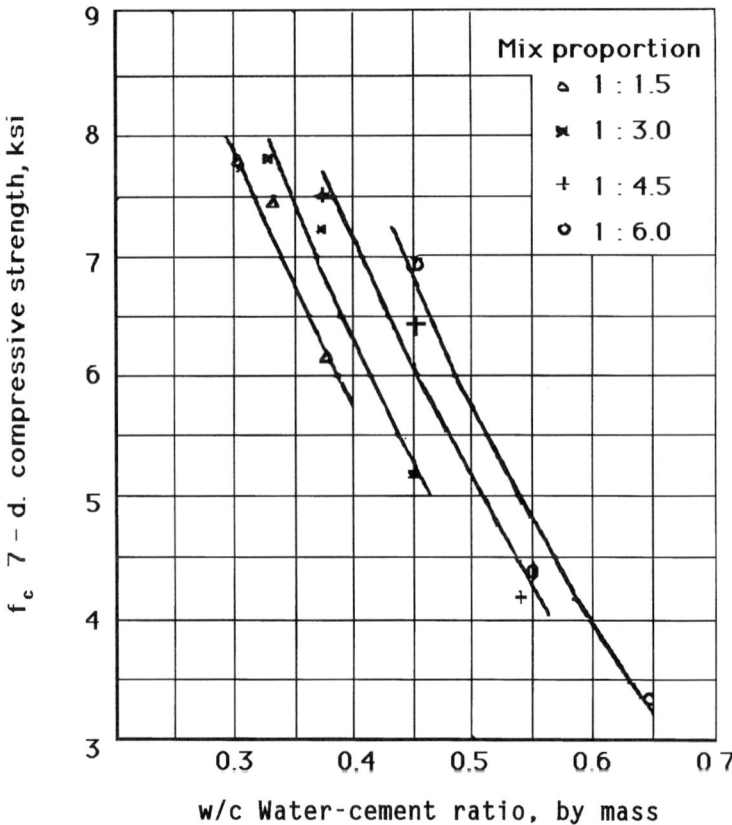

Figure 5.22 A richer concrete gives lower strengths than a leaner one of the same water-cement ratio. 1 ksi = 6.90 MPa. (From S. Popovics, 1990a. Copyright ACI. Reprinted with permission.)

gate content is lower. These are various aspects of the same phenomenon. It is shown in this section that an increase in the cement content, water content etc., reduces the concrete strength if the water-cement ratio is kept unchanged. The reason for this strength reduction is not clear at present, especially since the opposite is expected from the role of the interface between the cement paste and aggregate (Section 4.5). One could surmise that the higher the cement or paste content of the concrete, the larger the shrinkage, and/or the larger the bleeding that weakens the interface between the cement paste and coarse aggregate particles, and/or the more intensive the development of heat of hydration, all weakening the bond between the paste and coarse aggregate surface. This, in turn, reduces the strength of the concrete. Also, the effect of interlocking of the aggregate particles on the strength may be greater in lean concretes than in concretes with high cement content because the particles are closer to each other in the former. In any case, this phenomenon requires and deserves further research.

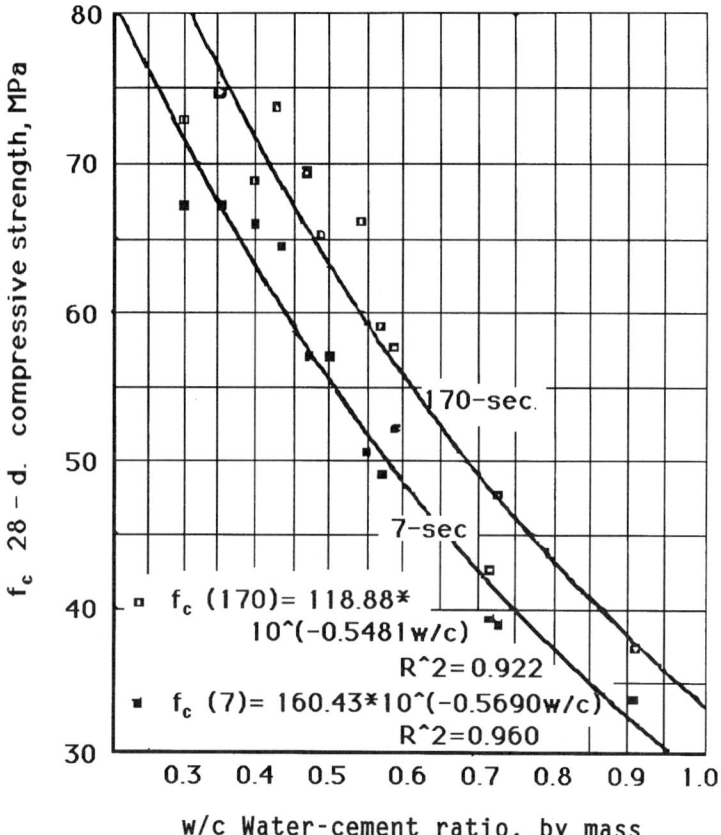

Figure 5.23 A wetter concrete, that is, concrete having lower Vebe time, gives lower strengths than a drier one of the same water-cement ratio. Squares represent experimental data; the lines are calculated values. 1 MPa = 145 psi (From S. Popovics, 1990a. Copyright ACI. Reprinted with permission.)

Recently, an effort was made to express this additional effect of the cement, water, and paste contents numerically for improvement of the accuracy of the strength versus water-cement ratio formulas. This is discussed below (S. Popovics, 1990a).

New Formulas for Concrete Strength Versus Composition Relation

The new formulas are based on the premise that the addition of, or augmentation with, a term containing the cement content to an f_c versus w/c formula, or basic model, can improve the accuracy of the relationship. One can also use the water, paste or, aggregate content, or, in certain cases, a consistency measure, or raising these terms to a higher power, or any combination of these for such augmentation. Then statistical analysis reveals if the augmented

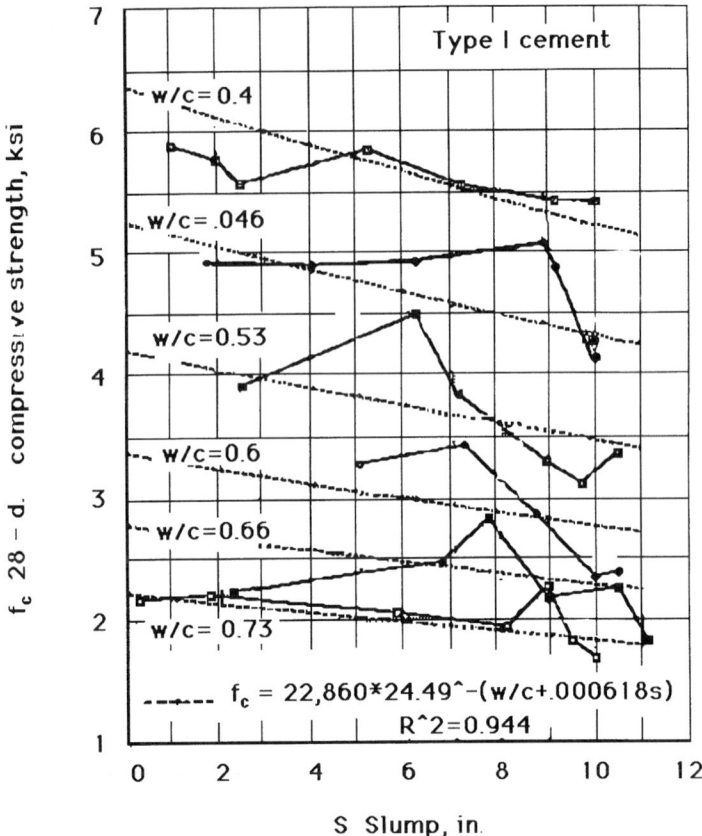

Figure 5.24 Effect of the wetness of concrete as measured by slump, on the strength versus water-cement ratio relationship. Points connected with solid lines represent Gruenwald's (1956a) experimental values, dotted lines represent values calculated by the Abrams' formula augmented by slump values. 1 in = 25.4 mm. 1 ksi = 6.90 MPA (From S. Popovics, 1990a. Copyright ACI. Reprinted with permission.)

formulas are supported by experimental data better than the basic model, and if so, which formulas approximate the experimental results best. This elementary statistical approach is justified because all the strength-predicting formulas are essentially empirical.

For specific illustration of the procedure, Abrams' formula is used here primarily as the basic model. However, the augmentation and general approach apply for other strength-predicting formulas based solely on the water-cement ratio, such as Eqs. 5.7 through 5.14.

Augmentation of Abrams' Formula (S. Popovics 1990a)

Abrams' formula, that is, Eq. 5.6 with $C_A = 0$, can be written in the form

$$\log f = \log A - (w/c) \log B = b_0 + b_1 \, w/c \qquad (5.42)$$

where f = concrete strength
w and c = water and cement content, respectively
w/c = water-cement ratio
A and B = experimental coefficients that depend on the type of strength, cement and aggregate type, admixtures, curing and testing conditions, age of concrete, and the units used
$b_0 = \log A$
$b_1 = \log B$.

The general form of the attempted augmentation of Eq. 5.42 is:

$$\log f = b_0 + b_1 w/c + b_2 c + b_3 w + b_4(w + c)$$
$$+ b_5 \log c + b_6 w^2/c + b_7(w/c)^2 + b_8 C + b_9 \log C \qquad (5.43)$$

where C is the measure of consistency (slump, VB time, etc.) and the b_i values are experimental coefficients. The other symbols are identical with the symbols of Eq. 5.42. Note that the consistency measure C is an acceptable variable in Eq. 5.43, and in similar strength equations presented later, if the changes in consistency are caused only by changes in the cement, water, and/or paste contents, and not by other factors, such as admixture, or aggregate.

By putting various coefficients equal to zero, various mathematical models were formed, the coefficients determined by multiple linear regression, and compared to experimental data, For example, if all the b_i values are equal to zero in Eq. 5.43 except b_o and b_1, the resulting model (model 1 in Table 5.4) is the original Abrams' formula. The coefficients obtained by linear regression for the compressive strength values published by Walker and Bloem (1961) for two cement contents, 12 results of each, and $D = 1.5$ in. (38.1 mm), are (Table 5.4)

$$\log f_c = 4.16 - 0.8586 \, w/c \qquad (5.44)$$

where f_c = compressive strength, psi
w/c = water-cement ratio by mass

The exponential form of Eq. 5.44 is represented in Figure 5.21 by dashed line.

When the nonzero coefficients are b_0, b_1 and b_2, the result is Abrams' formula augmented with the cement content, that is model 2 in Table 5.4. For the same group of experimental data as above for Eq. 5.44, the following was obtained:

$$\log f_c = 4.71 - 1.374 \, w/c - 0.00052c \qquad (5.45)$$

TABLE 5.4 Best-fit coefficients for various augmentaitons of Abrams' formula

Model	b_0	b_1	$b_2 \times 10^3$	$b_3 \times 10^3$	$b_4 \times 10^3$	b_5	$b_6 \times 10^3$	b_7	R^2
					Coefficients				
1	4.16	-0.859	0	0	0	0	0	0	0.84
2	4.71	-1.374	-0.52	0	0	0	0	0	0.93
3	4.43	-0.792	-0	-1.11	0	0	0	0	0.93
4	4.89	-1.766	-0.87[b]	0.77	0	0	0	0	0.93
5	4.62	-1.189	0	0	-0.36	0	0	0	0.93
6	6.26	-1.374	0	0	0	-0.672	0	0	0.93
7	4.10	0	0	0	0	0	-2.67	0	0.93
8	3.95	0	0	0	0	0	0	-0.846	0.87
9	4.11	-0.161[b]	0	0	0	0	-2.23	0	0.94
10	3.44	2.020	0	0	0	0	0	-2.798	0.89
11	3.41	1.177[b]	0	2.50	0	0	-6.90	0	0.95
12	2.04	3.940	1.40[b]	1.90[b]	0	0	-11.00[b]	0	0.95

[a]The specimens were 6 by 12-in. (150 by 300-mm) cylinders. Range of w/c: from 0.39 to 0.67 by weight. $D = 1.5$ in. (38.1 mm). $\log f_c = b_0 + b_1 w/c + b_2 c + b_3 w + b_4(c + w) + b_5 \log c + b_6 w^2/c + b_7 (w/c)^2$, where f is in psi, c and w in lb/cu yd. 1000 psi = 6.90 MPa. 1 lb/cu yd = 0.59 kg/m^3.

[b]Statistically not significant.

Source: Strength data from Walker and Bloem (1961). (Copyright ACI. Reprinted with permission.)

where c is the cement content in lb/yd^3 (0.593 kg/m^3). Note that Eq. 5.45 can also be written in exponential form:

$$f_c = \frac{51{,}290}{23.66^{w/c+0.000378c}} \tag{5.45a}$$

Equation 5.45a is represented in Figure 5.21 by the two solid lines, one for each cement content. When Abrams' formula is augmented by the water content (model 3), the result, for the same group of data as for Eqs. 5.44 and 5.45, is (Table 5.4)

$$\log f_c = 4.43 - 0.792w/c - 0.00111w \tag{5.46}$$

where w is the water content in lb/cu yd (0.593 kg/m^3).

Coefficients for further augmented Abrams' models are also given in Table 5.4 for the above-mentioned experimental results, that is, compressive strengths by Walker and Bloem for $D = 1.5$ in. (38.1 mm). These augmentations took place by the paste content, or squared terms, or slump, or combinations of some of these.

Similar augmented formulas were developed from Eq. 5.43 for the concretes by Walker and Bloem (1961) with $\frac{3}{4}$-in. (19.1 mm) maximum particle size, by Gruenwald (1956a) with two cement types, by Kamenski (1985) with two consistencies, and by Ujhelyi (1980). The improvements produced by the various augmentations in the fit to these test results are similar to those shown above for the data by Walker and Bloem (S. Popovics, 1990a).

Discussion of the Augmented Abrams' Formulas

The goodness of fit of Eq. 5.44 (i.e., Abrams' formula) to the results by Walker and Bloem can be judged visually from Figure 5.21. Here the single curve (i.e., the dashed line) represents Eq. 5.44. The fit is also characterized by $r^2 = 0.84$, where r is the correlation coefficient. The formula augmented by the cement content c, (i.e., Eq. 5.45 or 5.48) shows numerically to what extent an increase in the cement content decreases the concrete strength while the water-cement ratio is kept constant. These formulas have $R^2 = 0.93$, where R is the multiple correlation coefficient.

Figure 5.21 as well as the correlation coefficient $r^2 = 0.84$ indicate a decent fit of Abrams' formula to the 24 experimental data with two cement contents. The value of r^2 is much higher than the the minimum statistically significant value, which is 0.26 at the 0.01 level. Nevertheless, augmentation of this formula with the cement content (i.e., Eq. 5.45 or 5.45a, provides an improved fit, as shown both by the higher correlation coefficient and visually by the two solid lines in Figure 5.21. The goodness of fit of Eq. 5.46 is characterized by $R^2 = 0.93$. It can be seen from the correlation coefficients that the goodness of fit of Eq. 5.46 is better than that of Eq. 5.44 and is the same as that of Eq. 5.45.

Multiple correlation coefficients for other augmented Abrams' models are also presented in Table 5.4 for compressive strengths by Walker and Bloem (1961) for $D = 1.5$ in. (38.1 mm). These show that applied augmentations did improve the goodness of fit compared to Eq. 5.44, but the fits were not better than that of Eq. 5.45 or 5.46. Similarly, augmentation with more than one term, such as cement content *and* slump, did not produce better fit than augmentation with cement content or slump alone.

The same conclusions can be drawn from the analysis of test results by the other three investigators mentioned earlier (S. Popovics, 1990a). Also, Eqs. 5.45 and 5.46 reflect the fact that the change in concrete strength is more rapid with a change in the water-cement ratio than that indicated by Abrams' formula. Furthermore, concrete strength changes more rapidly when the change in the water-cement ratio causes an increase in the paste content than when the paste content decreases. For instance, increasing the water-cement ratio solely by adding water to the concrete increases the paste content. In contrast, when increases in the water-cement ratio are produced by reductions in the cement content (model 3, i.e., Eq. 5.46), simultaneous reductions occur in the paste content, thus, smaller strength changes take place. This is another demonstration that it is more efficient to increase the concrete strength by reducing the water-cement ratio through a reduction in the water content than with more cement. This observation further supports the proportioning of concretes for low water contents, including the use of water-reducing admixtures.

Linear Formulas

Figures 5.21 and 5.22 reveal that the graphs representing augmented formulas are very close to straight lines. It follows then that these relationships can be approximated within practical limits with linear formulas as well. The general forms of the attempted augmentations of such linear models are:

$$f = b_0 + b_1 w/c + b_2 c + b_3 w + b_4(w + c) + b_5 C + b_6 \log C \quad (5.47)$$

and

$$f = b_0 + b_1 c/w + b_2 c + b_3 w + b_4(w + c) + b_5 C + b_6 \log C \quad (5.48)$$

where the symbols are as in Eq. 5.43. Here again various mathematical models can be formed by making various b_i coefficients equal to zero. It was demonstrated that the augmented linear models are as good as the augmented Abrams formulas. Similarly, augmentation with more than one term did not produce better fit than augmentation with cement content or slump alone (S. Popovics, 1990a).

Example 5.4   Two models obtained from Eq. 5.47 for the Walker and Bloem compressive strength data that were used to calculate the coefficients in Table 5.4 are:

$$f_c = 15,890 - 15,094w/c - 5.15c \qquad (5.49)$$

and

$$f_c = 13,218 - 9319w/c - 11.34w \qquad (5.50)$$

where the symbols are as in Eq. 5.43.

Implications of the Augmented Formulas

The problems below focus on the case when the water-cement ratio is decreased solely by increasing the cement content, producing gradually diminishing strength increases.

PROBLEM 5.5   At constant water content, there should be a threshold cement content c_{th} beyond which further cement addition does not increase the concrete strength (f_{max}) despite the caused reduction in the water-cement ratio. Derive this cement content c_{th}.

Solution   From Eq. 5.45,

$$4.71 - 1.37w/c - 0.00052c = 4.71$$
$$- 1.374w/(c + \Delta c) - 0.00052(c + \Delta c) \qquad (5.51)$$

This equation reduces to:

$$w[1/c - 1/(c + \Delta c)] - 0.000378\Delta c = 0 \qquad (5.52)$$

the solution of which is the c_{th}. That is,

$$c_{th} = \frac{-\Delta c + \sqrt{(\Delta c)^2 + 10,580w}}{2} \qquad (5.53)$$

Since the term Δc is small, thus negligible in Eq. 5.53, the threshold cement content is, with good approximation,

$$c_{th} = 50\sqrt{w} \qquad (5.54)$$

Substituting Eq. 5.54 into Eq. 5.45, the top concrete strength f_{max} is

$$\log f_{max} = 4.71 - 0.0535\sqrt{w} \tag{5.55}$$

where w is the water content in lb/cu yd.

With the usual water contents, Eq. 5.54 gives around 800 lb/yd^3 (480 kg/m^3) for c_{th} and Eq. 5.55 about 7500 psi (50 MPa) for the maximum standard compressive strength achievable under the circumstances under which Walker and Bloem performed their tests. These values are in good agreement with the experience obtained from proportioning high-strength concretes (ACI, 1984), thus indirectly support Eq. 5.45. Note that according to Eq. 5.45, the limit f_{max} can be increased by reduction in the water content; this explains mathematically why superplasticizers are so effective in the production of high-strength concrete. The use of additional cementitious materials other than portland cement, such as fly ash or silica fume, can also raise f_{max}.

Otherwise, after c_{th}, further increases in the cement content usually do not cause sizable changes in the value of f_{max}. Exceptions have also been reported, however. For instance, the strength of a pure cement paste was higher at the age of 56 days than that of the comparable concrete of the same water-cement ratio (Hobbs, 1974).

PROBLEM 5.6   So, increasing a low cement content at constant water content is quite an efficient way to increase concrete strength but only up to a certain point. This point is the cement content at which the strength f_1 developed by 1 lb (kg, etc.) of cement reaches the maximum. This is obviously the most economical cement content for strength; therefore, it is designated as c_{eco}. The strength related to c_{eco} is f_{eco}. Calculate c_{eco} and f_{eco}.

Solution   An equation for this most economic compressive strength, from Eq. 5.49, for the experimental values by Walker and Bloem for $D = 1.5$ in. (38.1 m) is

$$f_1 = f/c = 15,890/c - 15,094w/c^2 - 5.15 \tag{5.56}$$

This equation reaches its maximum when $df_1/dc = 0$, which happens when

$$c = c_{eco} = 2w \text{ lb/cu yd} \qquad \text{that is,} \quad (w/c)_{eco} = 0.5 \quad \text{by weight} \tag{5.57}$$

Therefore,

$$f_{eco} = 8340 - 10.30w \quad \text{psi} \tag{5.58}$$

and

$$\left(\frac{f}{c}\right)_{eco} = \frac{4170}{w} - 5.15 \quad \text{psi/(lb/cu yd)} \tag{5.59}$$

This equation shows again that the most economical cement content as well as compressive strength can be increased by reduction in the water content in the concrete.

PROBLEM 5.7   Change the water-cement ratio in a concrete solely by changing the water content. Then change the same water-cement ratio of a similar concrete by changing solely the cement content. What is the difference between the changes of the two compressive strengths?

Solution   A comparison of Eqs. 5.45 and 5.46 shows that the compressive strength changes more rapidly when the water-cement ratio is changed solely by changing the water content than when the water-cement ratio is changed by the cement content.

Applications

Example 5.5   One can show that the augmented formulas can be used for proportioning concrete as simply as the original Abrams formula. For instance, the water content w for a given slump can be estimated in the usual manner, say, from ACI 211 Standard Practice (ACI, 1991). This value is substituted into Eq. 5.50, from which the water-cement ratio for the specified strength can be obtained:

$$w/c = 1.418 - 0.107 \times 10^{-3}f_c - 1.217 \times 10^{-3}w \qquad (5.60)$$

The rest of the procedure can then follow the traditional steps of concrete proportioning.

Models augmented with the cement content c, such as Eq. 5.45 or 5.49, are convenient because the composition of concrete is frequently characterized with the cement content. Applications of these formulas, for instance for proportioning, are made easy with computerization. Prop 21 on the accompanying disk is such a software. It appears that it is the paste content that directly influences the concrete strength in addition to the water-cement ratio. Therefore, models augmented with the paste content $w + c$, such as model 5 in Table 5.4, may be used advantageously for theoretical investigations.

5.10   OTHER EFFECTS ON B_A

Effects of Curing Temperature

Experimental data by Gruenwald (1956b) represented by the best-fitting lines in Figure 5.25 demonstrate that Abrams' rule is also valid for concretes cured

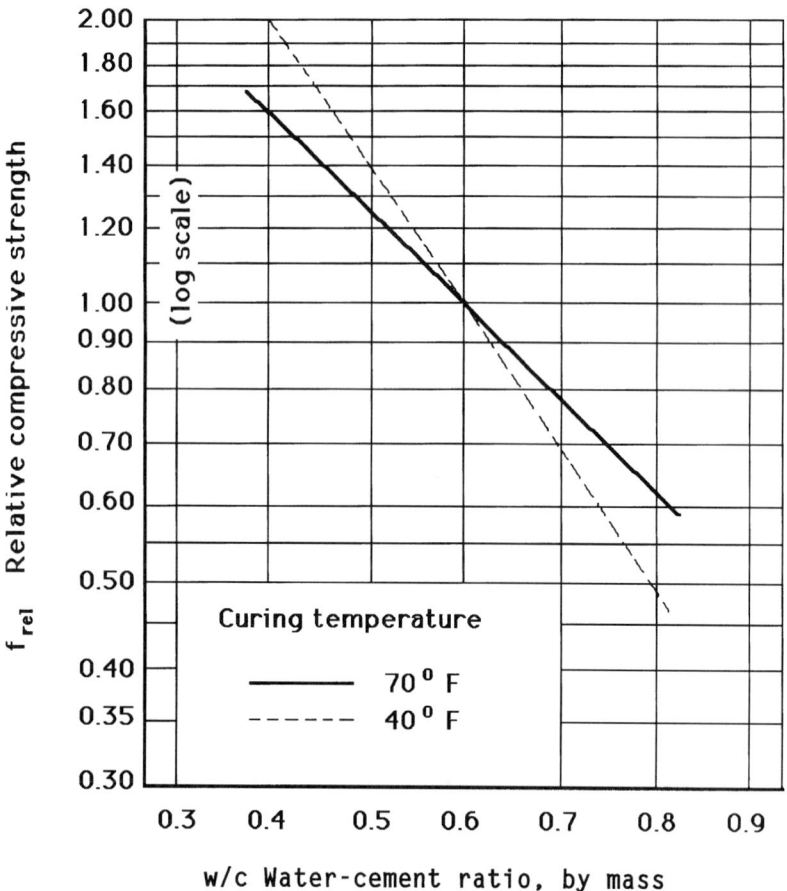

Figure 5.25 Effect of curing temperature on the compressive f_{rel} versus w/c relationship. Cement: Type III. The basic test results were published by Gruenwald (1956b). 70°F = 21.1°C; 40°F = 4.5°C.

at temperatures other than the standard 22.3°C. However, the lower the curing temperature, the steeper the log f_{rel} versus w/c line, that is, the higher the numerical value of B_A. This is particularly true for younger ages. Thus the effect of curing temperature on B_A, in general, follows the same pattern as some of the other effects (cement type, fineness, etc.) discussed earlier: namely, the lower the rate of strength development at early ages, the greater the magnitude of B_A for comparable concretes.

One would expect that the reverse of this statement is also true. That is, the strength of concretes cured at elevated temperature would show less water sensitivity than the same concrete after standard curing at the same age. The few available experimental data seem to support this expectation (Armuth et al., 1962).

Effects of the Type of Aggregate

Although the effects of aggregate on the strength of structural concretes are usually secondary in magnitude, there are exceptions. Figure 5.26 presents an example for this. The low strengths produced by the concretes with sand and gravel from Guntersville may be attributed to the weaker bonds between the cement paste and gravel particles at lower water-cement ratios. This bond strength probably did not increase as much with the reductions in the water-cement ratio as the strengths of the bonds to the crushed stone particles in the other concretes.

Another example, in Table 5.5, illustrates the reducing effects of structural lightweight aggregate on B_A calculated from 28-day strengths by Jindal

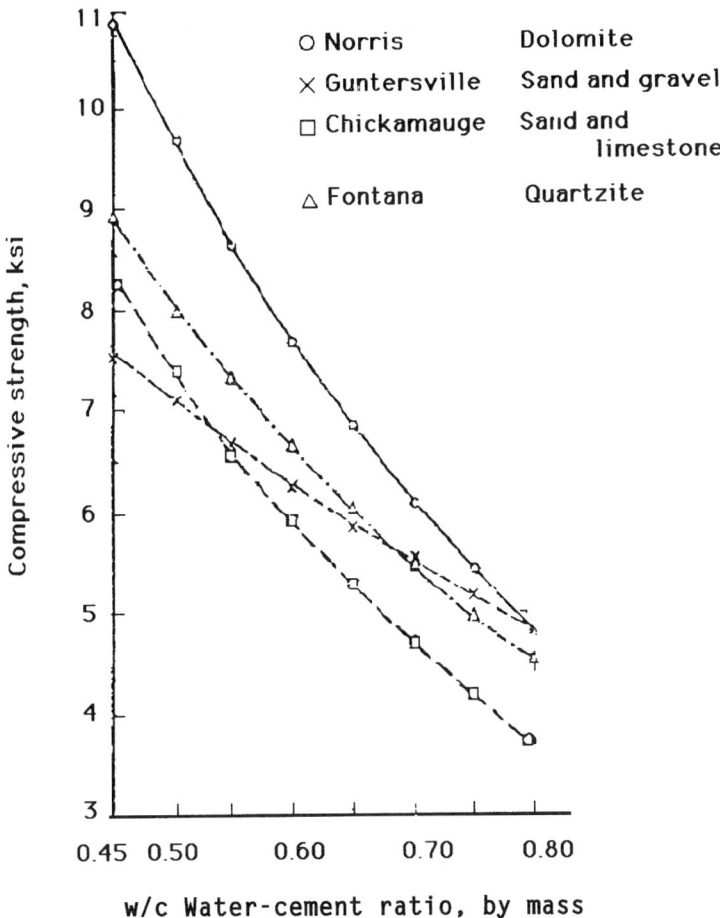

Figure 5.26 Effect of changing aggregate on the compressive strength versus water-cement ratio relationship at the age of 1 year. 1 ksi = 6.90 MPa. (From Tennessee Valley Authority, 1947.)

TABLE 5.5 Values of log B and B for various strengths of normal-weight and structural lightweight concretes

Type of Strength	Type of Concrete[a]		
	a	*b*	*c*
Compressive			
log *B*	1.30	0.90	0.85
B	20	8.0	7.1
Flexural			
log *B*	0.86	0.46	0.47
B	7.2	2.9	3.0
Splitting			
log *B*	0.93	0.51	0.51
B	8.5	3.2	3.2
Bond			
log *B*	1.06	0.70	0.79
B	11.5	5.0	6.1

Source: S. Popovics (1967b). (Copyright ASTM. Reprinted with permission.)
[a]*a*, concretes with natural sand and gravel; *b*, concretes with natural sand and lightweight coarse aggregate; *c*, concretes with lightweight fine and coarse aggregates.

(1964). That is, the water sensitivity of strengths of structural concretes made with lightweight coarse aggregate is less than that of normal-weight concretes (S. Popovics, 1967b). Experimental results published by Grieb and Werner (1962) and by Rothfuchs (1962) show a similar tendency. A contributing factor to this phenomenon is that the deformability of a typical structural lightweight aggregate is closer to the deformability of the hardened cement paste than that of a normal-weight aggregate. This reduces the inhomogeneity of the concrete and the stress concentrations resulting from this inhomogeneity (Bremmer and Helm, 1986; S. Popovics, 1987a; Kaplan, 1963; Hansen, 1968).

Although structural lightweight concretes do follow Abrams' rule, application of the formulas discussed for the prediction of their strengths from water-cement ratio is doubtful. This is so because the water absorption of lightweight-aggregate particles is usually large and difficult to estimate; thus the free water content of the fresh concrete is uncertain. However, once the water absorption of the lightweight aggregate is established (e.g., from experience), the effective water-cement ratio can be calculated in a reliable manner. Then the strength of the structural lightweight concrete can be estimated with the degree of accuracy associated with normal-weight concretes.

Effect of Maximum Particle Size and Grading

As Figs. 5.27 and 5.28 demonstrate, the strengths of comparable concretes with identical water-cement ratios usually increase as the *maximum particle*

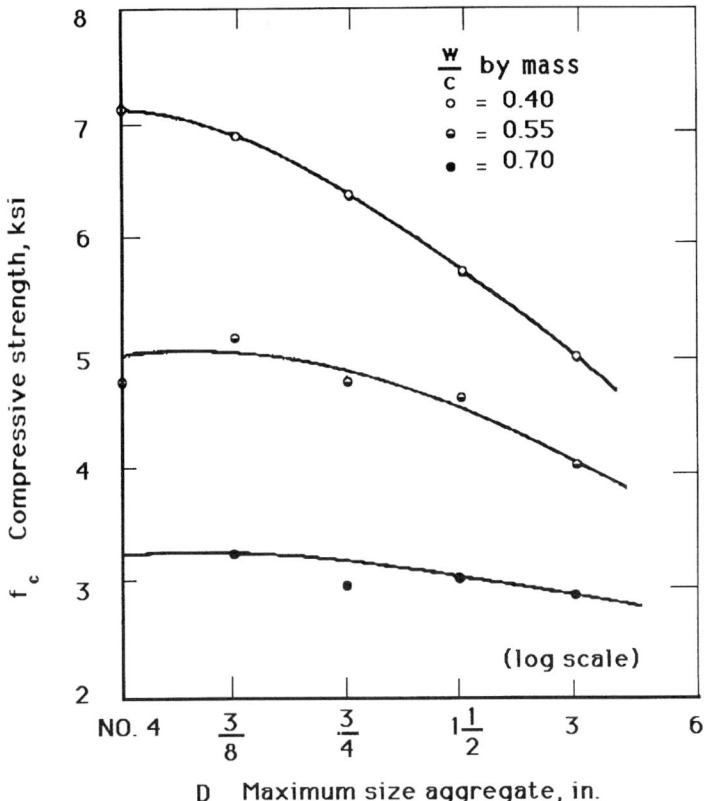

Figure 5.27 Effect of maximum size of aggregate on concrete strength at three w/c ratios; slump varied from 3.8 to 5.8 in. 1 in. = 25.4 mm. 1 ksi = 6.90 MPa. (From Cordon and Gillespie, 1963. Copyright ACI. Reprinted with permission.)

size decreases. Other pertinent experiments led to similar conclusion (Higginson et al., 1963; Walker et al., 1959, 1960; Cordon and Gillespie, 1963; Hughes and Chapman, 1966b), especially at lower water-cement ratios. A likely explanation for this is that the bond of the hardened cement paste to the aggregate particles of large sizes is weaker than the bond to the smaller sizes because of the smaller specific surface of the former.

This means that a more-or-less different f versus w/c curve prevails for each maximum particle size. In practical cases, however, the strength-reducing effect of an increase in the maximum particle size is less pronounced or even reversed. This is so because for concretes of *identical consistency* (Fig. 5.29), as the maximum size increases, the mixing water requirement for a specified consistency is reduced, particularly at lower maximum sizes (S. Popovics, 1982b). This lowers the water-cement ratio and improves the concrete strength, which is substantial in cases of low cement contents (Higginson, 1963).

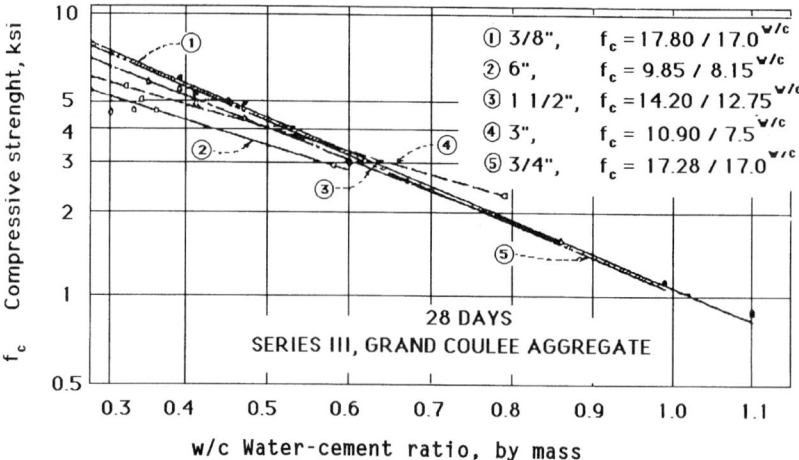

Figure 5.28 The 28-day compressive strength of concrete in ksi with various maximum size aggregates at different water-cement ratios. 1 ksi = 6.90 MPa. (From Higginson et al., 1963. Copyright ACI. Reprinted with permission.)

In addition to these opposing factors, the cement content also affects the relationship between concrete strength and maximum particle size, as discussed earlier in this section. The combined result of all this is that the strength-reducing effect of increasing the maximum size is strong in high-strength concretes, that is, when the water-cement ratio is low and the cement content is high; and when the maximum particle size is large (Fig. 5.27). Conversely, an increase in the maximum size does produce improved strength in comparable concretes of identical consistency when the cement content is low and/or the maximum size is small. Thus for each concrete there is an optimum maximum particle size that produces the highest compressive strength for a given consistency and cement content.

The conventional opinion has been that the *aggregate grading* does not have any direct effect on the strength of concrete (Lenhard, 1942). The effect is supposed to be indirect through the water need and workability of the fresh concrete (S. Popovics, 1982b). This view is also illustrated in Figure 5.30, where the effect of the coarseness of aggregate grading seems negligible on the strength versus water-cement ratio relationship within wide limits. Nevertheless, experimental evidence has also been presented implying that the coarser the grading up to a certain point, the higher the strength of comparable concretes with identical water-cement ratio (Graf et al., 1960; Singh 1958; Newman and Teychenne, 1954).

The expression of the concrete strength with other augmented formulas has also been attempted. Here a fictional water-cement ratio $(w/c)_f$ is calculated and used for strength calculation. There are formulas in the literature for

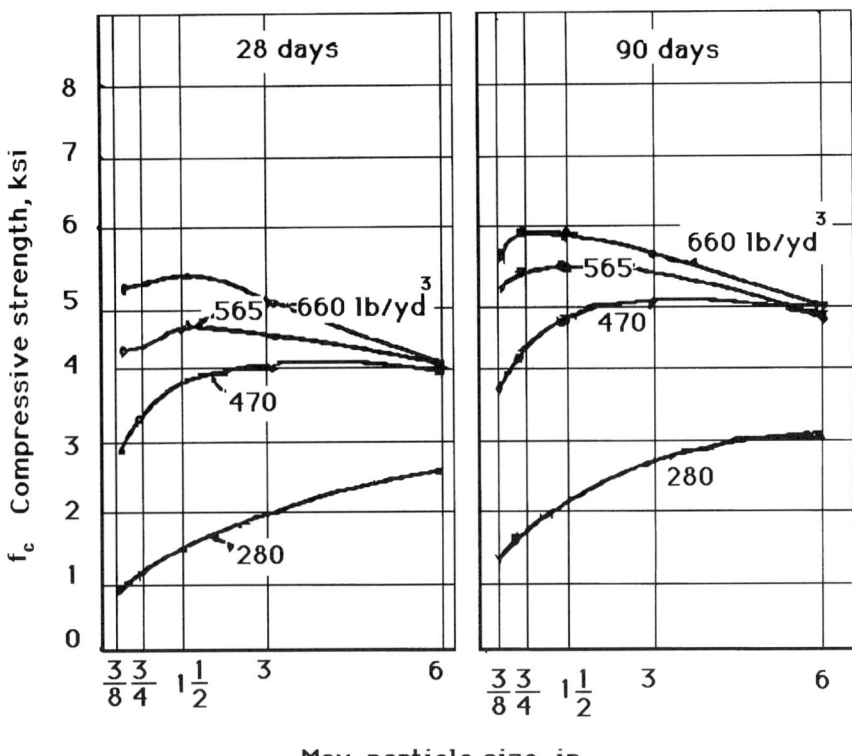

Figure 5.29   Effect of cement content and maximum particle size on compressive strengths of concretes of identical slumps at the ages of 28 and 90 days. 1 lb/cu yd = 0.59 kg/m³. 1 ksi = 6.90 MPa. (From Higginson et al., 1963. Copyright ACI. Reprinted with permission.)

calculation of such fictional water-cement ratios as a function of the effective water-cement ratio, grading, and mixture proportion (S. Popovics, 1971b; Singh 1958; Newman and Teychene, 1954).

Concrete Strength Versus Slump

When the water-cement ratio is changed by *changing the water content only,* that is, keeping the composition of the concrete otherwise unchanged, the concrete consistency also changes (Section 5.4). This gives the opportunity for such special cases, and only for such cases, to express the concrete strength for a measure of consistency, such as slump, along with the composition of the concrete. This can be done by the combination of any of the strength formulas in Section 5.5 with a water-predicting formula (S. Popovics, 1982b). For instance, the water-cement ratio can be expressed as

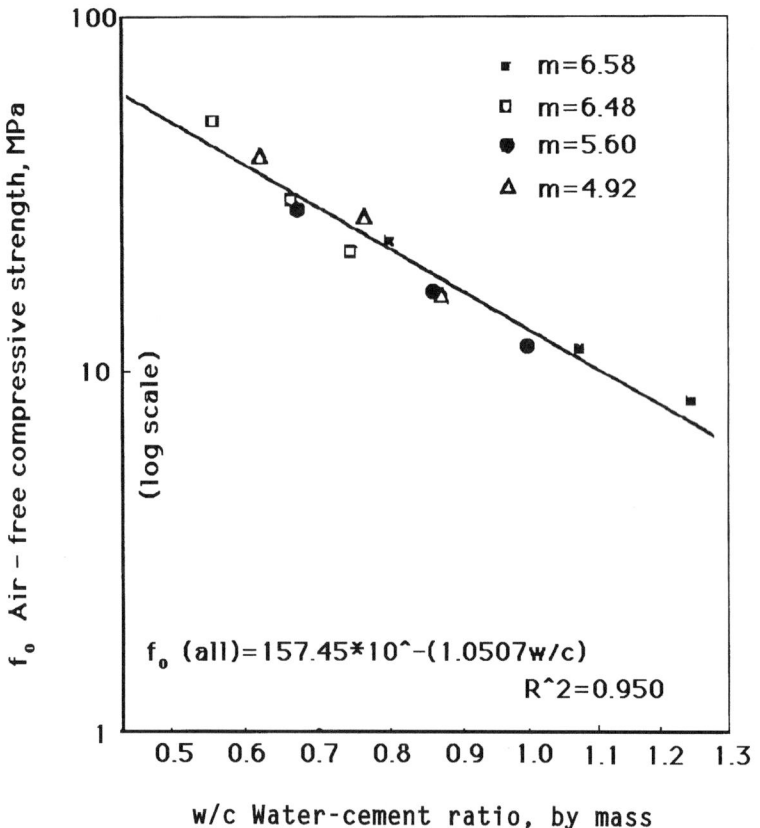

Figure 5.30 According to Ujhelyi's (1989) data, the aggregate grading does not have any direct effect on the strength versus water-cement ratio relationship. m = fineness modulus. 1 MPa = 145 psi.

$$w/c = \left(\frac{S}{4}\right)^{0.1} \left\{ \frac{14.5[(m - 5.9)^2 + 14.6]}{c - 100} + 0.1 \right\} \quad (5.61)$$

where w/c = water-cement ratio by mass
w and c = water and cement content, respectively, lb/cu yd
S = slump, in.
m = fineness modulus of the aggregate

Substituting this equation into Eq. 5.9 with $B_H = 1$, the following relationship is obtained for the concrete strength versus slump relationship (S. Popovics, 1966):

$$f_c = \left(\frac{4}{S}\right)^{0.1} \frac{A_H}{\{14.5[(m - 5.9)^2 + 14.6]/(c - 100)\} + 0.1} + C_H \quad (5.62)$$

A_H and C_H are coefficients of Eq. 5.9. The other symbols are the same as in Eq. 5.61. Equation 5.62 is still a strength versus water-cement ratio formula since the underlying condition for the validity of this formula is that the slump change be caused solely by a change in the water content.

Example 5.6   With a maximum particle size D of approximately 1 in. (25 mm), cement content c of 610 lb/cu yd (360 kg/m³), A_H of 3460 psi (23.9 MPa), and C_H of -950 psi (6.55 MPa), Eq. 5.62 provides the following for the 28-day strength versus slump relationship when the slump change is caused solely by changing the water content of the concrete:

1. When $m = 5.6$,

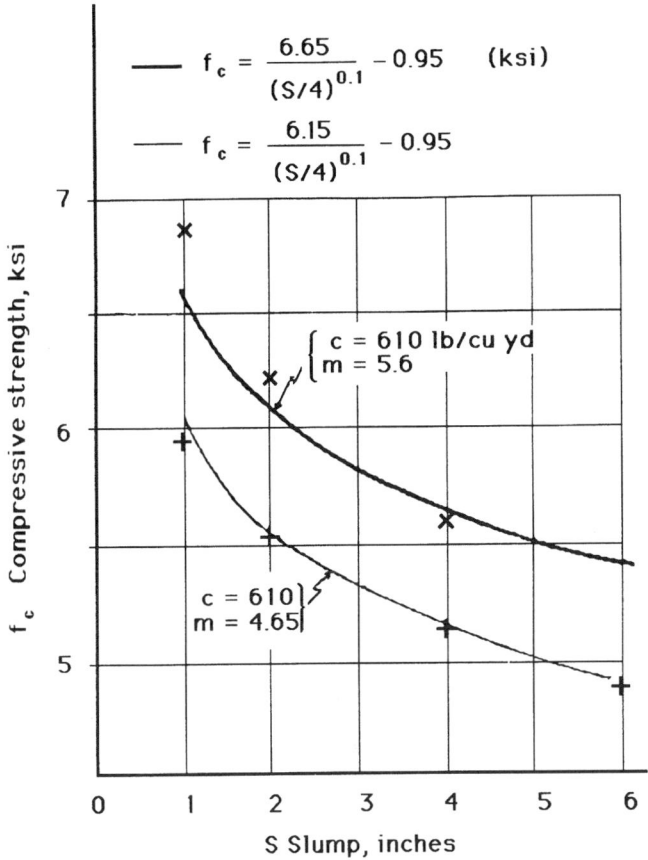

Figure 5.31   Compressive strength of concrete as a function of slump where the change in slump is due to the change in water content. 1 in. = 25.4 mm. 1 ksi = 6.90 MPa. 1 lb/cu yd = 0.59 kg/m³. (From S. Popovics, 1966. Copyright RILEM. Reprinted with permission.)

$$f_c = \left(\frac{4}{S}\right)^{0.1} 6650 - 950 = \frac{7640}{S^{0.1}} - 950$$

2. When $m = 4.65$,

$$f_c = \left(\frac{4}{S}\right)^{0.1} 6150 - 950 = \frac{7065}{S^{0.1}} - 950$$

where f_c is in psi. The graphs of these two formulas are presented in Figure 5.31 along with experimental data. Other experiments also seem to support these formulas (Nogula, 1956; Riley, 1975).

5.11 ABRAMS' FORMULAS FOR FLEXURAL AND OTHER STRENGTHS

Flexural Strength

Abrams' rule and the related formulas with the appropriate coefficients are also valid for concrete strengths other than the compressive strength. Numerous experiments have demonstrated this for flexural strength (Gonnerman and Shuman, 1928; Bonzel, 1964, 1965; Road Research Laboratory, 1950; Williams, 1962). Examples are shown in Figures 5.9, 5.18, and 5.32. Figure 5.32 is comparable to Figure 5.15. The best-fitting straight lines are also shown in these figures.

These fitted lines demonstrate, along with some of the figures as well as numerous other experiments presented, that the flexural strength is *less water sensitive*. It changes relatively less than the compressive strength, with a change in the water-cement ratio (Hummel, 1959; Portland Cement Association 1968; Bonzel, 1965; Abrams, 1922), at least for water-cement ratios higher than 0.3 by mass (Graf et al., 1960). This is also shown by Tables 5.3 and 5.5 as well as by Eqs. 5.28 through 5.31. The significance of these data is that they show these effects numerically, thus they contribute to a more reliable prediction of the flexural strength by, say, Abrams' formula. In addition, they clarify that part of the variation of the flexural strength–compressive strength ratio that is caused by the age, type of cement, and type of concrete (Section 2.1).

The smaller water sensitivity of the flexural strength may encourage the concrete engineer to use a little higher water-content in concretes where the flexural strength has primary importance. On the other hand, this also means that the flexural strength of concrete cannot be increased as effectively as the compressive strength of the same concrete by a reduction in the water-cement ratio. Thus the flexural strength–compressive strength ratio decreases with decreasing water-cement ratio, that is, with increasing compressive strength as shown, for instance, in Figure 2.1 (Section 2.1).

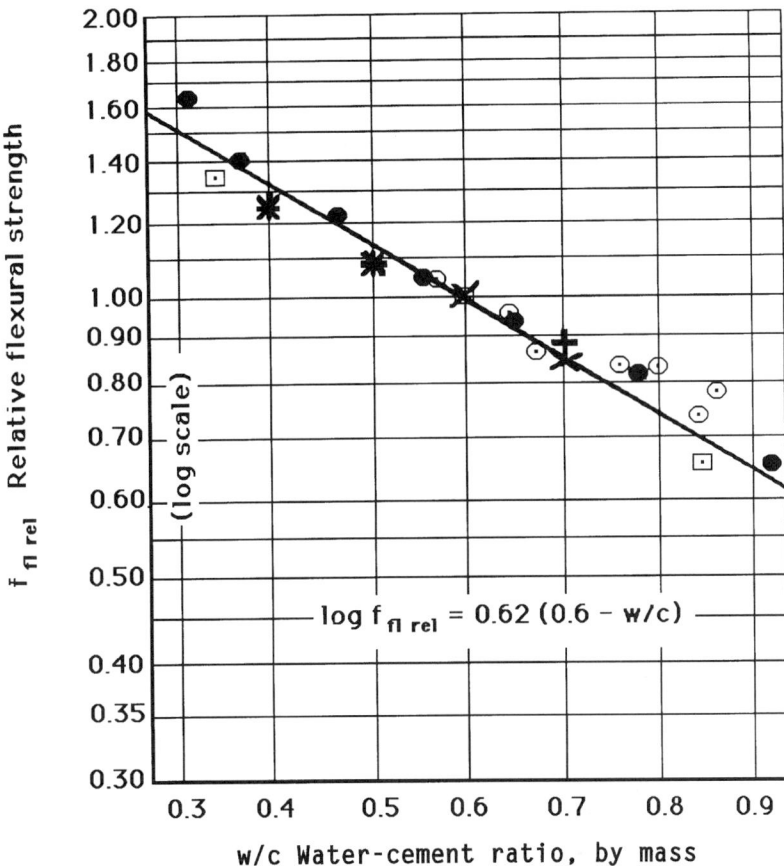

Figure 5.32 Relative flexural strength versus water-cement ratio relationship at the age of 28 days. The basic test results were taken from Gonnerman and Shuman (1928), Kaplan (1960a), Malhotra and Zoldners (1967), and Portland Cement Association (1955).

The coefficients of Eqs. 5.6 through 5.13 for flexural strength can be determined from appropriate test results in the same way as discussed in Section 5.6 for compressive strength. It is also possible to estimate approximate values of B_A for flexural strengths at various ages from Eqs. 5.37 and 5.39, respectively. A good example is Eq. 3.21, with the appropriate parameters b. Computerized examples are given in the User's Manual on the accompanying disk.

The f_{fl} versus w/c relationship is influenced by the same factors, by and large, to the same extent as was discussed about the compressive strength, except for the aggregate type. Most important, concretes with *crushed aggregate* have typically higher flexural strengths than comparable concretes with the same water-cement ratios but with round coarse aggregate particles (Fig. 5.26). The reason for this is assumed to be the stronger bond of the cement paste to the surface of the crushed stone particles, (i.e., a stronger interface).

This is attributed, at least partially, to the rougher surface texture of the crushed stone. The effect of the aggregate surface characteristics is greater on flexural strengths than on compressive strengths of concretes. That is, the strength of the paste–aggregate interface has a greater role in concrete under bending in resisting failure than under compression. It follows from this that the flexural strength–compressive strength ratio of a concrete is considerably greater when it is made with crushed stone aggregate than when made with gravel. Also, when the interface is weakened, for instance by a coating on the surface of the coarse aggregate particles, the flexural strength of the concrete made with such aggregate will suffer more than the compressive strength. Therefore, whenever the flexural (or tensile) strength of a concrete has a primary role, such as in concrete pavements, the use of clean crushed stone as coarse aggregate is a must.

Attempts to improve concrete strengths by making the aggregate surface rougher or by otherwise strengthening the interface have been unsuccessful so far (S. Popovics, 1987b). A hypothetical mechanism of lower water sensitivity of flexural and several other strengths was offered in Section 2.4.

Other Strengths

An analysis of test results (Malhotra and Zoldners, 1967; Bonzel, 1965) shows that the *splitting strength* of concrete is as sensitive to changes in water-cement ratio as the flexural strength; that is, less sensitive than the compressive strength (Figure 5.33). Direct uniaxial tensile *strength* shows somewhat less water sensitivity than is shown by the flexural strength (Johnston, 1970; Gonnerman and Shuman, 1928) (Fig. 5.34). Only a few data are available for the relationship between water-cement ratio and *bond* to reinforcement (Cong et al., 1992). These data do not show any considerable deviation from the sensitivity of other concrete strengths within the limit of 3 and 28 days. At the age of 1 day, however, the bond strength again appears to be very sensitive to changes in water-cement ratio (Narayanan and Rao, 1962). These strengths might be valid for the bond between cement pastes and aggregate as well (Alexander, 1959).

Experimental data concerning the water sensitivity of other concrete strengths are scarce. When such information is needed, it is usually assumed that it is similar to the water sensitivity of the compressive strength. This is a very rough approximation in most cases because the water sensitivity of the shear and torsional strengths is probably closer to that of the flexural strength. It can also happen that special strengths show water sensitivity higher than the compressive strength. Figure 5.35, for instance, demonstrates a very high water sensitivity for the *impact strength* of a concrete with B_A = approximately 6000 (!). There are not enough pertinent experimental data available to establish the general validity of this surprising trend. Note also that there are opposite statements both for the impact compressive strength and the impact splitting strength (Zielinski et al., 1980, 1981). This contradiction may

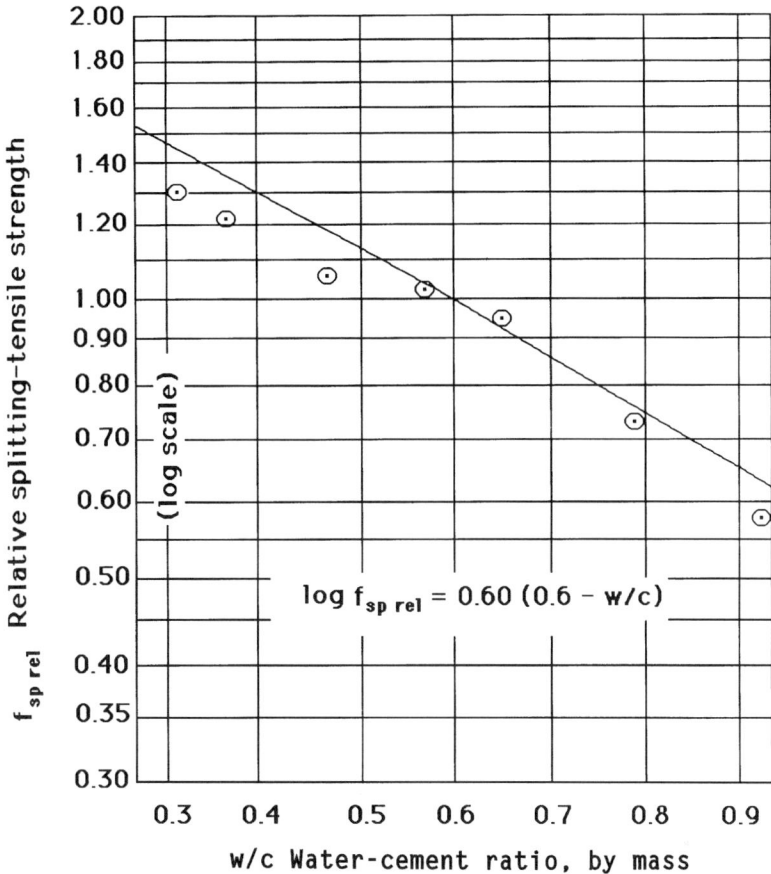

Figure 5.33 Relative splitting–tensile strength versus water-cement ratio relationship at the age of 28 days. The basic test results were published by Malhotra and Zoldners (1967).

be due to the type of aggregate (Green, 1958). That the impact strength increases much more at later ages than the compressive strength (Dahms, 1969) is another, indirect sign of the higher water sensitivity of the impact strength. Other data seem to indicate that the water sensitivity of the impact strength is independent of the age of concrete (Fulton and Davis, 1961). The reasons for these phenomena are not clear at present. Nevertheless, a pertinent hypothetical explanation can be offered.

The greater water sensitivity of the impact strength may be the consequence of the inhomogeneity of concrete. The higher the water-cement ratio, the greater the difference between the deformability of the hardening cement paste and that of the aggregate, especially at early ages; therefore, the greater the internal stresses produced by loading in the concrete. This, in turn, reduces the resistance of concrete against impact. The same hypothesis explains the

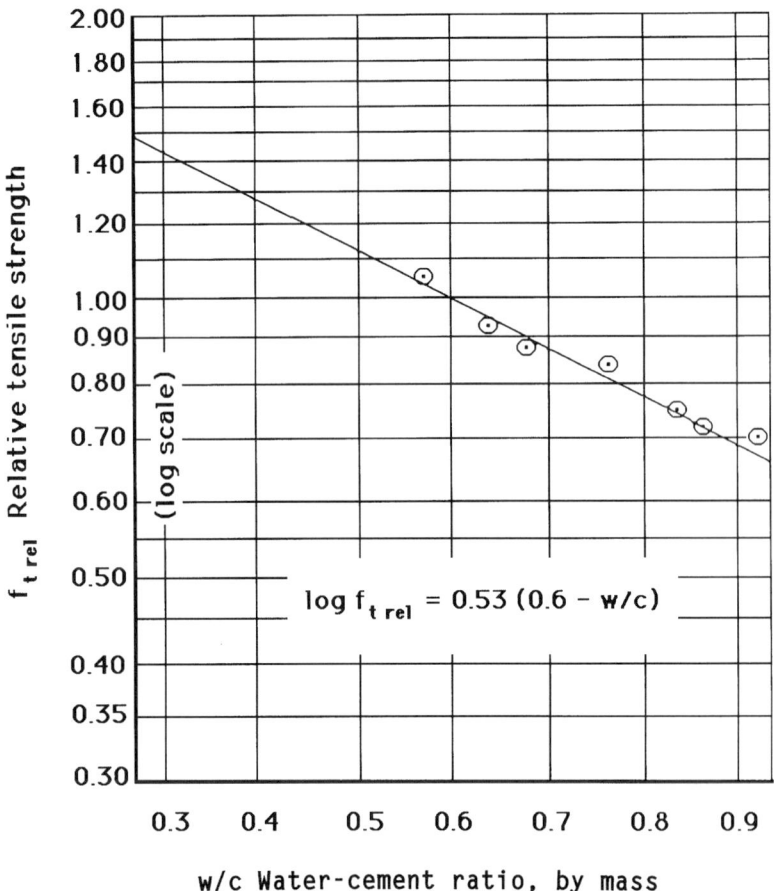

Figure 5.34 Relative direct tensile strength versus water-cement ratio relationship at the age of 28 days. The basic test results were published by Gonnerman and Shuman (1928).

strong increase of the impact strength with age. If this hypothetical explanation is true, the modulus of elasticity of the aggregate should also have a significant influence on the impact resistance of concrete.

The different water sensitivity values of the various mechanical properties of concrete are illustrated in Figure 5.36. This shows that the ultrasonic longitudinal pulse velocity has far the minimum, and the compressive and impact strengths far the maximum, water sensitivity.

5.12 EFFECT OF AIR CONTENT ON CONCRETE STRENGTH

Characteristics of Air Content in Concrete

Air content is that portion of the pores in the *fresh cement paste* portion of the concrete that is filled with *air*. The quantity of the liquid-filled pores can

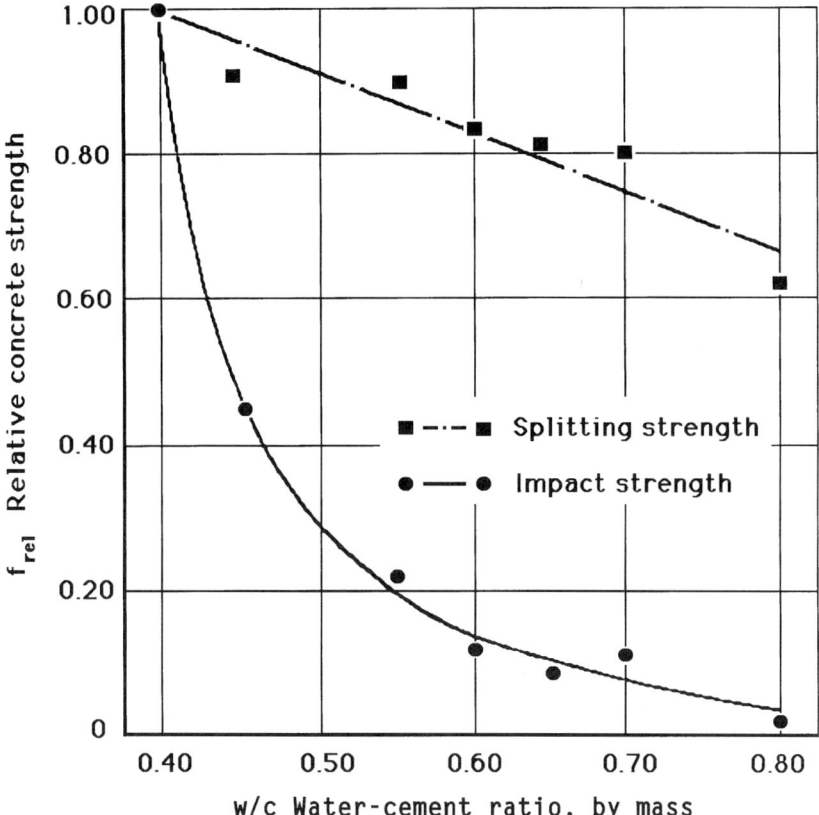

Figure 5.35 Relative splitting and impact strengths versus water-cement ratio relationships, (From Dahms, 1969. Copyright Beton Verlag. Reprinted with permission.)

be characterized by the water-cement ratio in the fresh cement paste, as discussed earlier.

A reduction of porosity in a solid material increases its strength in general, and the strength of concrete in particular. This was recognized long time ago. This is why adobe walls, bricks, and soils were compacted at early times, later the Roman builders compacted their concrete, and in the twentieth century various mechanical devices (vibrators, rollers, etc.) were developed for more efficient compaction of concrete. The strength-increasing effect of a reduction in the water-cement ratio is an example concerning the effect of the water-filled pores.

The strength-reducing effect of air content in the concrete is as drastic as that of the capillary porosity, although the magnitude is not the same. Therefore, Abrams' rule, and the other strength formulas presented earlier in this chapter, are valid for concretes only where the air contents in the specimens are practically the same (Section 5.4). When this condition is not fulfilled, the concretes are not comparable; therefore, the water-cement ratio alone is

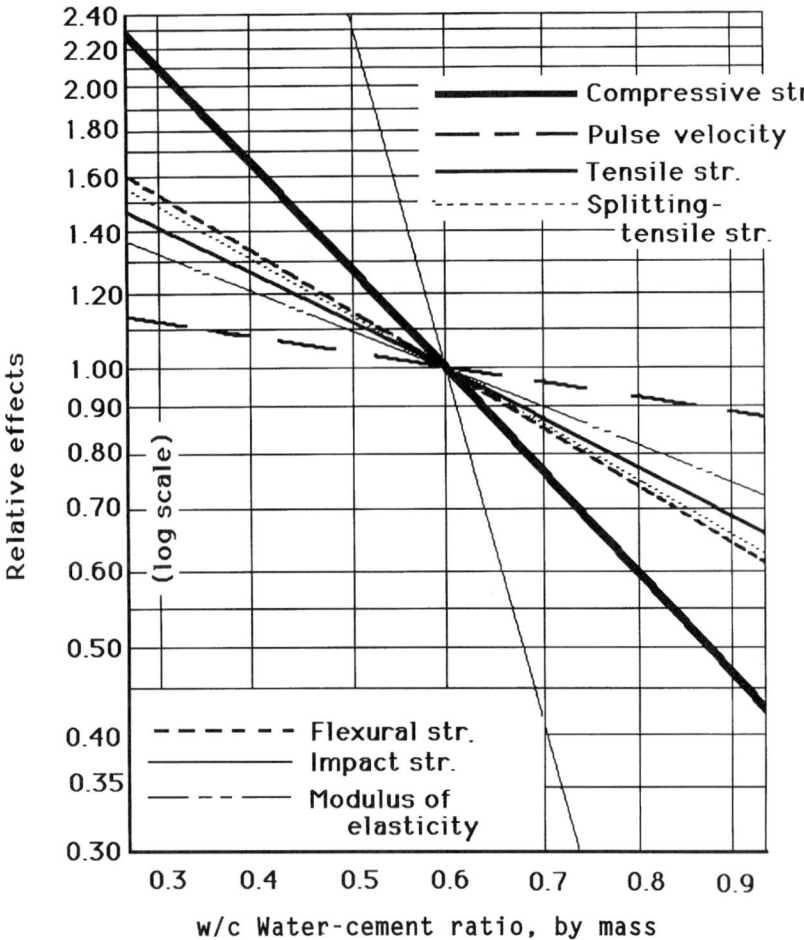

Figure 5.36 Relative effects of water-cement ratio on various mechanical characteristics of concrete at the age of 28 days.

not enough for the characterization of concrete strength. For this chapter, air voids are considered pores that are larger than the capillary pores, regardless whether they are filled with air, liquid, or other strengthless material (Section 4.4). The usual definition of the term air content of a concrete is given in Figure 4.6 as

$$a = \frac{100V_a}{V}$$

where a = air content, % of the volume of concrete

V_a = volume of air in the matrix of the compacted concrete

V = volume of the compacted concrete, including volume of air

Despite the simplicity and resulting popularity of the definition of air content in this equation, one can argue that a more meaningful way would be to express the air content as a percentage of the *cement paste volume*. After all, only the portion of the porosity influences the concrete strength that is in the paste; the porosity in the aggregate is usually negligible in normal-weight concretes. When the concrete consists of, say, 25% by volume of paste, 70% of aggregate, and 5% air, a change in the air content of the concrete from 5% to 6% means that the air content in the paste volume changes from 20% to 24%. This is the jump that influences the concrete strength.

The volume of the air voids remains essentially constant during the life of the concrete (Mielenz, 1969). That is, the air content in a concrete is practically the same at any age as the air content of the compacted concrete in the fresh state, apart from the changing volume of the pore liquid.

The following discussion is based on two conditions: (1) that the air voids are distributed in the matrix randomly but statistically uniformly, and (2), that none of the voids is too large for the size of the concrete specimen. It is also assumed at present that the various air voids (i.e., large and small pores, entrained air, entrapped air from bleeding, and air voids from inadequate amount of fines in the concrete) influence the concrete strength to the same extent. This assumption is probably not true, strictly speaking (Alford, 1982). For instance, experimental data appear to indicate that entrained air reduces concrete strengths less than the same amount of entrapped air. This may be so because of the differences in the shapes of voids (Cordon, 1946; Akutsu, 1969; Wright, 1953). A similar observation was reported by Schiller on the compressive strength of plaster of paris (Schiller, 1958) and by Spriggs (1962) on polycrystalline alumina containing open as well as closed pores. Hansen (1965), using different reasoning, reached a similar conclusion. It also follows from the Griffith criterion of fracture (Section 2.4) that larger pores produce less damage than the same total volume of smaller pores of the same voids shape. Therefore, the pore size distribution of the pores may also have an effect on the degree of strength reduction (Barovsky et al., 1979). For instance, one may expect that the reduction in load-bearing capacity of a concrete is less when there are a few large holes of controlled shape in it than when the same amount of air is present in the concrete in randomly distributed small pores of uncontrolled shape (Shanley, 1965). Spatial distribution of the pores can also be a factor. It is expected that two pores of the same size would reduce the concrete strength more when they are spaced closely to each other than when they are far apart.

Unfortunately, not enough data are available at present for the more definite establishment of these differences or to express them mathematically. It appears, however, that one can disregard these differences for practical purposes since the strength reductions approximate the same pattern in concrete regardless of the origin of the air voids. Thus this approach will be followed here.

For quantitative estimation of the effect of air content on the strength of concrete, most of the strength formulas offered in Section 5.5 may be applicable by substituting a for w/c, although the coefficients will be different. After all, the strength formulas presented are essentially strength versus capillary porosity formulas (Eq. 3.5). Nevertheless, only the exponential form has been developed successfully for this purpose.

Earlier Formulas

Earlier efforts for the numerical description of the effect of air content on the strength of concrete have been sporadic and the results less than satisfactory. The few available theoretical (Griffith, 1920; Wyllie, 1965; Biot, 1956; Mackenzie, 1950; Hashin, 1962) and semitheoretical (Coble, 1956; Spinner et al., 1963; Hansen, 1965; Martin, 1971) formulas on the effect of porosity are adequate only qualitatively at best.

Probably Feret was the first to offer an approach to this problem by introducing the *cement-paste ratio* as the major factor of composition influencing the mortar and concrete strengths (Feret, 1892). This ratio represents the proportion of cement in a cement paste:

$$r_F = \frac{C}{C + W + V_a} \tag{5.63}$$

where r_F = cement-paste ratio
C, W, and V_a = quantities of cement, water, and air in the compacted fresh concrete, in absolute volumes or in other suitable units

He then recommends the following formula for the estimation of strength

$$f = K \left(\frac{C}{C + W + V_a} \right)^2 = K r_F^2 \tag{5.64}$$

where f = mortar or concrete strength
       K = empirical coefficient that is dependent on the units used and on every factor that influences the concrete strength, except the cement-paste ratio

Talbot and Richart (1923) used a similar ratio in their method of proportioning under the name *cement-space ratio,* but they did not express the strength relationships mathematically. Equation 5.64 reflects the view that the strength is controlled by the quantity of cement. Despite its apparent logic, concrete strengths obtained experimentally do not show good enough correlations with the r_F ratio. Attempts to improve the Feret relationship have produced modifications of the original formula. Two of these are (1) taking only a portion of the water or a portion of the air into consideration as strength-affecting factors, and (2) changing the exponent or the structure of Eq. 5.64. However, these modifications have not been successful either. The modified gel–space ratio recommended by Powers for cement pastes (Eq. 5.4) is another example of the use of the Feret concept for consideration of the water and air contents together. This, however, has not been applied to concrete.

New developments have gradually replaced the cement–paste or cement–space ratio with the water–cement ratio. This is due to the simplicity of the w/c concept, the direct relation between porosity and water–cement ratio (Eq. 3.5), and the resulting better and more general correlation between water–cement ratio and concrete strength. Since the water–cement ratio does not take the air voids into consideration, use of the (water + air)–cement ratio was recommended for this purpose. This ratio, called the *space–cement ratio* (Talbot and Richart, 1923), is a form of the Feret concept. Its combination with the Abrams' formula gives us

$$f = \frac{A}{B^{(W+V_a)/C}} \tag{5.65}$$

When the water–cement ratio is expressed by mass, and the volume of air pores as a percentage, as customary, Eq. 5.65 takes the form

$$f = \frac{A}{B^{(w+10a)/c}} \tag{5.66}$$

where $(W + V_a)/C$ and $(w + 10a)/c$ are the (water + air)–cement ratio or space-cement ratio. The (water + air)-cement ratio as used in Eqs. 5.65 and 5.66 is the mathematical equivalent of the assumption that 1 unit volume of air in the fresh concrete has the same effect on the concrete strength as the same volume of mixing water. If this assumption (i.e., Eq. 5.65 or 5.66) is valid, strengths should form a straight line with -log B slope in the log f versus (water + air)-cement ratio in a semilog system of coordinates regardless of the cement content. To check this, a set of experimentally obtained compressive strengths published by Ujhelyi (1980) are plotted in Figure 5.37 against $(w + 10a)/c$. It can be seen that this requirement is not fulfilled; the strength results are spread all over instead of forming a single straight line. A closer look reveals that the results tend to position themselves according to their cement contents: The higher the cement content, the lower the concrete strength at the same $(w + 10a)/c$ ratio. Nevertheless, Figure 5.37 invalidates Eqs. 5.65 and 5.66.

Another rule of thumb recommended by ACI is that the compressive strengths of comparable concretes are reduced by approximately 5% of the air-free strength for every 1% of air in the fresh concrete (Wright, 1953; Gaynor, 1968). That is,

$$f = f_0(1 - 0.05a) \tag{5.67}$$

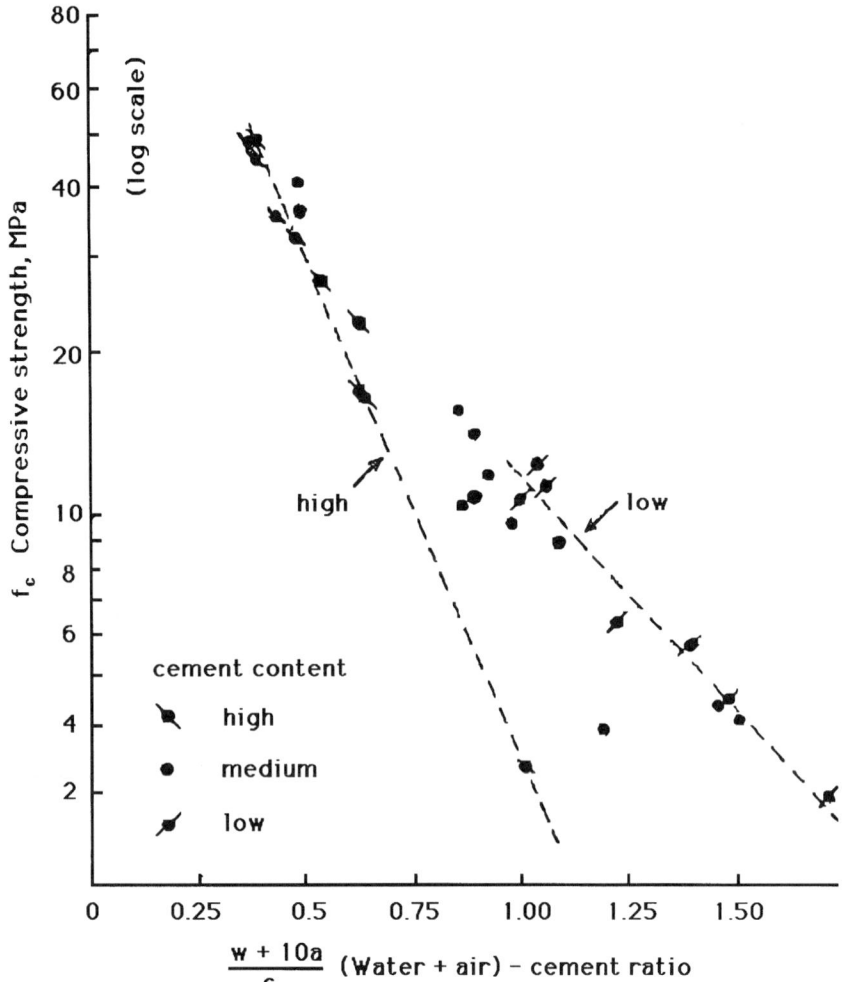

Figure 5.37 Compressive strength of concrete versus (water + air)–cement ratio of Feret. The experimental data were obtained by Ujhelyi (1980). 1 MPa = 145 psi. (From Popovics, 1985. Copyright ACI. Reprinted with permission.)

where f_0 = strength of concrete that is air-free in the fresh state
a = air content of the fresh concrete, %

Unfortunately, Eq. 5.67 is not satisfactory either; it under estimates the strength reductions at low air contents and overestimates them at high air contents. Nevertheless, it is probably better than Eq. 5.66 within narrow limits.

Exponential Formula

The following expression was recommended for the strength versus porosity relationship for polycrystalline bodies with air-filled voids (Ryskewitsch, 1953; Nurse, 1968):

$$f_{rel} = \frac{f}{f_0} = 10^{-\gamma a} \tag{5.68}$$

$$= \frac{1}{B_a^a} \tag{5.69}$$

where γ = experimental coefficient that is independent of the strength and age of the material within practical limits but depends on the material in question and type of strength

$B_a = 10^\gamma$.

The other symbols are as in Eq. 5.67. Note that the form of Eq. 5.69 is identical to the form of Abrams' formula for relative strength (i.e., Eq. 5.34). If Eq. 5.68 is applicable for a description of the effect of air content on concrete strength, the pertinent experimental data should form a straight line with a slope of $-\gamma$ in the system log f_{rel} versus a semilog system. It was found on this basis that $\gamma = 0.0384$ for the compressive strength of normal-weight concretes up to approximately 30% air content within the ages 7 and 90 days (S. Popovics, 1969a). Figure 5.38 demonstrates for one case, and Figure 5.39 in general, how well experimentally obtained compressive strengths support Eq. 5.68. This support is much stronger than any of the other pertinent equations discussed above. The data supporting this formula were obtained in various places at various times with different concrete compositions. Thus, not only the reliability but also the generality of Eq. 5.68 seem satisfactory. Similar generality has been noticed qualitatively by Kaplan (1960a) and Neville (1973). If measurable differences are established by further research on the effect of various types of air voids on the concrete strength, this improvement can be included in Eq. 5.68 as follows:

$$f_{rel} = 10^{-(\alpha a_1 + \beta a_2 + \gamma a_3)} \tag{5.70}$$

where a_1, a_2, and a_3 = quantities of the various types of air voids in the concrete

α, β, γ = experimental coefficients, each representing the effectiveness of a void type concerning strength reduction

Figure 5.39 illustrates the drastic reduction in compressive strength caused by air pores. For instance, an increase of 5% in air content reduces the com-

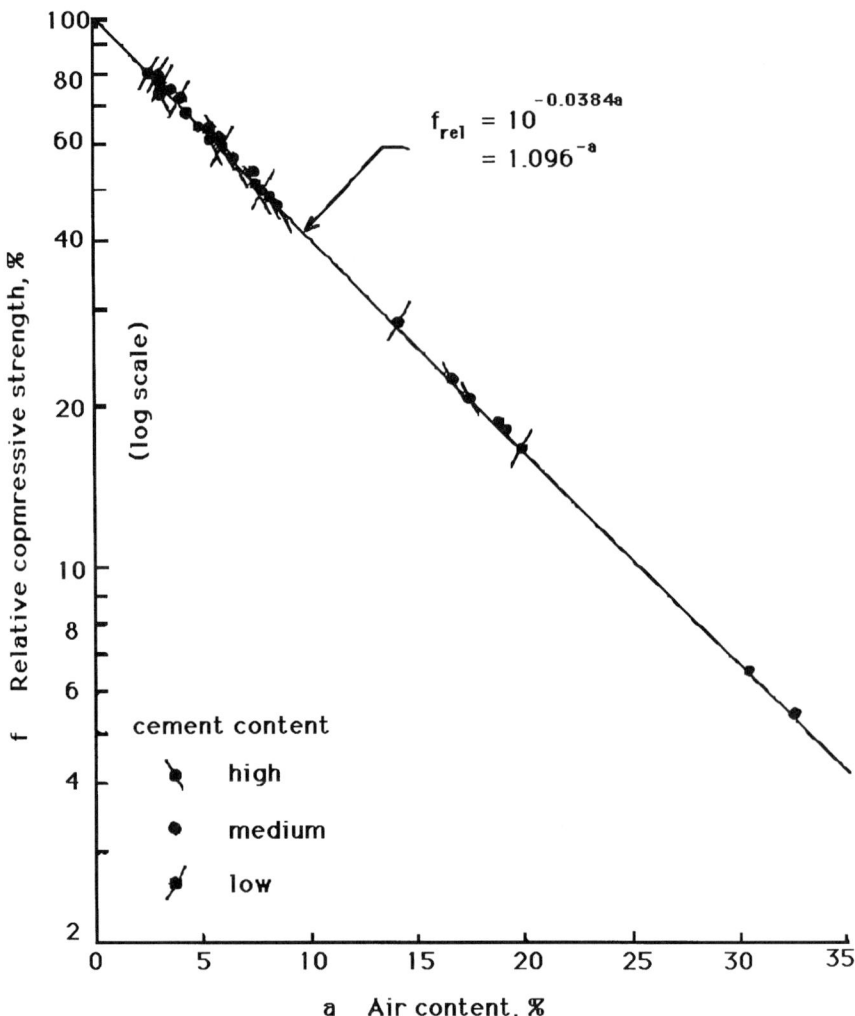

Figure 5.38 Effect of air content on the relative compressive strength of concrete. The experimental data represented by points are the same as in Figure 5.37. (From Popovics, 1985. Copyright ACI. Reprinted with permission.)

pressive strength to two-thirds of its original value. The restricted applicability of Eq. 5.67 is also demonstrated in Figure 5.39.

As shown in Figure 5.40, Eq. 5.68 can also be used for flexural strength but with $\gamma = 0.0232$. This equation is valid for several other mechanical properties of hardened concrete as well with different values of γ. A hypothetical explanation of the variation in γ values as well as the limitation of the present forms of the exponential models is offered in Section 5.13.

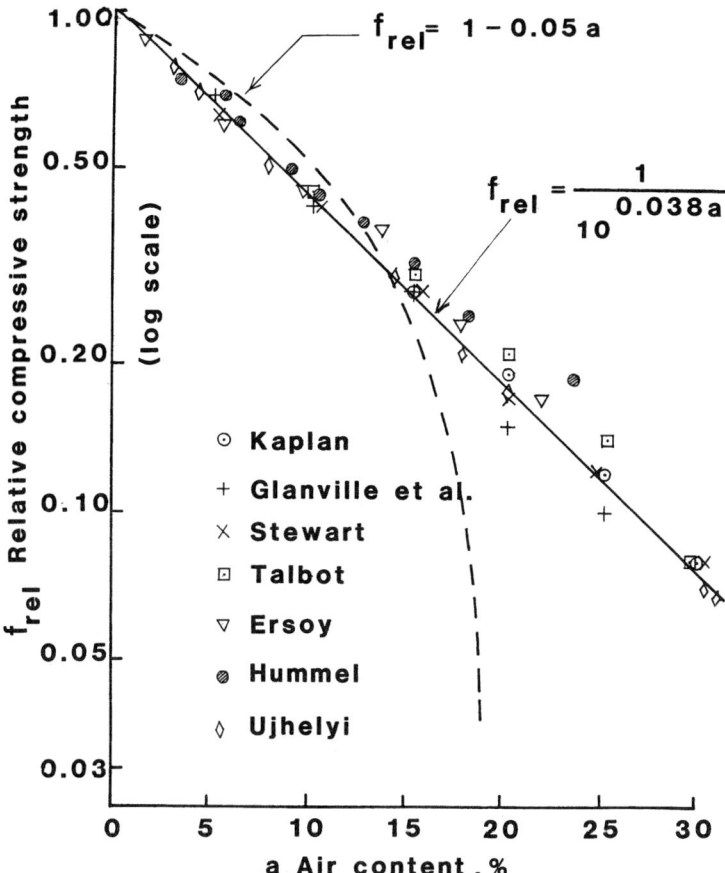

Figure 5.39 Comparison of the relative effects of air content *a* on the compressive strength of concrete obtained by various investigators. (From S. Popovics, 1969. Copyright ASTM. Reprinted with permission.)

Example 5.7 A given increase Δa in the air content produces a larger decrease, not only in absolute terms but also relatively, in the compressive strength f_c than in the flexural strength f_fl (Fig. 5.40). Therefore, the f_fl/f_c ratio is a function of the air content. The general numerical form of this relationship can be derived as follows:

$$\frac{f_\text{fl}}{f_c} = \frac{(A_f/B_f^{w/c}) \times 10^{-0.023a}}{(A_c/B_c^{w/c}) \times 10^{-0.038a}} = \frac{A_f}{A_c}\left(\frac{B_c}{B_f}\right)^{w/c} \times 10^{0.015a} \qquad (5.71)$$

$$= AB^{w/c} \times 10^{0.015a} \qquad (5.72)$$

where A and B are the pertinent coefficients for Abrams' formula (Eq. 5.6).

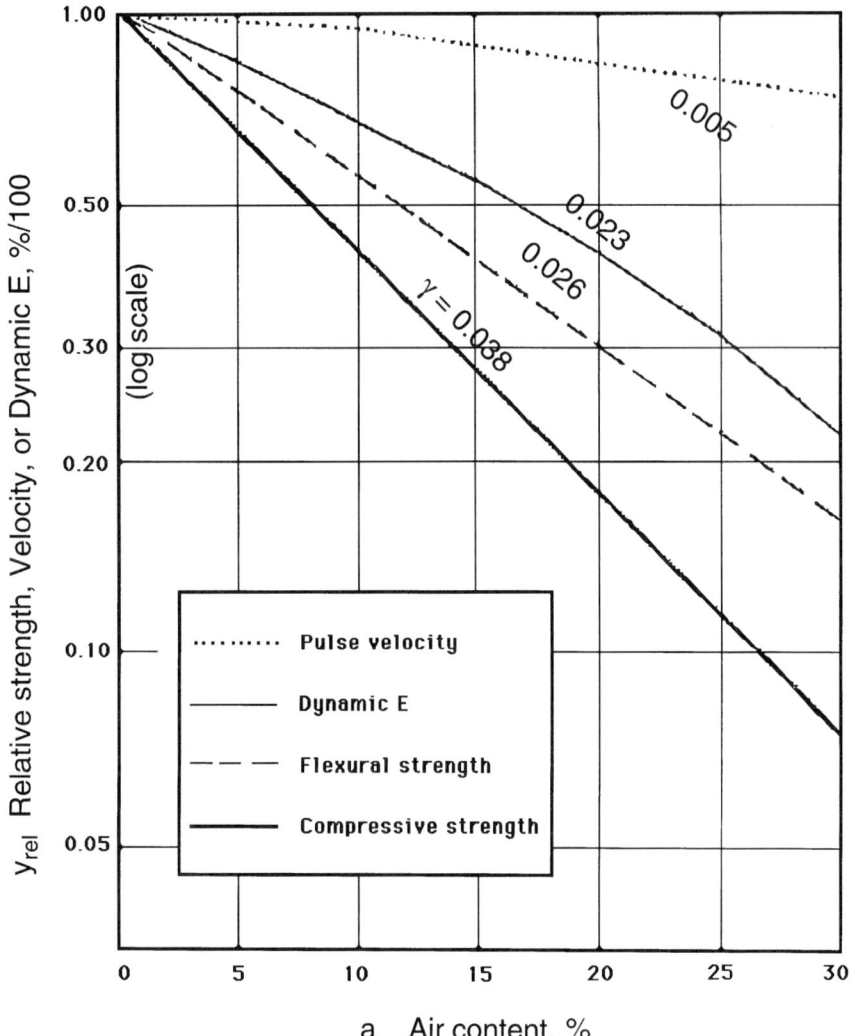

Figure 5.40 Relative effects of the entrapped air content on several mechanical characteristics of concrete. γ = slope. (From Popovics, 1969a. Copyright ASTM. Reprinted with permission.)

That is, the f_{fl}/f_c ratio increases exponentially with the air content, as noted in Section 2.1, along with the similar effect of the water–cement ratio.

In each of the cases above the effect of porosity is given only in relative terms, that is, as a percentage of the size of the pertinent property of the air-free concrete. Thus porosity can be used for practical calculations only when f_0 is known. Fortunately, the f_0 values for various strengths can be calculated from any of the formulas offered for the concrete strength versus water-cement ratio relationship, such as Eqs. 5.6 through 5.14. Thus the combination of, say, Abrams' formula with Eq. 5.68 provides the following equivalent equations for compressive strength (S. Popovics, 1985):

$$f = \frac{A_0}{B_0^{w/c}} \times 10^{-0.0384a} \tag{5.73}$$

$$= \frac{A_0}{B_0^{w/c} B_a^a} \tag{5.74}$$

$$= \frac{A_0}{10^{(\log B_0)(w/c)+0.384a}} \tag{5.75}$$

$$= \frac{A_0}{10^{(w \log B_0 + 0.0384ac)/c}} \tag{5.76}$$

$$= \frac{2.30 A_0}{\exp[(\log B_0)(w/c + 0.0384a/\log B_0)]} \tag{5.77}$$

where A_0 and B_0 are the coefficients of Eq. 5.6 for the air-free concrete with $C_A = 0$. The other symbols are the same as the symbols in Eqs. 5.6 and 5.68. The term $(w/c + \alpha a/\log B_0)$ is called the *porosity factor.*

The validity of Eqs. 5.73 through 5.77 can be checked experimentally for the description of the simultaneous effects of water-cement ratio and air content on strengths of concretes. If these equations are valid, the experimentally obtained strengths of these concretes should form a single straight line in the log f versus porosity factor semilog system for various compositions. The slope of this line is -log B within the limits of validity of these formulas. As Figure 5.41 demonstrates with Ujhelyi's data, this requirement is fulfilled for a variety of concrete compositions with a good approximation.

The goodness of fit of these formulas can be visualized further in Figure 5.42, where the calculated and experimental data of three test series are compared graphically. The average of the absolute values of deviations between the calculated and experimental values is less than 1 MPa (150 psi) in Glanville's series, and 1.6 MPa (230 psi) in the series by Ujhelyi. Further details of the comparison are reported elsewhere (S. Popovics, 1985).

Another possibility is a combination of Eqs. 5.68 and 3.20. When calibrated for a given portland cement, this provides, for example, the following for compressive and flexural strengths, respectively, at various ages (S. Popovics, 1985):

$$f_c = \frac{20{,}500}{6.4^{w/c}} \times 10^{-0.0384a} \frac{1 - 0.4e^{-b_1 t} - 0.6e^{-b_2 t}}{1 - 0.4e^{-90b_1} - 0.6e^{-90b_2}} \tag{5.78}$$

and

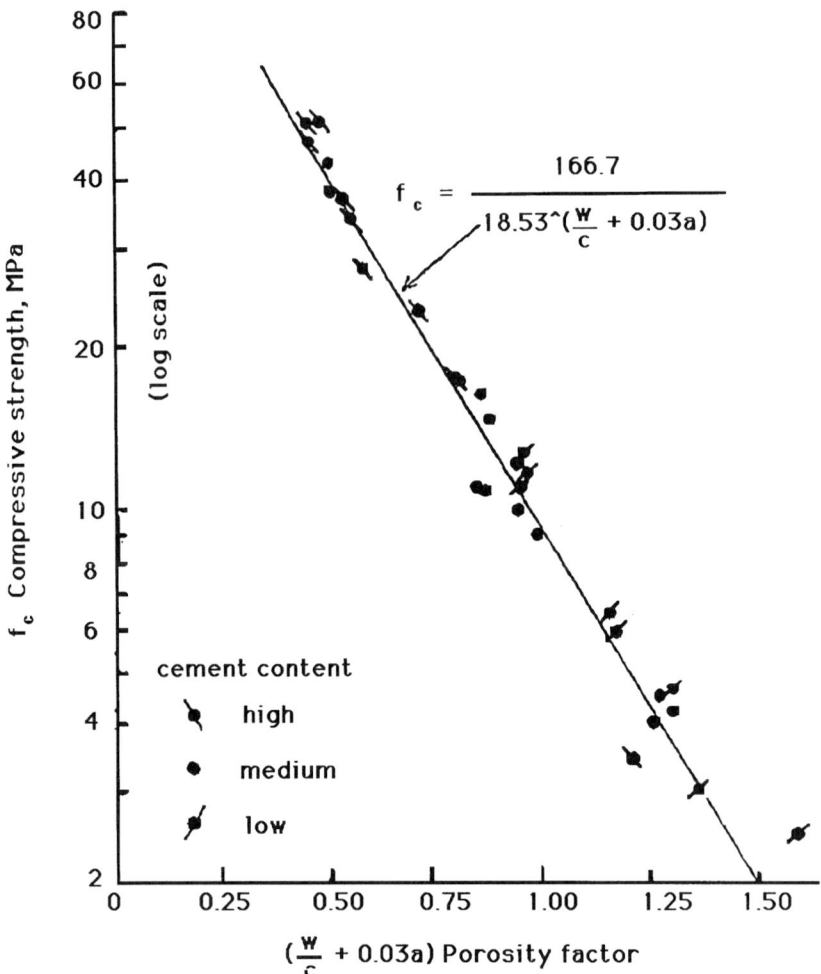

Figure 5.41 Compressive strength of concrete versus porosity factor. The experimental data represented by points are the same as in Figs. 5.37 and 5.38. 1 MPa = 145 psi. (From S. Popovics, 1985. Copyright ACI. Reprinted with permission.)

$$f_{fl} = \frac{1410}{2.55^{w/c}} \times 10^{-0.023a} \frac{1 - 0.4e^{-b_1t} - 0.6e^{-b_2t}}{1 - 0.4e^{-90b_1} - 0.6e^{-90b_2}} \qquad (5.79)$$

where f_c and f_{fl} are the compressive and flexural strengths in psi. The other symbols are as in Eqs. 3.20 and 5.68. The coefficients of -0.0384 and -0.0232 were taken from Figure 5.40.

Equations 5.78 and 5.79 show good fit to experimental data as illustrated in Fig. 5.42. The average of the absolute values of deviations between the

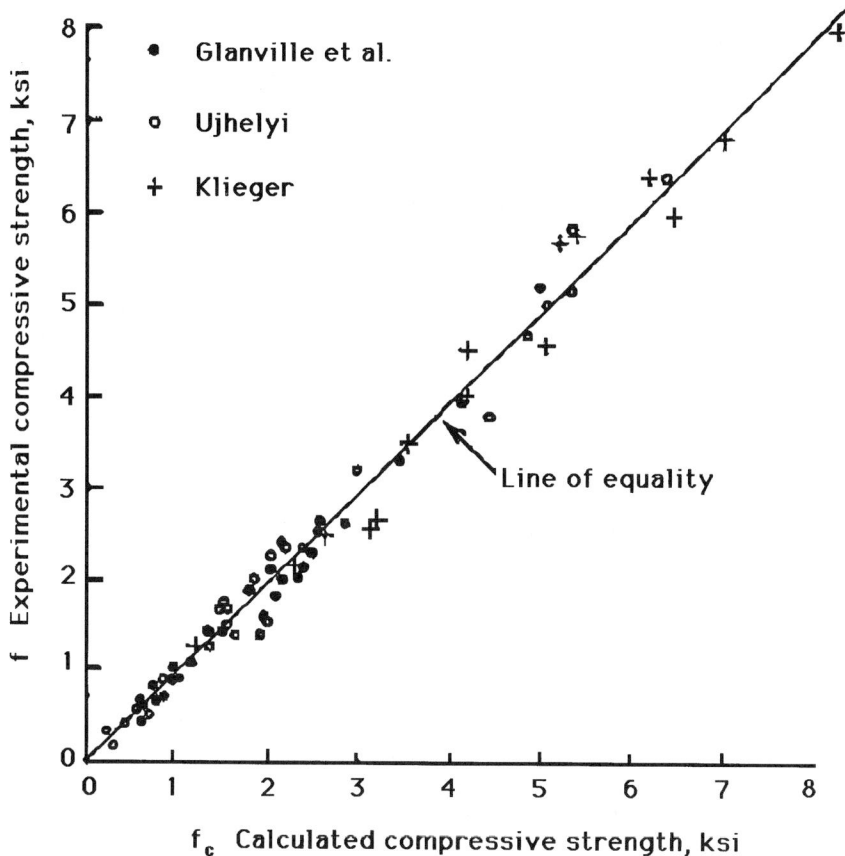

Figure 5.42 Comparison of calculated and experimentally obtained compressive strengths of concretes with different water-cement ratios and air contents. 1 ksi = 6.90 MPa. (From S. Popovics, 1985. Copyright ACI. Reprinted with permission.)

calculated and experimental values is 290 psi (2.0 MPa) for the compressive strength and 43.8 psi (0.30 MPa) for the flexural strength.

The primary advantage of Eqs. 5.78 and 5.79 is that the coefficients A_0 and B_0 are constant for every age and every standard cement type.

Difference Between the Effect of Water Content and That of Air Content on Strength

In hindsight, the superiority of Eqs. 5.73 through 5.79 to the earlier equations is not surprising. It is the total porosity that controls the concrete strength: the initial capillary portion characterized by the water-cement ratio, and the rest (i.e., the air voids), characterized by the air content. In this sense one can say that Eq. 5.65 or 5.66 uses the water content instead of the water-cement ratio, for characterization of capillary porosity, thus for the concrete strength at constant air content. This would be correct only if a unit volume of initial capillary pores had the same effect on concrete strength as the same total volume of air voids. This, however, is not true in general.

As shown by Eq. 5.76, the relative effect of air content on the concrete strength is influenced by the cement content. It is only around medium cement contents such as 330 kg/m³ (550 lb/cu yd) in the case of Figure 5.41 where the effects of water and air contents are the same on the strength of concrete. This means that it is only here where a decrease of, say, 1 unit volume of air content increases the concrete strength to the same extent as an identical decrease in the volume of mixing water, everything else remaining unchanged. At higher cement contents, the effect of air content will be greater on the strength than that of the water content, whereas at lower cement contents, this effect is reversed. Other experience also supports this statement (Reinsdorf, 1977; Armuth, 1971). Therefore, the Feret concept as represented by say, Eq. 5.65 has the tendency to overestimate concrete strengths at higher cement contents and underestimate them at low cement contents. This is also demonstrated by Figure 5.37.

This observation has several forms, as follows:

1. For concretes of high cement content, extra care should be paid to good compaction, whereas for of low cement contents control of the water-cement ratio is important,

2. Whereas air entrainment always reduces the strength of concrete with a high cement content, it may increase the strength of concrete with a low cement content when this concrete is reproportioned for equal slump (Klieger, 1952, 1956)

3. For mixtures of high cement content the optimum entrained air content is the minimum consistent with other requirements, such as frost resistance; however, for concretes of low cement content the optimum air content is the maximum that it is practical to obtain if the concrete composition is reproportioned for equal slumps (Mather, 1964; Tynes and Mather, 1969).

The difference in the effect of capillary porosity and the effect of air voids on concrete strength may be the consequence of the differences in the sizes and shapes of the two types of pores as a function of the cement content. Another contributing factor might be that the capillary porosity decreases with age, whereas the total volume of air voids remains practically unchanged. Thus the single factor B in Eq. 5.65 cannot serve adequately for both types of porosity.

5.13 EFFECTS OF POROSITY ON VARIOUS MECHANICAL PROPERTIES OF CONCRETE (S. Popovics, 1989)

The Phenomenon

The weakening effect of the presence of pores in the hardened matrix is not surprising. After all, a 1% increase in porosity means 1% less solid, load-carrying material in the same volume. However, the strength reduction by porosity is much greater than what would follow solely from reduction in the solid material. The larger-than-proportional reduction is also true for most other mechanical properties, although the rates of reduction are different. For instance, Figure 5.40 shows that an increase of 5% in the air content produces about a 33% reduction in the compressive strength, about a 25% reduction in the flexural strength. Only in the case of pulse velocity is this reduction 5%, which is approximately equal to the relative reduction of the solid content.

Therefore, the observed strength reduction by porosity is due primarily *not* to the decrease in solid material and the reduction of the number of bonds. Rather, it is due mostly to *tensile stress concentrations* developing around the pore tips during loading. This statement is based on Griffith's theory of fracture of brittle materials (Section 2.4). The stress concentrations crack the material progressively until it fails. It was pointed out in connection with Eq. 2.32 that the magnitude of the stress concentrations is proportional to the remote stress field in the specimen during loading. Thus, with the possible exception of the impact strength, the greatest stress concentrations develop in a concrete in the compressive strength test since the uniaxial average stress goes up to 30 to 50 MPa or more (several thousand psi). On the other hand, stress concentrations are practically missing when a pulse travels through a concrete because here the created remote stresses are not higher than about 0.01 MPa (1 psi). Since a larger stress concentration causes more cracking damage in the concrete, the pores produce the largest reduction in the uniaxial compressive strength and the least in the pulse velocity. The reductions in other mechanical properties are between these two extremes, according to the magnitudes of the prevailing uniaxial stress fields.

Model for the Effects of Air Content

On the basis of the qualitative discussion above, a mathematical model is offered for the *quantitative* description of the role of air content on the behavior of hardened concrete under stress. This new mathematical model is

$$f_{\text{rel}} = \left(1 - \frac{a}{a_{cr}}\right) \times 10^{-\gamma a/a_{cr}} \tag{5.80}$$

where f_{rel} = relative value of the mechanical property of a porous hardened concrete or some other brittle material, as a fraction of
the same property of the pore-free material, %/100

$a_{cr} < 100$ = critical porosity, that is, the air content at which the strength
or the characteristics of other mechanical properties become
zero, %

$a < a_{cr}$ = relative volume of the air content or other large pores in the
material, %

γ = experimental coefficient that is a function of the type of mechanical property in question

The first term on the right-hand side of the model in Eq. 5.80 reflects a
reduction in the quantity of solid material in the specimen. Thus it is independent of the mechanical property in question or the magnitude of the stress
field produced by the related test. The second term represents the additional
effects of air pores, primarily the effect of the tensile stress concentrations
around the pore tips. The model also reflects the fact that the strength or a
characteristic of a mechanical property of a hardened concrete becomes zero
when the porosity reaches a critical value a_{cr}. a_{cr} is less than 100% and is
probably a function of the mechanical property in question. More precisely,
it may increase with a decrease in the magnitude of the stress field in the
concrete. For instance, in the case of compressive strength it is around 60%;
in the case of pulse velocity it can be greater than 100%. In this section, for
the sake of simplicity, the constant value of 66.7% is assumed for concrete.

Equation 5.80 can be approximated quite well, at least for values of $a <$
30%, with the following simpler form:

$$f_{rel} = 10^{-\gamma a} \tag{5.81}$$

where the symbols are the same as in Eq. 5.80. Equation 5.81 is identical
with Eq. 5.68. An interesting application of Eq. 5.80 is when it was used for
the calculation of strength reductions caused by the substitution of chips of
rubber tires in concrete for traditional aggregate (Eldin, 1993a, 1993b). Since
the load-carrying capacity of these chips is very low, they can be considered
as pores with corresponding strength-reducing effects that can be calculated
from Eq. 5.80. It was found for this concrete that $\gamma = 0.20 \log(\bar{f}/0.007)$,
where $\bar{f}$ is the average uniaxial stress relatively far away from the pore,
causing the concentration of tensile stresses around the tip. With this parameter the experimentally observed strength reductions fit quite well the values
calculated in both the compressive and splitting strengths (Fig. 5.43).

The suggestion of using two terms for a description of the effect of air
content on strength, one for reduction of the solid content and the other for
the stress concentrations, probably came first from Millard (1958). There are,
however, differences between his approach and Eqs. 5.80 and 5.81. He used

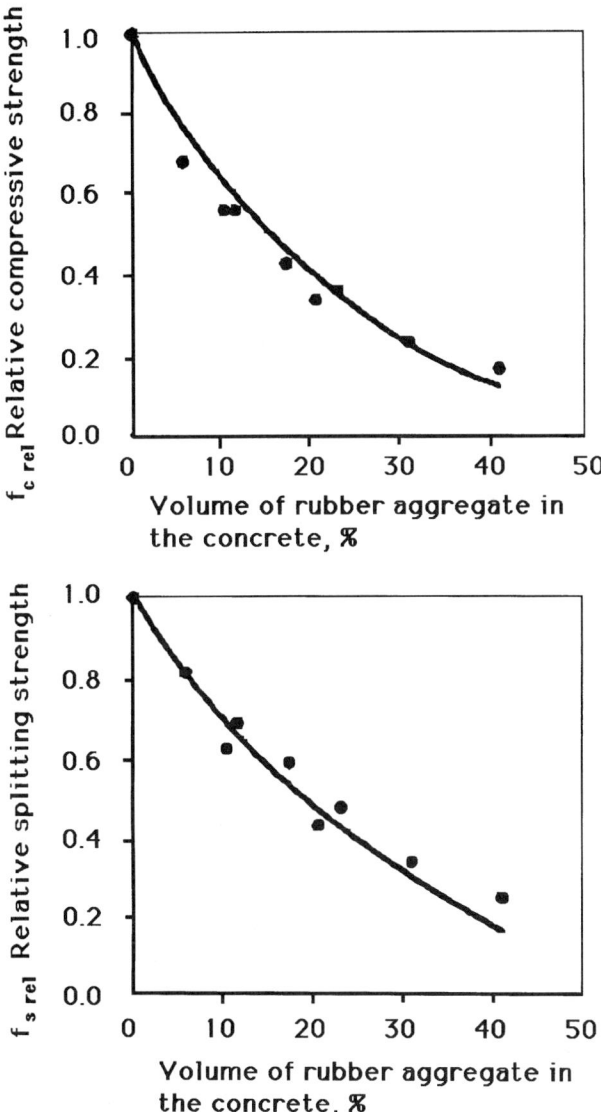

Figure 5.43 Comparison of experimental results (dots) to values calculated from Eq. 5.80 (lines) for compressive as well as splitting strength reductions of concretes as a function of the amount of rubber chips as aggregate. (From Eldin, 1993b. Copyright ASTM. Reprinted with permission.)

it only for compressive strength and for coal, and used a different formula, a power function, for stress concentrations.

Calibration of Eq. 5.80

The coefficient γ increases with increasing magnitude of the stress field. This means that at low stress concentrations, such as in pulse velocity measurements, the first term of Eq. 5.80 controls the value of f_{rel}. At high stress concentrations (compressive strength) the second term is in control.

Numerical values of γ can be obtained for each mechanical property from pertinent experimental data. For instance, several γ values are shown in Fig. 5.40 for Eq. 5.81. Experimental data support Eq. 5.81 with these γ values quite well, as shown in Fig. 5.39 for compressive strength. Far fewer pertinent experimental data are available for other concrete properties.

The γ values, of course, are different for Eq. 5.80 or 5.81. The following values were obtained for Eq. 5.80:

compressive strength	0.019
flexural strength	0.012
pulse velocity	−0.0027

These values of γ can be approximated with a simple continuous function. Assuming a value of 28 MPa (4000 psi) for the compressive strength, 2.8 MPa (400 psi) for the flexural strength, and a stress field of 0.01 MPa (1 psi) for pulse velocity, such a function is the following:

$$\gamma = 0.006 \log \overline{f} - 0.0027 \tag{5.82}$$

$$= \log 0.9938 \overline{f}^{\,0.006} \tag{5.83}$$

$$= 0.006 \log \frac{\overline{f}}{2.820} \tag{5.84}$$

where $\overline{f}$ is the average uniaxial stress relatively far away from the pore, causing the concentration of tensile stresses around the tip.

Equations 5.82 through 5.84 illustrate a method for the development of a mathematical model for a general description of the effects of air pores on various mechanical properties of a concrete. It will be simple in the future to refine coefficients as more pertinent experimental data become available. Further analysis of γ values may provide us with a better picture of stress concentrations and their role in the behavior of concrete under load.

Substitution of Eq. 5.82 into Eq. 5.80 provides the following, again for $a_{cr} = 66.7\%$:

$$f_{\text{rel}} = \left(1 - \frac{a}{a_{cr}}\right)\left(\frac{1.0063}{\overline{f}^{0.006}}\right)^{a/a_{cr}}$$ (5.85)

where the symbols are as in Eqs. 5.80 and 5.82. Equation 5.85 indicates that the weakening effect of porosity is greater in high-strength concretes than in low-strength concretes.

Since $10^{\gamma} = B$ (Eq. 5.69), a similar form for Eq. 5.80 is

$$f_{\text{rel}} = \frac{1 - a/a_{cr}}{B^a}$$ (5.86)

which is the same form as Abrams' formula has for the concrete strength versus water-cement ratio relationship.

This identity in forms also reveals a limitation of the present forms of the exponential formulas, such as Eqs. 5.80 and 5.81, and 5.82 through 5.86. This is, that these equations use the same γ, thus the same B parameter, within a strength type for concretes, regardless of the cement type and cement content, age, aggregate type, curing temperature, etc. That the B parameter of Abrams' formula is *not* independent of these factors has been demonstrated in the preceding sections for the water-cement ratio. One can surmise that the same is true for the air content. Therefore, the limits of validity of the exponential porosity formulas may be extended by the development of specific values for the γ, or B, parameters for various concretes, differing in age, cement type, and so on, in the same way as it was done for the water-cement ratio.

Interpretation of Eq. 5.81

Equation 5.81 can be written in the form (S. Popovics, 1969b)

$$f = f_0 \times 10^{-\gamma a}$$ (5.87)

$$= f_1^{0.01a} \times f_0^{(1-0.01a)}$$ (5.88)

where f = numerical measure of the mechanical property of porous concrete
f_0 = same measure for poreless concrete
f_1 = same measure attributed to the pores

The other symbols are as in Eq. 5.81. Equation 5.88 shows that the numerical measure of the mechanical property of a concrete with air pores is the weighted geometric average of the measure of the property of poreless material and of the pores. One can say that in the case of pulse velocity, f_0 is the velocity in the voidless material and f_1 is the velocity in the void. In the case of strength, the interpretation is not as clear, becaise Eq. 5.88 attributes a strength (i.e., a load-carrying capacity) to the voids as follows:

$$f_1 = f_0 \times 10^{-100\gamma} \qquad (5.89)$$

where the symbols are as in Eq. 5.87. That is, if the compressive strength of the voidless concrete is 28 MPa (approximately 4000 psi) and $\gamma = 0.038$, then from Eq. 5.89,

$$f_1 = 28 \times 10^{-3.8} = 0.0044 \quad \text{MPa} \quad (0.6 \text{ psi})$$

It is conceivable that this apparent strength of macropores is the result of a stress-relieving effect of the pores. It is also possible, however, that the assumption of $a_{cr} = 1$ in Eq. 5.80 for compressive strength is an oversimplification that produces the error f_1. Presumably, the findings above can be extended for both capillary porosity and for brittle materials other than concrete.

Porosity Versus Air Content

Whenever the concrete property in question is controlled by the matrix portion of the concrete, one can use the air content as the effective porosity influencing the property. Such is the case, for instance, for the strength of concrete, because the pores in the aggregate particles have scarcely any effect on strength in most normal-weight concretes. However, in other cases, such as the deformability of concrete (E_{st}, etc.), the property may be influenced significantly by the aggregate porosity. In such cases it may be better to use the sum of the air pores in the matrix and pores in the aggregate particles as effective porosity. Realization of this idea should, however, be left for future research.

6

ELASTIC DEFORMATIONS
OF CONCRETE

6.1 DEFORMATIONS OF CONCRETE

It is quite natural that when a *load* is applied to a concrete body, it deforms. There are, however, concrete deformations that can occur even *without* external loading. These are caused by the prevailing circumstances, such as changes in temperature (thermal expansion) or in moisture (shrinkage) or by internal chemical reactions (alkali aggregate reactions). These deformations can be quite large; thus they should be prevented, or reduced to acceptable size, by appropriate measures.

Another way to classify concrete deformations is by whether they are *elastic* or *inelastic,* or *plastic* or *permanent.* A deformation is said to be *elastic* if it appears and disappears upon application and removal of stress; otherwise, it is called *inelastic.* Concrete produces both kinds of deformation; that is why it is called a viscoelastic material. The inelastic part is greater for the first loading than for subsequent loadings but continues to a certain extent for each application of load. The curves for the repeated applications of load are more representative than the first curve of the actual conditions in a structure.

A third way to classify concrete deformations is by whether they are *instantaneous* or *time dependent.* A deformation is said to be *instantaneous* if it appears and disappears at the same time when stress is applied or removed. Since the loading (or unloading) of concrete, thus the development (elimination) of stresses, takes time, instantaneous deformation and time-dependent deformation are not entirely separable. It is known, however, that a major portion of the elastic deformations of concrete is instantaneous. Different concretes vary widely in their response to load. The amount and character of the deformation depend on the properties of the concrete, its strength, the particular environment, the magnitude of the load, how the load is applied, and the elapsed time after the load application when the observation is made.

Deformations under biaxial and triaxial loads follow similar, although more sophisticated, patterns. Details can be found in the literature (Gerstle, 1981a, 1981b; Stankovski and Gerstle, 1985; Tasuji et al., 1978; Kotsovos and Newman, 1980; Romstad et al., 1974).

Knowledge of the deformability of concrete is necessary to compute stresses from observed strains, to proportion sections of highway slabs and reinforced concrete members when certain design procedures are used, and to compute loss of prestress in prestressed structures. This chapter is restricted to a discussion of elastic deformations of concrete: primarily, instantaneous deformations under load.

6.2 STRESS–STRAIN DIAGRAM OF CONCRETE

Qualitative Aspects

The deformation of a specimen under increasing load can be described conveniently by a stress–strain or σ–ε or, in this book, f–ε diagram. The shape of such a diagram depends on the properties of the concrete and also, to a large extent, by the method of testing. Figure 6.1 shows two curves, for two different types of loading rates, for axial deformations of the same concrete

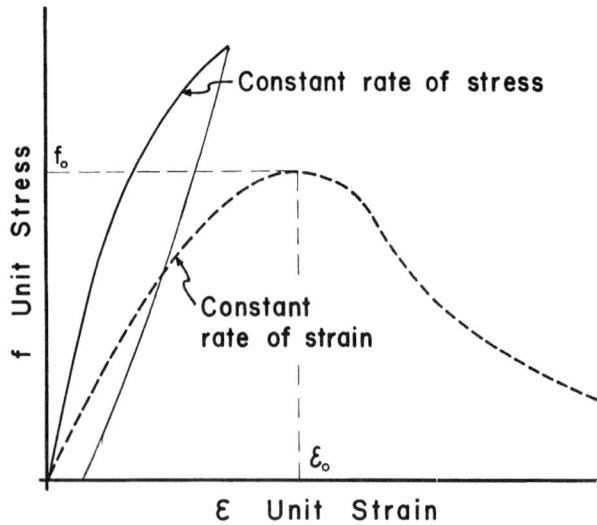

Figure 6.1 Two typical stress–strain curves for concrete under uniaxial load. The top curve is characteristic of a loading process where the rate of stress increase is kept constant during testing. The bottom curve is obtained by keeping the rate of strain increase constant.

under uniaxial compression or tension. The stresses are averages (i.e., load divided by cross-section area), as are the strains (S. Popovics, 1970a). Actually, the gross deformation measured is the result of many small, varying, discrete deformations occurring in the various constituents and parts of the concrete. Thus highly localized deformations, or strains, may be different from the average for the concrete as a whole (Slate et al., 1986).

In addition to the rate of loading, the diagram is influenced by the other testing conditions, such as the size and shape of the specimen, size and location of strain gages, and the character of the loading (Clark, 1967; Halasz, 1967; Rusch and Turk, 1959; Rasch 1962), as well as by the age and composition of concrete (Helmuth 1966), including the type and quantity of the aggregate. Strains at failure increase also with an increase in the confining pressure (Hobbs, 1970; Krahl et al., 1965; Iyengar et al., 1970; Rodrigez Cuevas, 1970), load repetition, and with the sustenance of the load (Mehmel and Verna, 1962; Rusch, 1960; Shah, 1970). The effect of the testing machine used may also be important since the magnitude of the machine deformations directly affects the time rates of stresses and strains in the specimen (Tegart, 1967). Note that these effects are similar, although not identical, to the way that concrete strength is influenced by most of these factors (S. Popovics, 1969a, 1970a). Moment-deflection diagrams and, to a lesser extent, torque-twist diagrams for concrete are similar to diagrams for compression or tension (Hughes and Ash, 1968) and serve a similar purpose.

Along with axial deformations, transverse or lateral deformations also take place under load. It is customary to characterize the lateral deformation by Poisson's ratio, which is the absolute value of the ratio of transverse strain to the corresponding axial strain resulting from a uniformly distributed axial stress below the proportional limit of the material.

The shapes of the stress–strain diagrams or moment deflection diagrams or torque–twist diagrams or stress–Poisson's ratio diagrams for concrete can be explained in qualitative terms by the fracture mechanism of concrete, discussed in Section 2.4. The stress–strain diagram starts out with a nearly linear portion that stretches to about 30% of the ultimate load. Beyond this point, the curve deviates gradually from the straight line toward the horizontal. This curving of the diagram needs explanation. As pointed out earlier, the two components of concrete, the hardened cement paste and the aggregate, produce individually linear stress–strain curves with a fair approximation (Johnson, 1928b; Larue, 1956; Johnston, 1970; Hobbs, 1969), yet when combined, the curve becomes nonlinear. The reason for this is the presence of interfaces between paste and aggregate (Section 4.5). The cracks on the interface of coarse aggregate and matrix increase continuously; this is the start of discontinuity of the specimen leading to failure. A subsequent contribution comes from intensive mortar cracking, although some creep probably also takes place (Ross et al., 1968; L'Hermite, 1962). Thus strain increases at a faster rate than applied stress, which causes the curving of the σ–ε diagram. In accordance with this mechanism, the stress–strain diagram is usually more curved for concretes than for comparable mortars, and these usually have more curved diagrams than those of comparable cement pastes. This failure mechanism also implies that:

1. The cement type and water-cement ratio influence the stress–strain diagram primarily through their effects on concrete strength.
2. The quantity, type, and grading of the aggregate can have greater effects on elastic deformations than on the concrete strength (Hummel, 1935; Graf et al., 1960).
3. The curvature of the diagram increases with the amount and size of the aggregate in the concrete.
4. The stress–strain diagram of a concrete becomes steeper and more nearly linear with increase in the rate of loading.
5. The shape of the diagram obtained by tensile load is similar (Hughes and Chapman, 1966a; Evans and Marathe, 1968) to, although more nearly linear than, the shape of the comparable diagram obtained by compressive load.
6. The major part of the concrete deformation caused by internal cracking and creep is permanent.

Also, when concrete and mortar specimens are subjected to increasing uniaxial compression, Poisson's ratio remains constant for a while but it begins to increase on attaining a certain stress level. After that, the volume of the specimen also begins to increase, resulting in an apparent increase of the Poisson's ratio (Fig. 2.13). This can again be explained by the appearance and propagation of cracking, because hardened paste specimens continue to consolidate at an increasing rate with increasing load, and stone specimens show only a slight volume expansion at stresses near failure (Shah and Chandra, 1968). Experimental evidence support all these implications.

Numerical Approximations for the Stress–Strain Diagram of Concrete

Stress–Strain Formulas   Despite the significant progress in the fracture mechanics of concrete, the best that the theory can do at present is to describe the stress–strain relationship in qualitative terms. Other theories are even less suitable for this purpose. Thus, only mathematical models or empirical formulas obtained from curve fitting and/or from boundary conditions, respectively, are available for numerical approximation of the stress–strain diagram of concrete. Some of these, for constant change of strain rate, are presented below.

Bach was probably the first to propose a formula, a simple power function, for approximation of the stress–strain diagram of concrete. Since then, many other formulas have been offered for the description of the ascending or ascending-and-descending branches of these curves (S. Popovics, 1970c, 1972b). Several of them are shown in Table 6.1 as well as in Figure 6.2 for

TABLE 6.1 Formulas for the ascending branch of the stress–strain diagram of concrete

Authority	f	Eq. No.	Remarks
Saenz	$E\varepsilon\left[1 + \left(\dfrac{3E_0}{E-2}\right)\dfrac{\varepsilon}{\varepsilon_0} + \dfrac{1-2E_0}{E}\left(\dfrac{\varepsilon}{\varepsilon_0}\right)^2\right]$	a	$E/E_0 = 2$ transforms this into Eq. c
Saenz	$\dfrac{E\varepsilon}{1 + (E/E_0 - 2)(\varepsilon/\varepsilon_0) + (\varepsilon/\varepsilon_0)^2}$	b	$E/E_0 = 2$ transforms this into Eq. e
European Concrete Committee (CEB)	$E\varepsilon\left(1 - \dfrac{\varepsilon}{2\varepsilon_0}\right)$	c	Implies that $E/E_0 = 2$
Torroja	$0.43E\varepsilon_0\left[1 - \left(1 - \dfrac{\varepsilon}{\varepsilon_0}\right)^{7/3}\right]$	d	Implies that $E/E_0 = 7/3$
Desayi and Krishnan	$\dfrac{E\varepsilon}{1 + (\varepsilon/\varepsilon_0)^2}$	e	Implies that $E/E_0 = 2$
Tulin and Gerstle	$\dfrac{E\varepsilon}{1 + (\varepsilon/\varepsilon_0)^n}$	f	Implies that $E/E_0 = 2$
Baumann	$E\varepsilon\left[1 - \left(1 - \dfrac{E_0}{E}\right)\dfrac{\varepsilon}{\varepsilon_0}\right]$	g	$E/E_0 = 2$ transforms this into Eq. c
Popovics	$(n - 1)\dfrac{E\varepsilon}{n - 1 + (\varepsilon/\varepsilon_0)^n}$	h	Implies that $E/E_0 = n/(n-1)$
Liebenberg	$E\varepsilon\left[1 - \dfrac{1}{n+1}\left(\dfrac{\varepsilon}{\varepsilon_0}\right)^n\right]$	i	Implies that $E/E_0 = 1 + 1/n$
Shah and Winter	$E\varepsilon\dfrac{E_0}{E}\left(\dfrac{E\varepsilon - 2}{E\varepsilon_0 - 2}\right)^m$	j	
Smith and Young	$E\varepsilon e^{-\varepsilon/\varepsilon_0}$	k	Implies that $E/E_0 = e$
Young	$\dfrac{2E\varepsilon_0}{\pi}\sin\left(\dfrac{\pi}{2}\dfrac{\varepsilon}{\varepsilon_0}\right)$	l	Implies that $E/E_0 = \pi/2$
Bach	$C_1\varepsilon^n$	m	Implies that E is infinite
Sturman et al.	$C_2\varepsilon(1 + C_3\varepsilon^{n-1})$	n	Implies that $E/E_0 = n/(n-1)$
Ritter	$C_4(1 - e^{-\varepsilon})$	o	
Alexander	$\dfrac{C_5\varepsilon}{(\varepsilon + C_6)^2 + C_7} - C_8\varepsilon$	p	
Kriz and Lee	$-0.5(A_8\varepsilon + A_9) \pm \sqrt{0.25(A_8\varepsilon + A_9)^2 - A_7\varepsilon^2 - A_{10}\varepsilon}$	q	See Eq. 6.4.

Source: S. Popovics (1972). (Copyright Materials Research Society. Reprinted with permission.)

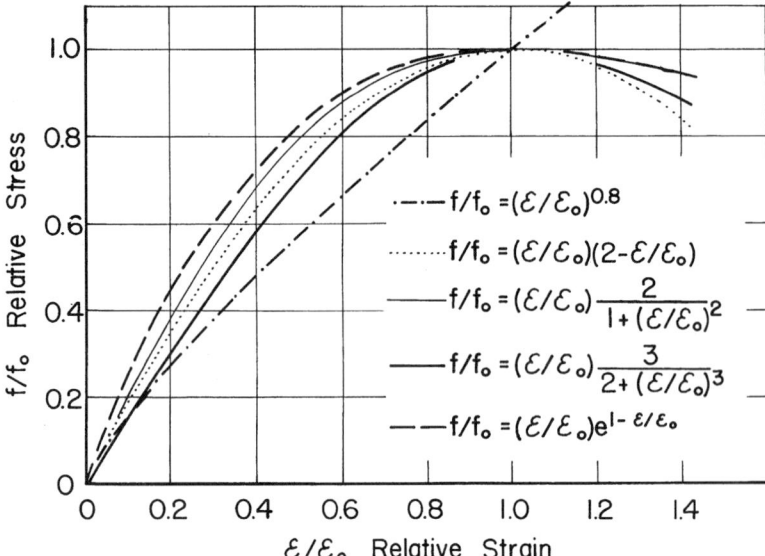

Figure 6.2 Comparison of several formulas for the stress–strain diagram of concrete. (From S. Popovics, 1970c. Copyright TRB. Reprinted with permission.)

normal-weight structural concretes under short-term loading. Many of these formulas are expressed in terms of relative stress f/f_0 and relative strain $\varepsilon/\varepsilon_0$, where f_0 is the ultimate stress, that is, the compressive strength, of the concrete, and ε_0 is the strain related to f_0. Other formulas for the stress–strain curve that have mathematical forms different from those in Table 6.1 have also been recommended:

1. Alexander (1965):

$$f = \frac{A_1\varepsilon}{(\varepsilon + A_2)^2 + A_3} - A_4\varepsilon \tag{6.1}$$

2. Hognestad et al. (1955):

$$f = \varepsilon\frac{df_a}{d\varepsilon} + f_a \tag{6.2}$$

where the $df_a/d\varepsilon$ term can be approximated by measured finite-element differences at any point a as $\Delta f_a/\Delta\varepsilon$.

3. Sinha et al. (1964a) for the *unloading* branches:

$$f = \frac{A_5}{X}(\varepsilon - X)^2 + A_6 \tag{6.3}$$

where the parameter X is a function of the number of load repetitions, applied repeated stress, strength of concrete, and so on. For the *reloading* branches straight lines were assumed.

4. Kriz and Lee (1960) for the stress–strain curve from *flexure* (Table 6.1, q):

$$f^2 + A_7\varepsilon^2 + A_8 f\varepsilon + A_9 f + A_{10}\varepsilon = 0 \tag{6.4}$$

5. Terzaghi for ε:

$$\varepsilon = \frac{f}{E} + A_{11}f^n \tag{6.5}$$

6. Ros (1950):

$$\varepsilon = \frac{f}{E} + \frac{A_{12}f}{A_{13} - f} \tag{6.6}$$

The descending branch of the diagram is important because the ultimate strength design calculation takes a portion of this branch into consideration. For instance, ACI 318 fixes the maximum concrete strain under flexure as 0.003 μin./in. for design purposes, which is in the case of the usual structural concretes, greater than the value of ε_0. Also, the cracked concrete specimen on the descending branch still has considerable load-bearing capacity, although this decreases with increasing deformation. This ability ensures relief of peak stresses, thus reducing the danger of sadden and violent failure (Swamy, 1970).

As Figure 6.2 shows, the curvatures of the various diagrams may differ considerably. This property can be used for the selection of a suitable stress–strain formula for a given case. Fortunately, the design of reinforced concrete members for simple flexure is hardly influenced by the assumed shape of the stress–strain diagram of concrete. This and the resulting simplicity justify the use of such fictitious stress distribution as the rectangular block (Kazinczy, 1938; Whitney, 1942).

In many of these formulas the E/E_0 ratio is a fixed number (E is the initial tangent modulus of the concrete; E_0 is f_0/ε_0). This restricts the limits of validity of these formulas because experimental data show that the E/E_0 ratio varies from near 4 for normal-weight concretes of 1000 psi (7 MPa) compressive strength to about 1.3 for concretes of 10,000 psi (700 MPa). Consequently, when a formula with fixed E/E_0 ratio fits a concrete of medium strength, it will overestimate the stress for a given strain in the ascending branch of the σ–ε diagram for high-strength concretes, and underestimate it for low-strength concretes (Smith, 1956b). Therefore, formulas with variable E/E_0 ratios are preferred. Such a formula is Eq. h in Table 6.1 (S. Popovics, 1970c, 1973c)

$$f = E\varepsilon\frac{n-1}{n-1+(\varepsilon/\varepsilon_0)^n} \tag{6.7}$$

or, considering that at $\varepsilon = \varepsilon_0$

$$E = \frac{n}{n-1}\frac{f_0}{\varepsilon_0} \tag{6.8}$$

Eq. 6.1 becomes

$$f = f_0\frac{\varepsilon}{\varepsilon_0}\frac{n}{n-1+(\varepsilon/\varepsilon_0)^n} \tag{6.9}$$

where f_0 = compressive strength of the concrete
f = axial stress in the specimen at unit strain ε
ε = unit strain in concrete caused by f
ε_0 = unit strain in concrete at ultimate stress f_0
n = experimental parameter

Parameter n can be expressed as an approximate function of the compressive strength of normal-weight concretes. From the data by Gilkey (1938) and Rusch (1955), the following expressions were obtained:

$$n_{concrete} = 0.4 \times 10^{-3}f_0 + 1 \tag{6.10}$$

$$n_{mortar} = 0.15 \times 10^{-3}f_0 + 1.5 \tag{6.11}$$

$$n_{paste} = 12 \tag{6.12}$$

Equation 6.7 is not recommended for practical purposes, although it is an interesting formula because for a given n value, it expresses the stresses purely in terms of deformations. The long fractions in Eqs. 6.7 and 6.9 represent deviation from the linear elasticity. This might be utilized in the future for the analysis of crack propagation in concrete.

Figure 6.3 illustrates Eq. 6.9 in relative terms for pastes and normal-weight concretes. The formulas are valid only when standard concrete cylinders are used and when the uniaxial compressive load is a short-term load applied at a rate that produces constant rate of strain in the specimen. Note that Eqs. 6.7 and 6.9, combined with Eq. 6.10, differ from the other formulas offered in the literature for similar purposes in that they provide more relative curvature in the diagram for concretes of lower strengths.

Stress–strain relations can also be obtained from *flexure* or from *eccentric loading*. However, the peak points of such curves are located at materially higher strains and stresses than the peak points of the corresponding curves for concentric compression. This is due primarily to differences in the time

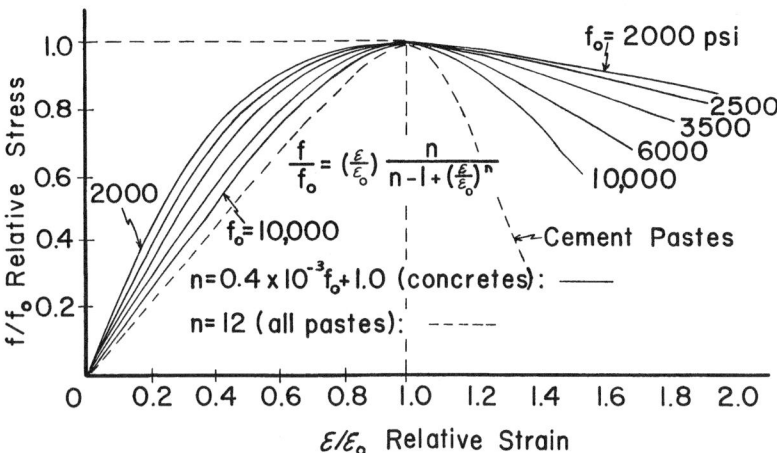

Figure 6.3 Calculated relative stress-strain diagrams for normal-weight concretes of various compressive strengths and for cement pastes. (From S. Popovics, 1973c. Copyright Cement and Concrete Research. Reprinted with permission.)

rate of application of strain in the two kinds of loading (Sturman et al., 1965; Neville, 1968; Clark, 1967; Halasz, 1967; Ghosh and Handa, 1970; Imbert, 1970).

Comparison with Experimental Data Equation 6.9 with Eqs. 6.10 through 6.12 are examined in two different situations: first, when both f_0 and ε_0 values are available; and second, the more practical situation, when only the value of f_0 is available. In the first situation the corresponding value of n is calculated from Eq. 6.10, 6.11, or 6.12 and used with Eq. 6.9. The σ–ε diagrams calculated fit not only the corresponding diagrams by Gilkey and Rusch but also the diagrams of Hognestad et al. (1955). A comparison was made with 20 pairs of diagrams representing five water-cement ratios (from 0.33 to 1.0 by weight) and four ages (from 7 to 90 days). As Figures 6.4a through d show, the diagrams calculated fit the experimental curves quite well.

In the second situation, when only f_0 is given, the value of ε_0 can be estimated from one of the formulas in the literature (Table 6.2) and used with Eq. 6.9 again, as shown above. Another empirical formula is (S. Popovics, 1972b, 1973c)

$$\varepsilon_0 = k \times 10^{-4} f_0^{0.25} \tag{6.13}$$

where k is a function of the type of mineral aggregate used and the test method. Equation 6.13 is well supported by experimental data by Hognestad et al. (1955) and by Watanabe (1972). Additional support comes from the observation that a combination of Eq. 6.9 with Eqs. 6.10 and 6.13 fits Hog-

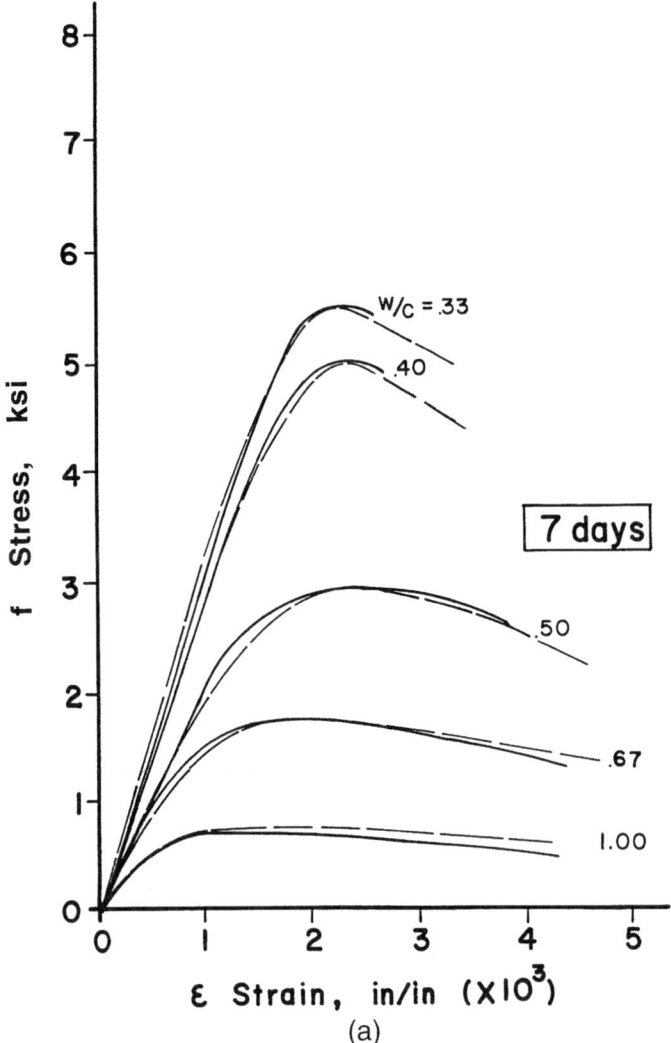

Figure 6.4 Comparison of experimental stress–strain diagrams of various concretes (solid lines) to those plotted from Eq. 6.9 with Eq. 6.10 (dashed lines). 1 ksi = 6.90 MPa. (From S. Popovics, 1973c. Copyright Cement and Concrete Research. Reprinted with permission.)

nestad curves in Figure 6.4a through d with $k = 2.7$ in.$^{0.5}$/lb$^{0.25}$ almost as well as do curves calculated with f_0 and obtained ε_0 experimentally. A similar good fit was obtained with data by Smith and Young (1956) not only on normal-weight but also on lightweight concretes. So it appears that the approximation of Eqs. 6.9 through 6.13 is better within wide ranges of water–cement ratio, compressive strength, and age than approximations of other formulas recommended in the literature for the same purpose (Smith

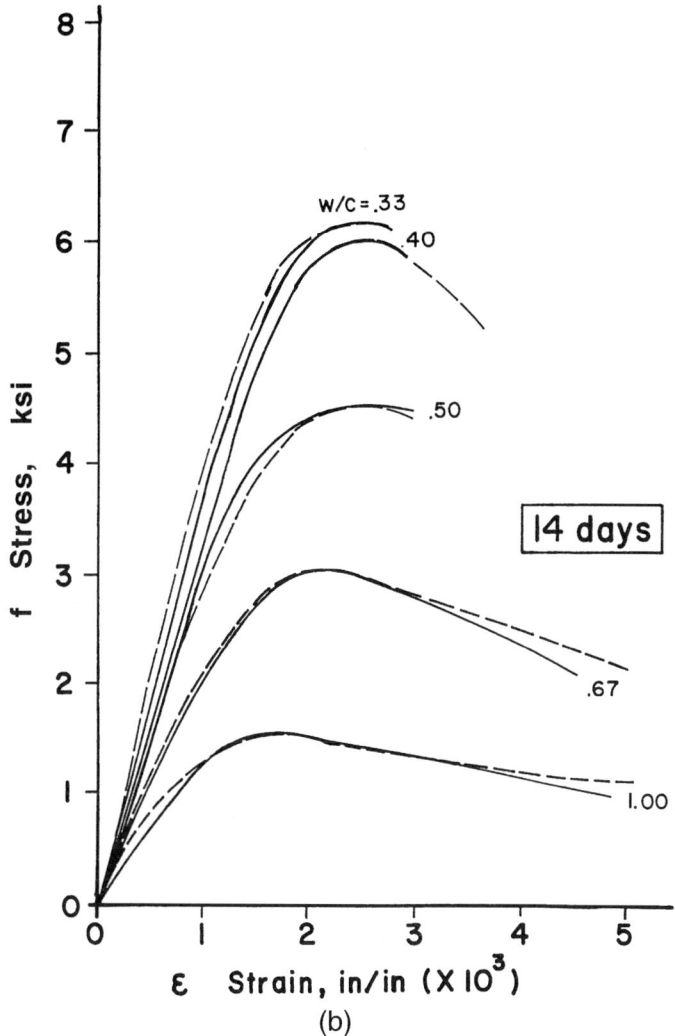

Figure 6.4   (Continued).

and Young, 1956; Liebenberg, 1962; Alexander, 1965; Desayi and Krishnan, 1964, 1970).

The stress–strain diagram for concrete under short-term uniaxial tension is similar to the diagram produced by compression (Hughes and Chapman, 1966a; Hughes and Ash, 1968; Komlos, 1967–1969; Evans and Marathe, 1968; Welch, 1966), except that the same curve is applicable for all relative tensile stress–strain diagrams regardless of the strength (Johnston, 1970). It is probably more than coincidence that this single tension curve is close to the compression curve presented in Figure 6.3 for pastes.

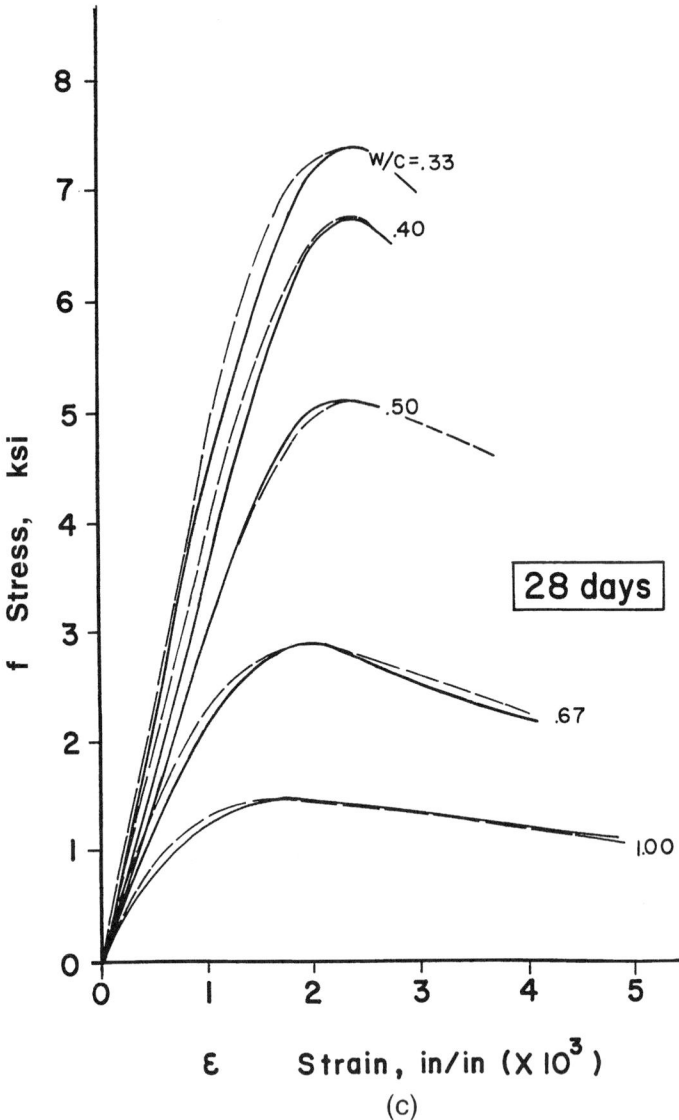

Figure 6.4 (Continued).

The stress–strain diagrams discussed in this section are valid for the case when the concrete specimen, with or without capping of the loaded surfaces, is in direct contact with the steel platens of the testing machine. The shape of such a stress–strain curve can change with the use of interlays. For instance, the use of Teflon interlayers drastically reduces the value of f_0 and

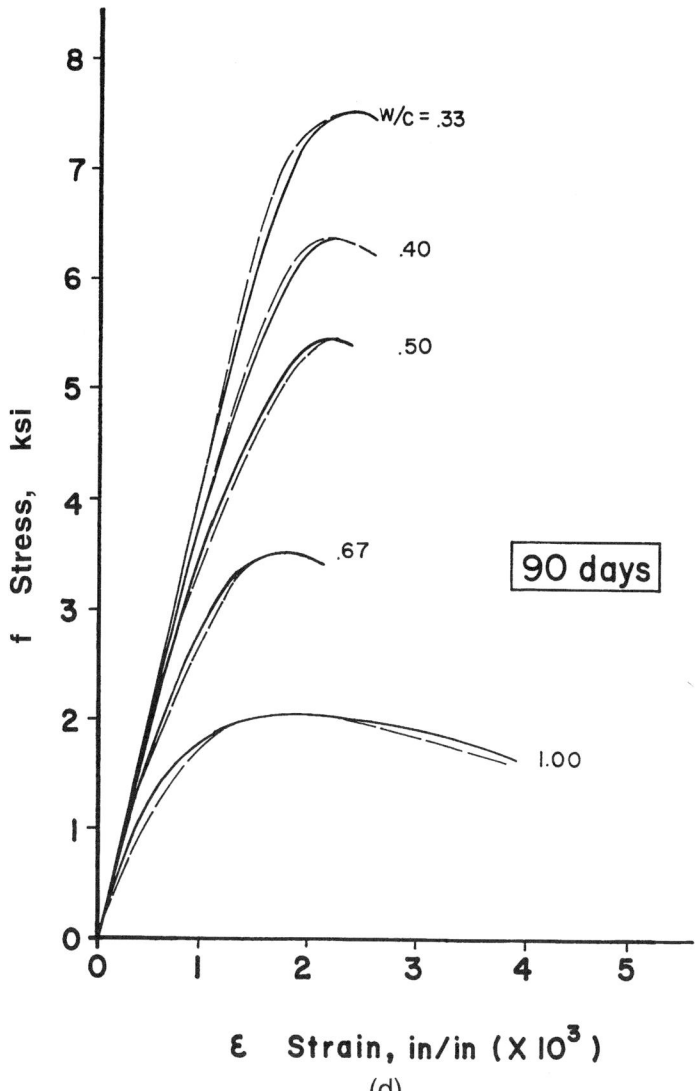

Figure 6.4 (Continued).

makes the descending portion of the curve steeper. Also, the effect of slenderness on f_0 is reduced by this interlayer (RILEM, 1997). These effects are attributed to the low friction coefficient of Teflon.

The area under any of the stress–strain diagrams can be determined by numerical integration. Differentiation of any f–ε (that is, σ–ε) formula provides the E_t tangent modulus of elasticity of the concrete (Section 6.3).

TABLE 6.2 Formulas for estimating the ε_0 and E_0 values of concrete under compression[a]

Authority	(a) ε_0 (%) (f in kg/cm²)	(b) ε_0 (%) (f in psi)	(c) E_0 (psi) (f in psi)	Eq. No.
Ros	$0.0546 + 3.64 \times 10^{-4} f_{pr}$	$0.0546 + 2.56 \times 10^{-5} f_0$	$\dfrac{f_0 \times 10^6}{546 + 0.256 f_0}$	a
Emperger	$0.014\sqrt{f_{pr}}$ [b]	$3.7 \times 10^{-3} \sqrt{f_0}$ [b]	$0.27 \times 10^5 \sqrt{f_0}$ [b]	b
Emperger	$0.0082 \sqrt{f_{pr}}$ [c]	$2.2 \times 10^{-3} \sqrt{f_0}$ [c]	$0.455 \times 10^5 \sqrt{f_0}$ [c]	c
Brandtzaeg	$\dfrac{f_{pr}}{478 + 2.6 f_{pr}}$	$\dfrac{f_0}{6800 + 2.6 f_0}$	$680{,}000 + 260 f_0$	d
Jäger	$0.014 \sqrt{f_{pr}}$	$3.7 \times 10^{-3} \sqrt{f_0}$	$0.27 \times 10^5 \sqrt{f_0}$	e
Hungarian Code	$\dfrac{f_{cu}}{790 + 3.95 f_{cu}}$	$\dfrac{f_0}{7900 + 3.95 f_0}$	$790{,}000 + 395 f_0$	f
Saenz	$10^{-3} \sqrt{f_{pr}} \, (61.1 - 3.76 \sqrt[4]{f_{pr}})$	$10^{-3} \sqrt[4]{f_0} \, (31.5 - \sqrt[4]{f_0})$	$\dfrac{10^5 \sqrt[4]{f_0^3}}{31.5 - \sqrt[4]{f_0}}$	g

Source: S. Popovics (1970c). (Copyright TRB. Reprinted with permission.)

[a] In the conversion from kg/cm² to psi it was assumed that the f_{pr} prism strength is identical with the f_0 cylinder strength and that the cylinder strength in psi is equal to 10 times the f_{cu} cube strength in kg/cm².

[b] For a high sand content.

[c] For a low sand content.

Concrete Deformations under Repeated Loads

The significance of this topic is that concrete structures are almost always subject to repeated load. The effect of load repetition is important when the live load–dead load ratio in the structure is high.

The effects of repeated loading on concrete can be divided into two classes (Sinha, 1964):

1. Fatigue effects that are caused by a large number of cycles of reloading–unloading of relatively low stress levels.
2. Incremental deformations that are caused by a relatively small number of load cycles of rather high stress, each of which produces additional deformation.

Only the second class of deformations is discussed here. There is considerable amount of literature on this topic, such as a RILEM symposium (RILEM, 1966), going back at least to 1907 (Bresztovszky, 1907).

The process of repeated, or cyclic, loading is to load the concrete up to a certain strain with monolitically increasing strain. This causes progressive crack propagation in the concrete (Shah, 1970). The resulting deformation can be presented in the form of a *loading curve* (Fig. 6.1). If, subsequently, the strain is gradually decreased at a given point, by reducing the load, descending curve will be obtained in the stress–strain diagram, which is called the *unloading curve*. The unloading curve is not identical with the loading curve; it is always under it looking from the strain axis. The reason for this is that upon the complete removal of the load, only a portion, the *elastic* portion, of the observed strain will disappear. The nonelastic portion of the strain, which is permanent, will remain. Therefore, repeated reloading–unloading cycles from various, increasing strain values produce a set of different reloading–unloading curves. In the case of a low-carbon steel the situation is different; the reloading–unloading curves are identical with a fair approximation and parallel with each other.

As shown in Figure 6.5, and elsewhere (Ros, 1950; Karsan, 1969; Shah and Winter, 1966; Linger, 1966; Halasz, 1967, 1970; Mander, 1989; Martinez-Rueda and Elnashai, 1997), the concrete deformations under repeated compressive loading have the following characteristics:

- The observed strain at failure is much larger than at failure caused by loading without repetition.
- The loading and reloading curves form a locus of *common points,* which are defined as the point where the reloading portion of any cycle crosses the pertinent unloading portion. Stresses above the common points produce additional strains, whereas stresses at or below these points will result in a stress–strain path going into a hysteresis loop.

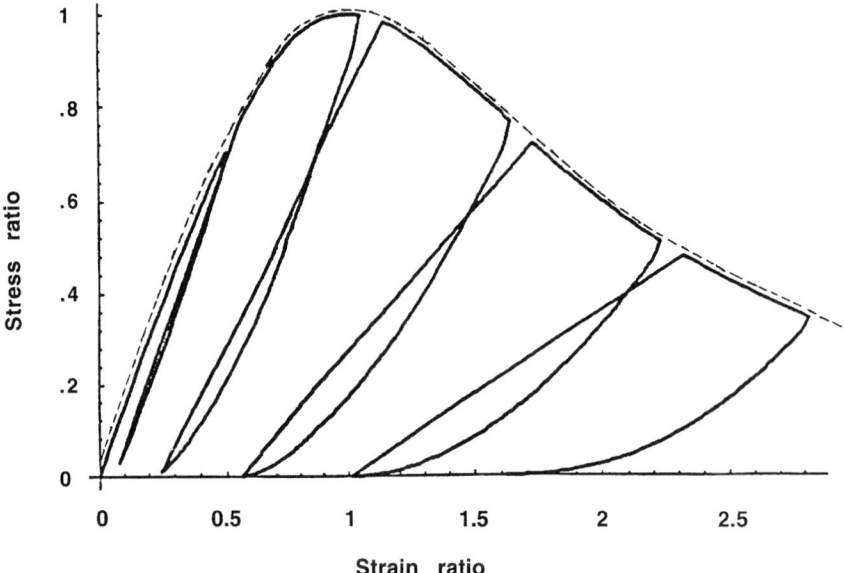

Figure 6.5 Normalized stress–strain curve of concrete under cyclic compressive loading. The envelope curve (dashed line) represents Eq. 6.9. (From Bahn, 1998. Copyright ACI. Reprinted with permission.)

- The area of such a loop represents an irreversible energy of deformation spent on temperature rise, crack formation, plastic deformation, etc. The area of the hysteresis loop first decreases with successive load cycles, but later, close to the failure, begins to increase.
- Repetition of the same loading cycle adds to the strain, but each repetition causes smaller and smaller strain increase up to a limit.
- The reloading curves are much closer to a straight line than the unloading curves. They are even straighter than the curve obtained with the first, unrepeated loading, represented, for instance, by the ascending portion of Eq. 6.9. That is, the relationship between stress and strain during reloading is practically linear.
- The slope of the reloading straight line, that is the modulus of elasticity of concrete at that strain section, decreases with an increase in the strain.
- The Poisson ratio decreases with load repetition.

Some of these characteristics have been expressed mathematically.

It was also observed that the set of unloading–reloading curves may possess an envelope curve. This curve is defined as the limiting curve within which all stress–strain curves lie regardless of the load pattern. It has been proposed (Sinha et al., 1964a, Otter and Naaman, 1988) that, for practical purposes, this envelope is identical with the stress–strain curve obtained under

monolitically increasing strain applied without repetition. Figure 6.5, for instance, has Eq. 6.9 as the envelope curve (Bahn and Hsu, 1998). This characteristic of the envelope curve also means that failure under constant stress
range cyclic loading can be predicted since the failure occurs when the strain
at the maximum load level reaches the envelope.

6.3 ELASTIC CONSTANTS OF CONCRETE

Although, as has been shown, the load–deformations diagrams of concrete
are nonlinear over their entire length, the assumption of linear stress–strain
relationships is frequently used because this simplifies engineering calculations. Since at zero stress the deformation is also zero, the characteristic of
such a stress–strain relationship is the slope of the straight line, which is
called the *modulus of elasticity,* which is an *elastic constant.* (See Section
1.5.) There are quite a few test methods for the determination of elastic constants. These can be divided into two classes: (1) static methods, where
stresses and corresponding strains are measured, and (2) dynamic methods
which are based on vibration or stress wave measurements.

Static Methods

Static elastic constants may be measured in compression, tension, and shear.
The modulus of elasticity in compression or tension is frequently called
Young's modulus and designated by E. The shear modulus, sometimes called
the *modulus of rigidity* or *torsional modulus,* is designated by G. These two
moduli describe completely the instantaneous elastic behavior of a linearly
elastic, homogeneous, and isotropic material; thus the third common elastic
constant, Poisson's ratio, μ, can be expressed in terms of E and G as

$$\mu = \frac{E}{2G} - 1 \tag{6.14}$$

Modulus of Elasticity The most commonly used elastic constant is the
modulus of elasticity. It is used to measure the instantaneous elastic deformability of a material under uniaxial compression of tension. The term is defined, in general, by ASTM E6-89 (1994) as the ratio of stress to
corresponding strain below the proportional limit. Like stress, this modulus
is expressed, in force per unit area, that is in psi, or MPa, or kg/cm^2. Different
methods produce different modulus values for the same material. For concrete
as well as for other materials where the stress–strain relationship is nonlinear,
one of the four following terms may be used for the static modulus of elasticity, as recommended by Stanton Walker in 1919 (Fig. 6.6):

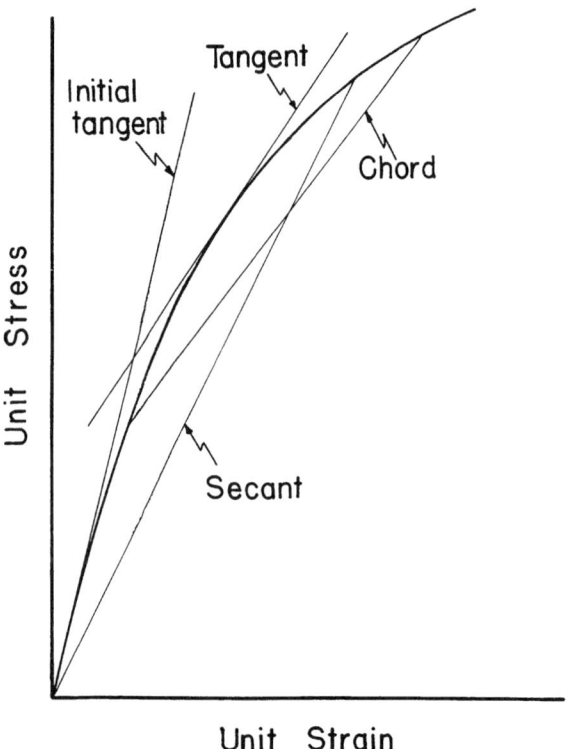

Figure 6.6   Four definitions for the static Young's modulus of elasticity of concrete. (From S. Popovics, 1970a. Copyright TRB. Reprinted with permission.)

1. *Initial tangent modulus:* Slope of the stress–strain curve at the origin (Section 6.2)
2. *Tangent modulus:* slope of the stress–strain curve at any specified stress or strain (Section 6.2)
3. *Secant modulus:* slope of the secant drawn from the origin to any specified point on the stress–strain curve
4. *Chord modulus:* slope of the chord drawn between any two specified points on the stress–strain curve

As was mentioned in the previous section, the E moduli can be calculated from any f–ε equation. For instance, differentiation of Eq. 6.9 results in the following:

$$E_t = \frac{df}{d\varepsilon} = n(n-1)\frac{f_0}{\varepsilon_0}\frac{1-(\varepsilon/\varepsilon_0)^n}{[n-1+(\varepsilon/\varepsilon_0)^n]^2} \tag{6.15}$$

E_t becomes E, the initial tangent modulus, at $\varepsilon = 0$, that is,

$$\frac{E}{E_0} = \frac{n}{n-1} \tag{6.16}$$

which, with Eq. 6.10, provides the following relationship for normal-weight concretes:

$$\frac{E}{E_0} = 1 + \frac{2500}{f_0} \tag{6.17}$$

where f_0 is in psi. This means that E/E_0 is 3.5 when $f_0 = 1000$ psi, and is 1.25 when $f_0 = 10,000$ psi. These figures are in accordance with experimental data.

Since $E_0 = f_0/\varepsilon_0$, the combination of Eq. 6.17 with Eqs. 6.10 and 6.13 provides the following relationship between the compressive strength and E modulus:

$$E = \frac{10^4}{k} \frac{f_0 + 2500}{f_0^{0.25}} \tag{6.18}$$

Equation 6.18 fits reasonably well, within 2000 and 10,000 psi compressive strength limits, the traditional empirical formula for normal-weight concretes:

$$E = Kf_0^{0.5}$$

It also follows from Eqs. 6.16 and 6.10 that

$$\varepsilon_0 = \frac{1}{E}(f_0 + 2500) \tag{6.19}$$

The only ASTM method for the determination of elastic constants is the "Standard Method of Test for Static Young's Modulus of Elasticity and Poisson's Ratio in Compression," C469-94. It stipulates a chord modulus between two points on the stress–strain curve defined as follows:

- The lower point corresponds to a stress corresponding to a longitudinal strain of 50 millionths, psi.
- The upper point corresponds to a stress equal to 40% of the strength of concrete at the time of loading.

The lower point is near the origin but far enough removed to be free of possible irregularities in the strain reading caused by seating of the testing machine platens and strain measuring devices. The upper point is taken near the upper end of the working stress range that was assumed in design. Thus the modulus determined is approximately the average modulus of elasticity in compression throughout the working stress range.

The modulus of elasticity can also be determined, at least in principle, on specimens *loaded as beams.* Deflection measurements, curvature measurements, or strain measurements in the extreme fibers can be used for this purpose. The usual approach is to measure deflections caused by known loads and to calculate the modulus of elasticity from a suitable beam-deflection formula. Unfortunately, application of the usual simple deflection formulas provides unreliable *E* values for concrete, for two reasons. First, these formulas are based on the assumption that the material follows Hooke's law, which in the case of concrete is not true. Second, the depth-to-span ratios of unreinforced concrete beams normally used for such tests are so large that shear deflection composes a significant part of the total deflection, and the usual beam-deflection formulas cannot take this into consideration. A formula corrected for shear can be derived (Philleo, 1966; Seewald, 1927; Klieger, 1957), but the *E* values obtained are still greater than the true value, due to the deviation of the concrete from the Hooke's law (S. Popovics, 1970a). A static modulus of elasticity can also be calculated from the *splitting tensile* test if principal strains are measured at the center of the concrete cylinder (Hondros, 1959).

The approximate range of *E* values has been from 10^6 to 6×10^6 psi (7000 to 42,000 MPa) for structural concrete.

Other Static Elastic Constants   Static determination of *Poisson's ratio* is made by measuring both the axial strain and the corresponding transverse strain. Details are described in ASTM C469-94 and in Yoshida's (1930) book. It can also be determined from the splitting tensile test (Hondros, 1959). The *static shear modulus* is rarely determined. When it is needed, a torsion test is generally used. Its value is usually somewhat less than $E/2$ (Andersen, 1935).

Dynamic Elastic Constants

Resonance Frequency Methods   These are the most frequently used dynamic methods for the determination of elastic constants. The basic principle was described briefly in Section 1.5. Here we show its application for the determination of elastic constants of concrete. The procedure is specified in ASTM C215-91, which is based on the theory developed by Picket (1945).

The standard forms of the resonance frequency method are described in ASTM C215-91. The test specimens recommended are beams made in accordance with the standard procedure prescribed for flexural specimens, but other suitable specimens, such as cylinders, can also be used. The elastic constants can be calculated from the formulas

$$E_r = CWn^2 \tag{6.20}$$

$$E_r = DW(n')^2 \tag{6.21}$$

$$G_r = BW(n'')^2 \tag{6.22}$$

where E_r = dynamic modulus of elasticity calculated from the fundamental longitudinal or transverse resonance frequency of the concrete specimen

G_r = dynamic shear modulus calculated from the fundamental torsional frequency

n = fundamental longitudinal frequency, Hz

n' = fundamental transverse frequency, Hz

n'' = fundamental torsional frequency, Hz

The values of the factors C, D, and B depend on the shape and size of the specimen tested, on the units used, and on Poisson's ratio of the concrete. Detailed instructions for calculation of these factors can be found in ASTM C215 or in the literature (Whitehurst, 1966; Jones, 1962; Malhotra et al., 1968; Spinner and Valore, 1958; Spinner and Tefft, 1961; Krautkramer and Krautkramer, 1983). Equation 6.21 can be written in the following form when E_r is in psi and the specimen is a rectangular prism:

$$E_r = 6 \times 10^{-6}L^2d_1(n')^2 \tag{6.23}$$

where L = length of the prism, in.

d_1 = unit weight of the concrete, lb/cu. ft

Because the true value of Poisson's ratio is rarely known, it has become popular to use the longitudinal resonance frequency for the calculation of E_r so that the effect of an error in the assumed value of Poisson's ratio may be minimized (Spinner and Teft, 1961). This method is also specified in the British Standard 1881. The dynamic Poisson's ratio can be calculated from Eq. 6.14 with E_r and G_r.

Pulse Velocity Methods   The fundamentals of these methods have also been discussed in Section 1.5. It was pointed out that there are theoretical equations for the pulse velocity versus elastic constant relationships, such as Eq. 1.19. When the units of the variables are in the inch-pound system, this equation, with $\mu = 0.24$, takes the form:

$$E_p = 0.000183d_1v^2 \tag{6.24}$$

where E_p = dynamic (pulse) modulus of elasticity calculated from longitudinal pulse velocity, psi

d_1 = unit weight of concrete, lb/cu ft

v = longitudinal pulse velocity, ft/sec

Examples for other pertinent relationships are Eqs. 1.20 through 1.30 and elsewhere (J.S. Popovics, 1996b). Nevertheless, this method is not recommended for determination of the dynamic (pulse) modulus of concrete because the moduli they produce are usually unrealistically high. A combination of longitudinal pulse velocity and shear pulse velocity has also been tried (e.g., Eq. 1.24) because this eliminates the need for the Poisson's ratio, as Eq. 1.28 shows (Yoshida, 1930; Leslie and Cheesman, 1949; Hanke, 1969). Unfortunately, such equations did not work well for concrete either.

Comparison of the Various Moduli of Elasticity

The moduli of elasticity of a concrete obtained in various manners are usually more or less different. The size of this difference depends on several factors, as described below.

Static Moduli of Elasticity Walker (1919) demonstrated that the relation of the tangent modulus to the secant modulus at the same load is 0.8 to 0.9 within practical stress limits. Davis and Troxell (1929) found that the secant modulus of a granite concrete of about 2500 psi (17 MPa) compressive strength decreases almost linearly from 2.5×10^6 psi (17,250 MPa) to 1.7×10^6 psi (11,730 MPa) at the age of 28 days as the applied stress is increased from 200 to 1000 psi (1.4 to 6.9 MPa). Data by Jones and Richart (1936) show that the secant modulus of elasticity is about 10^6 psi (7 MPa) less at $0.9f_c'$ than at $0.5f_c'$. Shideler (1957) reported that values of the secant modulus of structural lightweight and normal-weight concretes obtained at $0.3f_c'$ are almost identical with values obtained at $0.45f_c'$. Klieger's pertinent results reported by Philleo (1955) show a similar tendency.

Data by Klieger (1957) demonstrated that the static modulus of elasticity determined by the compression of 6×12-in. (150×300-mm) cylinders is about three-fourths of the comparable modulus obtained by the flexure of $6 \times 6 \times 30$-in. ($150 \times 150 \times 750$-mm) beams on an 18-in. (450-mm) span using correction for shear in the calculation. On the other hand, Witte and Price reported that the secant moduli determined by compression at the ages of 2 and 3 years were practically the same as the comparable moduli obtained by the flexure of $3 \times 3 \times 16\frac{1}{4}$-in. ($75 \times 75 \times 406$-mm) beams and using, presumably, a simple deflection formula without correcting for the shear of a nonlinear stress distribution. Measurements by Vile (1968) show a practical equality of elastic modulus values obtained under uniaxial compressive and tensile states of stresses.

Dynamic Moduli of Elasticity   As Figure 6.7 shows, the moduli of elasticity calculated from longitudinal resonance frequencies, E_{rl}, and those from transverse frequencies, E_{rt}, are practically identical (Batchelder, 1953; Woods and McLaughlin, 1959). Data by Obert and Duvall (1941) support this finding. The agreement between the resonance modulus and the modulus calculated from the pulse velocity, E_p, is less satisfactory. As a rule, the pulse moduli are greater. This is shown in Figure 6.8, where pertinent values obtained on the same specimens that provided data for Figure 6.7 are plotted (Batchelder, 1953; Woods and McLaughlin, 1959). Data published by other investigators (Leslie and Cheesman, 1949; Cheesman, 1949) show that E_p is about 10% greater than E_{rt} of the same concrete. Similarly, according to Philleo (1955), Whitehurst obtained pulse moduli that were up to 47% greater than the corresponding resonance moduli, with an average difference of 15%. Philleo's own data indicate a similar tendency.

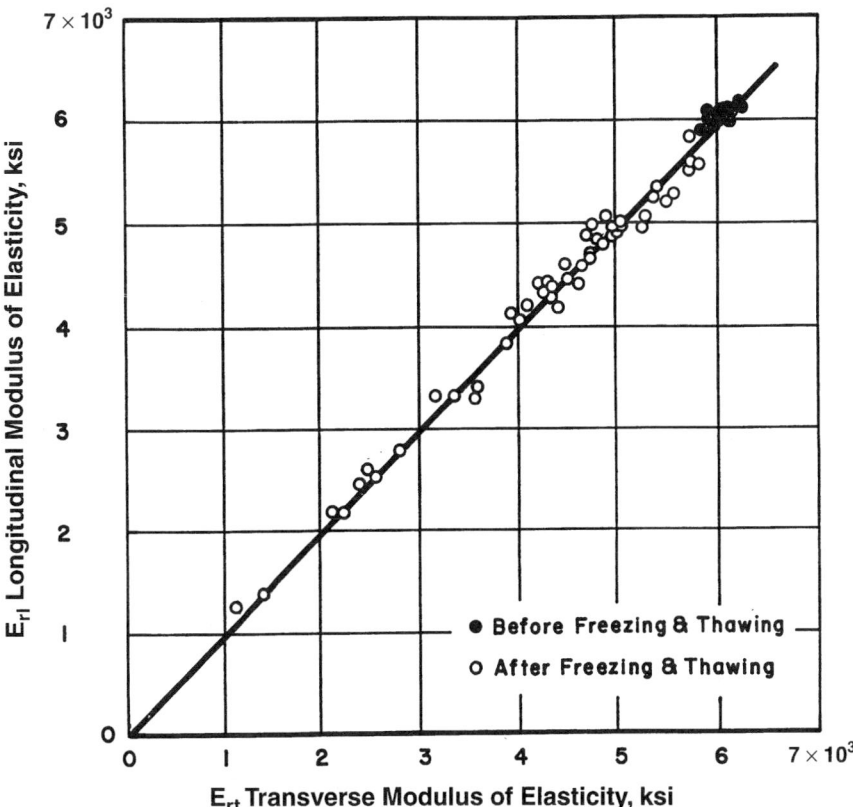

Figure 6.7   Comparison of moduli of elasticity from transverse resonance frequencies to moduli from longitudinal resonance frequencies. 1 ksi = 6.90 MPa. (From S. Popovics, 1970a, and Woods and McLaughlin, 1959. Copyright TRB. Reprinted with permission.)

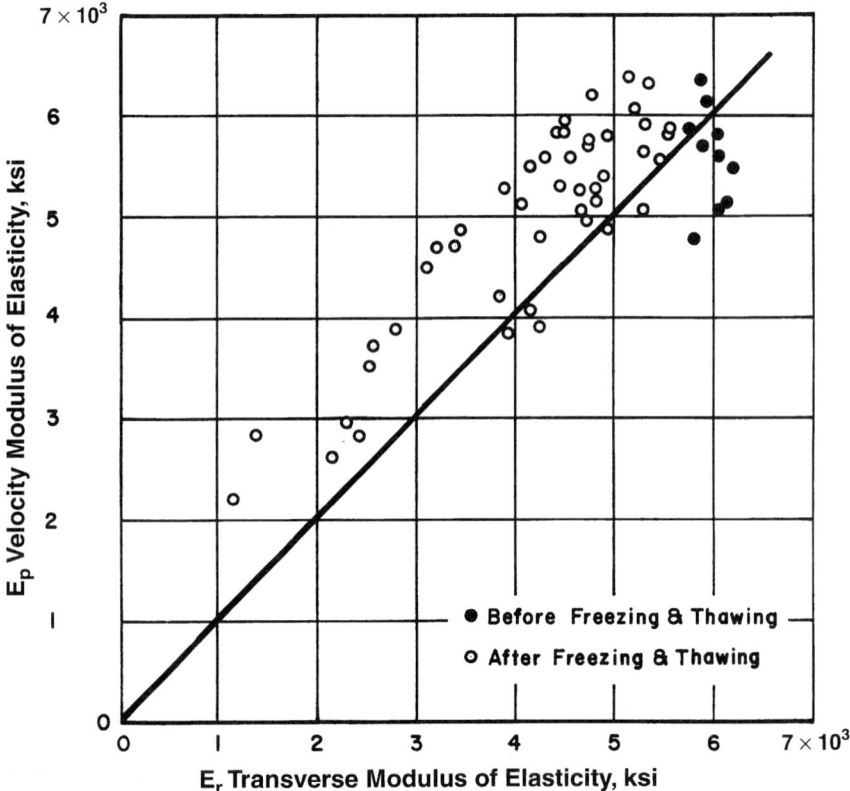

Figure 6.8 Comparison of moduli of elasticity from transverse resonance frequencies to moduli from pulse velocity. 1 ksi = 6.90 MPa. (From S. Popovics, 1970a, and Woods and McLaughlin, 1959. Copyright TRB. Reprinted with permission.)

Klieger (1957) determined the moduli of elasticity of a series of concretes by the resonance method both on 6 × 12-in. (150 × 300-mm) cylinders and on comparable 6 × 6 × 30-in. (150 × 150 × 750-mm) beams, having found the latter values to be about 8% higher. A similar series by Stanton (1944), however, showed close agreement between the moduli of the two types of specimens. The possible reasons for the tendency of E_p to be greater is that:

1. The pulse velocity is less affected by the porosity in concrete than the resonance frequency (Fig. 5.40). This opinion is supported by the observation in Figure 6.8 that the loosening of the internal structure of concrete caused by repeated freezing and thawing reduced the value of E_p less than that of E_{rt},
2. The transverse vibration produces higher stresses in the concrete specimen than the ultrasonic pulses, which makes the effect of porosity more pronounced (Section 5.13).

Static Versus Dynamic Moduli   The dynamic moduli are, as a rule, higher than the corresponding static values. Powers was the first to publish comparisons between the static and resonance moduli of concrete (Powers, 1938). He found good agreement between these two types of moduli. However, he used an oversimplified formula for calculation. Had he used the formula corrected by Pickett (1945), he would have obtained 7 to 10% higher values for E_{rt}. Later investigators also found that the resonance moduli are regularly higher than comparable static secant moduli. This difference can again be attributed to the two sources mentioned above for the E_r versus E_p difference.

It is shown in Section 6.4 that the relationship between the static and resonance moduli can be expressed as (S. Popovics, 1973b, 1975)

$$E_{st} = \frac{0.24E_r^{1.4}}{d} \qquad (6.25)$$

where E_{st} and E_r = static and dynamic moduli of concrete, respectively, psi
$\phantom{where E_{st} and E_r =}$ d = unit weight of the hardened concrete, lb/cu ft

A simpler formula is recommended in the British Code CP 110:1972:

$$E_{st} = 1.25E_r - 19 \qquad (6.26)$$

where the moduli are expressed in GPa.

Stanton (1944) reported that the secant modulus of 6 × 12-in. (150 × 300-mm) cylinder at 1000 psi (6.9 MPa) was found to range from 62% of the resonance modulus at 28 days to approximately 75% at later periods. Data by Shideler (1957) showed that the values of the resonance moduli of his moist-cured structural lightweight concretes are about 350,000 psi (2400 MPa) greater than the secant moduli, whereas this difference for sand-and-gravel concretes is about 1,500,000 psi (10,400 MPa). He also demonstrated that as the concretes dry, the differences between the secant moduli and resonance moduli are reduced. Similarly, Reichard (1964) found that the average ratios of the resonance to the secant modulus at the ages of 1 day, 28 days, and 1 year are 1.13, 1.11, and 1.10, respectively, for lightweight concretes, and 1.37, 1.32, and 1.16, respectively, for normal-weight concretes. Data by Hirsch (1962) demonstrated that the secant moduli are approximately 10% less than the resonance moduli. Chefdeville (1953) suggested that the E_{st}/E_r ratio increases to unity as the stress related to E_{st} decreases. Klieger's (1957) pertinent comparison showed that for 6 × 12-in. (150 × 300-mm) cylinders made of a wide range of concrete composition, the static moduli are in all cases lower than the related resonance moduli, the difference being smaller at 1 and 3 years than at 28 days. Other experimental evidence also indicates (Takabayashi, 1954; Elvery, 1954; Sharma and Gupta, 1960) that the higher the modulus of elasticity, the closer the agreement between resonance and static moduli.

A few comparisons are also available between values of static secant modulus corrected for shear from deflection measurements to the related resonance modulus. Philleo (1955) noticed a tendency for such E_r/E_{st} ratios to decrease as the modulus of elasticity increased. Klieger (1957) reported that such static moduli are slightly lower at 28 days than the resonance moduli, but at 1 and 3 years these differences are small and irregular. Thus it seems that on the average, the static secant moduli from deflection formulas, corrected for shear, agree fairly well with the resonance moduli.

A numerical relationship between the modulus E_p calculated from pulse velocity and the static modulus E_{st} is

$$\frac{E_p}{E_{st}} = \frac{2.52}{E_{st}^{0.5}} \tag{6.27}$$

where the moduli are expressed in psi. Other experimental evidence seems to indicate that the ratio of the pulse modulus to the static modulus is close to unity when the static modulus is about 10^6 psi (7000 MPa) and increases rapidly with decrease in the static modulus (Philleo, 1955). A group of aerated concrete in the compressive strength range 450 to 1000 psi (3 to 7 MPa) also provided practically unity for the E_p/E_{st} ratio (Morschtschichin, 1969).

There is nothing surprising in the fact that the static modulus is less than the corresponding dynamic modulus. In addition to the two sources mentioned above for the E_r versus E_p difference, another source is that in the static test, there is always an inelastic portion of the measured strain at a particular stress. This portion is due to the high stresses prevailing during the static test and is missing in the dynamic test. For this reason, several investigators consider the dynamic modulus as equivalent to the static initial tangent modulus. If unaccounted for, the inelastic deformations reduce the value calculated for the modulus. The magnitude of this inelastic portion varies with age and composition of concrete, curing and testing conditions, and other factors. There are also experimental data indicating that E_{st} is more dependent on the elastic properties of the hardened paste and less on the properties of the aggregate than is the dynamic modulus (Philleo, 1955).

Static tests for the modulus of elasticity involve the application of stresses of the same order as those in practice, whereas the stresses included in dynamic tests are very small. Thus the static modulus is of greater significance to the design engineer than are the corresponding dynamic values. On the other hand, determination of the dynamic values is easier.

Experimental data appear to indicate that the *reproducibility* of the static elastic constants is poorer than that of the dynamic constants. Witte and Price (1944) reported average coefficient of variations in the laboratory of 2.3, 7.4, and 10.3% for the resonance modulus of elasticity, compressive secant, and flexural secant static moduli, respectively. The variations in concrete in situ can be much higher. Neville (1968) reported cases after Elvery, where the ratio of the maximum to minimum value of the modulus of elasticity was as high as 2.

Poisson's Ratio   Several investigators have reached the conclusion that dynamic Poisson's ratios, as calculated from Eq. 6.19 with dynamic moduli, are consistently higher than the corresponding static values. Also, there seem to be differences between dynamic ratios obtained by different methods.

Anson and Newman (1966) are of the opinion that dynamic Poisson's ratios calculated from Eq. 6.19 may be 0.03 to 0.05 higher than the corresponding dynamic ratios calculated from pulse velocity and longitudinal resonance frequency tests. Simmons (1955) reported that dynamic Poisson's ratios calculated from Eq. 6.19 are consistently about 0.04 lower than the corresponding values from pulse velocity and longitudinal resonance frequency measurements, which, in turn, are about 0.08 higher than the corresponding static ratios. Data by Shideler (1957) show that in the case of wet concretes, the dynamic Poisson's ratio calculated from Eq. 6.19 is on the average 0.02 higher than the static ratio for structural lightweight concretes and 0.03 higher for sand-and-gravel concretes. For dry concretes this difference is negligible.

The source of differences between various Poisson's ratios is similar to that for the various moduli of elasticity. For instance, a significant portion of the difference between static and dynamic ratios is probably due to the inelastic deformations that occur when loads are applied in the static test. Also, the dynamic ratio appears to decrease with an increase in aggregate content and age of concrete, but increase as the water-cement ratio is increased. In contradistinction, the static ratio is hardly affected by the water-cement ratio and tends to increase with age approaching the dynamic value (Anson, 1964).

Simmons (1955) has shown that the standard deviation of static Poisson's ratio measurements, that of dynamic ratios calculated from Eq. 6.14, and that of dynamic ratios calculated from pulse velocity and resonance frequency measurements are 0.014, 0.007, and 0.009, respectively. These values correspond to a coefficient variation of 8.9, 3.6, and 3.8%, respectively. McCoy and Mather (1954) have reported that the coefficients of variation of dynamic Poisson ratios calculated from Eq. 6.14 were 11% at the age of 14 days and 9% at 180 days.

Other Effects on Elastic Constants (S. Popovics, 1972b)

Modulus of Elasticity   The concrete composition has a major effect on the modulus of elasticity. The higher the E value of the *coarse aggregate* and the larger its quantity in the concrete, the greater the E modulus of comparable concretes. This effect can be modeled mathematically as shown in Section 6.4. The higher the *water-cement ratio,* the lower the E value. However, this effect is less pronounced on the modulus than on the compressive strength (Walker, 1919; Graf et al., 1960; Koenitzer, 1935). It is surprising that no numerical relationship has been found in the literature for the E versus w/c relationship.

The effect of *air content* of concrete on the resonance modulus can be estimated using the formula (S. Popovics, 1969a)

$$E_r = E_{r0} \times 10^{-0022a} \qquad (6.28)$$

where E_r and E_{r0} = resonance modulus of elasticity of concrete with a and zero air content, respectively
a = air content, %

This is a sizable effect, which explains why reductions in the value of E_r can be used so conveniently to measure the damage in concrete caused by freezing and thawing, fire, or chemical attacks. Yet it is only approximately half of the relative effect of porosity on the compressive strength (Fig. 5.40). Note that the use of air content in Eq. 6.28 may not be correct. It is possible, for instance, that the effective porosity for E is not the air content of the concrete, as it is for the strength, but the total porosity, that is, the sum of the air content and the porosity in the aggregate particles. The reason for this distinction is that the pores in the stone have a measurable effect on the deformability of aggregate particles but have hardly any effect on concrete strength. Nevertheless, there is no effort here to differentiate between the two kinds of porosity. This simplification, or perhaps oversimplification, is inevitable at present because not enough data are available for meaningful distinction.

The value of E increases only slightly with an increase in the *rate of loading* (Jones and Richert, 1936; Yoshida, 1930; Atchley and Furr, 1967; Watstein, 1953; McHenry and Echideler, 1956). However, a sustained load has two effects on the modulus of elasticity (Ross 1968; Elvery and Furst, 1957; Balazs, 1960; Linger, 1963; Washa and Fluck, 1950):

1. Its value drops immediately after loading.
2. It increases while the load is sustained, probably due to an increasing denseness in the specimen. Thus, after awhile, the modulus of elasticity of the concrete under load will surpass the value of the original modulus of the unloaded concrete.

Age affects the modulus of elasticity less than it affects the compressive strength (Walker, 1919; Koenitzer, 1935; Davis and Troxell, 1929; Noble, 1931; Preece, 1948; Stanton, 1944). The following empirical formulas have been recommended for the E versus age relationship within 7 days and 1 year (Chefdeville, 1953; Harig, 1966; Erzen, 1956):

$$E_t = E_{28} + a \log(t - 27) \tag{6.29}$$

and

$$E_t = E_{28}\frac{8}{7 + 28/(t - 27)} \tag{6.30}$$

where E_t and E_{28} = modulus of elasticity at the age of t and 28 days, respectively

$t \geq 28$ = age, days

In contrast to strength, the static modulus of elasticity is less for concrete when *dry* then when *wet* (Davis and Troxell, 1929; Chefdeville, 1953; Plowman, 1963). From this, the influences of various normal curing methods on the modulus can be derived (Graf et al., 1960; Kesler and Higuchi, 1953; Koenitzer, 1936; Obert and Duvall, 1941). Also, as an extrapolation of this principle, one would expect that *heat treatment* of concrete would further decrease the modulus without damaging the strength, whereas cooling would increase it. Experimental data indeed support this expectation (Noble, 1931; Harig, 1966; Elvery, 1954; Elvery and Evans, 1964; Monfore and Lentz, 1962; Richards and Radjy, 1966). Both static and dynamic values for the moduli of elasticity of concrete are less than half at 1000°F (538°C) than at room temperature (Cruz, 1966; Philleo, 1958). Conversely, the moduli are usually about 10% higher at 0°F (−18°C) and 20 to 50% higher at −70°F (−57°C) than at room temperature (Saemann and Washa, 1957).

Covering the surface of coarse aggregate particles with a thin film of bituminous material or polymer emulsion can reduce the modulus of elasticity of concrete. However, they usually reduce the concrete strength as well (Harig, 1966; Zivica, 1965).

Other Elastic Constants   The effects of *porosity* on elastic constants are discussed by Spinner et al. (1963). Beyond this, very little is known about the effects on the *shear modulus* of concrete. It is customary to assume that the value of G is affected by the same factors and to the same extent as E is affected. Experimental data are also scarce about the compressibility of concrete (Yoshida, 1930; Krahl et al., 1965), although this can be estimated from other measurements (e.g., with Eq. 1.20).

Yoshida analyzed the influences of several factors on *Poisson's ratio* (Yoshida, 1930). He came to the conclusion that variations in this ratio are relatively small for a wide range of structural concretes compared with corresponding variations of the modulus of elasticity with similar changes in conditions. Thus one may assume in many cases that the static Poisson's ratio for such concretes is approximately 0.15 to 0.18 up to about 500 psi (3.5 MPa) compressive stresses. In present-day engineering practice, a value of

0.2 is accepted. There are, however, special cases, such as certain stress wave measurements (J.S. Popovics, 1997), where this approximation is not acceptable.

Under high loads Poisson's ratio increases rapidly. This is, however, not a true Poisson's ratio anymore because the increase is due to volume increase caused by cracking within the specimen (Neville, 1971). Other experiments support essentially these observations both for normal-weight (Davis and Troxell, 1929; Noble, 1931; Plowman, 1963; Richart and Roy, 1930) and lightweight concretes (Shideler, 1957; Reichard, 1964; Hanson, 1958, 1964), although more recent measurements do show relationships between Poisson's ratio and the mix proportion (Anson, 1964; Anson and Newman, 1966; Ishai, 1961). For instance, Poisson's ratio decreases with an increase in the aggregate content. Under biaxial stress, Poisson's ratio does not change significantly. A value of 0.20 has been measured in compression–compression, 0.18 in tension–tension, and between 0.18 and 0.20 in compression–tension (Kupfer et al., 1969).

The effects of porosity on elastic constants are further discussed by Spinner et al. (1963). The modulus of elasticity of concrete is discussed further in Sections 6.4 and 6.5.

6.4 RELATIONSHIP BETWEEN MODULUS OF ELASTICITY, CONCRETE STRENGTH, AND OTHER MECHANICAL PROPERTIES

Relationship Between Strength and Modulus of Elasticity

It is reasonable to expect, as Schule did in 1912 according to Ros (1950), that a stronger concrete has a higher resistance to deformations, that is, a higher modulus of elasticity. Eighty years ago Walker (1919) suggested a power function for this relationship whose general form

$$E = kf_c^n \qquad (6.31)$$

where E = modulus of elasticity
f_c = compressive strength of the concrete
k and n = experimental parameters that depend on testing conditions

When psi units and cylindrical specimens were used, Walker obtained the following values for normal-weight structural concretes having a unit weight of close to 145 lb/cu yd (2300 kg/m^3):

Initial tangent modulus	$k = 33,000$	$n = 0.63$
tangent modulus at 25% of the compressive strength	$k = 66,000$	$n = 0.5$

Walker's recommendation was followed by quite a few similar formulas

(Hummel, 1959; L'Hermite, 1950; RILEM, 1954; Chefdeville, 1953). The form of Eq. 6.31 in psi, as recommended by ACI, for similar concretes is (ACI, 1983b)

$$E = 57,000f_c^{0.5} \tag{6.32}$$

which is very close to Eq. 6.31 for tangent modulus. When E is expressed in GPa and f_c in MPa, Eq. 6.32 has the form

$$E = 4.73f_c^{0.5} \tag{6.33}$$

When the unit weight of the concrete is less than 145 lb/cu ft (2300 kg/m³), the E_{st} values calculated by Eq. 6.32 or 6.33 are typically much higher than the actual values. This is demonstrated in Figure 6.9 where pertinent experimental data published by Shideler (1957) are presented. Comparable data published by Reichard (1964) show a similar trend. Figure 6.9 also shows that the E_{st} versus f_c relationship is multivariable in nature.

It was probably Bolomey (1939) first, and later Schaffler (1954), who recommended that the modulus of elasticity of concrete be expressed in terms of compressive strength *and* its unit weight d. However, it was only after a reintroduction by Pauw (1960) that this multivariable concept obtained general acceptance. Pauw's empirical formula, as included in the ACI 318 Code, (ACI 1983b)

$$E_{st} = 33d^{1.5}f_c^{0.5} \tag{6.34}$$

where the stress units are psi and the unit weight d is lb/cu ft. When SI units are used, the formula becomes

$$E = 43d^{1.5}f_c^{0.5} \times 10^{-6} \tag{6.35}$$

It is obvious from the comparison of Figure 6.9 to Figure 6.10 that Eq. 6.34 provides a much better fit to a wide range of experimental data than any single-variable formula could. Both Shideler's results and Reichard's results show a standard error of estimates of only approximately 0.5×10^6 psi ($3.5 \times$ MPa) from the formula.

Additional, mostly European, formulas are presented by Neville (1981). The tangent modulus E in GPa versus cube strength f_{cu} in MPa relationship can be expressed according to the British Code of Practice for the Structural Use of Concrete (CP 110:1972) as

$$E = 9.1f_{cu}^{0.33} \tag{6.36}$$

when the density of concrete is about 145 lb/cu ft (about 2300 kg/m³) or

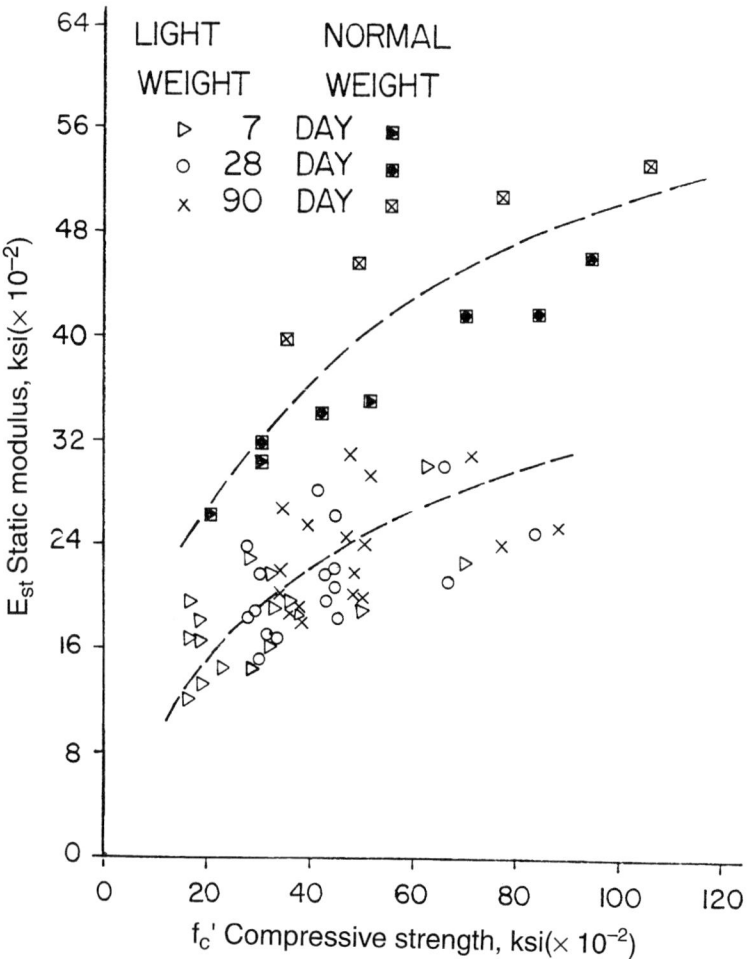

Figure 6.9 Experimental values of E_{st} static modulus of elasticity plotted against related values of f_c compressive strength as observed by Shideler on various concretes. 1 ksi = 6.90 MPa. (From S. Popovics, 1975. Copyright RILEM. Reprinted with permission.)

greater. The corresponding formula of the Comité Européen du Beton for cylinder strength at 28 days is

$$E = 9.5(f_c + 8)^{0.33} \qquad (6.37)$$

When the unit weight d is between 90 and 150 lb/cu ft (1400 and 2300 kg/m³), Eq. 6.36 becomes

$$E = 1.7d^2 f_{cu}^{0.33} \times 10^{-6} \qquad (6.38)$$

and Eq. 6.37 becomes

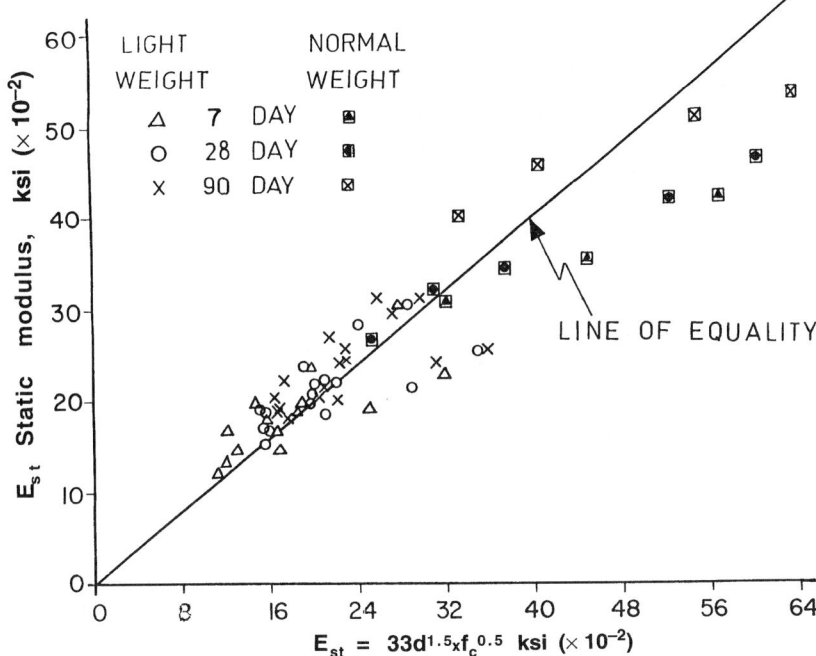

Figure 6.10 Comparison of experimental values of E_{st} obtained by Shideler on various concretes to related E_{st} values calculated by the ACI formula, in psi. *d* is in lb/cu ft, *f* is in psi, *E* values are in MPa. 1 lb/cu ft = 16 kg/m³. 1 ksi = 6.90 MPa. (From S. Popovics, 1975. Copyright RILEM. Reprinted with permission.)

$$E = 1.6d^2(f_c + 8)^{0.33} \times 10^{-6} \qquad (6.39)$$

Another type of formula, attributed to Gehler, is (Hummel, 1959)

$$E = \frac{555,000 f_{cu}}{f_{cu} + 200} \qquad (6.40)$$

where the stress unit is kg/cm².

The *E* versus *f* relationship is also influenced by the concrete composition. There is a decrease in elastic modulus of concrete with increasing paste content, given equal concrete strengths. This is illustrated by the following two formulas for concretes in the strength range of approximately 500 psi (3.5 MPa) to 11,000 psi (77 MPa) and unit weights ranging from about 135 lb/cu ft (2160 kg/m³) to 155 lb/cu ft (2480 kg/m³) (Hansen, 1986): For concretes containing 32% by volume paste,

$$E_{st} = 6,424 f_c'^{0.72} + c \qquad (6.41)$$

and for concretes containing 50% by volume paste,

$$E_{st} = 10,483 f_c'^{0.62} + c \qquad (6.42)$$

where c is an experimental parameter depending on the aggregate type, strength of concrete, and so on. However, given equal paste content, the E_{st} versus f_c relationship is independent of the water-cement ratio. Further analysis of some of these formulas is presented below.

Relationships Between Other Mechanical Properties
(S. Popovics, 1973b, 1975)

This is again a section where the emphasis is not so much on the presentation of formulas as on the demonstration of a method for rational development of numerical relationships between mechanical properties of brittle materials in general, and those of concrete in particular.

Basic Premise   The basic premise of the approach is that some mechanical properties of a brittle material may be affected more than other properties by a change in the porosity. In case of concrete, porosity means air content (Section 5.13). Therefore, a relationship between two mechanical properties, such as modulus of elasticity and compressive strength of a brittle material, is influenced by the porosity of the material. This means that such a relationship is multivariable in nature. Unfortunately, the present state of the art is insufficient to formulate this premise into an exact theory; therefore, approximation is necessary. Mathematical modeling is one such possibility, selecting porosity as one phase of the model.

The idea that porosity, or any function of it, can be a significant factor in relationships between mechanical properties of brittle materials is also supported by early equations such as Eqs. 1.19 through 1.21. These theoretically justified equations contain the density, or unit weight, as an extra variable in the relationships between the velocity of stress waves passing through a material and the elastic constants of this material. There are, however, unexplained phenomena that can be clarified by the consideration of the role of porosity, such as the following:

- Woods and McLaughlin (1959) compared the values of modulus of elasticity of concrete calculated from resonance frequencies to moduli calculated from pulse velocities before and after submitting the concrete to repeated freezing and thawing. They reported, without explanation, that whereas before the test the two kinds of modulus were more or less equal, after the test the pulse moduli were greater (Fig. 6.8).
- Whitehurst (1966) reported a similar trend for the effect of fire damage on concrete. The pulse modulus of the damaged concrete was found to be greater than the resonance modulus, but not explanation was offered.

· As discussed earlier, a normal-weight concrete has a greater modulus of elasticity and a greater flexural strength than does a structural lightweight concrete having the same compressive strength.

The damage caused by either freezing or fire essentially loosens up the internal structure of concrete, which suggests that the pulse modulus is less affected than the resonance modulus by a change in the porosity of concrete. For normal-weight versus lightweight concretes, again the explanation seems to be that a change in porosity affects the compressive strength more than it does either the modulus of elasticity or the flexural strength. So if porosity is included in a relationship between, say, the two types of moduli, the limits of validity as well as the accuracy of the relationship are improved.

The quantitative form of the explanation offered is the following: If the effects of porosity on two mechanical properties of a concrete are known numerically, from these and from a relationship between these two properties for a given initial state of porosity ("zero state"), an improved relationship between the two properties can be obtained mathematically by including the porosity, or a function of it, as one of the independent variables.

The mathematical development of this statement could be quite general and rigorous, but this would lead to complicated formulations. This complication can be avoided by the use of special functions. For instance, let the relationship between m_1 and m_2 mechanical properties in the zero state be given as

$$m_{1,0} = F_0(m_{2,0}) \tag{6.43}$$

Also let the relative effects of porosity p on these properties as well as on the density d be given:

$$\frac{m_1}{m_{1,0}} = f_1(p) \tag{6.44}$$

$$\frac{m_2}{m_{2,0}} = f_2(p) \tag{6.45}$$

$$\frac{d}{d_0} = f_3(p) \tag{6.46}$$

Then, $m_{1,0}$ and $m_{2,0}$ can be substituted from Eqs. 6.44 and 6.45 into Eq. 6.43, which, after rearrangement, provides the formula:

$$m_1 = f(p)F(m_2) \tag{6.47}$$

or by considering Eq. 6.46,

$$m_1 = g(d)F(m_2) \qquad (6.48)$$

Therefore, the explanation offered is correct if experimental results support Eq. 6.48 better than they support a similar formula without the porosity term. As shown below, this is indeed the case.

Underlying Equations Two groups of approximate formulas are used as points of departure for the development of relationships between various mechanical properties of concrete. The first group of these equations, any two of which can be used as Eqs. 6.44 and 6.45 in the general derivation, refers to the effects of porosity on mechanical properties of concrete. For instance, the following formulas have been obtained from published results by curve fitting (Fig. 5.40):

$$f_c = f_{c0} \times 10^{-0.038p} \qquad (6.49)$$

$$f_f = f_{f0} \times 10^{-0.023p} \qquad (6.50)$$

$$E_{st} = E_{st0} \times 10^{-0.025p} \qquad (6.51)$$

$$E_r = E_{r0} \times 10^{-0.022p} \qquad (6.52)$$

$$E_p = E_{p0} \times 10^{-0.014p} \qquad (6.53)$$

$$V = V_0 \times 10^{-0.0045p} \qquad (6.54)$$

$$d = d_0 \times 10^{-0.005p} \qquad (6.55)$$

where p = porosity
f_f = flexural strength
V = longitudinal ultrasonic pulse velocity
E_p = pulse modulus of elasticity
d = unit weight

The zero subscripts in these equations indicate properties obtained on a concrete in the zero state. Note that the term *porosity* here is not well defined. It is likely, for instance, that the effective porosity is the air content of the concrete for the *strength* equations, such as Eqs. 6.49 and 6.50, whereas in the *moduli* equations the effective porosity is the total porosity, that is, the sum of the air content and the porosity in the aggregate particles. Nevertheless, there is no effort here to differentiate between the two types of porosity. This simplification, or perhaps oversimplification, also makes it possible to use the unit weight of the concrete as a measure of the total porosity.

The second group of underlying equations consists of formulas that provide the relationship between two of the mechanical properties of a concrete in a zero state, for instance, when the porosity in macroscopic sense is zero in the dry material. Any of these formulas can be used as Eq. 6.43 in the general derivation. Several such formulas are

$$E_{st0} = K_1(f'_{c0})^{0.5} \tag{6.56}$$

$$E_{st0} = K_2(E_{r0})^{1.4} \tag{6.57}$$

$$f_{f0} = K_3(f'_{c0})^{0.7} \tag{6.58}$$

Again, these formulas are empirical because no acceptable theoretically justified formulas are available. On the other hand, these formulas, especially Eq. 6.56, are supported by a considerable amount of evidence, as shown above. Equation 6.57 reflects the observation that the static modulus of elasticity is smaller than the modulus calculated from the resonance frequency but that the differences are small when the moduli either very small or large (Whitehurst, 1966; Klieger, 1957; Takabayashi, 1954; S. Popovics, 1970a; Elvery, 1954; Sharma and Grupta, 1960; Stanton, 1944; Malhotra et al., 1968). The same tendency was observed about the relationship between the static modulus and the pulse modulus (Philleo, 1955; Morschtschichin, 1959; Leslie and Cheesman, 1949). Equation 6.58 is similar to the compressive strength versus flexural strength formulas in Section 2.1 and encountered in the literature (Hummel, 1959; Sen and Bharara, 1961; Road Research Laboratory, 1955; ACI Committee 435, 1963).

Presentation of New Relationships Numerous formulas can be obtained from the equations of the two groups presented above for relationships between certain mechanical properties of a concrete. Sample derivations are shown in Problems 6.1 through 6.3. It is not claimed that all the formulas obtained are supported well by experimental results because of the restricted validity of the underlying equations. Of the acceptable relationships, Eqs. 6.25, 6.27, and 6.34 have already been discussed. Several other formulas are also presented below.

$$E_{st} = \frac{0.23 E_r^{1.4}}{d} \tag{6.25}$$

$$\frac{E_p}{E_{st}} = \frac{2.52}{E_{st}^{0.5}} \tag{6.27}$$

$$E_{st} = 33 d^{1.5} f'^{0.5}_c \tag{6.34}$$

$$\frac{E_r}{E_{r0}} = \left(\frac{f'_c}{f'_{c0}}\right)^{0.6} \tag{6.59}$$

$$\frac{E_r}{E_{r0}} = \left(\frac{f_f}{f_{f0}}\right)^{0.8} \tag{6.60}$$

$$\frac{V}{V_0} = \left(\frac{f'_c}{f'_{c0}}\right)^{0.12} \tag{6.61}$$

$$\frac{V}{V_0} = \left(\frac{f}{f_0}\right)^{0.15} \tag{6.62}$$

$$f_f = 0.15d^{0.5}f'^{0.7}_c \tag{6.63}$$

$$f_f = 0.08E_r^{0.2}f'^{0.7}_c \tag{6.64}$$

The limits of validity of these equations are determined by the limits of validity of Eqs. 6.49 through 6.58.

Evaluations of Eq. 6.25 and 6.63   In Figure 6.11 more than 200 trios of experimentally obtained E_r values are plotted against values calculated by Eq. 6.25 with E_{st} and d. The test results were published by Shideler (1957), Reichard (1964), and Manns (1969), respectively, on various structural light-weight and normal-weight concretes. The compositions and ages of the

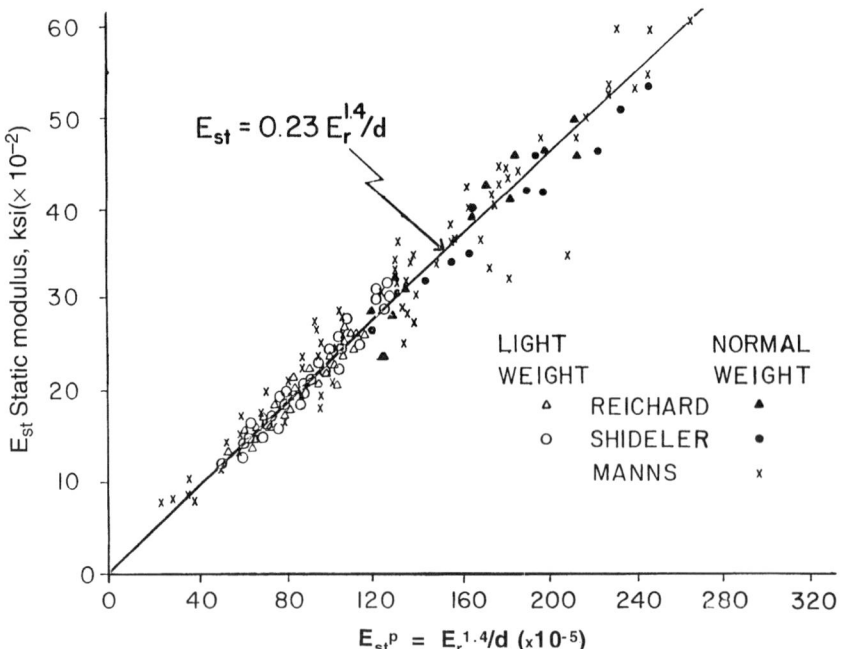

Figure 6.11 Comparison of experimental values of E_{st} obtained by Shideler, Reichard, and Manns, respectively, on various concretes to related E_{st} values calculated by Eq. 6.25, in psi. d is in lb/cu ft. 1 lb/cu ft = 16 kg/m³. 1 ksi = 6.90 MPa. (From S. Popovics, 1973b, 1975. Copyright ACI. Reprinted with permission.)

concretes covered a wide range. The test methods were similar to those in ASTM C215 and C469.

When the experimental E_r values are plotted against E_{st} values, the light-weight concrete results clearly separate from the results of normal-weight concretes, similar to Figure 6.9. In Figure 6.10, however, the two types of concrete results form a single group with a good fit. The coefficients of the equation were determined by the least-squares method. With these, the average square-root deviation of the values calculated by Eq. 6.25 is approximately 0.15×10^6 psi (2300 MPa), which is quite good. A few results on stones and hardened cement pastes also support Eq. 6.25.

Equation 6.63 is compared in Figure 6.12 to experimental flexural strengths f_f by Shideler (1957) with the same lightweight and normal-weight concretes that were used in Figure 6.9. Both the flexural and compressive strengths were determined in the standard way. Note that when the experimental f_f values are plotted against f_c values, the lightweight concrete results are again clearly distinguishable from the results of normal-weight concretes, similar to Figure 6.9. In Figure 6.12 however, the two types of concrete results form a single group with a good fit. The average of the f_f/f_c ratios was 0.113. Least-

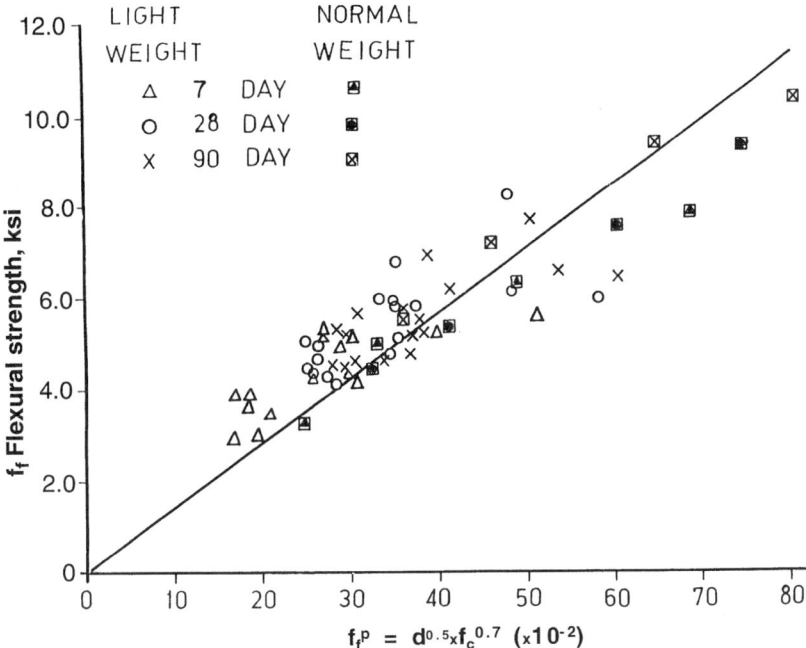

Figure 6.12 Comparison of experimental f_f flexural strengths obtained by Shideler on various concretes to related flexural strength values calculated by Eq. 6.63 in psi. d is in lb/cu ft, f_c is in psi, f_f is in ksi. 1 lb/cu ft = 16 kg/m³. 1 ksi = 6.90 MPa. (From S. Popovics, 1975. Copyright RILEM. Reprinted with permission.)

squares analysis produced an average square-root deviation of 130 psi (0.9 MPa) between experimental flexural strengths and f_f calculated from Eq. 6.63, which again demonstrates a good fit.

Analysis of Reichard's pertinent test results produced a similarly good fit for Eq. 6.63 (Fig. 2.2b). Further evaluation of some of the equations presented above can be found in the literature (S. Popovics, 1975).

PROBLEM 6.1   Derive Eq. 6.34 from the underlying equations.

Solution   Substitution of Eqs. 6.49 and 6.51 into Eq. 6.56 yields

$$E_{st} \times 10^{0.025p} = K_a(f_c \times 10^{0.035p})^{0.5}$$

that is,

$$E_{st} = K_a f_c^{0.5} \times 10^{-0.0075p} \tag{6.65}$$

Also, from Eq. 6.55,

$$10^{-0.0075p} = K_b d^{1.5}$$

The substitution of this into Eq. 6.65 provides Eq. 6.34.

PROBLEM 6.2   Derive Eq. 6.27 from the underlying equations.

Solution   Utilizing Eq. 6.57 for pulse modulus of elasticity, as well as Eqs. 6.51 and 6.53, gives us

$$E_{st} \times 10^{0.025p} = K_c E_p^{1.4} \times 10^{0.019p}$$

that is,

$$\frac{E_p}{E_{st}} = K_c E_p^{-0.4} \times 10^{0.006p} \tag{6.66}$$

But from Eq. 6.53,

$$E_p^{-0.4} = K_d \times 10^{0.006p}$$

Thus

$$E_p^{-0.4} \times 10^{0.006p} = K_d \times 10^{0.012p} = K_e E_{st}^{-0.5}$$

The substitution of this into Eq. 6.66 provides Eq. 6.27.

PROBLEM 6.3   Derive Eq. 6.59 from the underlying equations.

Solution   From Eq. 6.52,

$$\frac{E_r}{E_{r0}} = 10^{-0.022p} \qquad (6.67)$$

But from Eq. 6.49,

$$10^{-0.022p} = \left(\frac{f_c}{f_{c0}}\right)^{0.6}$$

The substitution of this into Eq. 6.67 provides Eq. 6.59.

Based on support from the wide range of experimental evidence presented, it appears that many of the equations derived from the underlying equations are valid for structural lightweight as well as normal-weight concretes under the circumstances upon which the underlying equations were based. They are as well justified as are many of the accepted formulas in concrete technology or materials science and show better fit to experimental results than do corresponding formulas without porosity as an additional variable. Thus it seems that the concept of derivation of relationships between mechanical properties of concrete based on the differing effects of porosity on these properties is valid.

6.5   ESTIMATION OF THE MODULUS OF ELASTICITY BY MODELS

Another approach is to use simplified structures, or *models,* to calculate properties of concrete from the properties and amount of ingredients. Such attempts have been made to describe the deformability of concrete by *physical models* and by their corresponding mathematical forms, that is, by *mathematical models.* Various laminated models (Hansen and Erikkson, 1966; Hirsch, 1962; Dougill, 1962) and lattice models (Reinius, 1956; Baker, 1959; Roy, 1963) are mentioned primarily for elastic deformations (S. Popovics 1969b) and as viscoelastic or rheological models for creep (Reiner, 1969, 1971; Flugge, 1967; Mase, 1970; Neville, 1970; Bazant and Panula, 1978; Bazant, 1982; Bazant and Chuan Charn, 1985). These model approaches are simple but have a number of weaknesses. For instance, laminated models predict that a composite material will be linearly elastic if the component materials are linearly elastic, which is not true for concrete (Section 6.2). Nevertheless, models can serve useful purposes within practical limits.

Modeling the Modulus of Elasticity

It has been customary to model the modulus of elasticity of composite materials by mathematical formulas with or without their structural equivalents, such as a matrix containing laminas and/or particles in different arrangements; or by the mechanical equivalents of the formulas, such as a set of connected springs; or by their electrical equivalents, such as a set of ohmic resistances or capacitors (Brooks and Newman, 1968). In this section the physical models of hardened concrete are laminated models (Fig. 6.13) to represent concrete as a two-phase composite. It is assumed that:

- The Poisson ratios of the two phases do not differ so much as to cause major changes in the stress distribution under load (Mills and Ono, 1972).
- The bond between phases is strong enough to prevent discontinuity as well as primary failure on the interfaces.
- The two phases do not interact with each other.

Under these circumstances one would expect that the deformability of this model is the weighted average of the deformabilities of the two phases. Nu-

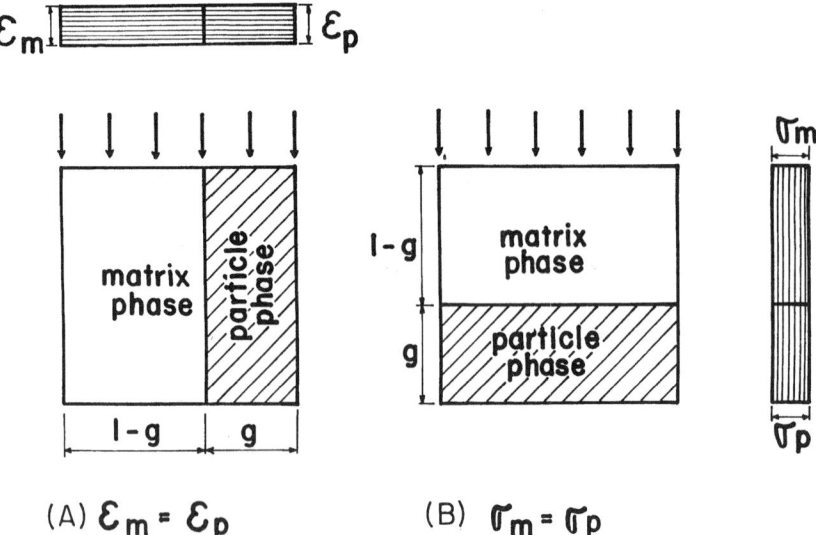

Figure 6.13 Simple laminated models of composite two-phase solids. (From S. Popovics, 1969b. Copyright American Ceramic Society. Reprinted with permission.)

merous attempts have been reported for the applications of the arithmetic and harmonic averages, respectively, for this purpose (Hansen, 1969). The mathematical forms of these averages are

$$E_a = g_1E_1 + g_2E_2 \qquad (6.68)$$

and

$$E_h = \frac{1}{g_1/E_1 + g_2/E_2} \qquad (6.69)$$

where E_a and E_h = weighted arithmetic and harmonic (i.e., "simple") average, respectively, of E_1 and E_2

g_1 and g_2 = fractional volume of phase 1 and phase 2, respectively, that is, $g_1 + g_2 = 1$

E_1 and E_2 = modulus of elasticity of phase 1 and phase 2, respectively

Spring analogs corresponding to Figure 6.13 as well as Eqs. 6.68 and 6.69, respectively, are given in Figure 6.14a and b.

Equation 6.68 can be interpreted as the mathematical model of a two-phase laminated composite where the *strains* in the two phases under a uniaxial load are equal, whereas in the composite represented by Eq. 6.69 the *stresses* are equal in the two phases. Unfortunately, experimental data did not support adequately these formulas, representing "simple averages." Equation 6.68, known as the *rule of mixes,* regularly overestimates the actual E values, whereas Eq. 6.69 underestimates them. As a matter of fact, it has been proved

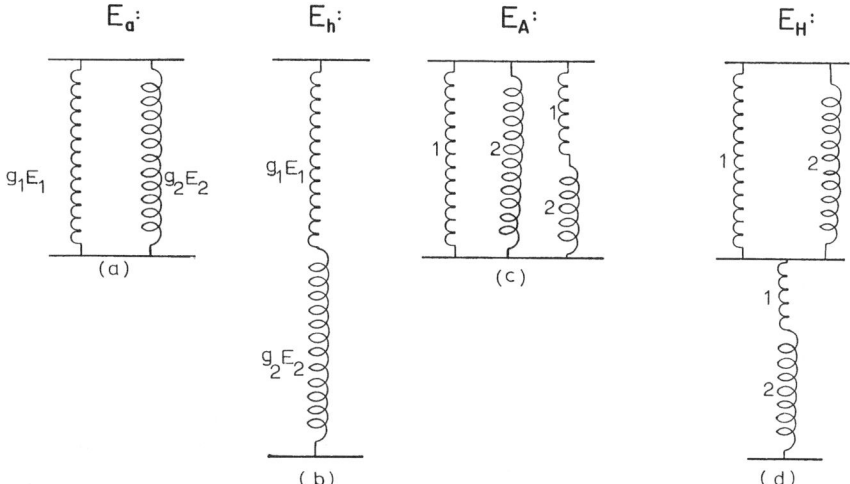

Figure 6.14 Various spring analogs for the elastic deformations of two-phase composite solids. (From S. Popovics, 1970d. Copyright RILEM. Reprinted with permission.)

that Eq. 6.68 represents the minimum potential energy, thus it is the upper limit, and Eq. 6.69 represents the least work, thus it is the lower limit of possible E values for combinations of the two phases provided that the three assumptions mentioned above for laminated models are fulfilled (Paul 1960).

Composite Average Concept

It logically follows from the limit nature of Eqs. 6.68 and 6.69 that the average of the limits (i.e., the average of the two averages, a *composite average*) would provide a better approximation for the modulus of elasticity in terms of composition. The question is then: What kind of composite average is suitable for the improved estimation of the modulus of elasticity of a two-phase composite? The answer (to be justified later) is that

- It is the *arithmetic composite average, E_A* or ACA (Eqs. 6.70, 6.71) for hard composites, that is, where the matrix portion is less deformable than the disperse phase. Examples of hard composites are lightweight-aggregate concrete, and tungsten carbide as matrix and cobalt as particles.
- It is the *harmonic composite average, E_H* or HCA (Eqs. 6.72, 6.73) for soft composites, that is, where the matrix is the more deformable phase. Examples of soft composites are normal-weight concrete, fiber-reinforced polymer, and boron-reinforced epoxy.

The mathematical forms of the two composite averages are (S. Popovics, 1969b)

$$E_A = A_1 E_a + (1 - A_1)E_h \tag{6.70}$$

$$= A_1[g_1 E_1 + (1 - g_1)E_2] + \frac{1 - A_1}{g_1/E_1 + (1 - g_1)/E_2} \tag{6.71}$$

and

$$\frac{1}{E_H} = \frac{A_1}{E_a} + \frac{1 - A_1}{E_h} \tag{6.72}$$

$$= \frac{A_1}{g_1 E_1 + (1 - g_1)E_2} + (1 - A_1)\left(\frac{g_1}{E_1} + \frac{1 - g_1}{E_2}\right) \tag{6.73}$$

where E_A and E_H = weighted arithmetic and harmonic composite average, respectively, of E_a and E_h as calculated by Eqs. 6.70 through 6.73

A_1 and A_2 = fractional volumes of the simple model elements represented by Eqs. 6.68 and 6.69, respectively; that is, $A_1 + A_2 = 1$

E_a and E_h = simple arithmetic and harmonic average of E_1 and E_2 as calculated by Eqs. 6.68 and 6.69, respectively

The other symbols are as in Eqs. 6.68 and 6.69. Note that Eqs. 6.69, 6.72, and 6.73 are inapplicable for the limit case when E_1 or E_2 is zero, or close to zero, that is, when one of the phases is air.

The following example illustrates the considerable differences among the four averages.

Example 6.1

1. If one wants to obtain a two-phase composite of, say, $E = 200$ with a matrix of $E_m = 100$, this can be achieved either by adding approximately 5% of particles of $E_p = 100,000$ or 10% of particles of $E_p = 10,000$ or 30% of particles of $E_p = 1000$.
2. If $E_1 = 10$, $E_2 = 1$, $A_1 = A_2 = 0.5$, and $g_1 = 0.4$, then $E_a = 4.6$, $E_h = 1.6$, $E_A = 3.1$, and $E_H = 2.4$.

Calculation of the Blending Proportions

When the E_1 and E_2 moduli are given, the blending proportions g_1 and g_2 can be calculated for a specified value of the modulus. From Eq. 6.68,

$$g_{1a} = \frac{E_a - E_2}{E_1 - E_2} \tag{6.74}$$

From Eq. 6.69,

$$g_{1h} = \frac{E_1(E_h - E_2)}{E_h(E_1 - E_2)} = \frac{E_1}{E_h}g_{1a} \tag{6.75}$$

From Eq. 6.71 when $A_1 = 0.5$,

$$g_{1A} = \frac{(2E_A + E_1 - E_2) \pm \sqrt{(2E_A + E_1 - E_2)^2 + 8E_1(E_2 - E_A)}}{2(E_1 - E_2)} \tag{6.76}$$

From Eq. 6.73 when $A_1 = 0.5$,

$$\begin{aligned} &g_{1H} \\ &= \frac{E_H(E_1 - E_2) - 2E_1E_2 \pm \sqrt{[E_H(E_1 - E_2) - 2E_1E_2]^2 + 8E_HE_1E_2(E_H - E_2)}}{2E_H(E_1 - E_2)} \end{aligned} \tag{6.77}$$

where g_{1a}, g_{1h}, etc. = fractional volume of phase 1 for the model of arithmetic average, harmonic average, etc., respectively

E_a, E_h, etc. = specified modulus of elasticity for simple arithmetic
average, simple harmonic average, etc.

For a two-phase composite and a fixed specified value of E, the solution
is always a unique g_1 value. That is, in such cases there is only one composition that can produce the specified value of E. This uniqueness no longer
exists when and $E \pm \Delta E$ range is specified and/or when the composite is
made up of more than two phases. In such cases there are usually infinitely
many different blending proportions that can provide the specified E, as
shown in Problem 6.4. This leaves a certain freedom to the engineer to introduce additional technical or economic conditions for selection of the composition, as shown below.

PROBLEM 6.4 (S. Popovics and Erdey, 1970)   A composite material should
be manufactured from the combination of three materials. The moduli of
elasticity of the three components are $E_1 = 10$, $E_2 = 4$, and $E_3 = 1$. Determine
the *totality* of the possible phase combinations with these components so that
the arithmetic composite average of the composite is

$$E_A = 5.0 \pm 0.5$$

Additional restrictions concern the composition of the composite:

amount of phase 1 $g_1 \leq 60\%$
amount of phase 3 $10\% \leq g_3 \leq 25\%$

Solution   Probably the simplest way to present the solution is in a triangular
system of coordinates (S. Popovics, 1964b). The totality of the phase proportions complying with the restrictions is shown in Figure 6.15 by the points
of the crosshatched area. The construction of the limits of this area is indicated
in the figure. One can see directly that g_{1A} may vary within 27 and 60%
limits, g_{2A} may vary within 15 and 63% limits, and g_{3A} may vary within 10
and 25% limits with the condition that $g_{1A} + g_{2A} + g_{3A} = 100$.

Internal Structure of the Composite Average Models

From a structural standpoint, these two-phase composite models represented
by Eqs. 2.70 through 2.73 are made up of closely packed two-phase composite
elements whose basic type for ACA is shown in Figure 6.13a. Its spring-
model form is presented as Figure 6.14c (S. Popovics, 1980b).

A composite element has a characteristic structure and a characteristic
composition. The structure is characterized by the simultaneous presence of
simple two-phase elements represented by Eq. 6.68, called the *Voigt element,*

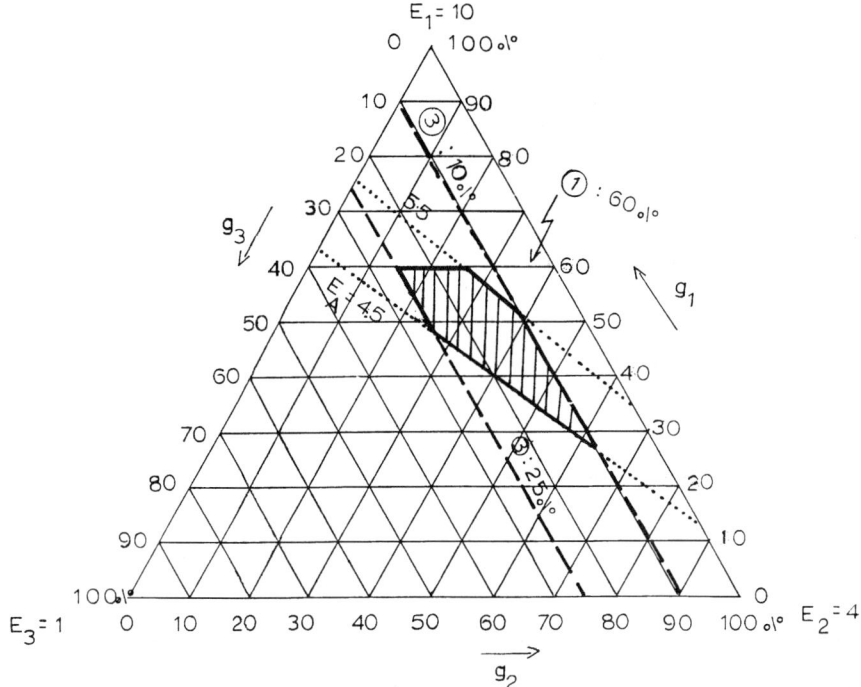

Figure 6.15 The crosshatched area represents the totality of the proportions for three phases which provide $E_A = 5.0 \pm 0.5$ composite modulus, with the restrictions that phase 1 be not more than 60% and phase 3 be within 10 and 25%. (From S. Popovics, 1970d. Copyright RILEM. Reprinted with permission.)

and other simple two-phase elements represented by Eq. 6.69, called the *Reuss element*. This arrangement is a schematic form of the concept that concrete is a *composite of composites* (Bates, 1963). The orientation of all these composite elements in ACA represented by Eq. 6.71 is identical relative to the direction of the uniaxial load; namely, the Voigt elements should be oriented parallel to the direction of the load. In the HCA model (Eq. 6.73) the orientation of all the composite elements is again identical, but here the Voigt elements are perpendicular to the load direction. The volume fraction of phase 1 in both the Voigt and Reuss elements is g_1 and that of phase 2 is $g_2 = 1 - g_1$. The fractional volume of the Voigt elements in the composite material A_1 and that of the Reuss element is $A_2 = 1 - A_1$. Otherwise, the distribution of the simple two-phase elements in the composite model is irrelevant.

Several equivalent geometric arrangements of the internal structure of the two phases for the arithmetic composite average model (Eq. 6.70) under a uniformly distributed load are presented in Figure 6.16*b–p*. It can be seen that arrangement *e* is similar to the model recommended by Counto (1964), and arrangement *p* is the same as the model by Hirsch (1962). Similar phase

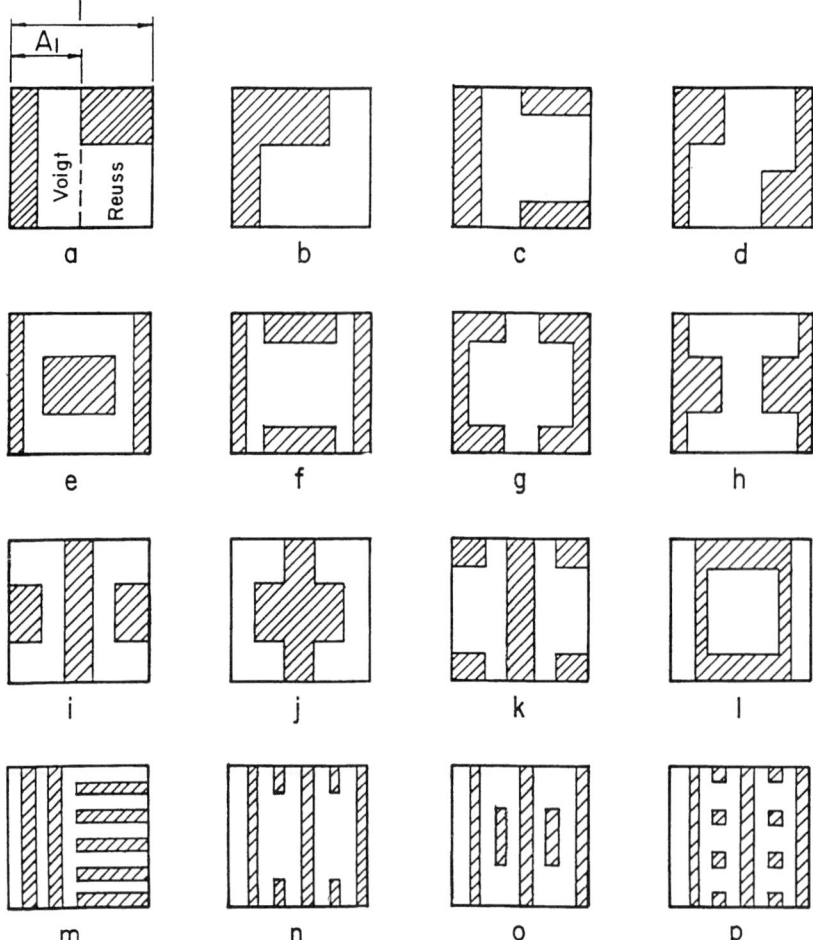

Figure 6.16 Various equivalent forms of the arithmetic composite average model. Mode of loading: uniformly distributed, vertical. (From S. Popovics, 1980c. Copyright Materials Research Society. Reprinted with permission.)

arrangements can be obtained, again for a uniformly distributed load, for the harmonic composite average (Eq. 6.71) by the rotation of the models in Figure 6.16 by 90° relative to the load direction. The spring equivalent of HCA is shown in Figure 6.14*d*.

The fact that composite models of such different internal structures can produce identical deformabilities greatly simplifies the utilization of material models since it makes it unnecessary to reproduce in the model the actual, usually complex phase distribution of the composite. In these composite average models neither the stresses nor the strains are equal in the two phases under a uniaxial load, although they still follow a simple, well-defined pattern.

For instance, in a ACA model the strains are equal (ε') in the Voigt elements (i.e., in the A_1 portion of the model under uniaxial load), whereas in the $1 - A_1$ portion the two phases in Reuss elements have the same stress σ'. The magnitude of A_1 can be used for the representation of the effect of anisotropy or the effect of the preferred orientation of the particle phase. In the case of random orientation, $A_1 = A_2 = 0.5$.

The graphs of Eqs. 6.68 through 6.73 with $A_1 = 0.5$, $E_1 = 10$, and $E_2 = 1$ are presented in Figure 6.17 for various g values. Graph E_g represents the geometric average. It can be seen that E_H is always smaller than E_A under identical circumstances. This means that the HCA, that is, the soft composite model, always provides a smaller modulus of elasticity for a given composition than does the ACA, the hard model.

Experimental Justification

The applicability of composite average models for the prediction of modulus of elasticity of normal-weight and lightweight concretes is demonstrated in Figures 6.18 through 6.22, using test results by Hirsch (1962), Counto (1964),

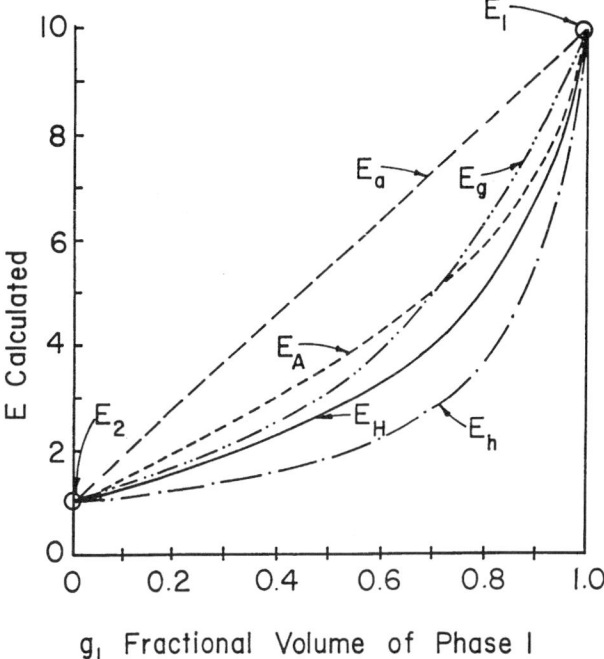

Figure 6.17 Various averages of the E_1 and E_2 moduli of elasticity of two phases as a function of the fractional volume of phase 1. (From S. Popovics, 1980c. Copyright Materials Research Society. Reprinted with permission.)

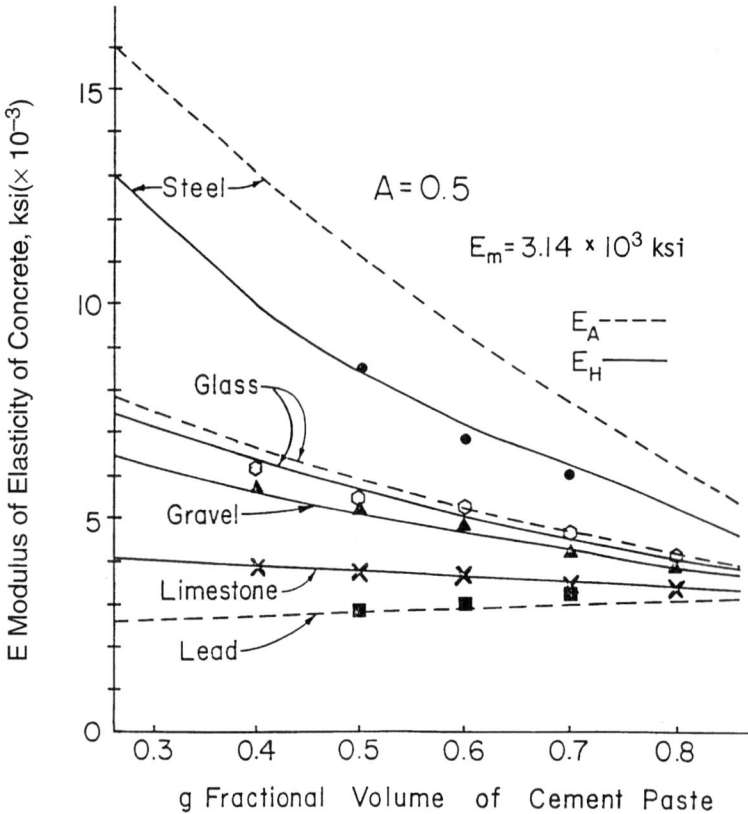

Figure 6.18 Composite averages for the moduli of elasticity of concrete at the age of 28 days as a function of the quantities and moduli of elasticity of aggregate. E_m is modulus of the matrix. Points represent experimental data reported by Hirsch. 1 ksi = 6.90 MPa. (From S. Popovics, 1980c. Copyright Materials Research Society. Reprinted with permission.)

and Stock et al. (1979), respectively. In Figures 6.18 through 6.20, the modulus of elasticity of concrete is plotted against the cement paste content with various aggregates, comparing the measured values to values calculated by composite averages. In Figures 6.21 and 6.22 the effect of the modulus of elasticity of the cement paste and that of the aggregate, respectively, are illustrated on the modulus of elasticity.

The good fit between experimental results with *normal-weight* concretes and Eq. 6.73 was demonstrated earlier (Hirsch, 1962; Dougill, 1962). Several years later it was shown (S. Popovics, 1969b) that the modulus of elasticity of *lightweight aggregate* concretes can be estimated reasonably well by the ACA model even in such an extreme case as polythene aggregate, as shown in Figure 6.19. This revelation is useful from a practical standpoint because it provides a simple model and a corresponding formula for lightweight ag-

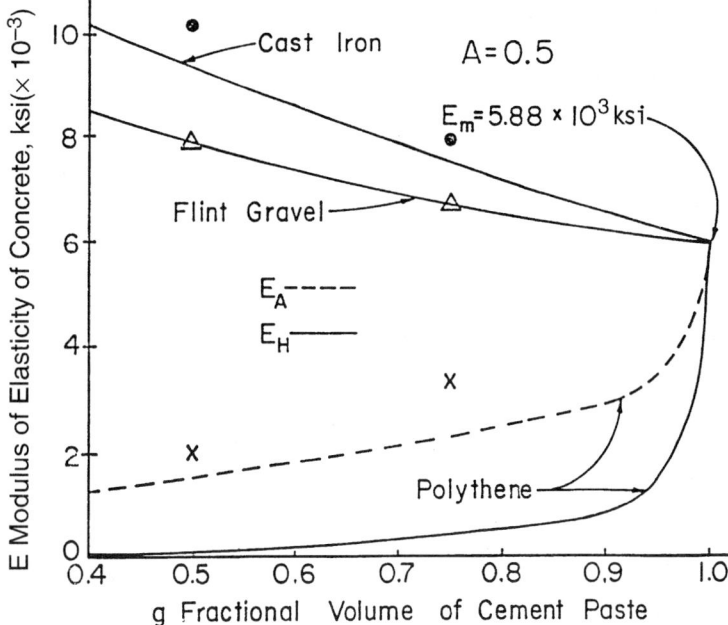

Figure 6.19 Comparison of measured values and composite averages for the modulus of elasticity of concrete as a function of the quantities and moduli of elasticity of aggregates. Solid lines represent the harmonic composite average; the dashed line represents the arithmetic composite average. The experimental data represented by points were published by Counto. 1 ksi = 6.90 MPa. (From S. Popovics, 1980c. Copyright Materials Research Society. Reprinted with permission.)

gregate concretes, and conceptually because the arithmetic composite average is a counterpart of the harmonic composite average. Thus the two elastic deformation models for normal-weight and lightweight concrete, respectively, are synthesized into a single composite average concept.

It also seems significant that the applicability of the two composite average models goes beyond portland cement concretes. For example, Figure 6.23 compares measured values (points) and values calculated by Eqs. 6.71 and 6.73 (lines) for the modulus of elasticity of composites consisting of tungsten carbide and cobalt (Paul, 1960). The figure not only demonstrates a reasonable fit of the estimated values to the measured values but also shows that the composites have higher moduli at identical cobalt contents when the tungsten carbide is the matrix (hard composite) than when the cobalt is the matrix (soft composite).

Figure 6.24 compares experimental values calculated by Eq. 6.73 for boron-reinforced epoxy. Figure 6.25 shows a similar comparison for glass-reinforced epoxy with randomly oriented particle phases using experimental data from Broutman (1967).

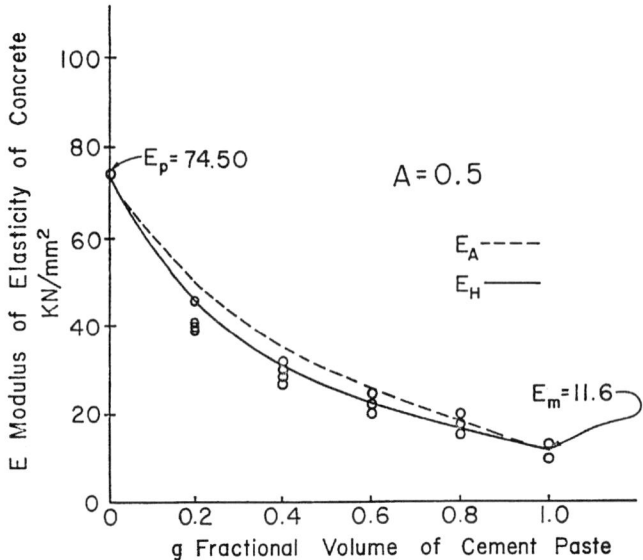

Figure 6.20 Comparison of measured values and composite averages for the modulus of elasticity of a concrete as a function of the quantities and moduli of elasticity of the cement paste and aggregate. Circles represent measured values published by Stock. (From S. Popovics, 1980c. Copyright Materials Research Society. Reprinted with permission.)

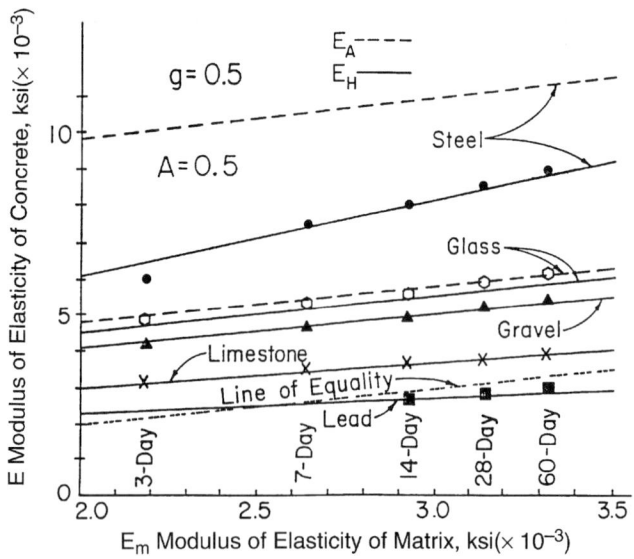

Figure 6.21 Composite average for the moduli of elasticity of concrete when 50% of the concrete volume is matrix, as a function of moduli of elasticity of aggregates and modulus of elasticity of matrix. Points represent experimental data reported by Hirsch. 1 ksi = 6.90 MPa. (From S. Popovics, 1980c. Copyright Materials Research Society. Reprinted with permission.)

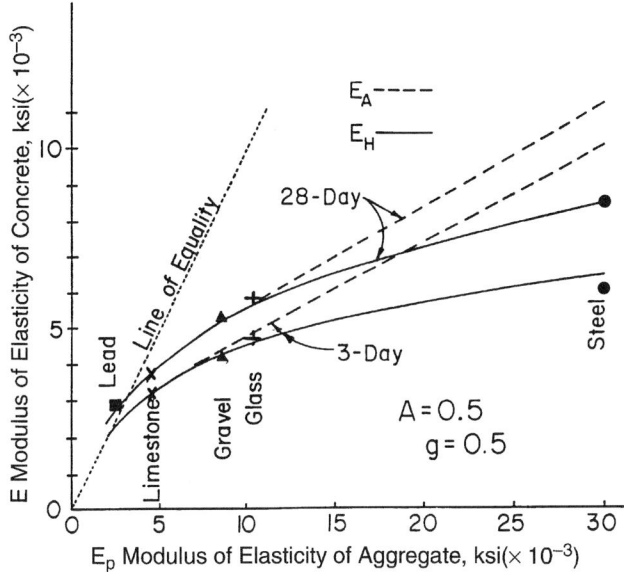

Figure 6.22 Composite averages for the moduli of elasticity of concrete when 50% of the concrete volume is matrix, as a function of the age of concrete and moduli of elasticity of aggregates. Points represent experimental data reported by Hirsch. 1 ksi = 6.90 MPa. (From S. Popovics, 1980c. Copyright Materials Research Society. Reprinted with permission.)

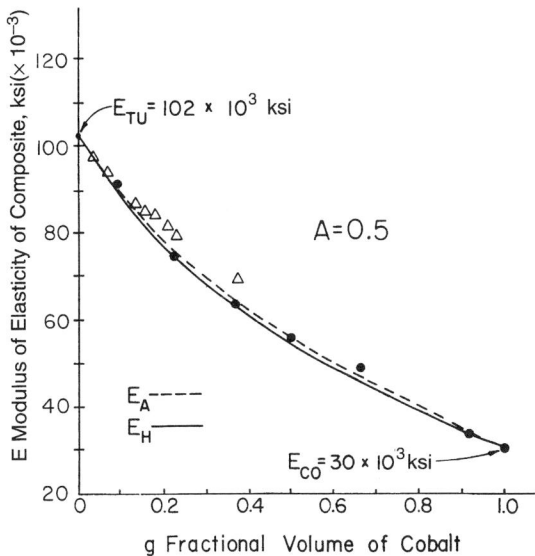

Figure 6.23 Comparison of measured values and composite averages for the modulus of elasticity of a composite as a function of the quantities and moduli of elasticity of tungsten carbide and cobalt. Circles represent measured values with cobalt as the matrix, triangles measured values with tungsten carbide as the matrix. The experimental data were taken from Paul. 1 ksi = 6.90 MPa. (From S. Popovics, 1980c. Copyright Materials Research Society. Reprinted with permission.)

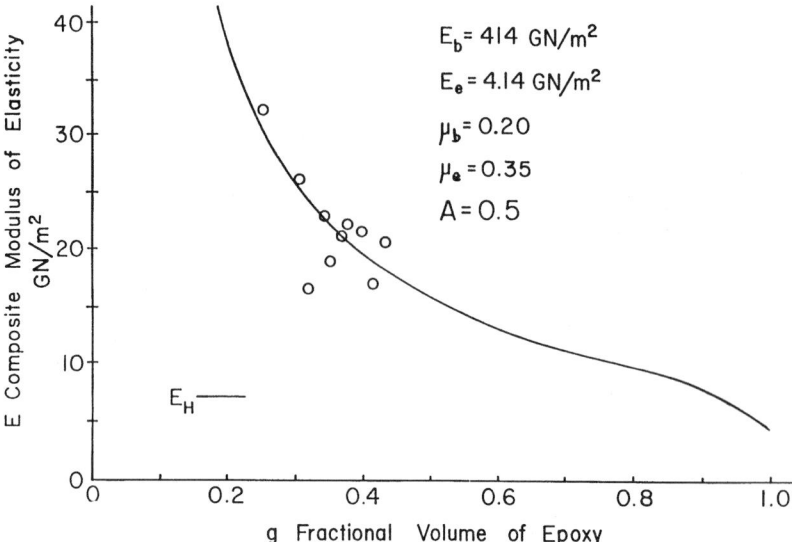

Figure 6.24 Comparison of measured values and composite averages for the modulus of elasticity of boron-reinforced epoxy. The experimental data represented by points were obtained by Whitney and Riley. (From S. Popovics, 1980c. Copyright Materials Research Society. Reprinted with permission.)

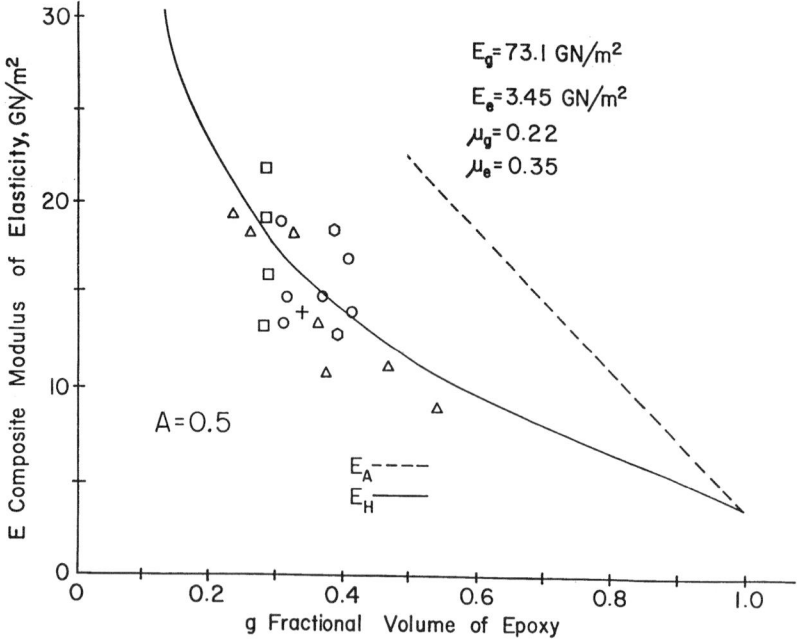

Figure 6.25 Comparison of measured values and composite averages for the modulus of elasticity of glass-reinforced epoxy. The experimental data represented by points were obtained by Tsai. (From S. Popovics, 1980c. Copyright Materials Research Society. Reprinted with permission.)

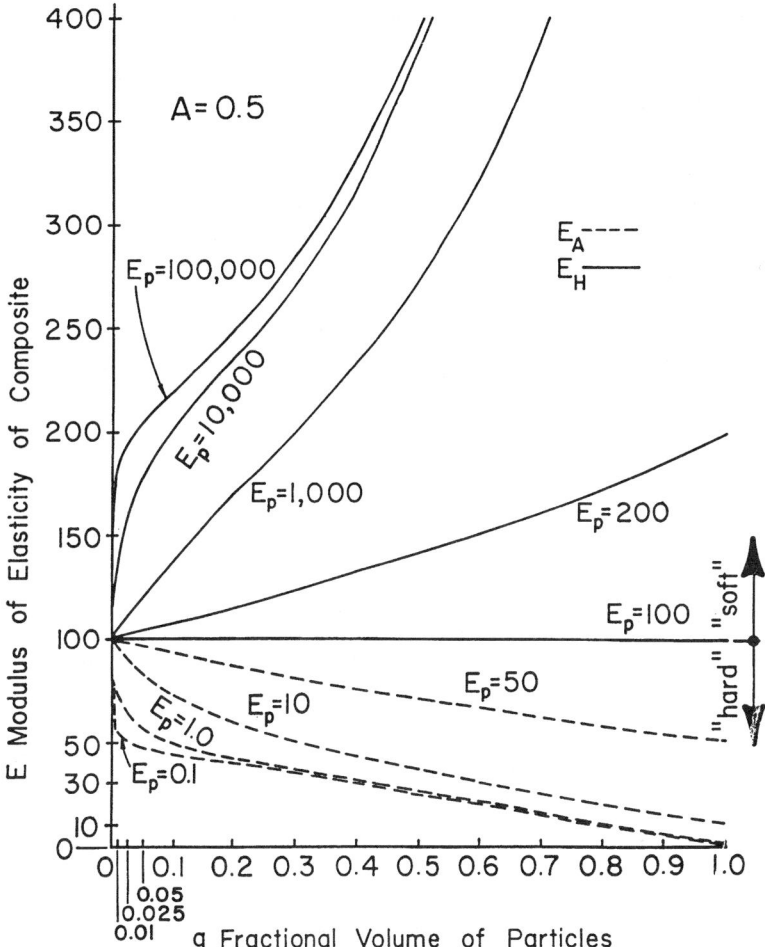

Figure 6.26 Moduli of elasticity of composite average models for soft and hard two-phase composites as a function of the modulus of elasticity and fractional volume of the dispersed phase (particles). (From S. Popovics, 1980c. Copyright Materials Research Society. Reprinted with permission.)

The goodness of fit in Figures 6.16 through 6.25 is just about the best one can expect from a model approach. The discrepancies between calculated and measured values in these figures are not greater than the usual inherent variations of modulus of elasticity measurements. This is at least as good as the fit of much more complicated models based on the theory of elasticity or other complex methods of stress–strain analysis.

Properties of the Composite Average Models

Despite its heuristic nature, the composite average models are attractive because they reproduce the modulus of elasticity of two-phase composites quite

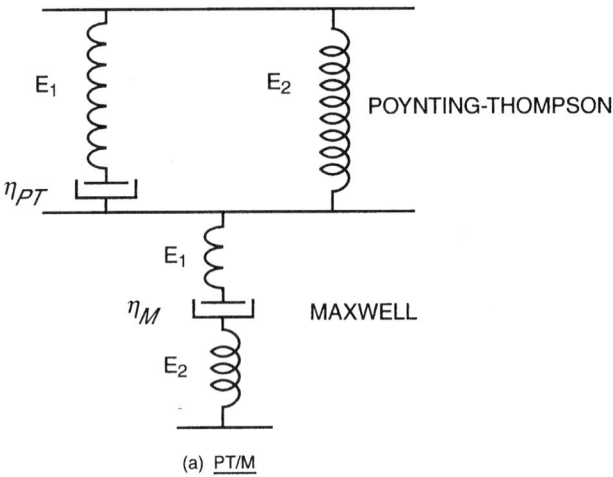

(a) <u>PT/M</u>

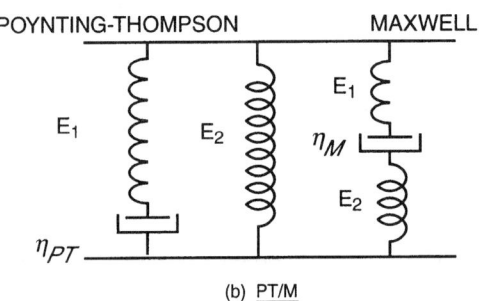

(b) <u>PT/M</u>

Figure 6.27 Rheological models of the generalized composite average concept for instantaneous and time-dependent deformations of two-phase viscoelastic composites. (From S. Popovics, 1987e. Copyright RILEM. Reprinted with permission.)

well and because the concept is simple. Therefore, it may be worthwhile to illustrate some of the special features of these models.

1. The estimated effects of the modulus of elasticity and fractional volume of particles on the modulus of elasticity of a two-phase composite are shown in Figure 6.26. Since the portion of the figure over the $E_m = E_p = 100$ line represents the soft composites and the lower portion the hard composites, the upper part contains only the graphs of HAC and the lower part only the graphs of ACA. It can be seen that the addition of relatively stiff particles, stiffer by, say, three orders of magnitude than the matrix, will rapidly increase the modulus of the composite. This increase is more than the reduction that a similar addition of very weak particles, weaker by three orders of magnitude than the matrix, would cause.

2. A somewhat different picture emerges when the same graphs are replotted in a log E versus g semilog system of coordinates. This shows a symmetry about the $E_m = E_p = 100$ line, indicating that increasing the modulus of elasticity of the particles by three orders of magnitude will produce a change of the same *percentage* in the modulus of the composite as a reduction of E_p.

3. When the E_m/E_p ratio is between approximately 0.5 and 2, the HAC is practically the same as the corresponding ACA.

4. Doubling (or halving) the amount of particles influences the modulus of the composite less when the quantity of the particles is between 1 and 20% than at higher or lower particle contents.

5. The effect of preferred orientation of the particles can be reproduced by proper selection of A_1 in Eqs. 6.70 through 6.73.

6. It is conceivable that the method can be generalized to mechanical properties of composites other than the modulus of elasticity, such as the estimation of shear modulus G, deformability under biaxial and triaxial stresses, and so on. The extension of the model to include load-induced time-dependent deformations (creep, relaxation, delayed elasticity) is illustrated in Figure 6.27 which is a modification of the composite models for E shown in Figure 6.14c and d. In this way, the models can reproduce both the instantaneous and the time-dependent deformations of the composites. Beyond this, however, time-dependent deformation aspects of these rheological models are not discussed in this book. Interested readers can find pertinent details elsewhere (S. Popovics, 1986d, 1987e; Paven et al., 1983; Paven and Popovics, 1991, 1992; 1994).

7. Application of the composite average concept can be extended easily to three phases.

REFERENCES

Abdun-Nur, E. A., Adapting Statistical Methods to Concrete Production, *Proceedings of National Conference on Statistical Quality Control Methodology in Highway and Airfield Construction,* Charlottesville, VA, Nov. 1966, pp. 483–513.

Abdun-Nur, E. A., Accelerated, Early, and Immediate Evaluation of Concrete Quality, *Accelerated Strength Testing,* ACI Publication SP-56, American Concrete Institute, Detroit, MI, 1978, pp. 1–13.

Abeles, P. W., Notes on Materials for Prestressed Concrete, *Concrete and Constructional Engineering,* Vol. 56, Nov., 1960, pp. 19–29.

Abrams, D. A., *Design of Concrete Mixtures,* Bulletin 1, Structural Materials Research Laboratory, Lewis Institute, Chicago, Dec. 1918.

Abrams, D. A., Flexural Strength of Plain Concrete, *Proceedings of the American Concrete Institute,* Vol. 18, 1922, pp. 20–51.

ACI, *Temperature and Concrete,* ACI Publication SP-25, American Concrete Institute, Detroit, MI, 1971.

ACI Committee 209, Prediction of Creep Shrinkage and Temperature Effects in Concrete Structure, *ACI Manual of Concrete Practice,* Part I. 1982a.

ACI Committee 201, Guide for Making Condition Survey of Concrete in Service, ACI 201.1R.68 (revised in 1984), *ACI Journal,* Proc. Vol. 65, No. 11, Nov. 1968, pp. 905–918.

ACI Committee 209, Prediction of Creep, Shrinkage, and Temperature Effects in Concrete Structure, *Designing for Effects of Creep, Shrinkage, Temperature in Concrete Structures,* ACI Publication SP-27, American Concrete Institute, Detroit, MI, 1971a, pp. 51–93.

ACI Committee 211, Standard Practice for Selecting Proportions for Normal, Heavy-weight, and Mass Concrete, ACI Standard ACI 211.1-91, 1991.

ACI Committee 211, Standard Practice for Selecting Proportions for No-Slump Concrete, ACI Standard ACI 211.3-75, revised 1987, reapproved 1992b.

ACI Committee 211, Guide for Selecting Proportions for High-Strength Concrete with Portland Cement and Fly Ash, ACI Standard ACI 211.4R-93, 1993.

ACI Committee 214, Recommended Practice for Evaluation of Strength Test Results of Concrete, ACI Standard ACI 214-77, 1977.

ACI Committee 214, *Accelerated Strength Testing,* ACI Publication SP-56, American Concrete Institute, Detroit, MI, 1978.

ACI Committee 224, Control of Cracking in Concrete Structures, *Concrete International: Design and Construction,* Vol. 2, No. 10, Oct. 1980, pp. 35–76.

ACI Committee 224, Causes, Evaluation, and Repair of Cracks in Concrete Structures, *ACI Journal,* Proc. Vol. 81, No. 3, May–June 1983a, pp. 211–229.

ACI Committee 225, Guide to the Selection and use of Hydraulic Cements, *ACI Journal,* Vol. 82, No. 6, Nov.–Dec. 1985, pp. 901–929.

ACI Committee 228, In-Place Methods for Determination of Strength of a Concrete, *ACI Materials Journal,* Vol. 85, No. 5, Sept.–Oct. 1988a, pp. 446–471.

ACI Committee 228, In-Place Methods to Estimate Concrete Strength, ACI Standard ACI 228.1R-95, 1996.

ACI Committee 305, Hot Weather Concreting, *ACI Journal,* Proc. Vol. 74, No. 8, Aug. 1982, pp. 317–332.

ACI Committee 305, Making Good Concrete in Hot Weather, *Concrete International,* Vol. 14, No. 4, Apr. 1992, pp. 55–57.

ACI Committee 306, Proposal Standard Specification for Cold Weather Concreting, *ACI Journal,* Vol. 83, No. 6, 1986, pp. 1043–1047.

ACI Committee 306, Cold Weather Concreting, *ACI Journal,* Vol. 85, No. 4, 1988b, pp. 280–302.

ACI Committee 318, Building Code Requirements for Reinforced Concrete, ACI Standard ACI 318-83, American Concrete Institute, Detroit, MI, 1983b.

ACI Committee 318, Building Code Requirements for Reinforced Concrete, ACI Standard ACI 318-89, revised 1992c.

ACI Committee 363, State-of-the-Art Report on High-Strength Concrete, *ACI Journal,* Proc., Vol. 81, No. 4, July–Aug. 1984, pp. 364–411.

ACI Committee 435, Deflection of Prestressed Concrete Members, *ACI Journal,* Proc. Vol. 60, No. 12, Dec. 1963, pp. 1697–1728.

ACI Committee 437, Strength Evaluation of Existing Concrete Buildings, ACI Standard ACI 437R-67 (revised in 1982), American Concrete Institute, Detroit, MI, 1982.

ACI Committee 516, *Menzel Symposium on High Pressure Steam Curing,* ACI Publication SP-32, American Concrete Institute, Detroit, MI, 1972, p. 282.

ACI Committee 517, Accelerated Curing of Concrete at Atmospheric Pressure—State of the Art, *ACI Journal,* Proc. Vol. 77, No. 6, Nov.–Dec. 1980, pp. 429–448.

Ackroyd L. W., and Rhodes, F. G., An Investigation of the Crushing Strengths of Concrete Made with Three Different Cements in Nigeria, *Proceedings, Institution of Civil Engineers,* Vol. 27, Feb. 1964, pp. 325–340.

Adler, L., Rose, J. H., and Molbey, C., Ultrasonic Method to Detect Gas Porosity in Aluminum Alloy Casting: Theory and Experiment, *Journal of Applied Physics,* Vol. 59, No. 2, 1986, pp. 36–347.

Agbabian, M. S., and Masri, S. F. (Eds.), Nondestructive Evaluation for Performance of Civil Structures—1988, *Proceedings of the International Workshop,* University of Southern California, Los Angeles, Feb. 1988, p. 431.

Akashi, T., On the Measurement of Velocity and Loss of Ultrasonic Pulse in Concrete, *Proceedings of the Third International Conference on Nondestructive Testing,* Tokyo and Osaka, Mar. 1960.

Akashi, T., and Amaski, S., Study of the Stress Waves in the Plunger of a Rebound Hammer at the Time of Impact, *In Situ/Nondestructive Testing of Concrete,* V. M. Malhotra (Ed.), ACI Publication SP-82, American Concrete Institute, Detroit, MI, 1984.

Akazawa, T., Tension Test Method for Concretes, *RILEM Bulletin,* No. 16 (Old Series), Nov. 1953, pp. 13–23.

Akroyd, T. N. W., The Accelerated Curing of Concrete Test Cubes, *Journal of the Institution of Civil Engineers,* Proc. Vol. 19, May 1961a, pp. 1–22.

Akroyd, T. N. W., Concrete Under Triaxial Stress, *Magazine of Concrete Research,* Vol. 13, No. 39, Nov. 1961b, pp. 111–118.

Aktan, A., Farhey, D., Hunt, V., Helmicki, A., Brown, D., and Shelley, S., Objective Bridge Condition Assessment, International Symposium, *Non-destructive Testing in Civil Engineering* (NDT-CE), Vol. 1, G. Schickert and H. Wiggenhouser (Eds.), BAM, Berlin, Sept. 26–28, 1995, pp. 51–59.

Akutsu, K., Study of the Admixture for Aerated Concretes Including Maleic Anhydride Modified, *Proceedings of the Fifth International Symposium on the Chemistry of Cement,* Part IV, Admixtures and Special Cements, Tokyo, Dec. 1969, pp. 65–73.

Alexander, K. M., Effect of Water/Cement Ratio on the Contribution Made by Powdered Admixtures to the Strength of Blended Portland Cement Mortars, *Australian Journal of Applied Science,* Vol. 6, No. 1, Mar. 1955, pp. 61–77.

Alexander, K. M., Strength of the Cement–Aggregate Bond, *ACI Journal,* Proc. Vol. 56, No. 11, Nov. 1959, pp. 377–390.

Alexander, K. M., The Relationship Between Strength and the Composition and Fineness of Cement, *Cement and Concrete Research,* Vol. 2, No. 6, Nov. 1972, pp. 663–680.

Alexander, K. M., Taplin, J. H., and Wardlaw, J., Correlation of Strength and Hydration with Composition of Portland Cement, *Proceedings of the Fifth International Symposium on the Chemistry of Cement,* Part III, Properties of Cement Paste and Concrete, Tokyo, Dec. 1969, pp. 152–166.

Alexander, S., A Single Equation for the Stress–Strain Curve of Concrete, *Indian Concrete Journal,* Vol. 39, No. 7, July 1965, pp. 274–277.

Alford, N. M., Physical Properties of High Strength Cement Pastes, *Cement and Concrete Research,* Vol. 12, No. 3, May 1982, pp. 349–358.

Alford, N. M., and Rahman, A. A., An Assessment of Porosity and Pore Sizes in Hardened Cement Pastes, *Journal of Materials Science,* Vol. 16, 1981, pp. 3105–3114.

Alldred, J., Chua, J., and Chamberlain, D., Determination of Reinforcing Bar Diameter and Cover by Analysing Travers Profiles from a Cover Meter, *International Symposium Non-Destructive Testing in Civil Engineering* (NDT-CE), G. Schickert, H. Wiggenhauser, (Eds.), Berlin, 1995, pp. 721–728.

Alongi, A. V., Cantor, T. R., and Alongi, A. V., Jr., Concrete Evaluation by Radar—Theoretical Analysis, *Concrete Analysis and Deterioration,* Transportation Research Record No. 853, Transportation Research Board, Washington, DC, 1982, pp. 31–37.

Andenaes, E., Gerstle, K., and Ko, H. Y., Response of Mortar and Concrete to Biaxial Compression, *Journal of the Engineering Mechanics Division, ASCE,* Vol. 103, No. EM4, Proc. Paper 13115, Aug. 1977, pp. 515–526.

Andersen, J., and Nerenst, P., Wave Velocity in Concrete, *ACI Journal,* Proc. Vol. 48, Apr. 1952, pp. 613–636.

Andersen, P., Experiments with Concrete in Torsion, *Transactions, ASCE,* Vol. 100, 1935, pp. 949–960.

Anderson, O. L., The Griffith Criterion for Glass Fracture, *Fracture,* B. L. Averbach et al. (Eds.), Technology Press of MIT/John Wiley, New York, 1959, pp. 331–353.

Anon., *Proceedings, National Conference on Statistical Quality Control Methodology in Highway and Airfield Construction,* University of Virginia School of General Studies, Charlottesville, VA, Nov. 1966.

Anon., Anleitung fur die Bestimmung von luftporenkennwerten am Festbeton—Mikroskopische Luftporen-untersuchung, *Betontechnische Berichte 1980/81,* Belton-Verlag, Dusseldorf, 1982, pp. 180–188.

Anson, M., An Investigation into a Hypothetical Deformation and Failure Mechanism for Concrete, *Magazine of Concrete Research,* Vol. 6, No. 47, June 1964, pp. 73–82.

Anson, M., and Newman, K., The Effect of Mix Proportions and Method of Testing on Poisson's Ratio for Mortars and Concretes, *Magazine of Concrete Research,* Vol. 18, No. 56, Sept. 1966, pp. 115–130.

Antrim, J. D., The Mechanics of Fatigue in Cement Paste and Plain Concrete, *Highway Research Record No. 210,* Highway Research Board, Washington, DC, 1967, pp. 95–107.

Armuth, A., Kis-cementadagolasu betonok (Concretes of Low Cement Content), *ETI Kutatasi Jelentes,* Budapest, 1971.

Armuth, A., Szuk, G., and Kochis, B., A labatlani epuletelemgyarban keszult feszitett vasuti-aljak betontechnologiai ellenorzese (Concrete Technological Control of Pre-stressed Cross Ties), *Epitestudomanyi Intezet Jelentese,* No. 22, Budapest, 1962.

Arni, H. T., Impact and Penetration Tests of Portland Cement Concrete, Nondestructive Testing of Concrete, Highway Research Record No. 378, Highway Research Board, Washington, DC, 1972, pp. 55–67.

ASTM, *Symposium on Application of Statistics,* ASTM STP No. 103, Philadelphia, 1950.

ASTM, *Symposium on Nuclear Methods for Measuring Soil Density and Moisture,* ASTM STP No. 293, Philadelphia, 1960.

ASTM *Statistical Methods for Quality Control of Road and Paving Materials,* ASTM STP No. 362, Philadelphia, 1964.

ASTM Committee C-9, Bibliography on Mixing Concrete, *Proceedings, ASTM,* Vol. 30, Part I, 1930, pp. 598–626.

ASTM Committee E-11, *ASTM Manual on Presentation of Data and Control Chart Analysis,* ASTM STP 15D, Philadelphia, 1976.

Atchley, B. L., and Furr, H. L., Strength and Energy Absorption Capabilities of Plain Concrete Under Dynamic and Static Loadings, *ACI Journal,* Proc. Vol. 64, No. 11, Nov. 1967, pp. 745–756.

Avram, C., Facaoaru, I., Filtmon, I., Mirsu, O., and Tertea, I., *Concrete Strength and Strain,* Elsevier, Amsterdam, 1981.

Axon, E. O., A Method of Estimating the Original Mix Composition of Hardened Concrete Using Physical Tests, *Proceedings, ASTM,* Vol. 62, 1962, pp. 1068–1080.

Babic, B., Relationships Between Mechanical Properties of Cement Stabilized Materials, *Materials and Structures—Design and Construction,* RILEM, Vol. 20, No. 120, Nov. 1987, pp. 455–460.

Bache, H. H., and Nepper-Christensen, P., Observations on Strength and Fracture in Lightweight and Ordinary Concrete, *The Structure of Concrete, Proceedings of an International Conference,* London, Sept. 1968, pp. 93–108.

Bahn, B. Y., and Hsu, Ch-T. T., Stress-Strain Behaviour of Concrete Under Cyclic Loading, *ACI Materials Journal,* Proc. Vol. 95, No. 2, Mar.–Apr. 1998.

Baker, A. L. L., An Analysis for Deformation and Failure Characteristics of Concrete, *Magazine of Concrete Research,* Vol. 11, No. 33, Nov. 1959, pp. 119–128.

Baker, W. M., and McMahon, T. F., Quality Assurance in Highway Construction: Part 3, Quality Assurance of Portland Cement Concrete, *Public Roads,* Vol. 35, No. 8, June 1969, pp. 184–189.

Balaguru, P., and Ramakrishnan, V. (Eds.), *Computer Use for Statistical Analysis of Concrete Test Data,* ACI Publication SP-101, American Concrete Institute, Detroit, MI, 1987a.

Balaguru, P. N., and Ramakrishnan, V., Criteria for Estimating the Required Average Strength f'_{cr} to Comply with the Specified f'_c, *ACI Materials Journal,* Vol. 84, No. 1, Jan.–Feb. 1987b, pp. 35–41.

Balazs, Gy., A tartosterheles hatasa a beton toroszilardsagara es rugalmassagi modulusara (The Effect of Sustained Load on the Compressive Strength and Modulus of Elasticity of Concrete), *Minosegi Beton Konferencia,* Budapest, 1960, pp. 257–261.

Balazs, G. L., Eroatadas betonban (Force transfer in Concrete), Thesis, Budapest, 1991.

Barker, M. G., and Ramirez, J. A., Determination of Concrete Strengths with Break-off Tester, *ACI Materials Journal,* Vol. 85, No. 4, July–Aug. 1988, pp. 221–228.

Barovsky, N., Bozhinov, G., and Encheva, Z. H., Experimental Design for Investigation of Pore Structure of Cement Stone, Mechanics and Technology of Composite Materials, *Proceedings of the Second National Conference,* Varna, Oct. 1–3, 1979, Publishing House of the Bulgarian Academy of Sciences, Sofia, 1979, pp. 534–537.

Barros, R. T., Weed, R. M., and Willenbrock, J. H., Software Package for Design and Analysis of Acceptance Procedures Based on Percent Defective, *Transportation Research Record No. 924,* Transportation Research Board, Washington, DC, 1983, pp. 85–92.

Bartlett, F. M., and MacGregor, J. G., Equivalent Specified Concrete Strength from Core Test Data, *Concrete International,* Vol. 17, No. 3, Mar. 1995, pp. 52–58.

Bartlett, F. M., and MacGregor, J. G., Statistical Analysis of the Compressive Strength of Concrete in Structures, *ACI Materials Journal,* Vol. 93, No. 2, Mar.–Apr. 1996, pp. 158–168.

Barton, W. R., Cement, *Mineral Facts and Problems,* 1965 ed., Bureau of Mines Bulletin 630, U.S. Department of the Interior, Washington, DC, 1965.

Bartos, P. (Ed.), *Bond in Concrete,* Applied Science Publishers, London, 1982.

Basalla, A., *Baupraktische Betontechnologie* (Construction Concrete Technology), Bauverlag, Wiesbaden, 1965.

Batchelder, G. M., and Lewis, D. W., Comparison of Dynamic Methods of Testing Concretes Subjected to Freezing and Thawing, *Proceedings ASTM,* Vol. 53, 1953, pp. 1053–1065.

Bates, A. A., Composites in Construction, Stanton Walker Lecture Series on the Materials Sciences, No. 1, University of Maryland, Nov. 1963.

Bates, P. H., and Klien, A. A., *Properties of the Calcium Silicates and Calcium Aluminate Occurring in Normal Portland Cement,* Technologic Paper No. 78, National Bureau of Standards, Washington, DC, June 9, 1917.

Bauer, L. A., and Olivan, I. L., Use of Accelerated Tests for Concrete Made with Slag Cement, *Accelerated Strength Testing,* ACI Publication SP-56, American Concrete Institute, Detroit, MI, 1978, pp. 117–128.

Bazant, Z. P., Instability, Ductility, and Size Effect in Strain Softening Concrete, *Journal of the Engineering Mechanics Division, ASCE,* Vol. 102, No. EM2, Proc. Paper 12042, Apr. 1976, pp. 331–344.

Bazant, Z. P., Mathematical Models for Creep and Shrinkage of Concrete, *Creep and Shrinkage in Concrete Structures,* Z. P. Bazant and F. H. Wittmann (Eds.), John Wiley, London, 1982, pp. 163–255.

Bazant, Z. P., and Chuan Charn, J., Log Double Power Law for Concrete Creep, *ACI Journal,* Proc. Vol. 82, No. 5, Sept.–Oct. 1985, pp. 665–675.

Bazant, Z. P., and Panula, L., Practical Prediction of Time-Dependent Deformations of Concrete, *Materials and Structures—Research and Testing,* RILEM, Vol. 11, No. 65, Sept.–Oct. 1978, pp. 307–328.

Bazant, Z. P., and Sener, S., Size Effect in Pullout Tests, *ACI Materials Journal,* Vol. 85, No. 5, Sept.–Oct. 1988, pp. 347–351.

Bazant, Z. P., and Sun, H.-H., Size Effect in Diagonal Shear Failure: Influence of Aggregate Size and Stirrups, *ACI Materials Journal,* Vol. 84, No. 4, July–Aug. 1987, pp. 259–272.

Bazant, Z. P., Kazemi, M. T., Hasegawa, T., and Mazars, J., Size Effect in Brazilian Split-Cylinder Tests: Measurements and Fracture Analysis, *ACI Materials Journal,* Vol. 88, No. 3, May–June 1991, pp. 325–331.

Bazant, Z. P., Gu, W., and Faber, K. T., Softening Reversal Another Effect of a Change in Loading Rate on Fracture of Concrete, *ACI Material Journal,* Vol. 92, No. 1, Jan.–Feb. 1995, pp. 3–9.

Beaudoin, J. J., and MacInnis, C., The Effect of Admixtures on the Strength–Porosity Relationship of Portland Cement Paste, *Cement and Concrete Research,* Vol. 1, No. 1, Jan. 1971, pp. 3–11.

Beauzee, M. C., Errors of Measurement in the Determination of the Modulus of Elasticity by the Sonic Method, *RILEM International Symposium on Nondestructive Testing of Materials and Structures, Paris,* Vol. 1, 1954, pp. 120–136.

Bellamy, C. J., Strength of Concrete Under Combined Stress, *ACI Journal,* Proc. Vol. 58, No. 10, Oct. 1961, pp. 367–381.

Bentz, D. P., and Garboczi, E. J., Simulation Studies of the Effects of Mineral Admixtures on the Cement Paste–Aggregate Interfacial Zone, *ACI Materials Journal,* Vol. 88, No. 5, Sept.–Oct. 1991, pp. 518–529.

Bentz, D. P., Garboczi, E. J., and Coverdale, R. T., *Computational Materials Science of Cement-Based Materials—An Educational Module,* NIST, Technical Note 1405, 1993.

Bentz, D. P., and Clifton, J. R., and Synder, K. A., Predicting Service Life of Chloride-Exposed Steel-Reinforced Concrete, *Concrete International,* Vol. 18, No. 12, Dec. 1996, pp. 42–47.

Beres, L., Investigation of Structural Loosening of Compressed Concrete, *RILEM Bulletin,* No. 36 (New Series), Sept. 1967, pp. 185–190.

Beres, L., Relationship of Deformational Processes and Structure Changes in Concrete, Structure, Solid Mechanics and Engineering Designs, *Proceedings of the Southampton 1969 Civil Engineering Materials Conference,* Part 1, Wiley-Interscience, London, 1971, pp. 643–652.

Berg, O. Ya., K Voprosu O. Prochnosti I Plastichnosti Betona (On the Problem of Strength and Plasticity of Concrete), *Dokladi Akademii Nauk USSR,* Vol. 70, No. 4, 1950, pp. 617–620.

Berg, O. Ya., Research on the Concrete Strength Theory, *Building Research and Documentation, Contributions and Discussions,* First CIB Congress, Rotterdam, 1959, pp. 60–69.

Berio, A., General Report of the RILEM Symposium by Correspondence, Accelerated Hardening of Concrete with View to Rapid Control Tests, *RILEM Bulletin,* No. 31, June 1966, pp. 158–166.

Berhardt, C. J., Hardening of Concrete at Different Temperatures, RILEM Symposium, Winter Concreting Theory and Practice, Session BII, *Danish National Institute of Building Research, Special Report,* Copenhagen, 1956a.

Bernhardt, C. J., Damage Due to Freezing of Fresh Concrete, *ACI Journal,* Proc. Vol. 52, Jan. 1956b, pp. 537–580.

Berry, E. E., and Malhotra, V. M., Fly Ash for Use in Concrete—A Critical Review, *ACI Journal,* Proc. Vol. 77, No. 2, Mar.–Apr. 1980, pp. 59–73.

Berthier, R. M., Mesures comparées de traction sur ciments et mortiers (Comparative Measurements of Tensile Strength on Cements and Mortars), *Revue des Materiaux de Construction,* Nos. 431–432, 1951, pp. 241–247.

Best, C. H., A Universal Test Specimen for Concrete, *RILEM Symposium on the Experimental Research of Field Testing of Concrete,* Trondheim, Norway, Oct. 5–7, 1964, pp. 118–134.

Bickley, J. A., Practical Application of the Maturity Concept to Determine In Situ Strength of Concrete, *Recent Developments in Accelerated Testing and Maturity,* Transportation Research Record No. 558, Transportation Research Board, Washington, DC, 1975.

Bickley, J. A., Accelerated Concrete Strength Testing at the CN Tower, *Accelerated Strength Testing,* ACI Publication SP-56, American Concrete Institute, Detroit, MI, 1978, pp. 29–38.

Bickley, J. A., The Variability of Pullout Tests and In-Place Concrete Strength, *Concrete International: Design and Construction,* Vol. 4, No. 4, Apr. 1983, pp. 44–51.

Bickley, J. A., The Evaluation and Acceptance of Concrete Quality by In-Place Testing, *In Situ/Nondestructive Testing of Concrete,* V. M. Malhotra (Ed.), ACI Publication SP-82, American Concrete Institute, Detroit, MI, 1984, pp. 84–110.

Bickley, J. A., and Mukherjee, P. K., Cement-Content Measurements with the Rapid-Analysis Machine, *Superplasticizers in Concrete,* Transportation Research Record No. 720, Transportation Research Board, Washington, DC, 1979, pp. 40–44.

Bilgeri, P., Giesbrecht, P., and Rempe, M., Einfluss der Lagerungsbedingungen und der Zementart auf die Betondruckfestigkeit, *Beton Informationen,* May 1991, pp. 54–58.

Biot, M. A., Theory of Propagation of Elastic Waves in a Fluid-Saturated Porous Solid, *Journal of the Acoustic Society of America,* Vol. 28, 1956, pp. 168–191.

Bisaillon, A., Accelerated Strength Test Results from Expanded Polystyrene Molds with Emphasis on Initial Concrete Temperature, *Accelerated Strength Testing,* ACI Publication SP-56, American Concrete Institute, Detroit, MI, 1978, pp. 201–228.

Bisaillon, A., Frechette, G., and Keyser, J. H., Field Evaluation of Expanded Polystyrene Molds for Self-Cured, Accelerated Strength Testing of Concrete, *Recent Developments in Accelerated Testing and Maturity of Concrete,* Transportation Research Record No. 558, Transportation Research Board, Washington, DC, 1975, pp. 50–60.

Bischoff, P. H., and Perry, S. H., Compressuve Behaviour of Concrete at High Strain Rates, *Materials and Structures—Research and Testing,* RILEM, Vol. 24, No. 144, Paris, Nov. 1991, pp. 425–450.

Blackman, J. S., Smith, G. M., and Young, L. E., Stress Distribution Affects Ultimate Tensile Strength, *ACI Journal,* Proc. Vol. 55, No. 6, Dec. 1958, pp. 679–684.

Blaine, R. L., Arni, H. T., and Defore, M. R., Compressive Strength of Test Mortars, *Interrelations Between Cement and Concrete Properties,* Part 3, Building Science Series 8, National Bureau of Standards, Washington, DC, 1968a, pp. 1–65.

Blaine, R. L., Arni, H. T., and Defore, M. R., Compressive Strength of Steam-Cured Portland Cement Mortars, Section 8, *Interrelations Between Cement and Concrete Properties,* Part 3, Building Science Series 8, National Bureau of Standards, Washington, DC, 1968b, pp. 67–98.

Blakey, F. A., Mechanism of Fracture of Concrete, *Nature,* Vol. 170, No. 4339, Dec. 27, 1952, p. 1120.

Blakey, F. A., and Beresford, F. D., *Tensile Strains in Concretes,* Part 1, Report No. C2.2-1, Part I, Division of Building Research, Melbourne, 1953.

Blakey, F. A., and Beresford, F. D., Strain Distribution in Unreinforced Concrete Beams, *Civil Engineering and Public Works Review,* Vol. 50, Apr. 1955, pp. 415–416.

Bloem, D. L., Concrete Strength Measurement—Cores vs. Cylinders, *Proceedings, ASTM,* Vol. 65, 1965, pp. 668–696.

Bloem, D. L., Concrete Strength in Structures, *ACI Journal,* Proc. Vol. 65, No. 3, Mar. 1968, pp. 176–187.

Blondiau, L., Qualité et applications des ciments de haut fournaeu belges (Quality and Application of Belgian blast-Furnace Slag Cements), *Communication au Congres de la Production et des Application des Cements Siderugiques,* Faculte des Ingenieurs de I'Universite de Neaples, May 1960.

Bobrov, B. S., Mutual Influence of $3CaO–SiO_2$ and $4CaO–Al_2O_3–Fe_2O_3$ in Portland Cement Hydration, Supplementary Paper II-2, *VIth International Congress on the Chemistry of Cement,* Moscow, Sept. 1974.

Bogue, R. H., Calculations of Compounds in Portland Cement, *Industrial and Engineering Chemistry,* Anal. Ed., Vol. 1, Oct. 15, 1929, pp. 192–195.

Bogue, R. H., and Lerch, W., The Hydration of Portland Cement Compounds, *Industrial and Engineering Chemistry*, Vol. 26, Aug. 1934, pp. 837–847.

Bolomey, J., Bestimmung der Druckfestigdeit von Mortel und Beton (Determination of the Compressive Strength of Mortar and Concrete), *Schweitzerische Bauzeitung*, Vol. 88, Nos. 2 and 3, 1926, pp. 41–44, 55–59.

Bolomey, J., Module d'élasticité du beton (Modulus of Elasticity of Concrete), *Bulletin Technique de la Suisse Romand*, Nos. 17 and 18, 1939.

Bombled, J. P., and Kalvenes, O., Comportement rheologique des pâtes, mortiers et betons: mesure, evolution, influence de certains parametres (Rheological Behavior of Pastes, Mortars and Concretes: Measurement, Evolution, Influence of Certain Parameters), *Revue des Materiaux de Construction, Ciments et Betons*, Paris, No. 617, Feb. 1967, pp. 39–52.

Bonzel, J., Zur Gestaltsabhangigkeit der Betondruckfestigkeit (Form Dependence of the Compressive Strength of Concrete), *Beton- und Stahlbetonbau*, Vol. 54, Nos. 9 and 10, Sept. and Oct. 1959, pp. 223–228, 247–248.

Bonzel, J., Uber die Bigezugfestigkeit des Betons (About the Flexural Strength of Concrete), *Betontechnische Berichte 1963*, Beton-Verlag, Dusseldorf, 1964, pp. 59–83.

Bonzel, J., Uber die Spaltzugfestigkeit des Betons (Determination of the Tensile-Splitting Strength of Concrete), *Betontechnische Berichte 1964*, Beton-Verlag GmbH, Dusseldorf, 1965, pp. 59–96.

Bonzel, J., and Dahms, J., Der Einfluss des Zements, des Wasserzementwertes und der Lagerung auf die Festigkeitsentwicklung des Betons (The Effect of Cement, Water–Cement Ratio, and Curing on the Strength Development of Concretes), *Betontechnische Berichte 1966*, Beton-Verlag, Dusseldorf, 1967, pp. 115–138.

Bonzel, J., and Manns, W., Beurteilung der Betondruckfestigkeit mit Hilfe von Annahmekennlinien (The Assessment of Concrete Compressive Strength by Means of Assumed Characteristic Curves), *Betontechnische Berichte 1969*, Beton-Verlag, Dusseldorf, 1970, pp. 85–114.

Bonzel, J., and Siebel, E., Bestimmung von Luftporen-kennwerten am Festbeton (Determination of the Parameters of the Air-Void System in Hardened Concrete), *Betontechnische Berichte 1980/81*, Beton-Verlag, Dusseldorf, 1982, pp. 169–179.

Borjan, J., *Roncsolasmentes betonvizsgalatok* (Nondestructive Testing of Concrete), Maszaki Konyvkiado, Budapest, 1981.

Bossi, J., Recherches expérimentales sur le comportment des eprouvettes de beton a l'essai bresilien (Experimental Research on the Behavior of Concrete Specimens in the Brazilian Test), *RILEM Bulletin*, No. 22, Mar. 1964, pp. 9–35.

Boundy, C. A. P., and Hondros, G., Rapid Field Assessment of Strength of Concrete by Accelerated Curing and Schmidt Rebound Hammer, *ACI Journal*, Proc. Vol. 61, No. 1, Jan. 1964, pp. 77–84.

Brander, M. E., Cold Weather Early Age Strength Measurement by Duplicating Structure Temperatures in Remotely Cured Test Cylinders, *Concrete International: Design and Construction*, Vol. 2, No. 12, Dec. 1980, pp. 52–54.

Brandtzaeg, A., Failure of a Material Composed of Non-isotropic Elements, *Det. Kgl. Norske Videnskabers Selskabs Skrifter*, 1927, Nr. 2, Trondjem, 1927.

Branson, D. E., and Christianson, M. L. Time-Dependent Concrete Properties Related to Design-Strength and Elastic Properties, Creep and Shrinkage, *ACI Symposium on Creep, Shrinkage and Temperature Effect,* Vol. SP-27, 1971, pp. 257–277.

Bremner, T. W., and Holm, T. A., Elastic Compatibility and the Behavior of Concrete, *ACI Journal,* Proc. Vol. 83, No. 2, Mar.–Apr. 1986, pp. 244–250.

Bresler, B., and Pister, K. S., Failure of Plain Concrete Under Combined Stresses, *ASCE Transactions,* Vol. 122, 1957, pp. 1049–1059.

Bresler, B., and Pister, K. S., Strength of Concrete Under Combined Stresses, *ACI Journal,* Proc. Vol. 55, No. 3, Sept. 1958, p. 321–345.

Bresler, B., and MacGregor, J. G., Review of Concrete Beams Failing in Shear, *Journal of the Structural Division, ASCE,* Feb. 1967, pp. 343–372.

Bresztovszky, B., A Beton Mechanikai Tulajdonsagainak Megallapitasa Nyomokiser-letekkel, *Patria Iroc. Vall. Es Nyomadai R.-T. Nyomasa,* Budapest, 1907.

Brickett, E. M., A Plastic Mortar Compression Test for Cement, *Proceedings, ASTM,* Vol. 28, Part II, 1928, pp. 432–442.

Brink, R. H., Physical Tests for Investigating Performance of Concrete, *Observations of the Performance of Concrete in Service,* Special Report No. 106, Highway Research Board, Washington, DC, 1970, pp. 18–28.

Brisbane, J. J., Axial Loading of an Elliptical Disk, *Journal of Applied Mechanics,* June 1963, p. 306.

Brooks, A. E., and Newman, K. (Eds.), *The Structure of Concrete and Its Behaviour Under Load,* Proceedings of an International Conference, London, Sept. 1965, Cement and Concrete Association, London, 1968.

Broutman, L. J., Fiber-Reinforced Plastics, *Modern Composite Materials,* Broutman Krock (Ed.), Addison-Wesley, Reading, MA, 1967, pp. 337–411.

Brown, C. B., and Mostaghel, N., Modulus and Strength of Reinforced Matrices: Critical Values of Inclusion Concentration, *Journal of Materials,* Vol. 2, No. 1, Mar. 1967, pp. 120–130.

Brown, H. E., Application of Statistical Evaluation Techniques for Quality Control of Steam Cured Concrete, *Proceedings of National Conference on Statistical Quality Control Methodology in Highway and Airfield Construction,* Charlottesville, VA, Nov. 1966, pp. 343–373.

Brunauer, S., and Copeland, L. E., The Chemistry of Concrete, *Scientific American,* Vol. 210, No. 4, Apr. 1964, pp. 80–92.

Brunauer, S., Emmett, P. H., and Teller, E., Adsorption of Gases in Multimolecular Layers, *Journal of the American Chemical Society,* Vol. 60, 1938, pp. 309–319.

Brunauer, S., Older, I., and Yudenfreund, M., The New Model of Hardened Portland Cement Paste, *Concrete Durability, Cement Paste, Aggregates, and Sealing Compounds,* Highway Research Record No. 328, Highway Research Board, Washington, DC, 1970, pp. 89–101.

Budnikov, P. P., and Erschler, E. Ya., Studies of the Processes of Cement Hardening in the Course of Low-Pressure Steam Curing of Concrete, *Symposium on Structure of Portland Cement Paste and Concrete,* Special Report No. 90, Highway Research Board, Washington, DC, 1966, pp. 431–446.

Bungey, J. H., Determining Concrete Strength by Using Small Diameter Cores, *Magazine of Concrete Research,* Vol. 31, No. 107, June 1979, pp. 91–98.

Bungey, J. H., Testing by Penetration Resistance, *Concrete,* Vol. 15, No. 1, Jan. 1981, pp. 30–32.

Bungey, J. H., *The Testing of Concrete in Structures,* Surrey University Press, 1982.

Bungey, J. H., The Influence of Reinforcement on Ultrasonic Pulse Velocity Testing, *In Situ/Nondestructive Testing of Concrete,* V. M. Malhotra (Ed.), ACI Publication SP-82, American Concrete Institute, Detroit, MI, 1984, pp. 229–246.

Butt, Y. M., Kolbasov, V. M., and Timashev, V. V., High Temperature Curing of Concrete Under Atmospheric Pressure, *Proceedings of the Fifth International Symposium on the Chemistry of Cement,* Part III, Properties of Cement and Concrete, Tokyo, Dec. 1969, pp. 437–470.

Byfors, J., Properties of Concrete at Early Ages, *Studies on Concrete Technology,* Swedish Cement and Cement Research Institute, Stockholm, 1979, pp. 55–72.

Bynum, D., Jr., Agarwal, R., and Fleisher, H. O., Constitutive Relations for Split Cylinder Tests on Bituminous Concrete, *Materials and Structures—Research and Testing,* RILEM, Vol. 4, No. 21, May–June 1971, pp. 163–169.

Campbell, R., Sr., *Computer-Aided Analysis of Concrete Strength Test Results,* Final Report, Instruction Report SL-83-1, USAE Waterways Experiment Station, Vicksburg, MS, Oct. 1983.

Campbell, R. H., and Tobin, R. E., Core and Cylinder Strengths of Natural and Lightweight Concrete, *ACI Journal,* Proc. Vol. 64, No. 4, Apr. 1967, pp. 190–195.

Campus, F., Relation entre l'hydration des liants hydrauliques et les résistance mécaniques des conglomerats (Relation Between the Hydration of Hydraulic Cements and the Strength of the Mixes), *Silicates Indstriels,* Vol. 20, No. 1, 1955, pp. 9–12.

Campus, F., Dantinne, R., Dzulynski, M., and Pirotte, F., Essais sur les effets de l'adhérence du mortier aux agregates (Tests Concerning the Effects of the Bond of Mortar to the Aggregates), *Memoires du C.E.R.E.S.,* No. 16, Sept. 1966, pp. 59–74.

Cantor, T. R., Review of Penetrating Radar as Applied to Nondestructive Evaluation of Concrete, *In Situ/Nondestructive Testing of Concrete,* V. M. Malhotra (Ed.), ACI Publication SP-82, American Concrete Institute, Detroit, MI, 1984, pp. 581–602.

Cantor, T. R., and Kneether, C. P., Radar as Applied to Evaluation of Bridge Decks, *Concrete Analysis and Deterioration,* Transportation Research Record No. 853, Transportation Research Board, Washington, DC, 1982, pp. 37–42.

Carette, G. G., and Malhotra, V. M., In-Situ Tests: Variability and Strength Prediction of Concrete at Early Ages, *In Situ/Nondestructive Testing of Concrete,* V. M. Malhotra (Ed.), ACI Publication SP-82, American Concrete Institute, Detroit, MI, 1984, p. 11.

Carino, N. J., Laboratory Study of Flaw Detection in Concrete by the Pulse-Echo Method, *In Situ/Nondestructive Testing of Concrete,* V. M. Malhotra (Ed.), ACI Publication SP-82, American Concrete Institute, Detroit, MI, 1984, pp. 557–580.

Carino, N. J., Specification for Cold Weather Concreting, *Concrete International: Design and Construction,* Vol. 10, No. 10, Oct. 1988, pp. 50–58.

Carino, N. J., The Maturity Method, *CRC Handbook on Nondestructive Testing of Concrete,* V. M. Malhotra and N. J. Carino (Eds.), CRC Press, Boca Raton, FL, 1991a, pp. 101–146.

Carino, N. J., Pullout Test, *CRC Handbook on Nondestructive Testing of Concrete,* V. M. Malhotra, and N. J. Carino (Eds.), CRC Press, Boca Raton, FL, 1991b, pp. 39–82.

Carino, N. J., and Lew, H. S., Re-examination of the Relation Between Splitting Tensile and Compressive Strength of Normal Weight Concrete, *ACI Journal,* Proc. Vol. 79, No. 3, May–June 1982, pp. 214–219.

Carino, N. J., and Lew, H. S., Temperature Effects on Strength–Maturity Relations of Mortar, *ACI Journal,* Proc. Vol. 80, No. 3, May–June 1983, pp. 177–182.

Carino, N., Sansalone, M., Detection of Voids in Grouted Ducts Using the Impact-Echo Method, *ACI Materials Journal,* Vol. 89, No. 3, May–June 1992a, pp. 296–303.

Carino, N. J., and Slate, F. O., Limiting Tensile Strain Criterion for Failure of Concrete, *ACI Journal,* Proc. Vol. 73, No. 3, Mar. 1976, pp. 160–165.

Carino, N. J., and Tank, R. C., Maturity Functions for Concrete Made with Various Cements and Admixtures, *ACI Materials Journal,* Vol. 89, No. 2, Mar.–Apr. 1992, pp. 188–196.

Carlsson, M., *Economical Considerations Between the Break-off Tester and Traditional Laboratory Tests,* A/S Scancem, Slemmestad, Norway, 1983.

Carlsson, M., Eeg, I. R., and Jahren, P., Field Experience in the Use of the Break-off Tester, *In Situ/Nondestructive Testing of Concrete,* V. M. Malhotra (Ed.), ACI Publication SP-82, American Concrete Institute, Detroit, MI, 1984, pp. 277–292.

Carneiro, F. L. L. B., and Barcellos, A., *Resistance a la Traction des Betons,* Broschure des Instituto Nacional de Technologia, Rio de Janeiro, 1949.

Carneiro, F. L. L. B., and Barcellos, A., Tensile Strength of Concretes, *RILEM Bulletin,* No. 13 (Old Series), Mar. 1953, pp. 99–107.

Celani, A., Moggi, P. A., and Roa, A., The Effect of Tricalcium Aluminate on the Hydration of Tricalcium Silicate and Portland Cement, *Proceedings of the Fifth International Symposium on the Chemistry of Cement,* Part II, Hydration of Cements, Tokyo, 1968, pp. 592–603.

Cement and Concrete Association, *Symposium on Mix Design and Quality Control of Concrete,* London, 1954.

Cement and Concrete Association, *Hydraulic Cement Pastes: Their Structure and Properties,* Proceedings of a Conference Held at the University of Sheffield, Apr. 8–9, 1976, London.

Chabowski, A. J., and Bryden-Smith, D., A Simple Pull-out Test to Assess the In Situ Strength of Concrete, *Concrete International: Design and Construction,* Vol. 1, No. 12, Dec. 1979, pp. 35–40.

Chabowski, A. J., and Bryden-Smith, D., Assessing the Strength of In Situ Portland Cement Concrete by Internal Fracture Tests, *Magazine of Concrete Research,* Vol. 32, No. 112, Sept. 1980, pp. 164–172.

Chandra, D., Sereda, P. J., and Swenson, E. G., Hydration and Strength of Neat Portland Cement, *Magazine of Concrete Research,* Vol. 20, No. 64, Sept. 1968, pp. 131–136.

Chandrashekhara, K., and Krishnaswamy, K. T., Determination of the Tensile Strength of Concrete by the Cube Split Test, *Indian Concrete Journal,* Jan. 1964, pp. 16–18.

Cheesman, W. J., Dynamic Testing of Concrete with the Soniscope Apparatus, *Proceedings, Highway Research Board,* Vol. 29, Washington, DC, 1949, pp. 176–181.

Chefdeville, J., Application de la méthode a l'estimation de la qualité du beton (Application of a Method for Estimating the Quality of Concrete), *RILEM Bulletin,* No. 15, Aug. 1953, pp. 59–78.

Chen, W. F., Double Punch Test for Tensile Strength of Concrete, *ACI Journal,* Proc. Vol. 67, No. 12, Dec. 1970, pp. 993–995.

Chen, W. F., and Chang, T. Y. P., Plasticity Solutions for Concrete Splitting Tests, *Journal of the Engineering Mechanics Division, ASCE,* Vol. 104, No. EM3, Proc. Paper 13852, June 1978, pp. 691–704.

Chen, W. F., and Colgrove, T. A., Double-Punch Test for Tensile Strength of Concrete, *Portland Cement Concrete,* Transportation Research Record No. 504, Transportation Research Board, Washington, DC, 1974, pp. 43–50.

Chen, W. F., and Ting, E. C., (Eds.), *Fracture in Concrete,* American Society of Civil Engineers, New York, 1980a, p. 105.

Chen, W. F., and Yuan, R. L., Tensile Strength of Concrete: Double-Punch Test, *Journal of the Structural Division, ASCE,* Vol. 106, No. ST8, Proc. Paper 15593, August 1980b, pp. 1673–1693.

Chin, F. K., Relation Between Strength and Maturity of Concrete, *ACI Journal,* Proc. Vol. 68, No. 3, Mar. 1971, pp. 196–203.

Chomahidze, R. O., and Chikovani, H. S., On the Type of the Analytic Relation Between Strength and Porosity of the Cement Stone (in Russian), *Proceedings of the Third National Conference on Mechanics and Technology of Composite Materials,* Varna, Oct. 4–6, 1982, Publishing House of the Bulgarian Academy of Sciences, Sofia, 1982, pp. 508–511.

Chung, H. W., Effects of Embedded Steel Bars upon Ultrasonic Testing of Concrete, *Magazine of Concrete Research,* Vol. 30, No. 102, Mar. 1978a, pp. 19–25.

Chung, H. W., How Good Is Good Enough—A Dilemma in Acceptance Testing of Concrete, *ACI Journal,* Proc. Vol. 75, No. 8, Aug. 1978b, pp. 374–380.

Chung, H. W., and Law, K. S., Diagnosing In Situ Concrete by Ultrasonic Pulse Technique, *Concrete International: Design and Construction,* Vol. 5, No. 10, Oct. 1983, pp. 42–49.

Clark, L. E., Gerstle, K. H., and Tulin, L. G., Effect of Strain Gradient on the Stress-Strain Curve of Mortar and Concrete, *ACI Journal,* Proc. Vol. 64, No. 9, Sept. 1967, pp. 580–586.

Clark, S. P., Jr. (Ed.), *Handbook of Physical Constants,* rev. ed., Geological Society of America, New York, 1966.

Clayton, N., Fluid-Pressure Testing of Concrete Cylinders, *Magazine of Concrete Research,* Vol. 30, No. 102, Mar. 1978, pp. 26–30.

Clemena, G. G., Short-Pulse Radar Methods, *CRC Handbook on Nondestructive Testing of Concrete,* V. M. Malhotra, and N. J. Carino (Eds.), CRC Press, Boca Raton, FL, 1991, pp. 253–274.

Clifton, J. R., and Anderson, E. D., *Nondestructive Evaluation Methods for Quality Acceptance of Hardened Concrete in Structures,* NBSIR 80-2163, National Bureau of Standards, Washington, DC, Jan. 1981.

Clifton, J. R., Brown, P. W., and Frohnsdorff, G., Reactivity of Fly Ashes with Cement, *Cement Research Progress 1977,* Chap. 15, American Ceramic Society, Columbus, OH, pp. 321–341.

Coble, R. L., and Kingery, W. D., Effect of Porosity on Physical Properties of Sintered Alumina, *Journal of the American Ceramic Society,* Vol. 29, No. 11, 1956, pp. 377–385.

Concrete Society, *Concrete Core Testing for Strength,* Concrete Society Technical Report No. 11, Concrete Society, London, May 1976.

Concrete Society, *Manual on Concrete Cube Crushing,* Concrete Society, London, 1975.

Cong, X., Gong, S., Darwin, D., and McCabe, S. L., Role of Silica Fume in Compressive Strength of Cement Paste, Mortar, and Concrete, *ACI Materials Journal,* Vol. 89, No. 4, July–Aug. 1992, pp. 375–387.

Cook, R. A., and Hover, K. C., Mercury Porosimetry of Cement-Based Materials and Associated Correction Factors," *ACI Materials Journal,* Vol. 90, No. 2, Mar.–Apr. 1993, pp. 152–161.

Copeland, L. E., Bodor, E., Chang, T. N., and Weise, C. H., Reactions of Tobermorite Gel with Aluminates, Ferriates, and Sulfates, *Journal of the PCA Research and Development Laboratories,* Vol. 9, No. 1, Jan. 1967, pp. 61–74.

Copeland, L. E., and Kantro, D. L., Chemistry of Hydration of Portland Cement at Ordinary Temperature, *The Chemistry of Cements,* H. F. W. Taylor (Ed.), Vol. I, Chap. 8, Academic Press, London, 1964, pp. 313–370.

Copeland, L. E., and Kantro, D. L., Hydration of Portland Cement, *Proceedings of the Fifth International Symposium on the Chemistry of Cement,* Part II, Hydration of Cements, Tokyo, 1968, pp. 387–421.

Copeland, L. E., Kantro, D. L., and Verbeck, G., Chemistry of Hydration of Portland Cement, *Chemistry of Cement,* Proceedings of the Fourth International Symposium, NBS Monograph 43, Vol. I, Session IV, Washington, DC, 1960, pp. 429–465.

Copeland, L. E., and Schulz, E. G., Electron Optical Investigation of the Hydration Products of Calcium Silicates and Portland Cement, *Journal of the PCA Research and Development Laboratories,* Vol. 4, No. 1, Jan. 1962, pp. 2–12.

Copeland, L. E., and Verbeck, G. J., Structure and Properties of Hardened Cement Pastes, *VIth International Congress on the Chemistry of Cement,* Moscow, Sept. 1974.

Cordon, W. A., Entrained Air—A Factor in the Design of Concrete Mixes, *ACI Journal,* Proc. Vol. 42, June 1946, pp. 605–620.

Cordon, W. A., Size and Number of Samples and Statistical Considerations in Sampling, *Significance of Tests and Properties of Concrete and Concrete Making Materials,* ASTM STP No. 169-A, Philadelphia, 1966, pp. 21–31.

Cordon, W. A., *Properties, Evaluation, and Control of Engineering Materials,* McGraw-Hill, New York, 1979.

Cordon, W. A., and Gillespie, H. A., Variables in Concrete Aggregates and Portland Cement Paste Which Influence the Strength of Concrete," *ACI Journal,* Proc. Vol. 60, No. 8, Aug. 1963, pp. 1029–1052.

Cottrell, A. H., *The Mechanical Properties of Matter,* John Wiley, New York, 1964.

Counto, U. J., The Effect of the Elastic Modulus of the Aggregate on the Elastic Modulus, Creep and Creep Recovery of Concrete, *Magazine of Concrete Research,* Vol. 16, No. 48, Sept. 1964, pp. 129–138.

Cowan, H. J., The Strength of Plain, Reinforced and Prestressed Concrete Under the Action of Combined Stresses, with Particular References to the Combined Denting and Torsion of Rectangular Sections, *Magazine of Concrete Research,* Vol. 5, No. 14, Dec. 1953, pp. 75–86.

Creskoff, J. J., Estimating 28-Day Strength of Concrete from Earlier Strengths—Including the Probable Error of the Estimate, *ACI Journal,* Proc. Vol. 41, Apr. 1945, pp. 493–512.

Cruz, C. R., Elastic Properties of Concrete at High Temperatures, *Journal of the PCA Research and Development Laboratories,* Vol. 8, No. 1, Jan. 1966, pp. 37–45.

Czernin, W., *Cement Chemistry and Physics for Civil Engineers,* Crosby Lockwood & Son, London, 1962.

Dahl-Jorgensen, E., and Johansen, R., General and Specialized Use of the Break-off Tester, *Situ/Nondestructive Testing of Concrete,* V. M. Malhotra (Ed.), ACI Publication SP-82, American Concrete Institute, Detroit, MI, 1984, pp. 293–308.

Dahms, J., *Die Schlagfestigkeit des Betons* (Impact Resistance of Concrete), Beton-Verlag, Dusseldorf, 1968.

Dahms, J., Uber die Schlagfestigkeit des Betons fur Rammpfahle (The Impact Resistance of Concrete for Driven Piles), *Betontechnische Berichte 1968,* Beton-Verlag, Dusseldorf, 1969, pp. 49–82.

Daniels, H. E., The Statistical Theory of the Strength of Bundles of Threads I, *Proceedings, Royal Society of London,* Vol. 153, 1945, p. 405.

Danilow, N., Die Anwendung der Infrarotstrahlen bei der Herstellung vorgefertigter Stahlbetonkonstruktionen und-details (Application of Infra-Red Radiation in the Production of Precast Concrete Construction), *Wissenschaftiche Zeitschrift,* Heft 6, Hochschule fur Bauwesen Leipzig, 1960, pp. 3–14.

Darwin, D., and Pecknold, D. A., Nonlinear Biaxial Stress–Strain Law for Concrete, *Journal of the Engineering Mechanics Division, ASCE,* Vol. 103, No. EM2, Proc. Paper 12839, April 1977, pp. 229–241.

Date, C. G., and Schnormeier, R. H., Day-to-Day Comparison of 4- and 6-Inch Diameter Concrete Cylinder Strengths, *Concrete International: Design and Construction,* Vol. 6, No. 8, Aug. 1984, pp. 24–26.

Date, C. G., and Schnomeier, R. H., Use of Prediction Relations, *ACI Journal,* Proc. Vol. 82, No. 4, July–Aug. 1985, pp. 525–530.

David, J. H., *Quality Control of Construction by Statistical Tolerance,* Alabama Highway Research HPR Report No. 29, Alabama Highway Department, Montgomery, AL, May 1967.

Davies, J. D., A Modified Splitting Test for Concrete Specimens, *Magazine of Concrete Research,* Vol. 20, No. 64, Sept. 1968, pp. 183–186.

Davies, J. D., and Bose, D. K., Stress Distribution in Splitting Tests, *ACI Journal,* Proc. Vol. 65, No. 8, Aug. 1968, pp. 662–669.

Davis, R. E., and Troxell, G. E., Modulus of Elasticity and Poisson's Ratio for Concrete, and the Influence of Age and Other Factors upon These Values, *Proceedings, ASTM,* Vol. 29, Part II, 1929, pp. 678–701.

DeHoff, R. T., and Rhines, F. N., *Quantitative Microscopy,* McGraw-Hill, New York, 1968.

De Larrard, F., Belloc, A., Renwez, S., and Boulay, C., Is the Cube Test Suitable for High Performance Concrete?, *Materials and Structures—Research and Testing,* RILEM, Vol. 27, No. 174, Dec. 1994, pp. 580–583.

Desayi, P., Strength of Concrete Under Combined Compression and Tension—Determination of Interaction Curve at Failure from Cylinder Split Test,

Materials and Structures—Research and Testing, RILEM, Vol. 2, No. 9, May–June 1969, pp. 179–185.

Desayi, P., A Simple Single Equation for the Stress–Strain Curve of Concrete in Compression, Publication No. 30, Annual Report of the Department of Civil and Hydraulic Engineering, Indian Institute of Science, Banglore, 1970, pp. 79–82.

Desayi, P., and Krishnan, S., Equation for the Stress–Strain Curve of Concrete, *ACI Journal,* Proc. Vol. 61, Mar. 1964, pp. 345–350.

Desayi, P., and Veerappan, M., A New Indirect Tension Test for Concrete and Other Brittle Materials, *Materials and Structures—Research and Testing,* RILEM, Vol. 5, No. 30, Nov.–Dec. 1972, pp. 371–377.

Desayi, P., and Viswanatha, C. S., The True Ultimate Strength of Plain Concrete, *RILEM Bulletin,* No. 36 (New Series), Sept. 1967, pp. 163–173.

Deutsche Bundesbahn, *Zusatzliche Technische Vorschriften fur Beton* (Additional Specifications for Concrete), 1962.

Diamond, S., A Critical Comparison of Mercury Porosimetry and Capillary Consideration Pore Size Distributions of Portland Cement Pastes, *Cement and Concrete Research,* Vol. 1, No. 5, Sept. 1971, pp. 531–545.

Diamond, S., Identification of Hydrated Cement Constituents Using a Scanning Electron Microscope-Energy Dispersive X-Ray Spectrometer Combination, *Cement and Concrete Research,* Vol. 2, No. 5, Sept. 1972, pp. 617–632.

Diamond, S., A Review of Alkali-Silica Reaction and Expansion Mechanisms, I. Alkalies in Cements and in Concrete Pore Solutions, *Cement and Concrete Research,* Vol. 5, No. 4, July 1975, pp. 329–346.

Diamond, S., Microstructure and Microstructure Engineering, *Teaching the Materials Science, Engineering, and Field Aspects of Concrete,* NSF-ACBM Center, Evanston, IL, 1993, pp. 46–73.

DiLeo, A., Pascale, G., and Viola, E., Core Sampling Size in Nondestructive Testing of Concrete, *In Situ/Nondestructive Testing of Concrete,* V. M. Malhotra (Ed.), ACI Publication SP-82, American Concrete Institute, Detroit, MI, 1984, pp. 459–478.

Dilly, R. L., and Ledbetter, W. B., Concrete Strength Based on Maturity and Pullout, *Journal of Structural Engineering,* ASCE, Vol. 110, No. 2, Feb. 1984, pp. 354–369.

Di Maio, A., Giaccio, G., and Zerbino, R., The Use of Break-off Test to Evaluate High Strength Concrete, International Symposium, *Non-destructive Testing in Civil Engineering* (NDT-CE), Vol. 2, Berlin, 1995, pp. 963–968.

Dinsdale, A., and Wilkinson, W. T., Impact Testing, *Mechanical Properties of Non-metallic Brittle Materials,* W. H. Walton (Ed.), Interscience Publishers, New York, 1958, pp. 193–203.

Double, D. D., and Hellawell, A., The Solidification of Cement, *Scientific American,* Vol. 237, No. 1, July 1977, pp. 82–86, 88–90.

Dougill, J. W., Discussion of the paper "Modulus of Elasticity of Concrete Affected by Elastic Modulus of Cement Paste, Matrix and Aggregate," by T. J. Hirsch, *ACI Journal,* Proc. Vol. 59, No. 9, Sept. 1962, pp. 1363–1365.

Duriez, M., and Arrambide, J., *Nouveau traite des materiaux de construction* (New Treatise of the Materials of Construction), Vol. 1, Dunod, Paris, 1961.

Dutron, P., Etude de l'influence du ciment en nature, classe de qualité et teneur et du mode de conservation sur les résistances mécaniques du beton de route (Study

Concerning the Influence of the Type and Quality of Cement, and the Length and Type of Curing on the Strength of Highway Concrete), Rapport presente au Xe Congres Belge de la Route, Bruxelles, 1962.

Dutron, R., *Ciments et betons* (IIIe partie)—Compositions et résistances des betons (Cements and Concretes, Third Part—Compositions and Strengths of Concretes), GPF Bulletin Technique, Bruxelles, 1955.

Eldin, N. N., and Senouci, A. B., Rubber-Tire Particles as Concrete Aggregates, *Journal of Materials in Civil Engineering,* Vol. 5, No. 4, Nov. 1993a, pp. 478–496.

Eldin, N. N., and Senouci, A. B., Observations on Rubberized Concrete Behavior, *Cement, Concrete, and Aggregates,* CCAGDP, Vol. 15, No. 1, 1993b, pp. 74–84.

Elices, M., and Planas, J., Measurement of Tensile Strength of Concrete at Very Low Temperatures, *ACI Journal,* Proc. Vol. 79, No. 3, May–June 1982, pp. 195–200.

Elvery, R. H., Symposium on the Non-destructive Testing of Concrete, *RILEM International Symposium on Nondestructive Testing of Materials and Structures,* Vol. I, Paris, 1954, pp. 111–119.

Elvery, R. H., and Evans, E. P., The Effect of Curing Conditions on the Physical Properties of Concrete, *Magazine of Concrete Research,* Vol. 16, No. 46, Mar. 1964, pp. 11–20.

Elvery, R. H., and Furst, M., The Effect of Compressive Stress on the Dynamic Modulus of Concrete, *Magazine of Concrete Research,* Vol. 9, No. 27, Nov. 1957, pp. 145–150.

Elvery, R. H., and Haroun, W., A Direct Tensile Test for Concrete Under Long or Short-Term Loading, *Magazine of Concrete Research,* Vol. 20, No. 63, June 1968, pp. 111–116.

Epstein, B., Statistical Aspects of Fracture Problems, *Journal of Applied Physics,* Vol. 19, Feb. 1948, pp. 140–147.

Erdei, C. K., Finite Element Analysis and Tests with a New Load-Transmitting Medium to Measure Compressive Strength of Brittle Materials, *Materials and Structures—Research and Testing,* RILEM, Vol. 13, No. 74, Mar.–Apr. 1980, pp. 83–90.

Erntroy, H. C., *The Variation of Works Test Cubes,* Research Report No. 10, Cement and Concrete Association, London, Nov. 1960.

Erzen, C. Z., An Expression for Creep and Its Application to Prestressed Concrete, *ACI Journal,* Proc. Vol. 53, Aug. 1956, pp. 205–213.

Evans, R. H., Extensibility and Modulus of Rupture of Concrete, *Structural Engineer,* Vol. 24, London, Dec. 1946, pp. 630–659.

Evans, R. H., and Dongre, A. V., The Suitability of a Lightweight Aggregate (Aglite) for Structural Concrete, *Magazine of Concrete Research,* Vol. 25, No. 44, July 1963, pp. 93–100.

Evans, R. H., and Marathe, M. S., Stress Distribution Around Holes in Concrete, *Materials and Structures—Research and Testing,* RILEM, Vol. 1, No. 1, Jan.–Feb. 1968, pp. 57–60.

Faber, J. H., Capp, J. P., and Spencer, J. D., Fly Ash Utilization, *Proceedings: Edison Electric Institute—National Coal Association,* Bureau of Mines Symposium, Pittsburgh, PA, Mar. 1967, U.S. Department of the Interior, Bureau of Mines, Information Circular 8348, Washington, DC, 1967.

Faber, J. H., Eckard, W. E., and Spencer, J. D. (Eds.), Ash Utilization, *Proceedings: Third International Ash Utilization Symposium,* U.S. Department of the Interior, IC 8640, 1974.

Facaoaru, I., Comparison Between Recommendations Existing in Some East European Countries Concerning the Determination of Concrete Strength by Surface Hardness Methods, *Materials and Structures—Research and Testing,* RILEM, Vol. 9, No. 51, May–June 1976, pp. 207–210.

Facaoaru, I., Rumanian Achievements in Nondestructive Strength Testing of Concrete, *In Situ/Nondestructive Testing of Concrete,* V. M. Malhotra (Ed.), ACI Publication SP-82, American Concrete Institute, Detroit, MI, 1984, pp. 35–56.

Facaoaru, I., and Stamate, G., Kombinierte zerstorungsfreie Prufmethoden fur Beton in Rumanien (Combined Nondestructive Methods for Concrete in Rumania), Zerstorungsfreie Pruf- und Messtechnik fur Beton und Stahbeton, Proceedings of an International Conference, Leipzig, Apr. 1969, pp. 25–34.

Facaoaru, E. I., Dumitrescu, E. I., and Stamate, E. G., *New Developments and Experience in Applying Combined Non-destructive Methods for Testing Concrete,* Report for the RILEM Working Group for Non-destructive Testing Meeting, Bucharest, 1968.

Faury, J., *Le beton* (Concrete), 3rd Ed., Dunod, Paris, 1958.

Feldman, R. F., Sorption and Length-Change Scannings Isotherms of Methanol and Water on Hydrated Portland Cement, *Proceedings of the Fifth International Symposium on the Chemistry of Cement,* Part III, Properties of Cement Paste and Concrete, Tokyo, Dec. 1969, pp. 53–66.

Feldman, R. F., *Assessment of Experimental Evidence for Models of Hydrated Portland Cement,* Highway Research Record No. 370, Highway Research Board, Washington, DC, 1971a, pp. 8–24.

Feldman, R. F., The Flow of Helium into the Interlayer Spaces of Hydrated Portland Cement Paste, *Cement and Concrete Research,* Vol. 1, No. 3, May 1971b, pp. 285–300.

Feldman, R. F., Helium Flow and Density Measurement of the Hydrated Tricalcium Silicate–Water System, *Cement and Concrete Research,* Vol. 2, No. 1, Jan. 1972, pp. 123–136.

Feldman, R. F., and Sereda, P. J., Use of Compacts to Study the Sorption Characteristics of Powdered Plaster of Paris, *Journal of Applied Chemistry,* Vol. 13, No. 4, Apr. 1963, pp. 158–167.

Feldman, R. F., and Sereda, P. J., A Model for Hydrated Portland Cement Paste as Deduced from Sorption-Length Change and Mechanical Properties, *Materials and Structures—Research and Testing,* RILEM, Vol. 1, No. 6, Nov.–Dec. 1968, pp. 509–520.

Feret, R., Sur la compacité des mortiers hydrauliques (About the Denseness of Cement Mortars), *Annales des Ponts et Chaussees,* Paris, 1892, pp. 1–184.

Ferguson, P. M., and Thompson, J. N., Diagonal Tension in T-Beams Without Stirrups, *ACI Journal,* Proc. Vol. 49, No. 7, Mar. 1953, pp. 665–676.

Fifth International Symposium on the Chemistry of Cement, Proceedings, Part IV, Admixtures and Special Cements, Vol. IV, Cement Association of Japan, Tokyo, 1969.

Figg, J. W., and Bowden, S. R., *The Analysis of Concretes,* Building Research Station, Her Majesty's Stationery Office, London, 1971.

Flugge, W., *Viscoelasticity,* Blaisdell, Waltham, MA, 1967.

Forrester, J. A., *The Application of Gamma Radiography to Concrete,* Technical Report TRA/273, Cement and Concrete Association, London, Aug. 1957.

Forrester, J. A., The Use of Gamma Radiography to Detect Faults in Grouting, *Magazine of Concrete Research,* Vol. 11, No. 32, July 1959, pp. 93–96.

Forstie, D. A., and Schnormeier, R., Development and Use of 4 × 8-Inch Concrete Cylinders in Arizona, *Concrete International: Design and Construction,* Vol. 3, No. 7, July 1981, pp. 42–45.

Foster, B. E., and Blaine, R. L., A Comparison of ISO and ASTM Tests for Cement Strength, *Cement, Comparison of Standards and Significance of Particular Tests,* ASTM Special Technical Publication, No. 441, Philadelphia, 1968, pp. 33–60.

Franca, G. De C., and Pincus, G., The Distribution of Concrete Strains in the Split Cylinder Test, *Journal of Materials,* JMLSA, Vol. 4, No. 2, June 1969, pp. 393–407.

Frankel, J. P., Relative Strengths of Portland Cement Mortar Bending Under Various Loading Conditions, *ACI Journal,* Proc. Vol. 45, Sept. 1948, pp. 21–32.

Fritsch, J., Wovon hangt die Betonfestigkeit ab? (What Does the Strength of Concrete Depend On?), *Zement–Kalk–Gips,* No. 7, 1951, pp. 185–188.

Frydman, S., The Applicability of the Brazilian (Indirect Tension) Test to Soils, *Australian Journal of Applied Science,* Vol. 15, No. 4, Dec. 1964, pp. 335–343.

Fuller, W. B., and Thompson, S. E., The Laws of Proportioning Concrete, *Transactions, ASCE,* Dec. 1907, pp. 67–143.

Fulton, F. S., *Some Physical Aspects of the Hydration of Portland Cement,* Laboratory Report SF-5, Portland Cement Institute, Johannesburg, South Africa, Feb. 1963.

Fulton, F. S., *Concrete Technology, A South African Handbook,* 3rd ed., Portland Cement Institute, Johannesburg, South Africa, 1964.

Fulton, F. S., and Davis, D. E., *Normal Portland vs. Portland Blastfurnace Cement,* Laboratory Report DF-1, Portland Cement Institute, Johannesburg, South Africa, Aug. 1961.

Fung, Y. C., *Foundations of Solid Mechanics,* Prentice Hall, Englewood Cliffs, NJ, 1965, pp. 17–34, 45–49.

Gaede, K., Die Prufung der Betonfestigkeit in Bauwerk (Testing the Strength of Concrete in Structures), *Bauingenieur,* Vol. 22, No. 13–16, 1941, pp. 138–142.

Gaede, K., Die Kugelschlagprufung von Beton (The Ball Test of Concrete), *Deutscher Ausschuss fur Stahlbeton,* Heft 107, Wilhelm Ernst & Sohn, Berlin, 1952.

Gaede, K., Kugelschlagprufung von Porenbeton (The Ball Test of Concrete of Porous Beton), *Deutscher Ausschuss fur Stahlbeton,* Heft 117, Wilhelm Ernst & Sohn, Berlin, 1954a.

Gaede, K., Report of the Ball Test, *RILEM International Symposium on Nondestructive Testing of Materials and Structures,* Vol. II, Paris, 1954b, pp. 307–309.

Gaede, K., Kugelschlagprufung von Beton (The Ball of Test of Concrete), *Deutscher Ausschuss fur Stahlbeton,* Heft 128, Wilhelm Ernst & Sohn, Berlin, 1957.

Gaede, K., Zur Frage der Verwendung von Bolzensetzgeraten zur Ermittlung der Druckfestigkeit von Beton (Application of the Bolt Setting Procedure for the Determination of Compressive Strength of Concrete), *Deutscher Ausschuss fur Stahlbeton,* Heft 168, Wilhelm Ernst & Sohn, Berlin, 1965, pp. 71–79.

Gaede, K., and Schmidt, E., Ruckprallprufung von Beton mit dichtem Gefuge (Rebound Testing of Concrete of Dense Structure), *Deutscher Ausschuss fur Stahlbeton,* Heft 158, Wilhelm Ernst & Sohn, Berlin, 1964, pp. 1–37.

Galan, A., Estimate of Concrete Strength by Ultrasonic Pulse Velocity and Damping Constant, *ACI Journal,* Proc. Vol. 64, No. 10, Oct. 1967, pp. 678–684.

Galan, A., *Combined Ultrasonic Methods of Concrete Testing,* Elsevier, Amsterdam, 1990.

Garboczi, E. J., Computational Materials Science of Cement-Based Materials, *Materials and Structures—Research and Testing,* Vol. 26, No. 158, May 1993, pp. 191–195.

Garboczi, E. J., Schwartz, L. M., and Bentz, D. P., Modeling the Influence of the Interfacial Zone on the DC Electrical Conductivity of Mortar, *Advanced Cement Based Materials,* Vol. 2, No. 5, Sept. 1995, pp. 169–181.

Gaynor, R. D., Effect of Horizontal Reinforcing Steel on the Strength of Molded Cylinders, *ACI Journal,* Proc. Vol. 62, No. 7, July 1965, pp. 837–840.

Gaynor, R. D., *High Strength Air-Entrained Concrete,* Joint Research Laboratory Publication No. 17, NSGA and NRMCA, Mar. 1968.

Gaynor, R. D., *In-Place Strength of Concrete—A Comparison of Two Test Systems,* NRMCA Technical Information Letter No. 272, Silver Spring, MD, Nov. 1969.

Gaynor, R. D., *One Look at Concrete Compressive Strength,* NRMCA Publication No. 147, Silver Spring, MD, Nov. 1974.

Geiker, M., Studies of Portland Cement Hydration by Measurement of Chemical Shrinkage and a Systematic Evaluation of Hydration Curves by Means of the Dispersion Model, Ph.D. Thesis, Institute of Mineral Industry, Technical University of Denmark, Lyngby, Denmark, 1983.

Gerend, M. S., Steam Cured Cylinders Give 28-Day Concrete Strength in 48 Hours, *Engineering News-Record,* Vol. 98, No. 7, Feb. 17, 1927, pp. 282–283.

Gerstle, K. H., Simple Formulation of Biaxial Concrete Behavior, *ACI Journal,* Proc. Vol. 78, No. 1, Jan.–Feb. 1981a, pp. 62–68.

Gerstle, K. H., Simple Formulation of Triaxial Concrete Behavior, *ACI Journal,* Proc. Vol. 78, No. 5, Sept.–Oct. 1981b, pp. 382–387.

Ghosh, R. K., Chatterjee, M. R., and Lal, Ram, Accelerated Strength Tests for Quality Control of Paving Concrete, *Accelerated Strength Testing,* ACI Publication SP-56, American Concrete Institute, Detroit, MI, 1978, pp. 169–182.

Ghosh, R. S., Proportioning Concrete Mixes Incorporating Fly Ash, *Canadian Journal of Civil Engineering,* Vol. 3, 1976, pp. 68–82.

Ghosh, S. K., and Handa, V. K., *Strain Gradient and the Stress–Strain Relationship of Concrete in Compression,* Highway Research Record No. 324, Highway Research Board, Washington, DC, 1970, pp. 44–53.

Gilkey, H. J., The Moist Curing of Concrete, *Engineering News-Record,* Oct. 14, 1937, pp. 630–631.

Gilkey, H. J., Water–Cement Ratio Versus Strength—Another Look, *ACI Journal,* Proc. Vol. 57, Apr. 1961, pp. 1287–1312.

Gilkey, H. J., and Murphy, G., Stress–Strain Characteristics of Mortars and Concretes, *Proceedings, ASTM,* Vol. 38, Part I, 1938, pp. 318–326.

Glanville, W. H., Grime, G., Fox, E. N., and Davies, W. W., *Investigation of Stresses in Reinforced Concrete Piles During Driving,* Building Research Technical Paper No. 20, His Majesty's Stationary Office, London, 1938a.

Glanville, W. H., Collins, A. R., and Matthews, D. D., *The Grading of Aggregates and Workability of Concrete,* Road Research Technical Paper No. 5, Department of Scientific and Industrial Research and Ministry of Transport, London, 1938b, p. 42.

Glucklich, J., Fracture of Plain Concrete, *Journal of the Engineering Mechanics Division, ASCE,* Vol. 89, No. EM6, Dec. 1963, pp. 127–138.

Golis, J. M., Pavement Thickness Measurement Using Ultrasonic Pulses, *Mechanical Properties of Plastic Concrete and Pavement Thickness Measurement,* Highway Research Record No. 218, Highway Research Board, Washington, DC, 1968, pp. 40–48.

Gonnerman, H. F., Effect of End Condition of Cylinder in Compression Tests of Concrete, *Proceedings, ASTM,* Vol. 24, Part II, 1924, pp. 1036–1063.

Gonnerman, H. F., Study of Cement Composition in Relation to Strength, Length Changes, Resistance of Sulfate Waters and to Freezing and Thawing of Mortars and Concrete, *Proceedings, ASTM,* Vol. 34, Part II, 1934, pp. 244–295.

Gonnerman, H. F., and Lerch, W., *Changes in Characteristics of Portland Cement as Exhibited by Laboratory Tests Over the Period 1904 to 1940,* ASTM Special Publication No. 127, Philadelphia, 1952, pp. 1–56.

Gonnerman, H. F., and Shuman, E. C., Compression, Flexure, and Tension Tests of Plain Concrete, *Proceedings, ASTM,* Vol. 28, Part II, Philadelphia, 1928, pp. 527–552.

Goral, M. L., Empirical Time–Strength Relations of Concrete, *ACI Journal,* Proc. Vol. 53, Aug. 1956, pp. 215–224.

Gouda, G. R. (Ed.), *Proceedings of the Fourth International Conference on Cement Microscopy,* International Cement Microscopy Association, Duncanville, TX, 1982.

Graf, O., Versuche uber die Widerstandfahigkeit von Beton- und Eisenbetonrohren gegen Innendruck (Experiments Concerning the Resistance of Concrete and Reinforced Concrete Pipes Against Internal Pressure), *Der Bauingenieur,* Vol. 4, Nos. 15 and 16, Aug. 15 and 31, 1923, pp. 441–448, 474–481.

Graf, O., Einfluss der Kornug des Zements? (Effect of Particle Size of Cement?), *Zement,* Vol. 19, 1930, pp. 48–52.

Graf, O., Albrecht, W., and Schaffler, H., *Die Eigenschaften des Betons* (Properties of Concrete), 2nd ed., Springer-Verlag, Berlin, 1960.

Green, H., The Impact Testing of Concrete, *Mechanical Properties of Non-metallic Brittle Materials,* W. H. Walton (Ed.), Interscience Publishers, New York, 1958, pp. 300–315.

Greenberg, S. A., and Meyer, L. M., Rheology of Fresh Portland Cement Pastes: Influence of Calcium Sulfates, *Properties of Concrete,* Highway Research Record No. 3, Washington, DC, 1963, pp. 9–29.

Greene, G. W., Test Hammer Provides New Method of Evaluating Hardened Concrete, *ACI Journal,* Proc. Vol. 51, No. 3, Nov. 1954, pp. 249–256.

Greer, W. C., Jr., Variation of Laboratory Concrete, Flexural Strength Tests, *Cement, Concrete, and Aggregates,* CCAGDP, Vol. 5, No. 2, Winter 1983, pp. 111–122.

Grieb, W. W., Use of Swiss Hammer for Estimating Compressive Strength of Hardened Concrete, *Rapid Tests for Aggregates and Concrete,* Bulletin 201, Highway Research Board, Washington, DC, 1958, pp. 45–50.

Grieb, W. E., and Werner, G., Comparison of Splitting Tensile Strength of Concrete with Flexural and Compressive Strengths, *Proceedings, ASTM,* Vol. 62, Philadelphia, 1962, pp. 972–990.

Griffith, A. A., The Phenomena of Rupture and Flow in Solids, *Philosophical Transactions, Royal Society of London,* Ser. A, Vol. 221, 1920, pp. 163–198.

Griffith, A. A., The Theory of Rupture, *Proceedings, 1st International Congress for Applied Mechanics,* Delft, 1924, pp. 55–63.

Grudemo, A., Electron Microscopy of Portland Cement Pastes, *The Chemistry of Cements,* H. F. W. Taylor (Ed.), Vol. I, Chapter 9, Academic Press, London and New York, 1964, pp. 371–390.

Grudemo, A., Strength vs. Structure in Cement Pastes, Supplementary Paper II-3, II-4, II-5, VIth International Congress on the Chemistry of Cement, Moscow, Sept. 1974.

Grudemo, A., The Structures of Cement Hydration Components—Information from X-ray Diffractometric Data, *Studies on Concrete Technology,* Swedish Cement and Concrete Research Institute, Stockholm, 1979, pp. 9–31.

Gruenwald, E., Effect of Slump on Compressive Strength of Concrete of Constant Water–Cement Ratio, *ACI Journal,* Proc. Vol. 53, Aug. 1956a, pp. 230–231.

Gruenwald, E., Cold Weather Concreting with High-Early Strength Cement, *RILEM Symposium, Winter Concreting Theory and Practice,* Session B1, Danish National Institute of Building Research, Special Report, Copenhagen, 1956b.

Guedes, Q. M., and Souza, M. O. L., Study of and Operating Characteristic Function of the Sampling Inspection Plan of Concrete, *Materials and Structures—Research and Testing,* RILEM, Vol. 11, No. 66, Nov.–Dec. 1978, pp. 401–406.

Guttmann, A., Zur Frage der Zugfestigkeit unbewehrten Betons (Tensile Strength of Nonreinforced Concrete), *Zement,* Vol. 24, No. 35, Aug. 29, 1935, pp. 532–542.

Gyengo, T., Neue Grundlagen fur die Bestimmung der Betonfestigkeit (New Basis for the Determination of Concrete Strength), *Acta Technica Academiae Scientiarum Hungaricae,* Tomus XXVI, Fasciculi 1–2, Budapest, 1959.

Hadley, D. W., The Nature of the Paste-Aggregate Interface, Joint Highway Research Project, Interim Report No. 40, Purdue University, West Lafayette, IN, No. 9, 1972.

Halabe, B., Stoodehnia, A., Maser, K. R., and Kausel, E. A., Modeling the Electromagnetic Properties of Concrete, *ACI Materials Journal,* Vol. 90, No. 6, Nov.–Dec. 1993, pp. 552–563.

Halasz, I., Deformation in Concrete, *Proceedings of the Technical University of Building and Transport Engineering,* Vol. XII, No. 6, Budapest, 1967, pp. 125–154.

Halasz, I., A beton terhelesokozta alakvaltozasai es a tonkremenetel folyamata (Deformations of Concrete Under Load and the Process of Fracture), *Tudomanyos Kozlemenyek, No. 2,* Budapesti Muszaki Egyetem Epitomernoki Kar Epitonayagok Tanszek, Budapest, 1970, pp. 57–81.

Halasz, I., A terhelesi sebesseg hatasa a beton tonkrementeli folyamatara es ellenallasara (The Effect of Loading Speed on the Process of Fracture and Resistance of Concrete), *Tudomanyos Kozlemenyek,* No. 21, Kozlekedesi Dokumentacios Vallalat, Budapest, 1975, pp. 57–74.

Hale, K. F., and Brown, M. H., Application of High Voltage Electron Microscopy to the Study of Cement, *Structure, Solid Mechanics and Engineering Design,* The Proceedings of the Southampton 1969 Civil Engineering Materials Conference, Part 1, Wiley-Interscience, London, 1971, pp. 289–294.

Halstead, P. E., The Significance of Concrete Cube Tests, *Magazine of Concrete Research,* Vol. 21, No. 69, Dec. 1969, pp. 187–194.

Hanke, I., Methode zur Bestimmung des Transversalwelleneinsatzes bei Senkrechtdurchscallung von Beton (Methods for the Determination of Transverse Waves Generated by Vertical Sonic Vibration of Concrete), *Zerstorungsfreie Pruf- und Messtechnik fur Beton and Stahlbeton,* Proceedings of an International Conference, Leipzig, April 1969, pp. 57–60.

Hannant, D. J., Creep and Creep Recovery of Concrete Subjected to Multiaxial Compressive Stress, *ACI Journal,* Proc. Vol. 66, No. 5, May 1969, pp. 391–394.

Hannant, D. J., Buckley, K. J., and Croft, J., The Effect of Aggregate Size on the Use of the Cylinder Splitting Test as a Measure of Tensile Strength, *Materials and Structures—Research and Testing,* RILEM, Vol. 6, No. 31, Jan.–Feb. 1973, pp. 15–21.

Hansen, H., Kielland, A., Nielsen, K. E. C., and Thaulow, S., Compressive Strength of Concrete—Cube or Cylinder?, *RILEM Bulletin,* No. 17, Dec. 1962, pp. 23–30.

Hansen, T. C., Influence of Aggregate and Voids on Modulus of Elasticity of Concrete, Cement Mortar, and Cement Paste, *ACI Journal,* Proc. Vol. 62, Feb. 1965, pp. 193–216.

Hansen, T. C., Theories of Multi-Phase Materials Applied to Concrete, Cement and Cement Paste, *The Structure of Concrete,* Proceedings of an International Conference, London, Sept. 1965, Cement and Concrete Association, London, 1968, pp. 16–23.

Hansen, T. C., Physical Structure of Hardened Portland Cement Paste—A Classical Approach, *Materials and Structures—Research and Testing,* RILEM, Vol. 19, No. 114, Nov.–Dec., 1986, pp. 423–436.

Hansen, T. C., and Eriksson, L., Temperature Change Effect on Behavior of Cement Paste, Mortar, and Concrete Under Load, *ACI Journal,* Proc. Vol. 63, No. 3, Apr. 1966, pp. 489–504.

Hansen, W., Static and Dynamic Elastic Modulus of Concrete as Affected by Mix Composition and Compressive Strength, *Properties of Concrete at Early Ages,* J. F. Young (Ed.), ACI SP-95, 1986, pp. 115–137.

Hansen, W. C., Interactions of Organic Compounds in Portland Cement Pastes, *Journal of Materials,* Vol. 5, No. 4, Dec. 1970, pp. 842–855.

Hanson, J. A., Shear Strength of Lightweight Reinforced Concrete Beams, *ACI Journal,* Proc. Vol. 55, No. 3, Sept. 1958, pp. 387–403.

Hanson, J. A., Tensile Strength and Diagonal Tension Resistance of Structural Lightweight Concrete, *ACI Journal,* Proc. Vol. 58, No. 7, July 1961, pp. 1–39.

Hanson, J. A., Replacement of Lightweight Aggregate Fines with Natural Sand in Structural Concrete, *ACI Journal,* Proc. Vol. 61, July 1964, pp. 779–793.

Harig, S., Die Beeinflussung des E-Modulus von Beton durch Zemente mitunterschiedlichem mineralischem Aufbau und durch naturliche und kunstliche Zuschlagstoffe (How the Modulus of Elasticity of Concrete Is Affected by Cements of Differing in Mineralogical Composi), *Betonstein-Zeitung,* Vol. 32, Nos. 9 and 10, Sept. and Oct. 1966, pp. 510–520, 557–567.

Harris, D. H. C., Windsor, C. G., and Lawrence, C. D., Free and Bound Water in Cement Pastes, *Magazine of Concrete Research,* Vol. 26, No. 87, June 1974, pp. 65–72.

Hashin, Z., The Elastic Moduli of Heterogeneous Materials, *Journal of Applied Mechanics,* Vol. 29, No. 1, March 1962, pp. 143–150.

Hedstrom, R. O., Tensile Testing of Concrete Block and Wall Elements, *Journal of the PCA Research and Development Laboratories,* Vol. 8, No. 2, May 1966, pp. 42–52.

Heilmann, H. G., Hilsdorf, H., and Finsterwalder, K., Festigkeit und Verformung von Beton unter Zugspannungen (Strength and Deformation of Concrete Under Tensile Stresses), *Deutscher Ausschuss fur Stahlbeton,* Heft 203, Wilhelm Ernst & Sohn, Berlin, 1969.

Helmuth, R. A., and Turk, D. H., Elastic Moduli of Hardened Portland Cement and Tricalcium Silicate Pastes, *Symposium on Structure of Portland Cement Paste and Concrete,* Special Report 90, Highway Research Board, Washington, DC, 1966, pp. 135–144.

Henk, B., Zur Fruhfestigkeit von Beton bei naturlichen Erhartungsbedingungen (The Early Strength of Concrete under Natural Conditions of Hardening), *Betonstein-Zeitung,* Vol. 32, No. 8, Aug. 1966, pp. 461–468.

Hester, W. T., Field Testing High-Strength Concretes: A Critical Review of the State-of-the-Art, *Concrete International: Design and Construction,* Vol. 2, No. 12, Dec. 1980, pp. 27–38.

Higginson, E. C., Wallace, G. B., and Ore, E. L., Effect of Maximum Size Aggregate on Compressive Strength of Mass Concrete, *Symposium on Mass Concrete,* ACI Publication SP-6, 1963, pp. 219–246.

Himsworth, F. R., The Variability of Concrete and Its Effect on Mix Design, *Proceedings, Institution of Civil Engineers,* Vol. 3, No. 2, Mar. 1954, pp. 163–195.

Hindo, K. R., and Bergstrom, W. R., Statistical Evaluation of the In-Place Compressive Strength of Concrete, *Concrete International: Design and Construction,* Vol. 7, No. 2, Feb. 1985, pp. 44–48.

Hirsch, T. J., Modulus of Elasticity of Concrete Affected by Elastic Modulus of Cement Paste Matrix and Aggregate, *ACI Journal,* Proc. Vol. 59, No. 3, Mar. 1962, pp. 427–451.

Hobbs, D. W., The Dependence of the Bulk Modulus, Young's Modulus, Shrinkage and Thermal Expansion of Concrete Upon Aggregate Volume Concentration, Technical Report TRA 437, Cement and Concrete Association, London, December 1969.

Hobbs, D. W., *Strength and Deformation Properties of Plain Concrete Subject to Combined Stress*—Part 1: Strength Results Obtained on One Concrete, Technical Report, SBN 7210 0742 2, Cement and Concrete Association, London, Nov. 1970.

Hobbs, D. W., Strength of Concrete Under Combined Stress, *Cement and Concrete Research,* Vol. 1, No. 1, Jan. 1971, pp. 41–56.

Hobbs, D. W., The Compressive Strength of Concrete: A Statistical Approach to Failure, *Magazine of Concrete Research,* Vol. 24, No. 80, Sept. 1972, pp. 127–138.

Hobbs, D. W., *Strength and Deformation Properties of Plain Concrete Subject to Combined Stress,* Part 3, Results Obtained on a Range of Flint Gravel Aggregate Concrete, Technical Report 42.497, Cement and Concrete Association, London, July 1974.

Hobbs, D. W., Mix Design Quality of Mixing Water, *W/C* Ratio Homogeneity, *Le beton et l'eau* (Concrete and Water), Conseil International de la Langue Française, 1985, pp. 46–67.

Hofsoy, A., Comparison of Apparent Compression Strength of Concrete Cores, *Cubes and Cylinders,* RILEM Symposium on the Experimental Research of Field Testing of Concrete, Trondheim, Norway, Oct. 5–7 1964, pp. 176–188.

Hognestad, E., Hanson, N. W., and McHenry, D., Concrete Stress Distribution in Ultimate Strength Design, *ACI Journal,* Proc. Vol. 52, Dec. 1955, pp. 455–479.

Hognestad, E., Elstner, R. C., and Hanson, J. A., Shear Strength of Reinforced Structural Lightweight Aggregate Concrete Slabs, *ACI Journal,* Proc. Vol. 61, No. 6, Part 1, June 1964, pp. 643–656.

Holliday, L., Geometrical Considerations and Phase Relationships, *Composite Materials,* L. Holliday (Ed.), Elsevier, Amsterdam, 1966, pp. 1–27.

Hondros, G., The Evaluation of Poisson's Ratio and the Modulus of Materials of a Low Tensile Resistance by the Brazilian (Indirect Tensile) Test with Particular Reference to Concrete," *Australian Journal of Applied Science,* Vol. 10, No. 3, September 1959, pp. 243–268.

Horimatsu, K., Curing of Concrete in Winter Construction by Electric Heating, *RILEM Symposium, Winter Concreting, Theory and Practice,* Session BII, Danish National Institute of Building Research, Special Report, Copenhagen, 1956.

Howdyshell, P. A., Evaluation of a Chemical Technique to Determine Water and Cement Content of Fresh Concrete, *Recent Developments in Accelerated Testing and Maturity,* Transportation Research Record No. 558, Transportation Research Board, Washington, DC, 1975, pp. 104–113.

Howdyshell, P. A., Concrete Quality Control: 28 Days–24 Hours–15 Minutes, *Accelerated Strength Testing,* ACI SP-56, 1978, pp. 183–200.

Hsu, T. T. C., Mathematical Analysis of Shrinkage Stresses in a Model of Hardened Concrete, *ACI Journal,* Proc. Vol. 60, Mar. 1963, pp. 371–390.

Hsu, T. T. C., Torsion of Structural Concrete—A Summary of Pure Torsion, *Torsion of Structural Concrete,* ACI Publication SP-18, American Concrete Institute, Detroit, MI, 1968a, pp. 165–178.

Hsu, T. T. C., Torsion of Structural Concrete—Behavior of Reinforced Concrete Rectangular Members, *Torsion of Structural Concrete,* ACI Publication SP-18, American Concrete Institute, Detroit, MI, 1968b, pp. 261–306.

Hsu, T. T. C., Torsion of Structural Concrete—Plain Concrete Rectangular Sections, *Torsion of Structural Concrete,* ACI Publication SP-18, American Concrete Institute, Detroit, MI 1968c, pp. 203–238.

Hsu, T. C., A Study of the Compression Test for Ductile Materials, *Materials Research and Standards,* MTRSA, Vol. 9, No. 12, Dec. 1969, pp. 20–25, 47–53.

Hsu, T. T. C., and Slate, F. O., Tensile Bond Strength Between Aggregate and Cement Paste or Mortar, *ACI Journal,* Proc. Vol. 60, No. 4, Apr. 1963, pp. 465–485.

Hsu, T. T. C., and Slate, F. O., Sturman, G. M., and Winter, G., Microcracking of Plain Concrete and the Shape of the Stress–Strain Curve, *ACI Journal,* Proc. Vol. 60, Feb. 1963, pp. 209–224.

Hudson, J. A., and Fairhurst, C., Tensile Strength, Weilbull's Theory and a General Statistical Approach to Rock Failure, *Structure, Solid Mechanics and Engineering Designs,* Proceedings of the Southampton 1969 Civil Engineering Materials Conference, Part 2, Wiley-Interscience, London, 1971, pp. 901–914.

Hudson, S. B., and Steele, G. W., Prediction of Potential Strength of Concrete from the Results of Early Tests, *Concrete,* Highway Research Record No. 370, Highway Research Board, Washington, DC, 1971, pp. 25–35.

Hudson, S. B., and Steele, G. W., Developments in the Prediction of Potential Strength of Concrete from Results of Early Tests, *Recent Developments in Accelerated Testing and Maturity,* Transportation Research Record No. 558, Transportation Research Board, Washington, DC, 1975, pp. 1–12.

Hughes, B. P., and Ash, J. E., Short-Term Loading and Deformation of Concrete in Unaxial Tension and Pure Torsion, *Magazine of Concrete Research,* Vol. 20, No. 64, Sept. 1968, pp. 145–154.

Hughes, B. P., and Chapman, G. P., Direct Tensile Test for Concrete Using Modern Adhesives, *RILEM Bulletin,* No. 26, Mar. 1965, pp. 787–780.

Hughes, B. P., and Chapman, G. P., The Complete Stress–Strain Curve for Concrete in Direct Tension, *RILEM Bulletin,* No. 30, Mar. 1966a, pp. 95–98.

Hughes, B. P., and Chapman, G. P., The Deformation of Concrete and Micro-concrete in Compression and Tension with Particular Reference to Aggregate Size, *Magazine of Concrete Research,* Vol. 18, No. 54, Mar. 1966b, pp. 19–24.

Hulshizer, A. J., Edgar, M. A., Daniels, R. E., Suminsby, J. D., and Meyers, G. E., Maturity Concept Proves Effective in Reducing Form Removal Time and Winter Curing Cost, *In Situ/Nondestructive Testing of Concrete,* V. M. Malhotra (Ed.), ACI Publication SP-82, American Concrete Institute, Detroit, MI, 1984, pp. 351–376.

Hummel, A., Beeinflussung der Betonelastizitat (Effect on the Elasticity of Concrete), *Zement,* Vol. 24, Nos. 42 and 43, Oct. 17 and 24, 1935, pp. 665–669, 684–689.

Hummel, A., *Das Beton—ABC* (The Alphabet of Concrete), 12th ed., Wilhelm Ernst & Sohn, Berlin, 1959.

Hunt, J. G., *The Curing of Concrete Pavement Slabs in Hot Weather,* Technical Report TRA 435, Cement and Concrete Association, London, Nov. 1969.

Ichiki, Y., Some Experiments on Electrical Curing of Concrete in Cold Weather, *RILEM Symposium, Winter Concreting, Theory and Practice,* Session D, Danish National Institute of Building Research, Special Report, Copenhagen, 1956.

Imbert, I. D. C., The Effect of Holes on Tensile Deformations in Plain Concrete, *Symposium on Concrete Deformation,* Highway Research Record No. 324, Highway Research Board, Washington, DC, 1970, pp. 54–65.

Inglis, C. E., Stresses in a Plate Due to the Presence of Cracks and Sharp Corners, *Transactions, Institute of Naval Architects,* Vol. 60, London, 1913, p. 219.

Ishai, O., Influence of Sand Concentration on the Deformations of Mortar Beams Under Low Stresses, *ACI Journal,* Proc. Vol. 58, Nov. 1961, pp. 611–623.

Ivey, D. L., and Buth, E., Splitting Tension Test of Lightweight Concrete, *Journal of Materials,* ASTM, Vol. 1, No. 4, Dec. 1966, pp. 859–871.

Iyengar, Sundara Raja, K. T., On the Determination of True Tensile Strength of Concrete, *RILEM Bulletin,* No. 21, Dec. 1963, pp. 38–45.

Iyengar, Sundara Raja, K. T., Desayi, P., and Reddy, K. N., Stress–Strain Characteristics of Concrete Confined in Steel Binders, *Magazine of Concrete Research,* Vol. 22, No. 72, London, Sept. 1970, pp. 173–184.

Jaeger, J. C., and Cook, N. G. W., *Fundamentals of Rock Mechanics,* 2nd ed., Chapman & Hall, London/John Wiley, New York, 1978.

Jambor, J., Phase Composition Structure and Strength of Hardened Cement Pastes, Supplementary Paper II-3, II-4, II-5, The VIth International Congress on the Chemistry of Cement, Moscow, Sept. 1974.

Jambor, J., Plenary Report: Studying the Laws of Structure Forming of Hardening Cement Pastes and the Effect of the Laws Mentioned Over the Concrete Strength and Elastic–Plastic Properties, *Mechanics and Technology of Composite Materials,* Proceedings of the Second National Conference, Varna, Oct. 1–3, 1979, Publishing House of the Bulgarian Academy of Sciences, Sofia, 1979, pp. 477–490.

Javor, T., Testing of Concrete Structures in Situ, *Materials and Structures—Research and Testing,* RILEM, Vol. 11, No. 66, Nov.–Dec. 1978, pp. 457–475.

Jenkins, R. S., Non-destructive Testing—An Evaluation Tool, *Concrete International: Design and Construction,* Vol. 7, No. 2, Feb. 1985, pp. 22–26.

Jensen, A. D., and Chatterji, S., State of the Art Report on Micro-cracking and Lifetime of Concrete—Part 1, *Materials and Structures,* Vol. 29, No. 185, Jan.–Feb. 1996, pp. 3–8.

Jevtic, D., Influence de l'age sur la résistance—quelques essais effectues avec un ciment a haute résistance initiale (Influence of Age on Strength—Some Tests Carried Out with a Cement of High Early Strength), *RILEM Bulletin,* No. 5, Dec. 1959, pp. 41–48.

Jiang, D. H., Shah, S. P., and Andonian, A. T., Study of the Transfer of Tensile Forces by Bond, *ACI Journal,* Proc. Vol. 81, No. 3, May–June 1984, pp. 250–258.

Jindal, B. K., Properties of Structural Lightweight Concrete Using Sintered Fly Ash Aggregate, *Indian Concrete Journal,* Nov. 1964, pp. 413–418.

Johansen, R., In Situ Strength Evaluation of Concrete—The Break-off Method, *Concrete International: Design and Construction,* Vol. 1, No. 9, Sept. 1979, pp. 45–51.

Johnson, A. N., Strength Characteristics of Concrete, *Public Roads,* Vol. 9, No. 9, Nov. 1928a, pp. 177–181.

Johnson, A. N., The Modulus of Elasticity of Cores from Concrete Roads, *Public Roads,* Vol. 9, No. 8, Oct. 1928b, pp. 164–168.

Johnston, C. D., Strength and Deformation of Concrete in Uniaxial Tension and Compression, *Magazine of Concrete Research,* Vol. 22, No. 70, Mar. 1970, pp. 5–16.

Johnston, C. D., and Sidewell, E. H., Testing Concrete in Tension and Compression, *Magazine of Concrete Research,* Vol. 20, No. 65, Dec. 1968, pp. 221–228.

Jones, R., The Application of Ultrasonics to the Testing of Concrete, *Magazine of Concrete Research,* London, May 1948, p. 383.

Jones, R., Measurement of Thickness of Concrete Pavements by Dynamic Methods: Survey of Difficulties, *Magazine of Concrete Research,* London, No. 1, Jan. 1949, pp. 31–34.

Jones, R., A Method of Studying the Formation of Cracks in a Material Subjected to Stress, *British Journal of Applied Physics,* Vol. 3, No. 7, July 1952, pp. 229–232.

Jones, R., A Vibration Method for Measuring the Thickness of Concrete Road Slabs in Situ, *Magazine of Concrete Research,* Vol. 7, No. 20, July 1955.

Jones, R., The Effect of Frequency on the Dynamic Modulus and Damping Coefficient of Concrete, *Magazine of Concrete Research,* Vol. 9, No. 26, Aug. 1957, pp. 69–72.

Jones, R., *Non-destructive Testing of Concrete,* Cambridge University Press, 1962, 104 p.

Jones, R., Cracking and Failure of Concrete Test Specimens Under Uniaxial Quasistatic Loading, *Structure of Concrete,* Proceedings of an International Conference, London, Sept. 1965, Cement and Concrete Association, London, 1968, pp. 125–130.

Jones, R., and Facaoaru, I., An Analysis of Answers to a Questionnaire on the Ultrasonic Pulse Technique, *Materials and Structures—Research and Testing,* RILEM, Vol. 1, No. 5, Sept.–Oct. 1968, pp. 457–466.

Jones, R., and Facaoaru, I., Analyse des RILEM-Fragebogens uber Ultraschallprufung von Beton, *Zerstorungsfreie Pruf- und Messtechnik fur Beton und Stahlbeton,* Proceedings of an International Conference, Leipzig, Apr. 1969, pp. 5–6.

Jones, R., and Gatfield, E. N., *Testing Concrete by an Ultrasonic Pulse Technique,* Road Research Technical Paper No. 34, Department of Scientific and Industrial Research, Road Research Laboratory, London, 1955.

Jones R., and Kaplan, M. F., The Effect of Coarse Aggregate on the Mode of Failure of Concrete in Compression and Flexure, *Magazine of Concrete Research,* Vol. 9, No. 26, Aug. 1957, pp. 89–94.

Jones, R., and Mayhew, H. C., Zerstorungsfreie Messung der Dicke und der Qualitat von betonierten Oberflachen und Platten mit Oberflachenwellen, *Wissenschaftliche Zeitschrift,* Vol. 12, Nos. 1–2, Hochschule fur Bauwesen, Leipzig, 1966, pp. 37–42.

Jones, P. G., and Richart, F. E., The Effect of Testing Speed on Strength and Elastic Properties of Concrete, *Proceedings, ASTM,* Vol. 36, Part II, 1936, pp. 380–391.

Jons, E. S., and Osbaeck, B., The Effect of Cement Composition on Strength Described by a Strength–Porosity Model, *Cement and Concrete Research,* Vol. 12, No. 2, 1982, pp. 167–178.

Jumikis, A. R., *Rock Mechanics,* 2nd ed., Trans Tech Publications, Clausthal-Zellerfeld, West Germany, 1983.

Kadlecek, V., and Spetla, A., Effect of Size and Shape of Test Specimens on the Direct Tensile Strength of Concrete, *RILEM Bulletin,* No. 36, Sept. 1967, pp. 175–184.

Kalousek, G. L., High Temperature Steam Curing of Concrete at High Pressure, *Fifth International Symposium on the Chemistry of Cement,* Part III, Session 4b, Tokyo, 1968.

Kamenski, M. F., Bewertung von Betonverflussigern und Zementen mit Zumahlstoffen (Evaluation of Concrete Plasticizers and Cements with Finely Divided Mineral Admixtures), Dissertation for the degree Dr. Ing., submitted to the Faculty of Hochschule fur Architektur und Bauwesen Weimar, May 1985, p. 107.

Kani, G. N. J., Basic Facts Concerning Shear Failure, *ACI Journal*, Proc. Vol. 63, No. 6, June 1966, pp. 675–692.

Kantro, D. L., Brunauer, S., and Weise, C. H., Development of Surface in the Hydration of Calcium Silicates, *Solid Surfaces and the Gas–Solid Interface*, American Chemical Society, Advances in Chemistry, Series 33, 1961, pp. 199–219.

Kantro, D. L., Brunauer, S., and Weise, C. H., Development of Surface in the Hydration of Calcium Silicates, II. Extension of Investigations to Earlier and Later Stages of Hydration, *Journal of Physical Chemistry*, Vol. 66, No. 10, American Chemical Society, Washington, DC, Oct. 1962, pp. 1804–1809.

Kaplan, M. F., Flexural and Compressive Strength of Concrete as Affected by the Properties of Coarse Aggregates, *ACI Journal*, Proc. Vol. 55, May 1959a, pp. 1193–1208.

Kaplan, M. F., The Effects of Age and Water/Cement Ratio upon the Relation Between Ultrasonic Pulse Velocity and Compressive Strength of Concrete, *Magazine of Concrete Research*, Vol. 11, No. 32, July 1959b, pp. 85–92.

Kaplan, M. F., Effects of Incomplete Consolidation on Compressive and Flexural Strength, Ultrasonic Pulse Velocity and Dynamic Modulus of Elasticity of Concrete, *ACI Journal*, Proc. Vol. 56, Mar. 1960a, pp. 853–867.

Kaplan, M. F., The Relation Between Ultrasonic Pulse Velocity and the Compressive Strength of Concretes Having the Same Workability but Different Mix Proportions, *Magazine of Concrete Research*, Vol. 12, No. 34, Mar. 1960b, pp. 3–8.

Kaplan, M. F., The Reproducibility of Flexure and Compression Tests for Determining the Strength of Cement, *Proceedings, ASTM*, Vol. 60, Philadelphia, 1960c, pp. 1006–1017.

Kaplan, M. F., Crack Propagation and the Fracture of Concrete, *ACI Journal*, Proc. Vol. 58, Nov. 1961, pp. 591–610.

Kaplan, M. F., Strains and Stresses of Concrete at Initiation of Cracking and Near Failure, *ACI Journal*, Proc. Vol. 60, July 1963, pp. 853–880.

Karsan, I. D., and Jirsa, J. O., Behavior of Concrete Under Compressive Loadings, *Journal of the Structural Division*, Proc. of ASCE, December 1969, pp. 2569–2563.

Kazinczy, G., A vasbeton tarto meretezesenek alapveto kerdesei (Fundamental Problems of Reinforced Concrete Design), *Anyagvizsgalok Kozlonye*, Vol. 16, No. 2–3, Budapest, 1938.

Keiller, A. P., *A Preliminary Investigation of Test Methods for the Assessment of Strength of In Situ Concrete*, Technical Report 551, Cement and Concrete Association, Wexham Springs Slough, UK, 1982.

Keiller, A. P., Assessing the Strength of In Situ Concrete: An Investigation of Test Methods, *Concrete International: Design and Construction*, Vol. 7, No. 2, Feb. 1985, pp. 15–21.

Kelly, J. W., Polivka, M., and Best, C. H., A Physical Method for Determining the Composition of Hardened Concrete, *Cement and Concrete*, ASTM Special Technical Publication 205, Philadelphia, 1958, pp. 135–152.

Kesai, Y., Yokoyama, K., and Matsui, I., Tensile Properties of Early-Age Concrete, *Mechanical Behavior of Materials*, Proceedings of the International Conference on

Mechanical Behavior of Materials, Vol. IV, Society of Materials Science, Japan, 1972, pp. 288–299.

Kesler, C. E., Strength, *Significance of Tests and Properties of Concrete and Concrete Making Materials,* ASTM STP No. 169-A, Philadelphia, 1966, pp. 144–159.

Kesler, C. E., and Higuchi, Y., Determination of Compressive Strength of Concrete by Using Its Sonic Properties, *Proceedings, ASTM,* Vol. 53, 1953, pp. 1044–1052.

Khajuria, A., and Balaguru, P., Comparison of Field, Core, and Laboratory Cylinder Data for Rebound Hammer, *Structural Materials Technology—An NDT Conference,* R. J. Scancella, and M. E. Callahan (Eds.), Technomic, Atlantic City, NJ, 1994, pp. 272–276.

Khoo, L. M., Pullout Technique—An Additional Tool for In Situ Concrete Strength Determination, *In Situ/Nondestructive Testing of Concrete,* V. M. Malhotra (Ed.), ACI Publication SP-82, American Concrete Institute, Detroit, MI, 1984, pp. 143–160.

Kierkegaard-Hansen, P., Lok-strength, *Saertry K of Nordisk Betong,* Vol. 3, 1975, pp. 1–10.

King, J. W. H., *Concrete Quality Control—A Technique of Accelerated Testing Developed at Queen Mary College for the Port of London Authority,* Bulletin, Institution of Civil Engineers, London, Nov. 1955, pp. 46–48.

Kirchner, H., and Rishel, P. A., Measuring the Tensile Strength of a Brittle Material Using a Thermal Contraction Loading Device, *Journal of Materials,* JMLSA, Vol. 6, No. 1, Mar. 1971, pp. 39–47.

Kirtschig, K., Zur Frage der Bohrkernentnahme an Stahlbetonsaulen und-balken (Core Extraction from Reinforced Concrete Columns and Girders), *Betonstein-Zeitung,* Vol. 34, No. 8, Wiesbaden, Aug. 1968, pp. 424–431.

Kirtschig, K., and Dulgeroglu, S., Zur Prufung der Spaltzugfestigkeit von Beton und Bewehrten Proben (Determining the Tensile Splitting Strength of Concrete and Reinforced Specimens), *Betonstein-Zeitung,* Vol. 32, No. 8, Aug. 1966, pp. 471–479.

Klieger, P., Effect of Entrained Air on Strength and Durability of Concrete Made with Various Maximum Sizes of Aggregate, *Proceedings, Highway Research Board,* Vol. 31, 1952, pp. 177–201.

Klieger, P., *Effect of Entrained Air on Strength and Durability of Concrete with Various Sizes of Aggregates,* Bulletin 128, Highway Research Board, Washington, DC, 1956, pp. 1–19.

Klieger, P., Long-Time Study of Cement Performance in Concrete, Chap. 10, Progress Report on Strength and Elastic Properties of Concrete, *ACI Journal,* Proc. Vol. 54, Dec. 1957a, pp. 481–504.

Klieger, P., Early High Strength Concrete for Prestressing, *Proc., World Conference on Prestressed Concrete,* San Francisco, July 1957b, pp. A5 1–14.

Klieger, P., Effect of Mixing and Curing Temperature on Concrete Strength, *ACI Journal,* Proc. Vol. 54, June 1958, pp. 1063–1081.

Klieger, P., and Isberner, A. W., Portland Blast Furnace Slag Cements, *Journal of the PCA Research and Development Laboratories,* Vol. 9, No. 3, Sept. 1967, pp. 2–22.

Koenitzer, L. H., Determination of Modulus of Elasticity and Poisson's Ratio of Concrete at Ages of Fourteen Days to Four Years, *Proceedings, ASTM,* Vol. 35, Part II, 1935, pp. 399–409.

Koenitzer, L. H., Elastic and Thermal Expansion Properties of Concrete as Affected by Similar Properties of Aggregate, *Proceedings, ASTM,* Vol. 36, Part II, 1936, pp. 393–406.

Kokubu, M., Fly Ash and Fly Ash Cement, *Fifth International Symposium on the Chemistry of Cement,* Part IV, Session 2, Tokyo, 1968.

Kokobu, M., and Yamada, J., Fly Ash Cements, *VIth International Congress on the Chemistry of Cement,* Moscow, Sept. 1974.

Kolek, J., An Appreciation of the Schmidt Rebound Hammer, *Magazine of Concrete Research,* Vol. 10, No. 28, Mar. 1958, pp. 27–36.

Kolsky, H., *Stress Waves in Solids,* Dover Publications, New York, 1963.

Komlos, K., O ciniteloch ovplyvnujucich stanovenie pevnosti betonu v nepriamom tahu (Factors That Affect the Measured Strength of Concrete in Indirect Tension), *Stavebnicky Casopis,* Sav. XV, No. 6, Bratislava, 1967, pp. 343–358.

Komlos, K., Determination of the Tensile Strength of Concrete: 1–4, *Indian Concrete Journal,* Nov. 1967, pp. 429–436; Feb. 1968, pp. 68–76; Nov. 1968, pp. 473–482; Feb. 1969, pp. 42–54.

Komlos, K., Factors Affecting the Stress–Strain Relation of Concrete in Uniaxial Tension, *ACI Journal,* Proc. Vol. 66, No. 2, Feb. 1969, pp. 111–114.

Komlos, K., Comments on the Long-Term Tensile Strength of Plain Concrete, *Magazine of Concrete Research,* Vol. 22, No. 73, Dec. 1970, pp. 232–238.

Komlos, K., Popovics, S., Nurnbergerova, T., Bahal, B., and Popovics, J. S., Comparison of Five Standards on Ultrasonic Pulse Velocity Testing of Concrete, *Cement, Concrete, and Aggregates,* CCAGDP, June 1966, pp. 42–48.

Kondo, R., Daimon, M., and Okabayashi, S., Effects of Grain Size Distribution in Alite Cement on Pore Size Distribution and Strength of Hardened Mortar, *Mechanical Behavior of Materials,* Proceedings of the International Conference on Mechanical Behavior of Materials, Vol. IV, Society of Materials Science, Japan, 1972, pp. 193–202.

Kopf, R. J., Powder Actuated Fastening Tools for Use in the Concrete Industry, *Mechanical Fasteners for Concrete,* ACI Publication SP-22, Detroit, MI, 1969, pp. 55–68.

Kopf, R. J., Cooper, C. G., and Williams, F. W., In Situ Strength Evaluation of Concrete—Case Histories and Laboratory Investigations, *Concrete International: Design and Construction,* Vol. 3, No. 3, Mar. 1981, pp. 66–71.

Kotsovos, M. D., Effect of Stress Path on the Behavior of Concrete Under Triaxial Stress States, *ACI Journal,* Proc. Vol. 76, No. 2, 1979, pp. 213–223.

Kotsovos, M. D. and Newman, J. B., Mathematical Description of Deformational Behavior of Concrete Under Generalized Stress Beyond Ultimate Strength, *ACI Journal,* Proc. Vol. 77, No. 5, Sept.–Oct. 1980, pp. 340–346.

Koufopoulos, T., Acceptance Rules for Concrete Strength, *Materials and Structures—Research and Testing,* RILEM, Vol. 15, No. 89, Sept.–Oct. 1982, pp. 453–460.

Krahl, N. W., et al., The Behavior of Plain Mortar and Concrete Under Triaxial Stress, *Proceedings, ASTM,* Vol. 65, Philadelphia, 1965, pp. 697–709.

Kramer, W., Blast-Furnace Slags and Slag Cements, *Chemistry of Cement,* Proceedings of the Fourth International Symposium, NBS Monograph 43, Vol. II, Washington, DC, 1960, pp. 957–973.

Krause, M., and Barmann, R., Frielinghaus, R., Kretzschmar, F., Kroggel, O., Langenberg, K., Maierhofer, C., Muller, W., Neisecke, J., Schickert, M., Schmitz, V., Wiggenhauser, H., and Wollbold, F., Comparison of Pulse-Echo-Methods for Testing Concrete, International Symposium, *Non-destructive Testing in Civil Engineering* (NDT-CE), Berlin, Vol. 1, 1995, pp. 281–296.

Krautkramer, J., and Krautkramer, H., *Ultrasonic Testing of Materials,* 3rd rev. ed., Springer-Verlag, Berlin, 1983.

Krishnaswamy, K. T., Strength and Microcracking of Plain Concrete Under Triaxial Compression, *ACI Journal,* Proc. Vol. 65, No. 10, Oct. 1968, pp. 856–862.

Krishnaswamy, K. T., Strength of Concrete Under Combined Tensile–Compressive Stresses, *Materials and Structures—Research and Testing,* RILEM, Vol. 2, No. 9, May–June 1969, pp. 187–194.

Kriz, L. B., and Lee, S. L., Ultimate Strength of Over-Reinforced Beams, *Journal of the Engineering Mechanics Division, ASCE,* Vol. 86, No. EM3, June 1960.

Krokosky, E. M., Strength vs. Structure: A Study for Hydraulic Cements, *Materials and Structures—Research and Testing,* RILEM, Vol. 3, No. 17, Sept.–Oct. 1970, pp. 313–324.

Kuhl, H., Zement-Chemie, Band III, Die Erhartung und die Verarbiettel (Chemistry of Cement), Vol. III, *The Hardening and Application of Hydraulic Cements,* VEB Verlag, Berlin, 1961.

Kupfer, H. B., and Gerstle, K. H., Behavior of Concrete Under Biaxial Stresses, *Journal of the Engineering Mechanics Division, ASCE,* Vol. 99, No. EM 4, Proc. Paper 9917, Aug. 1973, pp. 853–866.

Kupfer, H., Hilsdorf, H. K., and Rusch, H., Behavior of Concrete Under Biaxial Stresses, *ACI Journal,* Proc. Vol. 66, No. 8, Aug. 1969, pp. 656–666.

La Course, J., Doucet, R., Vezina, D., and Zaikoff, P., Problems Inherent in the Application of a Unit Price Adjustment System, *Concrete International: Design and Construction,* Vol. 5, No. 2, Feb. 1983, pp. 41–45.

Ladanyi, B., and Nguyen, D., Perforated Beam Test for Determining Tensile Strength of Rock, *Journal of Materials,* Vol. 3, No. 3, Sept. 1968, pp. 483–495.

Lambotte, H., Compliance Control of Concrete Survey of the Present and a View to the Future. Baustoffe'85 by the Institut fur Bauforschung, RWTH, Aachen, Germany, pp. 99–104, 1985.

Lamond, J. F., Accelerated Strength Testing by the Warm Water Method, *ACI Journal,* Proc. Vol. 76, No. 4, Apr. 1979, pp. 499–512.

Lamond, J. F., Quality Assurance Using Accelerated Strength Testing, *Concrete International: Design and Construction,* Vol. 5, No. 3, Mar. 1983, pp. 47–51.

Landis, E., and Shah, S., An Investigation of Frequency-Dependent Ultrasonic Attenuation in Concrete, International Symposium, *Non-destructive Testing in Civil Engineering* (NDT-CE), Berlin, Vol. 1, 1995, pp. 1189–1196.

Lapinas, R. A., Accelerated Concrete Strength Testing by Modified Boiling Method: Concrete Producer's View, *Accelerated Strength Testing,* ACI Publication SP-56, American Concrete Institute, Detroit, MI, 1968, pp. 75–93.

Larue, H. A., Modulus of Elasticity of Aggregates and its Effect on Concrete," *Proceedings, ASTM,* Vol. 46, 1946, pp. 1298–1309.

Lauer, K., Magnetic/Electrical Methods, *CRC Handbook on Nondestructive Testing of Concrete,* V. M. Malhotra, and N. J. Carino (Eds.), CRC Press, Boca Raton, FL, 1991, pp. 203–226.

Lawrence, P., Majumdar, A. J., and Nurse, R. W., The Application of the Mackenzie Model to the Mechanical Properties of Cements, *Cement and Concrete Research,* Vol. 1, No. 1, 1971, pp. 75–99.

Lea, F. M., *The Chemistry of Cement and Concrete,* 3rd ed., Chemical Publishing Company, New York, 1971.

Le Châtelier, H., Recherches expérimentales sur la constitution des mortiers hydrauliques (Experimental Researches on the Constitution of Hydraulic Mortats), *Annales des Mines,* Ser. 8, Vol. 11, 1887, pp. 345–465 (English translation by J. L. Mack, published by McGraw Publishing Company, New York, in 1905).

Ledbetter, W. B., and Thompson, J. N., A Technique for Evaluation of Tensile and Volume Change Characteristics of Structural Lightweight Concrete, *Proceedings, ASTM,* Vol. 65, Philadelphia, 1965, pp. 712–726.

Lee, S. L., Tam, C. T., Swaddiwudhipong, S., Ong, K. C. G., and Quek, S. T., Upper and Lower Bound Values of In-Situ Concrete Strength, *Rehabilitation of Concrete Structures,* Proceedings of the International RILEM Conference, Melbourne, 1992, pp. 267–274.

Lenhard, H., Zur Frage der praktischen Bedeutung der vollkommenen Frischbetonverdichtung (Practical Significance of the Complete Compaction of Concrete), *Zement,* Vol. 31, Nos. 11/12 and 13/14, 1942.

Lerch, W. C., and Ford, C. L., Long-Time Study of Cement Performance in Concrete, Chap. 3, Chemical and Physical Tests of the Cements, *ACI Journal,* Proc. Vol. 44, Apr. 1948, pp. 743–795.

Leslie, H. R., and Cheesman, W. J., An Ultrasonic Method of Studying Deterioration and Cracking in Concrete Structures, *ACI Journal,* Proc. Vol. 46, Sept. 1949, pp. 17–36.

Lew, H. S., and Reichard, T. W., Prediction of Strength of Concrete from Maturity, *Accelerated Strength Testing,* ACI Publication SP-56, American Concrete Institute, Detroit, MI, 1978, pp. 229–248.

L'Hermite, R., Méthodes modernes d'essais de materiaux et de prototypes d'ouvrages (Modern Methods for Testing Materials and Products), *Annales de l'Institut Technique du Batiment et des Travaux Publics,* July–Aug. 1936, pp. 19–34.

L'Hermite, R., Etude d'un nouveau type d'eprouvette pour essais de traction, compression et torsion sur le beton (A New Type of Specimen for Testing the Tension, Compression, and Torsion of Concrete), *Annales de l'Institut Technique du Batiment et des Travaux Publics,* No. 6, 1937, pp. 4–9.

L'Hermite, R., La résistance du beton et sa mesure (The Strength of Concrete and Its Measurement) *Annales de l'Institut Technique du Batiment et des Travaux Publics,* Serie Beton, Beton arme, Nos. 5 and 12, Feb. 1949 and Jan. 1950.

L'Hermite, R., Idées actuelles sur la technologie de beton (Up-to-Date Ideas in Concrete Technology), *La Documentation Technique du Batiment et des Travaux Publics,* Paris, 1955.

L'Hermite, R., Volume Changes of Concrete, *Chemistry of Cement,* Proceedings of the Fourth International Symposium, NBS Monograph 43, Vol. II, Washington, DC, 1962, pp. 659–694.

Liebenberg, A. C., A Stress–Strain Function for Concrete Subjected to Short-Term Loading, *Magazine of Concrete Research,* Vol. 14, No. 41, London, July 1962, pp. 85–90.

Lieberum, K. H., and Reinhardt, H. W., Strength of Concrete on an Extremely Small Bearing Area, *ACI Structural Journal,* Vol. 86, No. 1, Jan.–Feb. 1989, pp. 67–78.

Linger, D. A., *Effect of Stress on the Dynamic Modulus of Concrete,* Highway Research Record No. 3, Highway Research Board, Washington, DC, 1963, pp. 62–69.

Linger, D. A., and Gillespie, H. A., A Study of the Mechanism of Concrete Fatigue and Fracture, *Highway Research News,* No. 22, Washington, DC, Feb. 1966, pp. 40–51.

Lisle, R. J., and Strom, C. S., Least-Squares Fitting of the Linear Mohr Envelope, *Quarterly Journal of Engineering Geology,* Vol. 15, No. 1, London, 1982, pp. 55–56.

Lloyd, J. P., Lott, J. L., and Kesler, C. E., *Fatigue of Concrete,* Engineering Experiment Station Bulletin 499, University of Illinois, Urbana, IL, Nov. 1968.

Locher, F. W., Die Festigkeit des Zements (Strength of Cement), *Beton,* Vol. 26, Nos. 7 and 8, Dusseldorf, July and Aug. 1976, pp. 247–249, 283–286. (Also: *Beton-technische Berichte 1976,* Beton-Verlag, Dusseldorf, Germany, 1977, pp. 107–122.)

Locher, F. W., and Richartz, W., Study of the Hydration Mechanism of Cement, *VIth International Congress on the Chemistry of Cement,* Moscow, Sept. 1974.

Long, A. E., and Murray, A. M., Pull-off Test for In Situ Concrete Strength, *Concrete,* Vol. 15, No. 12, London, Dec. 1981, pp. 23–24.

Long, A. E., and Murray, A. M., The Pull-off Partially Destructive Test for Concrete, *In Situ/Nondestructive Testing of Concrete,* V. M. Malhotra (Ed.), ACI Publication SP-82, American Concrete Institute, Detroit, MI, 1984, pp. 327–350.

Lorman, W. R., Verifying the Quality of Freshly Mixed Concrete, *Proceedings, ASTM,* Vol. 62, 1962, pp. 944–959.

Lott, J., and Kesler, C. E., Crack Propagation in Plain Concrete, *Symposium on Structure of Portland Cement Paste and Concrete,* Special Report 90, Highway Research Board, Washington, DC, 1966, pp. 204–218.

Lott, J. L., and Kesler, C. E., Service Behavior of Concrete for Radiation Shielding, *Materials Research and Standards,* Vol. 7, No. 9, Sept. 1967, pp. 375–382.

Luthi, T. M., Meier, H., Primas, R., and Zogmal, O., Infrared Inspection of External Bonded CFPR-Sheets, International Symposium, *Non-destructive Testing in Civil Engineering* (NDT-CE), Berlin, DGZfP, Vol. 1, 1995, pp. 689–696.

Lutz, L. A., Information on the Bond of Deformed Bars from Special Pullout Tests, *ACI Journal,* Proc. Vol. 67, No. 11, Nov. 1970, pp. 884–887.

Lutz, L. A., Bond with Reinforcing Steel, *Significance of Tests and Properties of Concrete and Concrete Making Materials,* ASTM STP 169B, Chap. 21, Philadelphia, 1978, pp. 320–331.

Lutz, L. A., and Gergely, P., Mechanics of Bond and Slip of Deformed Bars in Concrete, *ACI Journal,* Proc. Vol. 64, No. 11, Nov. 1967, pp. 711–721.

Luzhin, O. V., Volokhov, V. A., Shmakov, G. B., Pochtovik, Pohl, Z., and Weber, Z., *Nondestructive Methods for Testing Concrete* (in Russian), Miskva Stroiyizdat, 1985.

Lynn, I. L., and Palmer, K. E., Correlation of Flexural and Compressive Strength of Concretes and Mortars, *Proceedings, ASTM,* Vol. 61, Philadelphia, 1961, pp. 1180–1196.

Mackenzie, J. K., The Elastic Constants of a Solid Containing Spherical Holes, Proceedings, Physical Society, Vol. 63, No. 1, London, 1950, pp. 2–11.

Mailer, H., Pavement Thickness Measurement Using Ultrasonic Techniques, *Nondestructive Testing of Concrete,* Highway Research Record No. 378, Highway Research Board, Washington, DC, 1972, pp. 20–28.

Mailhot, G., Bissailon, A., Carette, G. G., and Malhotra, V. M., In-Place Concrete Strength: New Pullout Methods, *ACI Journal,* Proc. Vol. 76, No. 12, Dec. 1979, pp. 1267–1282.

Malhotra, V. M., Predicting Compressive Strength from Properties of the Fresh Concrete, *Materials, Research and Standards,* MTRSA, Vol. 3, No. 6, June 1963, pp. 483–485.

Malhotra, V. M., *Non-destructive Methods for Testing Concrete,* Department of Energy, Mines and Resources, Mines Branch, Ottawa, Internal Report MPI 66-55, Nov. 1966a.

Malhotra, V. M., Problems Associated with Determining the Tensile Strength of Concrete, presented at the Engineering Institute of Canada, Region II, Technical Conference, Saskatoon, Oct. 31 and Nov. 1, 1966b.

Malhotra, V. M., Discussion of the paper "Relation Between Various Strengths of Concrete" by S. Popovics, *A Symposium on Concrete Strength,* Highway Research Record No. 210, Highway Research Board, Washington, DC, 1967, pp. 90–93.

Malhotra, V. M., Non-Destructive Methods for Testing Concrete, Mines Branch Monograph 875, Department of Energy, Mines and Resources, Ottawa, 1968b.

Malhotra, V. M., Effect of Specimen Size on Tensile Strength of Concrete, *ACI Journal,* Proc. Vol. 67, No. 6, June 1970a, pp. 467–469.

Malhotra, V. M., Concrete Rings for Determining Tensile Strength of Concrete, *ACI Journal,* Proc. Vol. 67, No. 4, Apr. 1970b, pp. 354–357.

Malhotra, V. M., *Maturity Concept and the Estimation of Concrete Strength—A Review,* Information Circular IC 277, Department of Energy, Mines and Resources, Mines Branch, Ottawa, Nov. 1971.

Malhotra, V. M., *Are 4 × 8-Inch Concrete Cylinders as Good as 6 × 12-Inch Cylinders for Quality Control of Concrete?,* Mines Branch Investigation Report IR 72-35, Canada, May 1973a.

Malhotra, V. M., *Recent Developments in Test Methods and Equipment for Evaluation of In-Situ Concrete,* Department of Energy, Mines and Resources, Ottawa, Canada, 1973b.

Malhotra, V. M., Evaluation of the Windsor Probe Test for Estimating Compressive Strength of Concrete, *Materials and Structures—Research and Testing,* RILEM, Vol. 7, No. 37, Jan.–Feb. 1974, pp. 3–15.

Malhotra, V. M., Canadian Experience in the Use of the Modified Boiling Method, *Recent Developments in Accelerated Testing and Maturity of Concrete,* Transportation Research Record No. 558, Transportation Research Board, Washington, DC, 1975a, pp. 13–18.

Malhotra, V. M., Evaluation of the Pull-out Test to Determine Strength of In-Situ Concrete, *Materials and Structures—Research and Testing,* RILEM, Vol. 8, No. 43, Jan.–Feb. 1975b, pp. 19–31.

Malhotra, V. M., *Testing Hardened Concrete: Nondestructive Methods,* ACI Monograph No. 9, American Concrete Institute, Detroit, 1976.

Malhotra, V. M., *Testing Hardened Concrete: Nondestructive Methods,* ACI Monograph No. 9, American Concrete Institute, Detroit, MI, 1976.

Malhotra, V. M., Contract Strength Requirements—Cores Versus In Situ Evaluation, *ACI Journal,* Proc. Vol. 74, No. 4, Apr. 1977, pp. 163–173.

Malhotra, V. M., An Accelerated Method of Estimating the 28 Day Splitting-Tensile and Flexural Strength of Concrete, *Accelerated Strength Testing,* ACI Publication SP-56, American Concrete Institute, Detroit, MI, 1978, pp. 169–182.

Malhotra, V. M., Accelerated Strength Testing: Is It a Solution to a Contractor's Dilemma?, *Concrete International: Design and Construction,* Vol. 3, No. 11, Nov. 1981, pp. 17–21.

Malhotra, V. M., *Proceedings of the CANMET/ACI First International Conference on the Use of Fly Ash, Silica Fume, Slag and Other Mineral By-Products in Concrete,* Vol. 1, ACI Publication SP-79, American Concrete Institute, Detroit, MI, 1983.

Malhotra, V. M. (Ed.), *In Situ/Nondestructive Testing of Concrete,* ACI Publication SP-82, American Concrete Institute, Detroit, MI, 1984.

Malhotra, V. M., In Situ Strength Evaluation of Concrete, *Concrete International: Design and Construction,* Vol. 1, No. 9, Sept. 1987, pp. 40–42.

Malhotra, V. M., Resonant Frequency Methods, *CRC Handbook on Nondestructive Testing of Concrete,* V. M. Malhotra and N. J. Carino (Eds.), CRC Press, Boca Raton, FL, 1991, pp. 147–168.

Malhotra, V. M., and Berwanger, C., *Effect of Below Freezing Temperatures on Strength Development of Concrete,* Mines Branch Investigation Report IR 71-71, Dept. of Energy, Mines and Resources, Ottawa, Canada, Nov. 1971, 34 p.

Malhotra, V. M., and Carette, G. G., In-Situ Testing—A Review, *Progress in Concrete Technology,* V. M. Malhotra, (Ed.), Department of Energy, Mines and Resources, Ottawa, Canada, MRP/MSL 80-89 (TR), 1980a, pp. 749–796.

Malhotra, V. M., and Carette, G., Comparison of Pullout Strength of Concrete with Compressive Strength of Cylinders and Cores, Pulse Velocity, and Rebound Number, *ACI Journal,* Proc. Vol. 77, No. 3, May–June 1980b, pp. 161–170.

Malhotra, V. M., and Carette, G., Penetration Resistance Methods, *CRC Handbook on Nondestructive Testing of Concrete,* V. M. Malhotra and N. J. Carino (Eds.), CRC Press, Boca Raton, FL, 1991, pp. 19–38.

Malhotra, V. M., and Carino, N. J. (Eds.), *CRC Handbook on Nondestructive Testing of Concrete,* CRC Press, Boca Raton, FL, 1991.

Malhotra, V. M., and Zoldners, N. G., Comparison of Ring Tensile Strength of Concrete with Compressive, Flexural and Splitting-Tensile Strengths, *Journal of Materials,* ASTM, Vol. 2, No. 1, Mar. 1967, pp. 160–183.

Malhotra, V. M., and Zoldners, N. G., Some Field Experience in the Use of an Accelerated Method of Estimating 28-Day Strength of Concrete, *ACI Journal,* Proc. Vol. 66, Nov. 1969, pp. 894–897.

Malhotra, V. M., Zoldners, N. G., and Woodrooffe, H. M., Ring Test for Tensile Strength of Concrete, *Materials Research and Standards,* MTRSA, Vol. 6, No. 1, Jan. 1966, pp. 2–12.

Malhotra, V. M., Zoldners, N. G., and Woodrooffe, H. M., Ring Test for Determining the Tensile Strength of Cementitious Materials, *Materials Technology—An Inter-American Approach,* Inter-American Conference on Materials Technology, San Antonio, TX, May 20–24, 1968, pp. 196–205.

Malquori, G., Portland-Pozzolan Cement, *Chemistry of Cement,* Proceedings of the Fourth International Symposium, NBS Monograph 43, Vol. II, Washington, DC, 1960, pp. 983–1000.

Mander, J. B., Priestley, M. J. N., and Park, R., Theoretical Stress-Strain Model for Confined Concrete, *Journal of Structural Engineering,* ASCE, Vol. 114, No. 8, Aug. 1988, pp. 1804–1826.

Manning, D. G., *Detecting Defects and Deterioration in Highway Structures,* National Cooperative Highway Research Program, Synthesis 118, Transportation Research Board, Washington, DC, July 1985.

Manning, D. G., and Holt, F. B., Detecting Delamination in Concrete Bridge Decks, *Concrete International: Design and Construction,* Vol. 2, No. 11, Nov. 1980, pp. 34–41.

Manning, D. G., and Holt, F. B., The Development of Deck Assessment by Radar and Thermography, *Pavement and Bridge Maintenance,* Transportation Research Record No. 1083, Transportation Research Board, Washington, DC, 1986, pp. 13–20.

Manning, D. G., and Hope, B. B., The Effect of Porosity on the Compressive Strength and Elastic Modulus of Polymer Impregnated Concrete, *Cement and Concrete Research,* Vol. 1, No. 6, Nov. 1971, pp. 631–644.

Manns, W., Uber den Einfluss der elastischen Eigenschaften von Zementstein und Zuschlag auf die elastischen Eigenschaften von Mortel und Beton (Effect of the Elastic Properties of Hardened Cement Paste and Aggregate on the Elastic Properties of Mortar and Concrete), Dissertation submitted to the Technische Hochschule, Aachen, Germany, for the degree of Doctor of Engineering, 1969.

Marshall, W. T., Discussion of the paper "Review of Code Requirements for Torsion Design" by G. P. Fisher and P. Zia, *ACI Journal,* Proc. Vol. 61, No. 9, Sept. 1964, pp. 1164–1166.

Martin, B. G., Ultrasonic Attenuations Due to Voids in Fibre-Reinforced Materials, *NDT International,* Vol. 9, 1976, pp. 242–246.

Martin, R. B., and Haynes, R. R., Theoretical Analysis of the Effects of Air Voids in Concrete, *ACI Journal,* Proc. Vol. 68, Jan. 1971, pp. 36–41.

Martin, I., and Junces, J. A., It Pays to Core Test Suspicious Concrete!, *Concrete International: Design and Construction,* Vol. 4, No. 4, Apr. 1982, pp. 52–54.

Martinet, C., Augenblicklicher technischer and wirtschaftlicher Stand der elektrischen Hartung des Betons (Present Technological and Economic Outlook of Electrical Curing of Concrete), *Betonstein-Zeitung,* Vol. 29, No. 11, Nov. 1963, pp. 604–607.

Martinez-Rueda, J. E., and Elnashai, A. S., Confined Concrete Model Under Cyclic Load, *Materials and Structures,* RILEM, Vol. 30, No. 197, Apr. 1997, pp. 139–147.

Mase, G. E., *Continuum Mechanics,* Schaum's Outline Series, McGraw-Hill, New York, 1970.

Maser, K. R., Highway Radar for Pavement and Bridge Deck Evaluation, *Structural Materials Technology—An NDT Conference,* R. J. Scancella, and M. E. Callahan (Eds.), Technomic, Atlantic City, NJ, 1994, pp. 136–140.

Maso, J. C., The Bond Between Aggregate and Hydrated Cement Paste, 7th International Congress on the Chemistry of Cement, Editions Septima, Vol. 1, No. 4, Paris, 1980, pp. VII-1/3–VII-1/15.

Mather, B., Laboratory Tests of Portland Blast Furnace Slag Cements, *ACI Journal,* Proc. Vol. 54, No. 3, Sept. 1957, pp. 205–232.

Mather, B., *The Strength of Portland Cement Concrete as Affected by Air, Water, and Cement Content,* U.S. Army Engineer Waterways Experiment Station, Jackson, MS, 1964, p. 6.

Mather, B., *Stronger Concrete,* Highway Research Record No. 210, Highway Research Board, Washington, DC, 1967, pp. 1–28.

Mather, B., How Soon is Soon Enough?, *ACI Journal,* Proc. Vol. 73, No. 3, Mar. 1976, pp. 147–150.

Mather, B., and Tynes, W. D., Investigation of Compressive Strength of Molded Cylinders and Drilled Cores of Concrete, *ACI Journal,* Proc. Vol. 57, No. 1, Jan. 1961, pp. 767–778.

Mather, K., Petrographic Examination, *Significance of Tests and Properties of Concrete and Concrete Making Materials,* ASTM STP 169B, Chap. 11, Philadelphia, 1978, pp. 132–145.

Mather, K., Condition of Concrete in Martin Dam After 50 Years of Service, ASTM, *Cement, Concrete, and Aggregates,* CCAGDP, Vol. 3, No. 1, Summer 1981, pp. 53–62.

Mather, K., Effects of Accelerated Curing Procedures on Nature and Properties of Cement and Cement-Fly Ash Pastes, *Properties of Concrete at Early Ages,* J. F. Young (Ed.), ACI SP-95, American Concrete Institute, Detroit, MI, 1986, pp. 155–171.

Mathews, D. H., and Metcalf, J. B., Statistical Study of the Compliance with Specification Concrete Supplied for Highway Structures in the United Kingdom, *Nuclear Testing, Construction Variability, Materials Control and Acceptance,* Highway Research Record No. 290, Highway Research Board Washington, DC, 1969, pp. 50–57.

McClintok, F. A., and Argon, A. S., *Mechanical Behavior of Materials,* Addison-Wesley, Reading, MA, 1966.

McCoy, E. E., and Mather, B., Discussion of paper "Dynamic Testing of Materials" by L. J. Mitchell, *HRB Proceedings,* Vol. 33, 1954, pp. 256–258.

McDowell, P. W., and Millett, N., Surface Ultrasonic Measurement of Longitudinal and Transverse Wave Velocities through Rock Samples, *International Journal of Rock Mechanics, Mineral Science and Geomechanics Abstract,* Vol. 21, No. 4, 1984, pp. 223–227.

McHenry, D., and Karni, J., Strength of Concrete Under Combined Tensile and Compressive Stresses, *ACI Journal,* Proc. Vol. 54, No. 10, Apr. 1958, pp. 829–840.

McHenry, D., and Schideler, J. J., Review of Data on Effect of Speed in Mechanical Testing of Concrete, *Symposium on Speed of Testing,* ASTM STP No. 185, 1956, pp. 72–82.

McIntosh, J. D., The Selection of Natural Aggregates for Various Types of Concrete, *The Reinforced Concrete Review,* Vol. IV, No. 5, Mar. 1957, pp. 281–305.

McIntosh, J. D., *Concrete and Statistics,* CR Books, London, 1963.

McIntosh, J. D., *Concrete Mix Design,* Cement and Concrete Association, London, 1964.

McNeely, D. J., and Lash, S. D., Tensile Strength of Concrete, *ACI Journal,* Proc. Vol. 60, No. 6, June 1963, pp. 751–761.

Mehmel, A., and Kern, E., Elastische und Plastische Stauchungen von Beton infolge Druckschwell- und Standbelastung (Elastic and Plastic Strains of Concrete due to Cyclic and Static Loading), *Deutscher Ausschuss fur Stahlbeton,* Heft 153, Wilhelm Ernst & Sohn, Berlin, 1962.

Mehta, P. K., Influence of Different Crystalline Forms of C_3A on Sulfate Resistance of Portland Cement, *7th International Congress on the Chemistry of Cement,* Vol. IV, Editions Septima, Paris, 1980, pp. 575–579.

Mehta, P. K., and Monteiro, P. J. M., *Concrete—Structure, Properties and Materials,* 2nd ed., Prentice Hall, Englewood Cliffs, NJ, 1993.

Meininger, R. C., Effect of Core Diameter on Measured Concrete Strength, *Journal of Materials,* ASTM, Vol. 3, No. 2, June 1968, pp. 320–336.

Mena-Ferrer, M., Quality Control of Concrete by Means of Short-Termed Tests at La Angostura Hydroelectric Project, State of Chiapas, Mexico, *Accelerated Strength Testing,* ACI Publication SP-56, American Concrete Institute, Detroit, MI, 1978, pp. 51–73.

Mercer, L. B., Concrete Strength Variation—60 Contributing Causes, *ACI Journal,* Proc. Vol. 47, No. 9, May 1951, pp. 745–747.

Meyer, A., Die Biegezugfestigkeit als Gutemerkmal des Betons (Flexural Strength as the Characteristic of Quality for Concrete), *Der Bauingenieur,* Vol. 38, No. 2, Mar. 1963a, pp. 45–51.

Meyer, A., Uber den Einfluss des Wasserzementwertes auf die Fruhfestigkeit von Beton (On the Effect of the Water–Cement Ratio on the Early Strength of Concrete), *Betonstein-Zeitung,* Vol. 29, No. 8, Aug. 1963b, pp. 391–394.

Meyer, A., Normen fur die Festigkeitsprufung von Zement (Standards for Testing the Strength of Cement), *Betonstein-Zeitung,* Vol. 32, No. 2, Feb. 1966, pp. 68–80.

Meyer, A., Beton mit hocher Fruhfestigkeit (High Early Strength Concrete), *Betonstein-Zeitung,* Vol. 3, No. 5, May 1967, pp. 213–227.

Michigan Department of State Highways, *Highway Quality Control Program—Statistical Parameters,* Research Report No. R572, Research Laboratory Division, Office of Testing and Research, Project 63-G-124, 1966.

Midgley, H. G., Electron Microscopy of Set Portland Cement, Structure, *Structure, Solid Mechanics and Engineering Design,* The Proceedings of the Southampton 1969 Civil Engineering Materials Conference, Part 1, Wiley-Interscience, London, 1971, pp. 275–288.

Mielenz, R. C., Use of Surface-Active Agents in Concrete, *Proceedings of the Fifth International Symposium on the Chemistry of Cement,* Part IV, Admixtures and Special Cements, Tokyo, 1969, pp. 1–33.

Millard, D. J., Discussion of the paper "Porosity and Strength of Brittle Solids" by K. K. Schiller, *Mechanical Properties of Non-metallic Brittle Materials,* W. H. Walton (Ed.), Interscience Publishers, New York, 1968, pp. 45–47.

Mills, R. H., Influence of Water in Areas of Restricted Adsorption on Properties of Concrete, *Materials and Structures—Research and Testing,* Vol. 1, No. 6, Paris, Nov.–Dec. 1968, pp. 553–559.

Mills, R. H., and Ono, K., Elastic Modulus of Close-Packed Randomly Oriented Maxwell Elements, *Materials and Structures—Research and Testing,* RILEM, Vol. 5, No. 27, May–June 1972, pp. 127–133.

Mindess, S., and Young, J. F., *Concrete,* Prentice Hall, Englewood Cliffs, NJ, 1981.

Mironov, S. A., Some Generalizations in Theory and Technology of Acceleration of Concrete Hardening, *Symposium on Structure of Portland Cement Paste and Concrete,* Special Report 90, Highway Research Board, Washington, DC, 1966, pp. 465–474.

Mitchell, L. J., and Hoagland, G. G., Investigation of the Impact-Type Concrete Test Hammer, *Tests for Concrete and Durability of Concrete Aggregates,* Bulletin 305, Highway Research Board, Washington, DC, 1961, pp. 14–27.

Mitchell, N. B., Jr., The Indirect Tension Test for Concrete, *Materials Research and Standards,* MTRSA, Vol. 1, No. 10, Oct. 1961, pp. 780–788.

Mitchell, T. M., Radioactive/Nuclear Methods, *CRC Handbook on Nondestructive Testing of Concrete,* V. M. Malhotra, and N. J. Carino (Eds.), CRC Press, Boca Raton, FL, 1991, pp. 227–252.

Modry, S. (Ed.), *Proceedings of the RILEM/IUPAC International Symposium on Pore Structure and Properties of Materials,* Academie Prague, Prague, 1973.

Mohr, O., Die Scherfestigkeit des Betons (Shear Strength of Concrete), *Armierter Beton,* Vol. IV, July 1911, pp. 247–250.

Monack, A. J., Intrinsic Impact Strength of Ceramics: Charpy Tests, *Materials Research and Standards,* MTRSA, Vol. 11, No. 4, Apr. 1971, pp. 24–25.

Monfore, G. E., and Lentz, A. E., Physical Properties of Concrete at Very Low Temperatures, *Journal of the PCA Research and Development Laboratories,* Vol. 4, No. 2, May 1962, pp. 33–39.

Moore, R. W., Earth-Resistivity Tests Applied as a Rapid, Nondestructive Procedure for Determining Thickness of Concrete Pavements, *Mechanical Properties of Plastic Concrete and Pavement Thickness Measurement,* Highway Research Record No. 218, Highway Research Board, Washington, DC, 1968, pp. 49–55.

Morrison, G. L., Virmant, Y. P., Stratton, F. W., and William, J. G., *Chloride Removal and Monomer Impregnation of Bridge Deck Concrete by Electro-osmosis,* Report No. FHWA-Ks-RD. 74-1, Kansas Department of Transportation, Topeka, Kansas, 1976a.

Morrison, G. L., Virmoni, Y. P., Ramamurti, K., and Gilliland, W. J., *Rapid In Situ Determination of Chloride Ion in Portland Cement Concrete Bridge Decks,* Final Report, Report No. FHWA-Ks-RD 75-2, Kansas Department of Transportation, Topeka, Kansas, Nov. 1976b.

Morschtschichin, W. N., Bestimmung des Elastizitatsmoduls von Beton mit der Schallimpulsemethode (Determination of the Modulus of Elasticity of Concrete by Sonic Impulse Methods), *Zerstorungsfreie Pruf- und Messtechink fur Beton und Stahlbeton,* Proceedings of an International Conference, Leipzig, Apr. 1969, pp. 45–48.

Muenow, R., A Sonic Method to Determine Pavement Thickness, *Journal of the PCA Research and Development Laboratories,* Vol. 5, No. 3, Sept. 1963, pp. 8–21.

Mukherjee, P. K., Practical Application of Maturity Concept to Determine in Situ Strength of Concrete, *Recent Developments in Accelerated Testing and Maturity,* Transportation Research Record No. 558, Transportation Research Board, Washington, DC, 1975, pp. 87–92.

Mullen, W. G., and Bodvarsson, G. W., *Determination of Air Void Content and Mixing Water Void Content of Hardened Concrete Using Electron Microscope Techniques,* Part I, Report No. ERSD-110-75-2, Highway Research Program, North Carolina Department of Transportation and Highway Safety, June 1978.

Mullin, J. R., and Knoell, A. C., Basic Concepts in Composite Beam Testing, *Materials Research and Standards,* MTRSA, Vol. 10, No. 12, December 1970, pp. 16–20, 33.

Munday, J. G. L., and Dhir, R. K., Assessment of In Situ Concrete Quality by Core Testing, *In Situ/Nondestructive Testing of Concrete,* V. M. Malhotra (Ed.), ACI Publication SP-82, American Concrete Institute, Detroit, MI, 1984, pp. 393–410.

Murphy, G., *Advanced Mechanics of Materials,* McGraw-Hill, New York, 1964.

Murphy, W. E., Formwork Striking Times, *VTT Symposium 49,* Third International Conference on the Durability of Building Materials and Components, Vol. 2, Technical Research Center of Finland, Espoo, Finland, Aug. 12–15, 1984, pp. 220–228.

Murray, D. W., Octahedral Based Incremental Stress-Strain Matrices, *Journal of the Engineering Mechanics Division,* ASCE, Vol. 105, No. EM4, Proc. Paper 14734, Aug. 1979, pp. 501–513.

Murray, A. M., and Long, A. E., A Study of the In Situ Variability of Concrete Using the Pull-Off Method, *Proceedings, Institution of Civil Engineers,* Part 2, Vol. 83, Dec. 1987, pp. 731–745.

Murrell, S. A. F., The Strength of Coal Under Triaxial Compression, *Mechanical Properties of Non-metallic Brittle Materials,* W. H. Walton (Ed.), Interscience Publishers, New York, 1958, pp. 123–146.

Murrell, S. A. F., Micromechanical Basis of the Deformation and Fracture of Rocks, *Structure, Solid Mechanics and Engineering Design,* Proceedings of the Southampton 1969 Civil Engineering Materials Conference, Part 1, Wiley-Interscience, London, 1971, pp. 239–248.

Nadai, A., *Theory of Flow and Fracture of Solids,* 2nd ed., Vol. 1, McGraw-Hill, New York, 1950.

Nagaraj, T. S., Shashiprakash, S. G., and Kameswara, R. B., Generalized Abrams' Law, *Properties of Fresh Concrete* (RILEM Colloquim), H.-J. Wierig, (Ed.), Chapman & Hall, London, 1990, pp. 242–252.

Narayanan, R., The Tensile Strength of Concrete by the Split Test, *Indian Concrete Journal,* Aug. 1961, pp. 307–309.

Narayanan, R., and Rao, T. K., The Bond Strength and Maturity of Concrete, *Indian Concrete Journal,* Vol. 66, No. 4, Apr. 1962, pp. 138–141.

Narrow, I., and Ullberg, E., Correlation Between Tensile Splitting Strength and Flexural Strength of Concrete, *ACI Journal,* Proc. Vol. 60, Jan. 1963, pp. 27–38.

Nasser, K. W., and Kenyon, J. C., Why Not 3 × 6-Inch Cylinders for Testing Concrete Compressive Strength?, *ACI Journal,* Proc. Vol. 81, No. 1, Jan.–Feb. 1984, pp. 47–53.

Nasser, K. W., and Al-Manaseer, A. A., Comparison of Nondestructive Testers of Hardened Concrete, *ACI Materials Journal,* Vol. 84, No. 5, Sept.–Oct. 1987, pp. 374–388.

National Science Foundation, *Civil Infrastructure Systems Research: Strategic Issues,* Washington, DC, 1993.

Navaratnarajah, V., A New Approach to the Ultimate Strength of Concrete in Pure Torsion, *ACI Journal,* Proc. Vol. 65, No. 2, Feb. 1968, pp. 121–128.

Nelson, G. G., and Frei, O. C., Lightweight Structural Concrete Proportioning and Control, *ACI Journal,* Proc. Vol. 54, No. 9, Jan. 1958, pp. 605–621.

Nerenst, P., and Plum, N. M., Freezing and Thawing Tests on Green Concrete, *RILEM Symposium, Winter Concreting Theory and Practice,* Session BI, Danish National Institute of Building Research, Special Report, Copenhagen, 1956.

Neville, A. M., The Relation Between Standard Deviation and Mean Strength of Concrete Test Cubes, *Magazine of Concrete Research,* Vol. 11, No. 32, July 1959, pp. 75–84.

Neville, A. M., A General Relation for Strengths of Concrete Specimens of Different Shapes and Sizes, *ACI Journal,* Proc. Vol. 63, Oct. 1966, pp. 1095–1109.

Neville, A. M., Concrete—A Non-elastic Material in the Laboratory and in Structures, Stanton Walker Lecture Series on the Materials Sciences, No. 6, University of Maryland, Nov. 1968.

Neville, A. M., *Creep of Concrete: Plain, Reinforced, and Prestressed,* North-Holland Amsterdam/American Elsevier, New York, 1970.

Neville, A. M., Creep of Concrete in Hot Climate, *Concrete and Reinforced Concrete in Hot Countries,* Vol. I, Proceedings of International RILEM Symposium, Haifa, Aug. 1971, pp. 97–108.

Neville, A. M., *Properties of Concrete,* 2nd ed., John Wiley, New York, 1973.

Neville, A. M., *High Alumina Cement Concrete,* John Wiley, New York, 1975.

Neville, A. M., *Properties of Concrete,* 3rd ed., Pitman Publishing, London, 1981.

Neville, A. M., *Properties of Concrete,* 4th ed., John Wiley, New York, 1997.

Neville, A. M., and Kennedy, J. B., *Basic Statistical Methods,* International Textbook Co., Scranton, PA 1964.

Neville, A. M., and Staunton, M. M., Method of Estimating Creep of Concrete When the Stress–Strength Ratio Varies with Time, *ACI Journal,* Proc. Vol. 62, No. 10, Oct. 1965, pp. 1293–1312.

Newlon, H. W., Jr., Variability of Portland Cement Concrete, *Proceedings of National Conference on Statistical Quality Control Methodology in Highway and Airfield Construction,* Charlottesville, VA, Nov. 1966, pp. 259–299.

Newman, A. J., and Teychenne, D. C., A Classification of Natural Sands and Its Use in Concrete Mix Design, *Symposium on Mix Design and Quality Control of Concrete,* Cement and Concrete Association, London, May 1954, pp. 175–194 and Discussion.

Newman, K., The Effect of Water Absorption by Aggregates on the Water/Cement Ratio of Concrete, *Magazine of Concrete Research,* Vol. 1, No. 33, Nov. 1959, pp. 135–142.

Newman, K., Concrete Control Tests as Measures of the Properties of Concrete, *Proceedings of a Symposium on Concrete Quality,* Cement and Concrete Association, London, Nov. 1964, pp. 119–138.

Newman, K., Properties of Concrete, *Structural Concrete,* Vol. 2, No. 11, Sept.–Oct. 1965, pp. 451–483.

Newman, K., Criteria for the Behavior of Plain Concrete Under Complex States of Stress, *The Structure of Concrete,* Proceedings, International Conference, London, Sept. 1965, Cement and Concrete Association, London, 1968, pp. 255–274.

Newman, K., Failure Theories and Design Criteria for Plain Concrete, *Structure, Solid Mechanics, and Engineering Designs,* Proceedings, Southhampton 1969 Civil Engineering Materials Conference, Part 2, Wiley-Interscience, London, 1971, pp. 963–996.

Nicholls, R., *Composite Construction Materials Handbook,* Prentice-Hall, Englewood Cliffs, NJ, 1976.

Nikkanen, P., A Research on the Electrical Heating Methods for Winter Concreting, *RILEM Bulletin,* No. 33, Dec. 1966, pp. 371–380.

Nilsson, L. O., Hygroscopic Moisture in Concrete Drying, Measurements and Related Material Properties, Doctoral Dissertation, Lund University, Lund, Sweden, Apr. 1980.

Nilsson, S., The Tensile Strength of Concrete Determined by Splitting Test on Cubes, *RILEM Bulletin,* No. 11, June 1961, pp. 63–67.

Nilsson, S., Tensile Strength and Compressive Strength of Concrete Determined on the Same Cube, *RILEM Bulletin,* No. 17, Dec. 1962, pp. 35–36.

Noble, P. M., The Effect of Aggregate and Other Variables on the Elastic Properties of Concrete, *Proceedings, ASTM,* Vol. 31, Part I, 1931, pp. 399–421.

Nogula, K., A betonok vizadagolasanak es konzisztenciajanak szuresek alapjan torteno szamitasa (Calculation of the Water Need and Consistency of Concretes Based on Filtering), ETI Tudomanyos Muszaki Beszamolo, Epitestudomanyi Intezet, Budapest, 1956.

Novgordsky, M., *Testing of Building Materials and Structures,* Mir Publishers, Moscow, 1973 (revised from the 1971 Russian edition).

NRMCA, *In-Place Concrete Strength Evaluation—A Recommended Practice,* Publication No. 133-79, National Ready Mixed Concrete Association, Silver Spring, MD, 1979.

Nurse, R. W., Steam Curing of Concrete, *Magazine of Concrete Research,* Vol. 1, No. 2, June 1949, pp. 79–88.

Nurse, R. W., Cohesion and Adhesion in Solids, *The Structure of Concrete,* Proceedings of an International Conference, London, Sept. 1965, Cement and Concrete Association, London, 1968, pp. 49–58.

Obert, L., and Duvall, W., Discussion of Dynamic Methods of Testing Concrete with Suggestions for Standardization, *Proceedings, ASTM,* Vol. 41, 1941, pp. 1053–1070.

Oehlers, D. J., and Johnson, R. P., The Splitting Strength of Concrete Prisms Subjected to Surface Strip or Patch Loads, *Magazine of Concrete Research,* Vol. 33, No. 116, Sept. 1981, pp. 171–179.

Okada, K., Kobayashi, K., and Miyagawa, T., Corrosion Monitoring Method of Reinforcing Steel in Offshore Concrete Structures, *In Situ/Nondestructive Testing of Concrete,* V. M. Malhotra (Ed.), ACI Publication SP-82, American Concrete Institute, Detroit, MI, 1984, pp. 703–720.

Okajima, T., Failure of Plain Concrete Under Combined Stresses (Compression–Torsion, Tension–Torsion), *Mechanical Behavior of Materials,* Proceedings, International Conference on Mechanical Behavior of Materials, Vol. IV, Society of Materials Science, Japan, 1972, pp. 12–20.

Orchard, D. F., *Concrete Technology Practice,* 2nd ed., Vol. 2, Contractors Record, London/John Wiley, New York, 1962.

Orchard, S. E., Dealing with the Data, *ASTM Standardization News,* STDNA, Vol. 2, No. 4, Apr. 1974, pp. 23–26.

Orowan, E., Fracture and Strength of Solids, *Reports on Progress in Physics,* Vol. 12, Physical Society, London, 1948–1949, pp. 185–232.

Orr, D. M. F., and Haigh, G. F., An Apparatus for Measuring the Shrinkage Characteristics of Plastic Mortars, *Magazine of Concrete Research,* Vol. 23, No. 74, Mar. 1971, pp. 43–48.

Osinski, A., Application de la similitude mécanique dans les essais non destructifs du beton (Application of the Theory of Mechanical Similitude to the Non-destructive Testing of Concrete), *Materials and Structures—Research and Testing,* RILEM, Vol. 7, No. 40, July–Aug. 1974, pp. 273–281.

Osinski, A., Relation résistance—vitesse du son dans les betons exprimée par une constante universelle (Dependence Between the Strength and the Sound Velocity in Concretes Expressed by Means of One Criterion), *Materials and Structures—Research and Testing,* RILEM, Vol. 12, No. 71, Sept.–Oct. 1979, pp. 407–412.

Otter, D. E., and Naaman, A. E., Properties of Steel Fiber Reinforced Concrete Under Cyclic Loading, *ACI Materials Journal,* Vol. 85, No. 4, July–Aug. 1988, pp. 254–261.

Ottosen, N. S., A Failure Criterion for Concrete, *Journal of the Engineering Mechanics Division, ASCE,* Vol. 103, No. EM4, Aug. 1977, pp. 527–535.

Palaniswamy, R., and Shah, S. P., Fracture and Stress–Strain Relationship of Concrete Under Triaxial Compression, *Journal of the Structural Division, ASCE,* Vol. 100, No. ST 5, Proc. Paper 10547, May 1974, pp. 901–916.

Palotas, L., Vergleich der verschiedenen Formeln zur Vorausbestimmung der Wurfelfestigkeit von Beton (Comparison of the Various Formulas for Predetermining the Compressive Strength of Concrete), *Zement,* Vol. 24, Nos. 36 and 37, Sept. 5 and 12 1935, pp. 565–570, 579–581.

Palotas, L., A minosegi beton mertekado alapadatai (Fundamental Characteristics of Quality Concrete), Minosegi Beton Konferencia, Budapest, 1960, pp. 9–38.

Palotas, L., *A vasbeton elmelete* (Theory of Reinforced Concrete), Akademiai Kiado, Budapest, 1973.

Palotas, L., and Halasz, I., Bewertung der verschiedenen Festigkeitskennwerte und Untersuchungsmethoden des Betons (Evaluation of the Different Strength Characteristics and Test Methods for Concrete), *Wissenschaftliche Zeitschrift,* Heft 12, Hochschule fur Bauwesen, Leipzig, 1963, pp. 117–126.

Papadakis, M., Rheologie des suspensions de ciment (Rheology of Cement Suspensions), *Revue des Materiaux de Construction,* No. 476, Mai 1955, pp. 121–137.

Papadakis, M., Die Bedeutung der Feinstoffe in der Fliesskunde des Frischbetons (Role of Fine Particles in the Rheology of Fresh Concrete), *Betonstein-Zeitung,* Vol. 29, No. 11, Nov. 1963, pp. 555–558.

Park, S., Uomoto, T., and Yoshizawa, M., Analysis of Radar Response on Subsurface Objects in Concrete by Simulation Technique, International Symposium, *Nondestructive Testing in Civil Engineering* (NDT-CE), G. Schickert and H. Wiggenhouser (Eds.), Sept. 26–28, 1995, DGZfP Berlin, 1995, pp. 673–680.

Parker, W. E., Pulse Velocity Testing of Concrete, *Proceedings, ASTM,* Vol. 53, 1953, pp. 1033–1042.

Parsons, T. J., and Naik, T. R., Early Age Concrete Strength Determination by Pullout Testing and Maturity, *In Situ/Nondestructive Testing of Concrete,* V. M. Malhotra (Ed.), ACI Publication SP-82, American Concrete Institute, Detroit, MI, 1984, pp. 177–200.

Parsons, T. J., and Naik, T. R., Early Age Concrete Strength Determination by Maturity, *Concrete International: Design and Construction,* Vol. 7, No. 2, Feb. 1985, pp. 37–43.

Patch, D. G., An 8-Hour Accelerated Strength Test for Field Concrete Control, *Proceedings, ACI,* Vols. 4–5, Mar.–Apr. 1933, pp. 318–324.

Paul, B., Prediction of Elastic Constants of Multiphase Materials, *Transactions of the Metallurgical Society of AIME,* Vol. 218, Feb. 1960, pp. 36–41.

Pauw, A., Static Modulus of Elasticity of Concrete as Affected by Density, *ACI Journal,* Proc. Vol. 57, Dec. 1960, Part I, pp. 679–687.

Paven, H., and Popovics, S., Peculiarities in the Phenomenological Quantitative and Qualitative Rheological Effects in Binary Solid Composites with Linear Viscoelastic Behavior, *Annual Symposium of the Institute of Solid Mechanics,* Rumanian Academy, Bucharest, Nov. 28–29, 1991, pp. 69–72.

Paven, H., and Popovics, S., Rheology of Polymeric Composites, VII. Composite Averages Approach for Binary Materials with Linear Viscoelastic Behavior (in Rumanian), *Revista de Chimie,* Vol. 43, Nos. 11–12, 1992, pp. 716–722.

Paven, H., and Popovics, S., Morph-rheological Interactions and Selection Rules in Hybrid Phase in Phase Composites with Solid Linear Viscoelastic Components 1. Composite Couplings Approach, *The Annual Symposium of the Institute of Solid Mechanics,* 1994, pp. 281–290.

Peirce, F. T., Tensile Tests for Cotton Yarns: V. The Weakest Link—Theorems on the Strength of Long and of Composite Specimens, *Journal of Textile Institute,* Vol. 17, 1926, p. 355.

Peltier, R., Theoretical Investigation of the Brazailian Test, *RILEM Bulletin,* No. 19 (Old Series), Oct. 1954, pp. 31–69.

Petersen, C. G., LOK-Test and CAPO-Test—Development and Their Applications, *Proceedings, Institution of Civil Engineers,* Part 1, No. 76, May 1984, pp. 539–549.

Petersen, P. H., and Stoll, U. W., Relation of Rebound-Hammer Test Results to Sonic Modulus and Compressive-Strength Data, *Highway Research Board Proceedings,* Vol. 34, Washington, DC, 1955, pp. 387–395.

Petersons, N., Strength of Concrete in Finished Structures, *RILEM Symposium on the Experimental Research of Field Testing of Concrete,* Trondheim, Norway, Oct. 5–7, 1964a, pp. 214–232.

Petersons, N., *Strength of Concrete in Finished Structures,* Kungl, Tekniska Hogskolans Handlingar, NR 232, Stockholm, 1964b.

Petersons, N., Should Standard Cube Test Specimens Be Replaced by Test Specimens Taken from Structures?, *Materials and Structures—Research and Testing,* RILEM, Vol. 1, No. 5, Sept.–Oct. 1968, pp. 425–435.

Philleo, R. E., Comparison of Results of Three Methods for Determining Young's Modulus of Elasticity of Concrete, *ACI Journal,* Proc. Vol. 51, Jan. 1955, pp. 461–467.

Philleo, R. E., Some Physical Properties of Concrete at High Temperatures, *ACI Journal,* Proc. Vol. 54, Apr. 1958, pp. 857–864.

Philleo, R. E., Elastic Properties and Creep, *Significance of Tests and Properties of Concrete and Concrete Making Materials,* ASTM STP No. 169-A, Philadelphia, 1966, pp. 160–175.

Philleo, R. E., The Strength of Concrete—A Statistical View, Stanton Walker Lecture Series on the Materials Sciences, No. 5, University of Maryland, Nov. 16, 1967.

Philleo, R. E., A Need for In Situ Testing of Concrete, *Concrete International: Design and Construction,* Vol. 1, No. 9, Sept. 1979, pp. 43–44.

Philleo, R. E., Accelerated Testing—A Review, *Progress in Concrete Technology,* V. M. Malhotra (Ed.), Department of Energy, Mines and Resources, Ottawa, Canada, MRP/MSL 80-90 (TR), 1980, pp. 729–748.

Philleo, R. E., Increasing the Usefulness of ACI 214: Use of Standard Deviation and a Technique for Small Sample Sizes, *Concrete International: Design and Construction,* Vol. 3, No. 9, Sept. 1981, pp. 71–74.

Phillips, C. J., The Strength and Weakness of Brittle Materials, *American Scientist,* Vol. 53, No. 1, Mar. 1965, pp. 20–51.

Pickett, G., Equations for Computing Elastic Constants from Flexural and Torsional Resonant Frequencies of Vibration of Prism and Cylinders, *Proceedings, ASTM,* Vol. 45, 1945, pp. 846–863.

Pihlajavaara, S. E., *On the Interrelation of the Moisture Content and the Strength of Mature Concrete, and Its Reversibility,* State Institute for Technical Research, Finland, Reports, Series III, Building 76, Helsinki, 1964.

Pihlajavaara, S., A Review of the Research Carried Out in the State Institute of Technical Research (Finland), *RILEM Bulletin,* No. 27 (New Series), June 1965, pp. 61–64.

Pihlajavaara, S. E., Some Results of the Effect of Carbonation on the Porosity and Pore Size Distribution of Cement Paste, *Materials and Structures—Research and Testing,* RILEM, Vol. 1, No. 6, Nov.–Dec. 1968, pp. 521–527.

Pincus, G., and Gesund, H., Evaluating the Tensile Strength of Concrete, *Materials Research and Standards,* MTRSA, Vol. 5, No. 9, Sept. 1965, pp. 454–458.

Pineiro, M., Vasquez, R., and Vergara, A., Tension Strength of Concrete: Comparison Between Direct Tensile, Splitting and Flexure Strength, *An Inter-American Approach for the Seventies,* Materials Technology—I. Second Inter-American Conference on Materials Technology, Mexico City, Mexico, Aug. 24–27, 1970, pp. 24–42.

Platts, D. R., and Kirchner, H. P., Comparing Tensile and Flexural Strengths of a Brittle Material, *Journal of Materials,* JMLSA, Vol. 6, No. 1, Mar. 1971, pp. 48–59.

Plowman, J. M., Young's Modulus and Poisson's Ratio of Concrete Cured at Various Humidities, *Magazine of Concrete Research,* Vol. 15, No. 44, July 1963, pp. 77–82.

Pogany, A., Bestimmung der Zugfestigkeit des Betons mit Hilfe von mit Innerdruck belasteten zylindrischen Probekorpern (Determination of the Tensile Strength of Concrete by Means of Cylindrical Specimens Loaded with Internal Pressure), *Zement,* Vol. 26, No. 24, June 17, 1937, pp. 397–398.

Pohl, E., Prufung von Beton mit nieder- und hochfrequenten mechanischen Schwingungen (Testing Concrete with Mechanical Vibrations of Low and High Frequency), *Wissenschaftliche Zeitschrift,* Heft 1, Hochschule fur Bauwesen, Leipzig, 1962, pp. 19–64.

Pohl, E., Zerstorungsfreie Prufung im Stahlbetonbau mit geschlossenen Gammastrahlen, *Wissenschaftliche Zeitschrift,* Heft 1, Hochschule fur Bauweesen, Leipzig, 1964a, pp. 81–86.

Pohl, E., *Kerntechnik im Bauwesen,* VEB Verlag fur Bauwesen, Berlin, 1964b.

Pohl, E., *Zerstorungsfreie Pruf- und Messmethoden fur Beton* (Nondestructive Test- and Measuring Methods for Concrete), 2nd ed., VEB Verlag fur Bauwesen, Berlin, 1969a, 200 p.

Pohl, E., Kombination der Ultraschall- und Drehmomentmethode zur Bestimmung der Druckfestigkeit und der Homogenitat von verarbeitetem Beton (Combination of Ultrasonic and Torque Methods for the Determination of Compressive Strength and Uniformity of Concrete), *Zerstorungsfreie Pruf- und Messtechnik fur Beton und Stahlbeton* (Nondestructive Testing and Measuring Techniques for Concrete and Reinforced Concrete), *Proceedings of an International Conference,* Hochschule fur Bauwesen, Leipzig, Apr. 1969b, pp. 93–96.

Pohl, E., Bericht uber Forschungsarbeiten "Zerstorungsfreie Pruf- und Messmethoden im Bauwesen" in der UdSSR (Report About the Research Activity in USSR Concerning "Non-Destructive Testing and Measuring Methods in Buildings"), *Wissenschaftliche Zeitschrift,* Heft 3, Hochschule fur Bauwesen, Leipzig, 1970, pp. 127–132.

Pohl, E., Anderung der dynamischen elastischen Konstanten von Betonprufkorpern in Abhangigkeit von der Temperatur und der Feuchtigkeit, *Wissenschaftliche Zeitschrift,* No. 3, Hochschule fur Bauwesen, Leipzig, 1975, pp. 157–160.

Poijarvi, H., and Syrjala, H., Betonin Pursitusluguusmaarityksista Erilaisilla Koekappaleilla (On the Determination of the Compressive Strength of Concrete with Various Test Specimens), *Tiedotus,* Sarja III—Rakennus 91, State Institute for Technical Research, Finland, Helsinki, 1965.

Polivka, M., Kelly, J. W., and Best, C. H., A Physical Method for Determining the Composition of Hardened Concrete, *Cement and Concrete,* ASTM STP 205, Philadelphia, 1958, pp. 135–152.

Popov, E. P., *Mechanics of Materials,* 2nd ed., Prentice Hall, Englewood Cliffs, NJ, 1976.

Popovics, J. S., Feasibility Study Concerning the Application of Advanced Ultrasonics to Concrete, Thesis, submitted to the faculty of Drexel University in partial fulfillment of the requirements for the degree of Master of Science of civil engineering, Philadelphia, June 1990a.

Popovics, J. S., Are Advanced Ultrasonic Techniques Suitable for Concrete?—An Exploratory Investigation, *Proceedings, Nondestructive Evaluation of Civil Structures and Materials,* B. A. Suprenant et al. (Eds.), University of Colorado, Boulder, CO, Oct. 1990b, pp. 327–339.

Popovics, J. S., Discussion of the paper "Detecting Flaws in Concrete Beams and Columns Using the Impact Echo Method" by Y. Lin and M. Sansalone, *ACI Materials Journal,* Vol. 90, No. 3, May–June 1993, p. 293.

Popovics, J. S., Some Theoretical and Experimental Aspects of the Use of Guided Waves for the Nondestructive Evaluation of Concrete, Ph.D. Thesis in Engineering Science and Mechanics, Pennsylvania State University, College of Engineering, May 1994.

Popovics, J. S., Ultrasound and Sound Generation Alternatives for Concrete Structures, *Proceedings of the 2nd International Conference on Nondestructive Testing of Concrete in the Infrastructure,* Nashville, TN, Society for Experimental Mechanics, 1996a, pp. 108–117.

Popovics, J. S., Comments on Determination of Elastic Contents of a Concrete Specimen Using Transient Elastic Waves, *Journal of the Acoustical Society of America,* Vol. 100, No. 5, Nov. 1996b, pp. 3451–3453.

Popovics, J. S., Effects of Poisson's Ratio upon Impact-Echo Test Analysis, *Journal of Engineering Mechanics,* ASCE, Vol. 123, No. 8, Aug. 1997, pp. 843–851.

Popovics, J. S., and Achenbach, J. D., Airport Pavement NDE Research at CQEFP, *Proceedings of the SPIE Conference Nondestructive Evaluation Techniques for Aging Infrastructure & Manufacturing,* Vol. 2945, 1996, pp. 294–303.

Popovics, J. S., and Rose, J. L., A Survey of Developments in Ultrasonic NDE of Concrete, *IEEE Transactions on Ultrasonics, Ferroelectrics, and Frequency Control,* Vol. 41, No. 1, 1994, pp. 140–143.

Popovics, J. S., and Rose, J. L., A New Approach for the Analysis of Impact-Echo Data, *Review of Progress in Quantitative Nondestructive Evaluation,* Vol. 12, D. O. Thompson and D. E. Chimenti (Eds.), Plenum Press, New York, 1993, pp. 2223–2230.

Popovics, J. S., and Rose, J. L., An Approach for Wave Velocity Measurement in Solid Cylindrical Rods Subjected to Elastic Impact, *The International Journal of Solids and Structures,* Vol. 33, No. 26, Nov. 1996, pp. 3925–3935.

Popovics, J. S., Rose, J. L., and Pilarski, A., A Theoretical Approach to Characterize Reinforced Concrete Using Stress Waves, *Materials—Performance and Prevention of Deficiencies and Failures,* Proceedings of the Materials Engineering Congress, ASCE, Atlanta, Aug. 1992, pp. 492–504.

Popovics, J. S., Rose, J. L., Popovics, S., Newhouse, V. L., Lewin, P., and Bilgutay, N., A Survey of Existing and New Approaches for the Generation of Stress Waves in Concrete, *International Conference on Nondestructive Testing of Concrete in the Infrastructure,* SEM, Dearborn, MI, June 1993, pp. 335–352.

Popovics, J. S., Rose, J. L., Popovics, S., Newhouse, V. L., Lewin, P., and Bilgutay, N., Approaches for the Generation of Stress Waves in Concrete, *Experimental Mechanics,* Vol. 35, No. 1, Mar. 1995, pp. 36–41.

Popovics, S., *How to Improve the Flexural Strength of Concrete Made with Alabama Aggregates,* Research Report, Alabama State Highway Department and Auburn University, Montgomery, AL, 1961.

Popovics, S., A Method of Evaluating Gradings of Concrete Aggregates, Thesis submitted to the Faculty of Purdue University in partial fulfillment of the requirements for the degree of Doctor of Philosophy, January 1961b.

Popovics, S., *Investigation of the Grading Requirements for Mineral Aggregates of Concrete,* Final Report, Alabama Highway Research, HRP Report No. 6, Montgomery, AL, June 1964a.

Popovics, S., Theory and Application of Triangular Diagrams, *RILEM Bulletin,* New Series No. 22, Paris, Mar. 1964b, pp. 37–43.

Popovics, S., An Investigation of the Unit Weight of Concrete, *Magazine of Concrete Research,* Vol. 16, No. 49, Dec. 1964c, pp. 211–220.

Popovics, S., Concrete Consistency and Its Prediction, *RILEM Bulletin,* No. 31, June 1966, pp. 235–252.

Popovics, S., Relations Between Various Strengths of Concrete, *A Symposium on Concrete Strength,* Highway Research Record No. 210, Highway Research Board, Washington, DC, 1967a, pp. 67–89.

Popovics, S., Factors Affecting the Relationship Between Strength and Water–Cement Ratio, *Materials Research and Standards,* MTRSA, Vol. 7, No. 12, Dec. 1967b, pp. 527–534.

Popovics, S., A Model for the Kinetics of the Hardening of Portland Cement, *Cement Hydration,* Highway Research Record No. 192, Highway Research Board, Washington, DC, 1967c, pp. 14–35.

Popovics, S., Discussion of "General Relation for Strengths of Concrete Specimens of Different Shapes and Sizes" by A. M. Neville, *ACI Journal,* Proc. Vol. 64, No. 6, Part Two, June 1967d, pp. 1566–1568.

Popovics, S., A Method for Evaluating How Well Observed Data Fit the Line $Y = X$, *Materials, Research and Standards,* ASTM, Vol. 7, No. 5, May 1967e, pp. 195–202.

Popovics, S., What Should an Engineer Know About the Nature of Admixtures? *Concrete,* Vol. 2, No. 7, London, July 1968a, pp. 272–277.

Popovics, S., Examples for the Application of Mathematics in Concrete Technology, *Reports, IVth International Congress on the Application of Engineering,* Weimar, Germany, 1967, Vol. I, VEB Verlag fur Bauwesen, Berlin, 1968b, pp. 375–382.

Popovics, S., Berechnung der Festigkeitsentwicklung von Morteln und Betonen unter Berucksichtigung der Klinkerphasen des verwendeten Portlandzementes Referate (Calculation of the Strength Development of Mortars and Concretes from the Compound Composition of the Portland Cement Used), *Betonstein Zeitung,* Vol. 34, No. 11, Nov. 1968c, pp. 587–590.

Popovics, S., Effect of Porosity on the Strength of Concrete, *Journal of Materials,* JMLSA, Vol. 4, No. 2, June 1969a, pp. 356–371.

Popovics, S., Structural Model Approach to Two-Phase Composite Materials: State of the Art, *American Ceramic Society Bulletin,* Vol. 48, No. 11, Nov. 1969b, pp. 1060–1064.

Popovics, S., Comparison of Various Measurements Concerning the Kinetics of Hydration of Portland Cements, *Proceedings of the Fifth International Symposium on the Chemistry of Cement,* Part III, Properties of Cement Paste and Concrete, Tokyo, Dec. 1969c, pp. 129–137.

Popovics, S., The Fracture Mechanism in Concrete: How Much Do We Know?, *Journal of the Engineering Mechanics Division, ASCE,* EM3, June 1969d, pp. 531–544.

Popovics, S., Characteristics of the Elastic Deformations of Concrete, *Symposium on Concrete Deformation,* Highway Research Record No. 324, Highway Research Board, Washington, DC, 1970a, pp. 1–14.

Popovics, S., The Meaning and Determination of Flexural and Tensile Strengths of Concrete, *An Inter-American Approach for the Seventies,* Materials Technology—I, Second Inter-American Conference on Materials Technology, Mexico City, Mexico, Aug. 24–27, 1970b, pp. 43–54.

Popovics, S., A Review of Stress–Strain Relationships for Concrete, *ACI Journal,* Proc. Vol. 67, No. 3, Mar. 1970c, pp. 243–248.

Popovics, S., Strength Tests for Bridge Concrete—Their Meaning and Significance, *Second International Symposium on Concrete Bridge Design,* ACI Publication SP-26, American Concrete Institute, Detroit, MI, 1971a, pp. 1174–2000.

Popovics, S., Some Aspects of the Concrete Strength, *Structures, Solid Mechanics and Engineering Design,* Proceedings of the Southampton 1969 Civil Engineering Materials Conference, Part 2, Wiley-Interscience, London, 1971b, pp. 1031–1037.

Popovics, S., Effect of Kinetics on the Ultimate Strength of Portland Cement Pastes (published in Russian), *Beton i Zhelezobeton,* Moscow, Mar. 1972a, pp. 23–24.

Popovics, S., Factors Affecting the Elastic Deformations of Concrete, *Mechanical Behavior of Materials,* Proceedings, International Conference on Mechanical Behavior of Materials, Vol. IV, Society of Materials Science, Japan, 1972b, pp. 172–183.

Popovics, S., Segregation and Bleeding (General Report), *Fresh Concrete: Important Properties and Their Measurement,* RILEM Seminar held at the University of Leeds, Vol. 2, Paper 6.1, Mar. 22–24, 1973a, pp. 1–36.

Popovics, S., Method for Developing Relationships Between Mechanical Properties of Hardened Concrete, *ACI Journal,* Proc. Vol. 70, No. 12, Dec. 1973b, pp. 795–798.

Popovics, S., A Numerical Approach to the Complete Stress–Strain Curve of Concrete, *Cement and Concrete Research,* Vol. 3, No. 5, Sept. 1973c, pp. 583–599.

Popovics, S., Aggregate Grading and the Internal Structure of Concrete, *Highway Research Record,* Number 441, *Grading of Concrete Aggregates,* Highway Research Board, Washington, DC, 1973d, pp. 56–64.

Popovics, S., Strength Development of Portland Cement Paste, *VIth International Congress on the Chemistry of Cement,* Supplementary Paper, Section II, Moscow, Sept. 1974a.

Popovics, S., Proportioning Concrete for a Specified Cement Content and/or for a Specified Unit Weight, *Proportioning Concrete Mixes,* ACI Publication SP-46, American Concrete Institute, 1974b, pp. 46–63.

Popovics, S., Verification of Relationships Between Mechanical Properties of Concrete-Like Materials, *Materials and Structures—Research and Testing,* RILEM, Vol. 8, No. 45, May–June 1975, pp. 183–191.

Popovics, S., Phenomenological Approach to the Role of C_3A in the Hardening of Portland Cement Pastes, *Cement and Concrete Research,* Vol. 6, No. 3, May 1976a, pp. 343–350.

Popovics, S., Reef Shell—Beach Sand Concrete, *Living with Marginal Aggregates,* ASTM STP 597, Philadelphia, 1976b, pp. 97–113.

Popovics, S., Verification of Relationships Between Mechanical Properties of Concrete-Like Materials (in Japanese), *Concrete Journal,* JCI No. 143, Vol. 15, No. 8, Tokyo, Aug. 1977, pp. 35–38.

Popovics, S., Judging the Precision and Reliability of Standard Test Methods, *Cement, Concrete, and Aggregates,* CCAGDP, Vol. 1, No. 1, 1979, pp. 38–43.

Popovics, S., Possibility of a Catalytic Role of C_3A in the Hardening of Portland Cement, *7th International Congress on the Chemistry of Cement,* Vol. IV, Editions Septima, Paris, 1980a, pp. 602–606.

Popovics, S., Calculation of Strength Development from the Compound Composition of Portland Cement, *7th International Congress on the Chemistry of Cement,* Vol. III, Editions Septima, Paris, 1980b, pp. VI.47–VI.51.

Popovics, S., Composite Averages for the Estimation of the Moduli of Elasticity of Composite Materials, *Advances in Cement-Matrix Composites,* D. M. Roy, A. J. Majumdar, S. P. Shah, and J. A. Manson (Eds.), Materials Research Society, Proceedings, Symposium L, Boston, November 1980c, pp. 119–133.

Popovics, S., Generalization of the Abrams Law—Prediction of Strength Development of Concrete from Cement Properties, *ACI Journal,* Proc. Vol. 78, No. 2, Mar.–Apr. 1981a, pp. 123–129.

Popovics, S., Extended Model for Estimating the Strength Developing Capacity of Portland Cement, *Magazine of Concrete Research,* Vol. 33, No. 116, London, Sept. 1981b, pp. 147–153.

Popovics, S., Accelerated Determination of the Control Strength of Concrete, *Proceedings of the Third National Conference on Mechanics and Technology of Composite Materials,* Varna, Publishing House of the Bulgarian Academy of Sciences, Sofia, 1982a, pp. V553–555.

Popovics, S., *Fundamentals of Portland Cement Concrete: A Quantitative Approach,* Vol. 1, *Fresh Concrete,* Wiley, New York, 1982b, 477 pp.

Popovics, S., Strength Relationships for Fly Ash Concrete, *ACI Journal,* Proc. Vol. 79, No. 1, Jan.–Feb. 1982c, pp. 43–49.

Popovics, S., New Results with Epoxy Modification of Portland Cement Concrete, *Polymers in Concrete,* H. Schulz (Ed.), Fourth International Congress, Institut fur Spanende Technologie und Werkzeugmaschinen, Technische Hochschule, Darmstadt, Sept. 19–21 1984, pp. 369–373.

Popovics, S., New Formulas for the Prediction of the Effect of Porosity on Concrete Strength, *ACI Journal,* Proc. Vol. 82, No. 2, Mar.–Apr. 1985, pp. 136–146.

Popovics, S., Stato attuale della determinazione della resistenza del calcestruzzo mediante la velocita degli impulsi in America (Present State of the Determination of

Concrete Strength by Pulse Velocity in America), *Il Cemento,* Anno 83, No. 3, July–Sept. 1986a, pp. 117–128.

Popovics, S., Effect of Curing Method and Final Moisture Condition on Compressive Strength of Concrete, *ACI Journal,* Proc. Vol. 83, No. 4, July–Aug. 1986b, pp. 650–657.

Popovics, S., What Do We Know About the Contribution of Fly Ash to the Strength of Concrete? *Fly Ash, Silica Fume, Slag, and Natural Pozzolans in Concrete,* V. M. Malhotra (Ed.), Proceedings, Second International Conference, Madrid, Vol. I, ACI Publication SP-91, Detroit, MI, 1986c, pp. 313–331.

Popovics, S., A Model for Deformations of Two-Phase Composites Under Load," Reprints, *Fourth RILEM International Symposium on Creep and Shrinkage of Concrete: Mathematical Modeling,* Z. P. Bazant (Ed.), Northwestern University, Aug. 26–29, 1986d, pp. 733–742.

Popovics, S., Discussion of the paper "Elastic Compatibility and the Behavior of Concrete" by T. W. Bremner and T. A. Holm, *ACI Journal,* Mar.–Apr. 1987a, pp. 74–75.

Popovics, S., Attempts to Improve the Bond Between Cement Paste and Aggregate, *Materials and Structures—Research and Testing,* RILEM, Vol. 20, No. 115, Jan. 1987b, pp. 32–38.

Popovics, S., Model for the Quantitative Description of the Kinetics of Hardening of Portland Cements, *Cement and Concrete Research,* Vol. 17, No. 5, 1987c, pp. 821–838.

Popovics, S., Strength-Increasing Effects of a Chloride-Free Accelerator, *Corrosion, Concrete, and Chlorides,* F. W. Gibson (Ed.), ACI SP-102, American Concrete Institute, Detroit, MI, 1987d, pp. 79–106.

Popovics, S., Quantitative Deformation Model for Two-Phase Composites Including Concrete, *Materials and Structures—Research and Testing,* RILEM, Vol. 20, No. 117, Paris, May 1987e, pp. 171–179.

Popovics, S., Effects of a Chloride-free Accelerator on Concrete Strength, *Second International Colloquum on Concrete in Developing Countries,* Session 8, Mineral and Fibre Admixtures to Concrete, Bombay, India, Jan. 3–8, 1988.

Popovics, S., A Hypothesis Concerning the Effects of Macro Porosity on Mechanical Properties of Concrete, *Fracture of Concrete and Rock,* S. P. Shah and S. E. Swartz (Eds.), Springer-Verlag, New York, 1989, pp. 170–174.

Popovics, S., Analysis of the Concrete Strength Versus Water–Cement Ratio Relationship, *ACI Materials Journal,* Vol. 87, No. 5, Sept.–Oct. 1990a, pp. 517–529.

Popovics, S., Discussion of the paper "Effect of Temperature on the Early-Age Properties of Type I, Type III, and Type I/Fly Ash Concretes, *ACI Materials Journal,* Vol. 87, No. 6, Nov.–Dec. 1990b, pp. 654–655.

Popovics, S., A Study on the Use of a Chloride-Free Accelerator, *Improvement of Properties,* E. Vazquez (Ed.), Chapman and Hall, London, 1990c, pp. 197–208.

Popovics, S., *Concrete Materials Properties, Specifications and Testing,* Noyes Publications, Park Ridge, NJ, 1992.

Popovics, S., Effects of the Fineness of Fly Ash on the Flow and Compressive Strength of Portland Cement Mortars, *New Concrete Technology,* T. C. Liu and G. C. Hoff (Eds.), SP-141, American Concrete Institute, Detroit, MI, 1994, pp. 205–225.

Popovics, S., History of a Mathematical Model for Strength Development of Portland Cement Concrete, To be published in the *ACI Materials Journal* in 1998.

Popovics, S., and Erdey, M. R. A., Estimation of the Modulus of Elasticity of Concrete-Like Composite Materials, *Materials and Structures Research and Testing,* RILEM, Vol. 3, No. 16, Paris, July–Aug. 1970, pp. 253–260.

Popovics, S., and McDonald, W. E., *Underwater Inspection of the Engineering Condition of Concrete Structures,* Technical Report, REMR-CS-9, USAE Waterways Experiment Station, Vicksburg, MS, 1989, p. 74.

Popovics, S., and Pfeifer, R. P., An Evaluation of Three Standard Methods for the Accelerated Curing of Concrete, unpublished report submitted to the Department of Civil Engineering, Drexel University, Philadelphia, 1992.

Popovics, S., and Popovics, J. S., Improved Determination of the Locations and Sizes of Steel Rebars in Concrete, *Evaluation and Rehabilitation of Concrete Structures and Innovations in Design,* Proceedings, ACI International Conference, Hong Kong, 1991a, pp. 485–495.

Popovics, S., and Popovics, J. S., Potential Ultrasonic Techniques Based on Surface Waves and Attenuation for Damage Evaluation in Concrete—A Review, *Diagnosis of Concrete Structures,* T. Javor (Ed.), Proceedings of the International RILEM-IMEKO Conference, Expertcentrum, Bratislava, 1991b, pp. 101–104.

Popovics, S., and Popovics, J. S., Effect of Stresses on the Ultrasonic Pulse Velocity in Concrete, *Materials and Structures—Research and Testing,* RILEM, Vol. 24, No. 139, Paris, Jan. 1991c, pp. 15–23.

Popovics, S., and Popovics, J. S., A Critique of the Ultrasonic Pulse Velocity Method for Testing Concrete, *Nondestructive Testing of Concrete Elements and Structures,* F. Ansari and S. Sture (Eds.), Proc. ASCE, San Antonio, TX, Apr. 1992, pp. 94–10.

Popovics, S., and Popovics, J. S., Misapplications of the Standard Ultrasonic Pulse Velocity Method for Testing Concrete, *Structural Materials Technology—An NDT Conference,* R. J. Scancella and M. E. Callahan (Eds.), Technomic, Atlantic City, NJ, 1994a, pp. 214–246.

Popovics, S., and Popovics, J. S., The Foundation of a Computer Program for the Advanced Utilization of w/c and Air Content in Concrete Proportioning," *Concrete International,* Vol. 16, No. 12, Dec. 1994b, pp. 21–26.

Popovics, S., and Popovics, J. S., Computerization of the Strength Versus w/c Relationship," *Concrete International,* Vol. 17, No. 4, April 1995, pp. 37–40.

Popovics, S., and Popovics, J. S., Selection of Test Locations for In-Place Testing of Concrete Strength, *Diagnosis of Concrete Structures,* T. Javor (Ed.), Expertcentrum, Bratislava, 1996a, pp. 586–589.

Popovics, S., and Popovics, J. S., Novel Aspects in Computerization of Concrete Proportioning, *Concrete International,* Vol. 18, No. 12, Dec. 1996b, pp. 54–58.

Popovics, S., and Popovics, J. S., NDT Methods for the In-Situ Evaluation of Pavements in the USA—A Review, *Non-Destructive Testing in Civil Engineering NDT-CE '97,* J. H. Bungey (Ed.), Vol. 2, The British Institute of Non-Destructive Testing, 1997, pp. 675–688.

Popovics, S., and Popovics, J. S., Misses in Ultrasonic Research from Which We Learned, To be published in the *Cement, Concrete and Aggregate,* ASTM, in 1998.

Popovics, S., and Rajendran, N., Early Age Properties of Magnesium Phosphate–Based Cements Under Various Temperature Conditions, *Concrete and Construction, Transportation Research Record* No. 1110, Transportational Research Board, Washington, DC, 1987, pp. 34–35.

Popovics, S., and Ujhelyi, J., Az 1952. evben vegzett cementvizsgalatok eredmenyeirol (Results of the Cement Tests Conducted in 1952), *Epitoanyag,* Vol. 5, No. 10, Oct. 1953, pp. 315–320.

Popovics, S., and Ujhelyi, J., Az ETI 1953. evben vegzett cementvizsgalatainak kiertekelese (Evaluation of Cement Tests Made in the Year 1953), *Epitoanyag,* Vol. 6, No. 9, Sept. 1954, pp. 307–312.

Popovics, S., and Ujhelyi, J., A cementszilardsag szabvanyos hazai vizsgalati modszereinek osszehasonlitasa (Comparison of Several Standard Methods for Testing of Cement Strength), *ETI Tudomanyos Kozlemenyek,* No. 7, 1955, pp. 50–73.

Popovics, S., Rajendran, N., and Penko, M., Rapid Hardening Cement for Repair of Concrete, *ACI Materials Journal,* Vol. 84, No. 1, Jan.–Feb. 1987, pp. 64–73.

Popovics, S., Rose, J. L., and Popovics, J. S., The Behavior of Ultrasonic Pulses in Concrete, *Cement and Concrete Research,* Vol. 20, No. 2, 1990, pp. 259–270.

Popovics, S., Silva-Rodrigez, R., Popovics, J. S., and Martucci, V., Behavior of Ultrasonic Pulses in Fresh Concrete, *New Experimental Techniques for Evaluating Concrete Material and Structural Performance,* American Concrete Institute, Detroit, MI, 1994, pp. 207–225.

Popovics, S., Komlos, K., and Popovics, J. S., Comparison of DIN/ISO 8047 (Entwurf) to Several Standards on Determination of Ultrasonic Pulse Velocity in Concrete, International Symposium, *Non-destructive Testing in Civil Engineering* (NDT-CE), G. Schickert and H. Wiggenhouser (Eds.), Sept. 26–28, 1995, DGZfP, Berlin, 1995, pp. 673–680.

Portland Cement Association, *Concrete Technology Instructor's Guide,* U.S. Edition, PCA, Skokie, IL, 1965.

Portland Cement Association, *Concrete Pavement Manual,* PCA, Skokie, IL, 1955.

Portland Cement Association, *Design and Control of Concrete Mixtures,* 11th ed., PCA, Skokie, IL, 1968.

Powers, T. C., Measuring Young's Modulus of Elasticity by Means of Sonic Vibrations, *Proceedings ASTM,* Vol. 38, Part II, 1938, pp. 460–467.

Powers, T. C., Should Portland Cement Be Dispersed? *ACI Journal,* Proc. Vol. 42, Nov. 1945, pp. 117–140.

Powers, T. C., The Nonevaporable Water Content of Hardened Portland-Cement Paste—Its Significance for Concrete Research and Its Method of Determination, *ASTM Bulletin* No. 158, Philadelphia, May 1949, pp. 68–76.

Powers, T. C., The Physical Structure and Engineering Properties of Concrete, Bulletin 90, Research and Development Laboratories, Portland Cement Association, Skokie, IL, July 1958a, p. 24.

Powers, T. C., Structure and Physical Properties of Hardened Portland Cement Paste, *Journal of the American Ceramic Society,* Vol. 41, No. 1, Jan. 1958b, pp. 1–6.

Powers, T. C., Capillary Continuity or Discontinuity in Cement Pastes, *Journal of the PCA Research and Development Laboratories,* Vol. 1, No. 2, May 1959, pp. 38–48.

Powers, T. C., Physical Properties of Cement Paste, *Chemistry of Cement,* Proceedings of the Fourth International Symposium, NBS Monograph 43, Vol. II, Washington, DC, 1960, pp. 577–609.

Powers, T. C., A Hypothesis on Carbonation Shrinkage, *Journal of the PCA Research and Development Laboratories,* Vol. 4, No. 2, May 1962, pp. 40–50.

Powers, T. C., The Physical Structure of Portland Cement Paste, *The Chemistry of Cements,* H. F. W. Taylor (Ed.), Vol. I, Chap. 10, Academic Press, London, 1964, pp. 391–416.

Powers, T. C., *The Properties of Fresh Concrete,* Wiley, New York, 1968.

Powers, T. C., The Specific Surface Area of Hydrated Cement Obtained from Permeability Data, *Materials and Structures, Research and Testing,* Vol. 12, No. 69, May–June 1978, pp. 159–168.

Powers, T. C., and Brownyard, T. L., Studies of the Physical Properties of Hardened Portland Cement Paste—Part 6, *ACI Journal,* Proc. Vol. 43, Mar. 1947, pp. 845–864.

Preece, E. F., Determination and Use of the Dynamic Modulus of Elasticity of Concrete, *Proceedings, Highway Research Board,* Vol. 28, 1948, pp. 233–237.

Preiss, K., Measuring Concrete Density by Gamma Ray Transmission, *Materials Research and Standards,* Vol. 5, No. 6, June 1965, pp. 285–291.

Preiss, K., Measuring the Thickness of a Concrete Slab by Gamma Ray Transmission, *ACI Journal,* Proc. Vol. 63, July 1966, pp. 743–748.

Preiss, K., and Newman, K., An Improved Technique for the Measurement of Density of Concrete and Soils with Gamma Radiation, *Proceedings, Fourth International Conference on Non-destructive Testing,* Butterworth, London, 1964, pp. 135–141.

Price, W. H., Factors Influencing Concrete Strength, *ACI Journal,* Proc. Vol. 47, No. 6, Feb. 1951, pp. 417–432.

Raithby, K. D., Behavior of Concrete Under Fatigue Loading, *Development in Concrete Technology*—1, F. D. Lydon (Ed.), Applied Science Publishers, London, 1979, pp. 83–109.

Ramakrishnan, V., Accelerated Strength Testing—Annotated Bibliography, *Accelerated Strength Testing,* ACI Publication SP-56, American Concrete Institute, Detroit, MI, 1978, pp. 285–312.

Ramakrishnan, V., and Dietz, J., Accelerated Methods of Estimating the Strength of Concrete, *Recent Developments in Accelerated Testing and Maturity of Concrete,* Transportation Research Record No. 558, Transportation Research Board, Washington, DC, 1975, pp. 29–44.

Ramakrishnan, V., and Dietz, J., ASTM Accelerated Strength Tests for Quality Control of Concrete, *Accelerated Strength Testing,* ACI Publication SP-56, Detroit, MI, 1978, pp. 85–116.

Ramakrishnan, V., Ananthanarayan, Y., and Gopal, K. C., The Determination of the Tensile Strength of Concrete: A Comparison of Different Methods, *Indian Concrete Journal,* Vol. 41, No. 5, May 1967a, pp. 202–206.

Ramakrishnan, V., Anathanarayana, Y., and Sabapathi, P., A Standard Test Procedure for Tensile Strength of Concrete, *Indian Concrete Journal,* Vol. 41, No. 8, Aug. 1967b, pp. 322–327.

Ramesh, C. K., and Chopra, S. K., Determination of Tensile Strength of Concrete and Mortar by the Split Test, *Indian Concrete Journal,* Vol. 34, No. 9, Sept. 1960, pp. 354–357.

Raphael, J. M., Tensile Strength of Concrete, *ACI Journal,* Proc. Vol. 81, No. 2, Mar.–Apr. 1984, pp. 158–165.

Rasch, C. H. R., Spannungs-Dehnungs-Linien des Betons und Spannungsverteilung in der Biegdruckzone bei konstanter Dehngeschwindigkeit (Stress–Strain Curves of Concrete and Stress Distribution in the Compressed Zone at Constant Strain Velocity), *Deutscher Ausschuss fuer Stahlbeton,* Heft 154, Wilhelm Ernst & Sohn, Berlin, 1962.

Rascon Chavez, O. A., Stochastic Model to Fatigue, *Journal of the Engineering Mechanics Division, ASCE,* No. EM3, June 1967, pp. 147–155.

Rayleigh (Lord), *The Theory of Sound,* Vols. I and II, Dover, New York, 1945.

Reagel, F. V., and Willis, T. F., The Effect of Dimensions of Test Specimens on the Flexural Strength of Concrete, *Public Roads,* Vol. 12, No. 2, Apr. 1931, pp. 37–46.

Reeves, J. S., *The Strength of Concrete Under Combined Direct and Shear Stresses,* Technical Report, TRA/365, Cement and Concrete Association, London, Nov. 1962.

Reichard, T. W., *Creep and Drying Shrinkage of Lightweight and Normal-Weight Concretes,* National Bureau of Standards Monograph 74, U.S. Department of Commerce, Washington, DC, Mar. 1964.

Reiner, M., *Deformation, Strain and Flow,* 3rd ed., H.K. Lewis & Co., London, 1969.

Reiner, M., *Advanced Rheology,* H.K. Lewis & Co., London, 1971.

Reinhardt, H. W., Rosi, P., and Van Mier, J. G. M., Joint Investigation of Concrete at High Rates of Loading," *Materials and Structure—Research and Testing,* RILEM, Vol. 23, No. 135, Paris, May 1990, pp. 213–216.

Reinius, E., A Theory of the Deformation and the Failure of Concrete, *Magazine of Concrete Research,* Vol. 8, No. 24, Nov. 1956, pp. 157–160.

Reinsdorf, S., *Betontaschenbuch,* Band 1, Bentontechnologie (Pocketbook of Concrete, Vol. 1, Concrete Technology), 4th ed., VEB Verlag fur Bauwesen, Berlin, 1977, p. 380.

Richards, C. W., and Radjy, F., A New Application of Internal Friction to Concrete Research, *Materials, Research and Standards,* ASTM, Vol. 6, No. 8, Aug. 1966, pp. 386–391.

Richards, O., Pullout Strength of Concrete, *Reproducibility and Accuracy of Mechanical Testing,* ASTM STP 626, Philadelphia, 1977, pp. 32–40.

Richardson, D. N., Point-Load Test for Estimating Concrete Compressive Strength, *ACI Materials Journal,* Vol. 86, No. 4, July–Aug. 1989, pp. 409–416.

Richardson, D. N., Review of Variables That Influence Measured Concrete Compressive Strength, *Journal of Materials in Civil Engineering,* ASCE, Vol. 3, No. 2, May 1991, pp. 95–112.

Richart, F. E., and Roy, N. H., Digest of Test Data on Poisson's Ratio for Concrete, *Proceedings, ASTM,* Vol. 30, Part I, 1930, pp. 661–667.

Richart, F. E., Brandtzaeg, A., and Brown, R. L., *A Study of Failure of Concrete Under Combined Compressive Stresses,* Bulletin No. 185, Engineering Experiment Station, University of Illinois, Urbana, 1928.

Richart, F. E., Brandtzaeg, A., and Brown, R. L., *The Failure of Plain and Spirally Reinforced Concrete in Compression,* Bulletin No. 190, Engineering Experiment Station, University of Illinois, Urbana, 1929.

Richartz, W., Electron Microscopic Investigations about the Relations between Structure and Strength of Hardened Cement, Supplementary Paper No. III-1245, *Fifth International Symposium on the Chemistry of Cement,* Tokyo, 1968.

Riha, J., Ausgewahlte Probleme der Betontechnologie (Selected Problems of Concrete Technology), *Wissenschaftliche Zeitschrift,* Technischen Hochschule, Leipzig, Vol. 8, No. 1, 1984, pp. 1–12.

RILEM, *International Symposium on Nondestructive Testing of Materials and Structures,* Vols. 1 and 2, Paris, 1954.

RILEM, Symposium, *Winter Concreting Theory and Practice,* Danish National Institute of Building Research, Special Report, Copenhagen, 1956.

RILEM, Non-destructive Testing of Concrete—Report of the Bucharest Meeting, *RILEM Bulletin,* No. 27, June 1965, pp. 121–125.

RILEM, Accelerated Hardening of Concrete with a View to Rapid Control Tests, *RILEM Bulletin,* No. 31, June 1966a, pp. 156–213.

RILEM, International Symposium *Effects of Repeated Loading on Materials and Structures,* Vol. III, RILEM—Instituto de Ingenieria, Mexico, Sept. 1966b.

RILEM, Testing of Concrete by the Ultrasonic Pulse Method, RILEM Recommendation NDT1, *Materials and Structures—Research and Testing,* RILEM, Vol. 2, No. 10, July–Aug. 1969, pp. 253–293.

RILEM, Compressive Test, RILEM Recommendation CPC 4, Nov. 1975a.

RILEM, Flexural Test, RILEM Recommendation CPC 5, Nov. 1975b.

RILEM, Tension by Splitting, RILEM Recommendation CPC 6, Nov. 1975c.

RILEM, On the Terminology, Notations and Symbols for Loading Test of In Situ Structures, *Materials and Structures—Research and Testing,* RILEM Vol. 11, No. 63, May–June 1978, pp. 217–220.

RILEM, Final Report of Task Group 1,68-MMTH Technical Committee on Strength of Cement, *Materials and Structures—Research and Testing,* RILEM, Vol. 24, No. 140, Paris, Mar. 1991, pp. 143–158.

RILEM/CEB/CIB/FIP Joint Committee on the Statistical Control of the Quality of Concrete, Recommended Principles for the Control of Quality and the Judgment of Acceptability of Concrete, *Materials and Structures—Research and Testing,* RILEM, Vol. 8, No. 47, Sept.–Oct. 1975, pp. 387–403.

RILEM Technical Committee 43-CND, Draft Recommendation for In Situ Concrete Strength Determination by Combined Non-destructive Methods, *Materials and Structures—Research and Testing,* RILEM, Vol. 26, No. 155, Jan.–Feb. 1993, pp. 43–49.

RILEM Tentative Recommendation, Recommendation for Terminology, Notations and Symbols in Loadings Tests of Structures In Situ, Part II, *Materials and Structures—Research and Testing,* RILEM, Vol. 11, No. 66, Nov.–Dec. 1978, pp. 453–456.

RILEM Tentative Recommendation, Quick Routine Inspection of Structures In Situ, *Materials and Structures—Research and Testing,* RILEM, Vol. 14, No. 79, Jan.–Feb. 1981, pp. 65–73.

RILEM Working Group on the Non-destructive Testing of Concrete, RILEM Bulletin, No. 33, Dec. 1966, pp. 367–370.

RILEM, TC 148-SEC, Strain-Softening of Concrete in Uniaxial Compression, *Materials and Structures,* Vol. 30, No. 198, May 1997, pp. 195–209.

Riley, O., Inspection Requirements for Highway Construction, *ACI Journal,* Proc. Vol. 72, No. 6, June 1975, pp. 283–285.

Road Research Laboratory, *Design of Concrete Mixes,* Road Note No. 4, Department of Scientific and Industrial Research, Her Majesty's Stationary Office, London, 1950.

Road Research Laboratory, *Concrete Roads,* Her Majesty's Stationary Office, London, 1955.

Robins, P. J., The Point-Load Strength for Concrete Cores, *Magazine of Concrete Research,* Vol. 32, No. 111, June 1980, pp. 101–111.

Robins, P. J., The Point-Load Test for Tensile Strength Estimation of Plain and Fibrous Concrete, *In Situ/Nondestructive Testing of Concrete,* V. M. Malhotra (Ed.), ACI Publication SP-82, American Concrete Institute, Detroit, MI, 1984, pp. 309–326.

Robins, P. J., and Austin, S. A., Core Point-Load Test Steel-Fibre-Reinforced Concrete, *Magazine of Concrete Research,* Vol. 37, No. 133, Dec. 1985, pp. 238–244.

Rodrigez Cuevas, C. N., Effect of Confining Pressures on the Mechanical Behavior of Materials, *An Inter-American Approach for The Seventies,* Materials Technology—I. Second Inter-American Conference on Materials Technology, Mexico City, Mexico, Aug. 24–27, 1970, pp. 492–501.

Romstad, K. M., Taylor, M. A., and Herrmann, L. R., Numerical Biaxial Characterization for Concrete, *Journal of the Engineering Mechanics Division,* ASCE, Vol. 100, No. EM 5, Proc. Paper 10879, Oct. 1974, pp. 935–948.

Roper, H., and Bryden, J. G., A New Approach to the Determination of the Tensile Strength of Cement Paste by Centrifugal Force, *Magazine of Concrete Research,* Vol. 16, No. 49, Dec. 1964, pp. 211–224.

Roper, H., Cox, J. E., and Erlin, B., Petrographic Studies on Concrete Containing Shrinking Aggregate, *Journal of the PCA Research and Development Laboratories,* Vol. 6, No. 3, Sept. 1964, pp. 2–18.

Ros, M., Die materialtechnischen Grundlagen und Probleme des Eisenbetons im Hinblick auf die zukunftige Gestaltung der Stahlbeton-Bauweise (Material-Technological Foundation and Problems of Reinforced Concrete . . .), *Bericht Nr. 162,* Eidgenossische Materialprufungs- und Versuchsanstalt fur Industrie, Bauwesen und Gewerbe, Zurich, 1950.

Rose, J. L., and Goldberg, B. B., *Basic Physics in Diagnostic Ultrasound,* John Wiley, New York, 1979.

Rose, J. L., and Jeong, Y., *Manual for Ultrasonic Nondestructive Testing Laboratory Experiments,* Drexel University, Philadelphia, 1984.

Rose, J. L., Popovics, J. S., and Pilarski, A., Effects of the Viscoelastic Nature of Concrete on Ultrasonic Nondestructive Evaluation, *Review of Progress in Quantitative Nondestructive Evaluation,* Vol. 13B, D. O. Thompson and D. E. Chimenti (Eds.), Plenum Press, New York, 1994, pp. 2131–2138.

Rosenthal, D., and Asimow, R. M., *Introduction to Properties of Materials,* 2nd ed., Van Nostrand Reinhold Company, New York, 1971.

Ross, C. A., Effects of Strain Rate on Concrete Strength, *ACI Materials Journal,* Vol. 92, No. 1, Jan.–Feb. 1995, pp. 37–47.

Ross, A. D., Illston, J. M., and England, G. L., Short- and Long-Term Deformations of Concrete as Influenced by Its Physical Structure and State, *The Structure of Concrete,* Proceedings of an International Conference, London, Sept. 1965. Cement and Concrete Association, London, 1968, pp. 407–422.

Rossi, P., A Physical Phenomenon Which Can Explain the Mechanical Behavior of Concrete Under High Strain Rates, *Materials and Structures—Research and Testing,* RILEM, Vol. 24, No. 144, Paris, Nov. 1991, pp. 422–424.

Rossi, P., Van Mier, J. G. M., Boulay, F., and Le Maou, F., The Dynamic Behaviour of Concrete: Influence of Free Water, *Materials and Structures—Research and Testing,* Vol. 25, No. 153, Nov. 1992, pp. 509–514.

Rothfuchs, G., *Betonfibel,* Band 1 (Concrete Primer, Vol. 1), Bauverlag, Wiesbaden-Berlin, 1962.

Roup, R. P., and Fillmore, C. L., Tensile Testing of Ceramic Materials Using a Modified Cross-Breaking Technique, *Ceramic Bulletin,* Vol. 40, No. 11, Nov. 1961, pp. 694–697.

Roy, D. M., and Gouda, G. R., Optimization of Strength in Cement Pastes, Supplementary Paper II-3, II-4, II-5, *VIth International Congress on the Chemistry of Cement,* Moscow, Sept. 1974.

Roy, H. E. H., A Failure Theory for Concrete, Ph.D. Thesis, University of Illinois, 1963.

Rudnick, A., Hunter, A. R., and Holden, F. C., An Analysis of the Diametral-Compression Test, *Materials Research and Standards,* ASTM, Vol. 3, No. 4, Apr. 1963, pp. 283–289.

Ruijie, L., The Diameter-Compression Test for Small Diameter Cores, *Materials and Structures,* Vol. 29, No. 185, Jan.–Feb. 1996, pp. 56–59.

Rusch, H., Versuche zur Festigkeit der Biegedruckzone (Experiments Concerning the Strength of the Compression Zone in Bending), *Deutscher Ausschuss fur Stahlbeton,* Heft 120, Wilhelm Ernst & Sohn, Berlin, 1955.

Rusch, H., Betrachtungen zur Prufung der Betonfestigkeit (Considerations Concerning the Testing of Concrete Strength), *Beton- und Stahlbetonbau,* June 1956, pp. 135–138.

Rusch, H., Remarks Concerning the Mechanics of Testing Cubical Specimens, *RILEM Bulletin,* No. 34 (Old Series), 1957, pp. 75–83.

Rusch, H., Physikalische Fragen der Betonprufung (Physical Problems in the Testing of Concrete), *Zement–Kalk–Gips,* No. 12, Heft 1, 1959, pp. 1–9.

Rusch, H., Research Toward a General Flexural Theory for Structural Concrete, *ACI Journal,* Proc. Vol. 57, July 1960, pp. 1–28.

Rusch, H., Der Einfluss der Streuung bei der Betonkontrolle, *Der Bauingenieur,* Vol. 37, No. 10, Oct. 1962, pp. 373–377.

Rusch, H., Sell, R., and Rachwitz, R., Statistische Analyse der Betonfestigkeit (Statistical Analysis of the Strength of Concrete), *Deutscher Ausschuss fur Stahlbeton,* Heft 206, Wilhelm Ernst & Sohn, Berlin, 1969.

Ryskewitsch, E., Compressive Strength of Porous Sintered Alumina and Zirconia—9th Communication to Ceramography, *Journal of the American Ceramic Society,* Vol. 36, No. 2, Feb. 1953, pp. 65–68.

Sadgrove, B. M., Prediction of Strength Development of Concrete Structures, *Recent Developments in Accelerated Testing and Maturity,* Transportation Research Record No. 558, Transportation Research Board, Washington, DC, 1975, pp. 19–28.

Saemann, J. C., and Washa, G. W., Variation of Mortar and Concrete Properties with Temperature, *ACI Journal,* Proc. Vol. 54, Nov. 1957, pp. 385–395.

Samarai, M., Popovics, S., and Malhotra, V. M., Effects of High Temperatures on the Properties of Fresh and Hardened Concrete: A Bibliography (1915–1983), *Transportation Research Record No. 924,* Transportation Research Board, Washington, DC, 1983a, pp. 56–63.

Samarai, M., Popovics, S., and Malhotra, V. M., Effects of High Temperatures on the Properties of Fresh Concrete, *Transportation Research Record No. 924,* Transportation Research Board, Washington, DC, 1983b, pp. 42–50.

Samarai, M., Popovics, S., and Malhotra, V. M., Effects of High Temperatures on the Properties of Hardened Concrete, *Transportation Research Record No. 924,* Transportation Research Board, Washington, DC, 1983c, pp. 50–56.

Samarin, A., and Dhir, R. K., Determination of In Situ Concrete Strength: Rapidly and Confidently by Non-destructive Testing, *In Situ/Nondestructive Testing of Concrete,* V. M. Malhotra (Ed.), ACI Publication SP-82, American Concrete Institute, Detroit, MI, 1984, pp. 77–94.

Samarin, A., and Meynink, P., Use of Combined Ultrasonic Rebound Hammer Method for Determining Strength of Concrete Structural Members, *Concrete International: Design and Construction,* Vol. 3, No. 3, Mar. 1981, pp. 25–29.

Sanchez-Trejo, R., and Flores-Castro, L., Experience in the Use of the Accelerated Testing Procedure for the Control of Concrete During the Construction of Tunnel Emisor Central in Mexico City, *Accelerated Strength Testing,* ACI Publication SP-56, American Concrete Institute, Detroit, MI, 1978, pp. 15–28.

Sandhu, R. S., Resistance of Concrete to Impact Loading, *Indian Concrete Journal,* Vol. 37, No. 5, May 1963, pp. 169–173.

Sandstedt, C. E., Ledbetter, W. B., and Gallaway, B. M., Prediction of Concrete Strength from the Calculated Porosity of the Hardened Cement Paste, *ACI Journal,* Proc. Vol. 70, No. 2, Feb. 1973, pp. 115–116.

Sansalone, M., and Carino, N. J., Laboratory and Field Studies of the Impact-Echo Method for Flaw Detection in Concrete, *Nondestructive Testing,* H. S. Lew (Ed.), ACI Publication SP-112, American Concrete Institute, Detroit, MI, 1988, pp. 1–20.

Sansalone, M., and Carino, N. J., Detecting Delaminations in Concrete Slabs With and Without Overlays Using the Impact-Echo Method, *ACI Materials Journal,* Vol. 86, No. 2, Mar.–Apr. 1989, pp. 175–184.

Sansalone, M., and Carino, N. J., Stress Wave Propagation Methods, *CRC Handbook on Nondestructive Testing of Concrete,* V. M. Malhotra and N. J. Carino (Eds.), CRC Press, Boca Raton, FL, 1991, pp. 274–304.

Sansalone, M., and Carino, N. J., Use of the Impact-Echo Method and Field Instrument for Non-destructive Testing of Concrete Structures, International Symposium, *Nondestructive Testing in Civil Engineering* (NDT-CE), G. Schickert and H. Wiggenhouser (Eds.), Sept. 26–28, 1995, DGZfP, Berlin, 1995, pp. 495–502.

Saul, A. G. A., Principles Underlying the Steam Curing of Concrete at Atmospheric Pressure, *Magazine of Concrete Research,* No. 6, London, Mar. 1951, pp. 127–140.

Saul, A. G. A., *A Comparison of the Compressive, Flexural, and Tensile Strengths of Concrete,* Technical Report TRA/333, Cement and Concrete Association, London, June 1960.

Schaffler, H., Elastizitat und Prismendruck-festigkeit von Leichbeton, im besonderen von Gas-und Schaumbeton (Elasticity and Compressive Strength of Lightweight Concrete, Particularly Those of Gas and Foam Concrete), *Betonstein-Zeitung,* Vol. 20, No. 10, Oct. 1954, pp. 432–434.

Schickert, G., and Wiggenhauser, H. (Eds.), International Symposium, *Non-destructive Testing in Civil Engineering* (NDT-CE), Sept. 26–28, 1995, DGZfP, Berlin, 1995.

Schiller, K. K., Porosity and Strength of Brittle Solids (with Particular Reference to Gypsum), *Mechanical Properties of Non-metallic Brittle Materials,* W. H. Walton (Ed.), Interscience Publishers, New York, 1958, pp. 35–49.

Schiller, K. K., Strength of Porous Material, *Cement and Concrete Research,* Vol. 1, No. 4, July 1971, pp. 419–422.

Schmidt, E., The Concrete Sclerometer, *RILEM International Symposium on Nondestructive Testing of Materials and Structures,* Vol. II, Paris, 1954, pp. 310–319.

Schramli, W., An Attempt to Assess Beneficial and Detrimental Effects of Aluminate in the Cement on Concrete Performance, Parts 1 and 2, *World Cement Technology,* Vol. 9, Nos. 3 and 4, Mar. and Apr. 1978, pp. 35–42, 75–80.

Schuman, L., and Tucker, J., Jr., Tensile and Other Properties of Concretes Made with Various Types of Cements, *Journal of Research of the National Bureau of Standards,* Vol. 31, Aug. 1943, pp. 107–124.

Scrivener, K. L., and Gariner, E. M., Microstructural gradients in cement paste around aggregate particles, *Proc. Materials Research Symposium,* Vol. 114, 1988, pp. 77–85.

Sedlacek, R., and Halden, F. A., Method for Tensile Testing of Brittle Materials, *Review of Scientific Instruments,* Vol. 33, No. 3, Mar. 1962, pp. 298–300.

Seefried, K. J., Gesund, H., and Pincus, G., Experimental Investigation of the Strain Distribution in the Split Cylinder Test, *Journal of Materials,* ASTM, Vol. 2, No. 4, Dec. 1967, pp. 703–718.

Seewald, F., Die Spannungen und Formanderungen von Balken mit rechteckigem Querschnitt (Stresses and Deformations of Beams of Rectangular Cross Section), *Abhandlungen aus dem Aerodynamischen Institut an der Technischen Hochschule Aachen,* No. 7, Springer Verlag, Berlin, pp. 11–33.

Seki, S., Kasahara, K., Kuriyama, T., and Kawasumi, M., Relation Between Compressive Strength of Concrete and the Effective Cement–Water Ratio Calculated from the Hydration Rate of Cement, *ACI Journal,* Proc. Vol. 66, No. 3, Mar. 1969, pp. 198–202.

Sell, R., Einfluss der Zwischenlage auf Streuung und Grosse der Spaltzugfestigkeit von Beton (Effect of Bearing Strips on the Standard Deviation and Magnitude of Splitting Strength of Concrete), *Deutscher Ausschuss fur Stahlbeton,* Heft 155, Wilhelm Ernst & Sohn, Berlin, 1963, pp. 36–47.

Sen, B. R., Needed Research on Method of Testing Mechanical Strength of Portland Cement, *Indian Concrete Journal,* Oct. 15, 1955, pp. 341–342.

Sen, B. R., and Bharara, A. L., A New Indirect Tensile Test for Concrete, *Indian Concrete Journal,* Vol. 35, No. 3, Mar. 1961, pp. 85–89.

Sen, B. R., and Desayi, P., Determination of the Tensile Strength of Concrete by Splitting a Cube Along Its Diagonal Plane, *Indian Concrete Journal,* Vol. 36, No. 7, July 1962, pp. 249–252.

Sereda, P. J., Feldman, R. F., and Swenson, E. G., Effect of Sorbed Water on Some Mechanical Properties of Hydrated Portland Cement Pastes and Compacts, *Symposium on Structures of Portland Cement Paste and Concrete,* Special Report 90, Highway Research Board, Washington, DC, 1966, pp. 58–73.

Sereda, P. J., Feldman, R. F., and Ramachandran, V. S., Structure Formation and Development in Hardened Cement Pastes, *7th International Congress on the Chemistry of Cement,* Vol. I, Editions Septima, Paris, 1980, pp. VI. 1/3–VI.1/44.

Shacklock, B. W., and Keene, P. W., Comparison of the Compressive and Flexural Strengths of Concrete with and Without Entrained Air, *Civil Engineering and Public Works Review,* Vol. 54, No. 631, Jan. 1959, pp. 77–80.

Shah, S. P. (Ed.), Application of Fracture Mechanics to Cementitious Composites, Preprints of the Proceedings NATO Advanced Workshop, Northwestern University, Evanston, Sept. 1984.

Shah, S. P., and Chandra, S., Critical Stress, Volume Change, and Microcracking of Concrete, *ACI Journal,* Proc. Vol. 65, No. 9, Sept. 1968, pp. 770–781.

Shah, S. P., and Chandra, S., Fracture of Concrete Subjected to Cyclic and Sustained Loading, *ACI Journal,* Proc. Vol. 67, No. 10, Oct. 1970, pp. 816–825.

Shah, S. P., and McGarry, F. J., Griffith Fracture Criterion and Concrete, *Journal of the Engineering Mechanics Division, ASCE,* Vol. 97, No. M6, Proc. Paper 8597, Dec. 1971, pp. 1163–1676.

Shah, S. P., and Quyang, C., Fracture Mechanics, *Testing the Materials Science, Engineering, and Field Aspects of Concrete,* Evanston, NSF-ACBM Center, 1993, pp. 93–130.

Shah, S. P., and Swartz, S. E. (Ed.), *Fracture of Concrete and Rocks,* Springer-Verlag, New York, 1989, p. 449.

Shah, S. P., and Winter, G., Inelastic Behavior and Fracture of Concrete, *ACI Journal,* Proc. Vol. 63, Sept. 1966, pp. 925–930.

Shah, S. P., Swartz, and Ouyang, C., *Fracture Mechanics of Concrete: Applications of Fracture Mechanics to Concrete, Rock, and Other Quasi-brittle Materials,* John Wiley, New York, 1995.

Shanley, F. R., and Knapp, W. J., Ceramics as Structural Materials, *Journal of the Structural Division, ASCE,* Vol. 91, No. ST4, Aug. 1965, pp. 47–56.

Sharma, M. R., and Gupta, B. L., Sonic Modulus as Related to Strength and Static Modulus of High Strength Concrete, *Indian Concrete Journal,* Vol. 34, No. 4, Apr. 1960, pp. 139–141.

Shideler, J. J., Lightweight Aggregate Concrete for Structural Use, *ACI Journal,* Proc. Vol. 54, Oct. 1957, pp. 299–328.

Short, A., and Kinniburgh, W., *Lightweight Concrete,* C.R. Books, London/John Wiley, New York, 1963.

Simmons, J. C., Poisson's Ratio of Concrete: A Comparison of Dynamic and Static Measurements, *Magazine of Concrete Research,* Vol. 7, No. 20, July 1955, pp. 61–68.

Sinha, B. P., Gerstle, K. H., and Tulin, L. G., Stress–Strain Relations for Concrete Under Cyclic Loading, *ACI Journal,* Proc. Vol. 61, No. 2, Feb. 1964a, pp. 195–211.

Sinha, B. P., Gerstle, K. H., and Tulin, L. G., Response of Singly Reinforced Beams to Cyclic Loading, *ACI Journal,* Proc. Vol. 61, No. 8, Aug. 1964b.

Singh, B. G., Specific Surface of Aggregates Related to Compressive and Flexural Strength of Concrete, *ACI Journal,* Proc. Vol. 54, Apr. 1958, pp. 897–907.

Siviero, E., Evaluation of Early Concrete Strength, *Materials and Structures—Research and Testing,* Vol. 27, No. 169, June 1994, pp. 273–284.

Skalny, J., and Bajza, A., Properties of Cement Pastes Prepared by High Pressure Compaction, *ACI Journal,* Proc. Vol. 67, Mar. 1970, pp. 221–227.

Skalny, J., Jawed, I., and Taylor, H. F. W., Studies of Hydration of Cement—Recent Developments, *World Cement Technology,* Vol. 9, No. 6, Sept. 1978, pp. 183–195.

Skramtayev, B. G., Determining Concrete Strength for Control of Concrete in Structures, *ACI Journal,* Proc. Vol. 34, Jan.–Feb. 1938, pp. 285–303.

Skramtayev, B. G., and Leschtschinski, M. J., Complex Methods of Non-destructive Tests of Concrete in Constructions and Structural Works, *RILEM Bulletin,* No. 30, Mar. 1966a, pp. 99–106.

Skramtayev, B. G., and Leschtschinski, M. J., Grossere Genauigkeit bei der Bestimmung der Betonfestigkeit in Konstruktionen und Bauwerken durch Anwendung komplexer Methoden and Gerate, *Wissenschaftliche Zeitschrift,* Vol. 12, No. 1–2, Hochschule fur Bauwesen, Leipzig, 1966b, pp. 81–84.

Slanicka, S., The Influence of Fly Ash Fineness of the Strength of Concrete, *Cement and Concrete Research,* Vol. 21, Nos. 2 and 3, Aug.–Sept. 1991a, pp. 285–296.

Slanicka, S., The Influence of Condensed Silica Fume on the Concrete Strength, *Cement and Concrete Research,* Vol. 21, No. 4, Aug.–Sept. 1991b, pp. 462–470.

Slate, F. O., and Meyers, B. L., Deformations of Plain Concrete, *Proceedings of the Fifth International Symposium on the Chemistry of Cement,* Part III, Properties of Cement Paste and Concrete, Tokyo, Dec. 1969, pp. 142–151.

Slate, F. O., and Olsefski, S., X-Ray for Study of Internal Structures and Microcracking of Concrete, *ACI Journal,* Proc. Vol. 60, May 1963, pp. 575–588.

Slate, F. O., Nilson, A. H., and Martinez, S., Mechanical Properties of High-Strength Lightweight Concrete, *ACI Journal,* Proc. Vol. 83, No. 4, July–Aug. 1986, pp. 606–613.

Slater, W. A., and Lyse, I., Compressive Strength of Concrete in Flexure from Tests of Reinforced Beams, *Proceedings, ACI,* Vol. 26, 1930, pp. 831–874.

Smith, G. M., Physical Incompatibility of Matrix and Aggregate in Concrete, *ACI Journal,* Proc. Vol. 52, Mar. 1956.

Smith, G. M. and Young, L. E., Ultimate Flexural Analysis Based on Stress–Strain Curves of Cylinders, *ACI Journal,* Proc. Vol. 53, Dec. 1956, pp. 597–609.

Smith, J. R., Estimating Later Age Strengths of Concrete, *ACI Journal,* Vol. 81, No. 6, Nov.–Dec. 1984, pp. 609–612.

Smith, M. A., Review of Standard Specifications for Fly Ash for Use in Concrete, *Concrete,* BRE Building Research Series, Vol. 1, Construction Press, Lancaster, 1978, pp. 119–134.

Smith, P., Opening the Oyster of Concrete Quality, Stanton Walker Lecture Series on the Materials Sciences, No. 9, University of Maryland, Nov. 1971.

Smith, P., and Chojnacki, B., Accelerated Strength Testing of Concrete Cylinders, *Proceedings, ASTM,* Vol. 63, Philadelphia, 1963, pp. 1079–1104.

Smith, P., and Tiede, J., Earlier Determination of Concrete Strength Potential, *A Symposium on Concrete Strength,* Highway Research Record No. 210, Highway Research Board, Washington, DC, 1967, pp. 29–66.

Sneck, T., and Oinonen, H., *Measurement of Pore Size Distribution of Porous Materials,* Julkaisu 155, State Institute for Technical Research, Finland, Helsinki, 1970.

Snell, L. M., and Rutledge, R. B., Statistical Behavior of Testing Plans for Concrete, *Concrete International: Design and Construction,* Vol. 4, No. 2, Feb. 1982, pp. 36–40.

Sokolov, I. B., and Logunova, V. A., Strength of Brittle Materials in Non-uniform Triaxial Compression, *Advances in Research on the Strength and Fracture of Materials,* Vol. 3B, D. M. R. Taplin (Ed.), Pergamon Press, New York, 1978, pp. 1193–1195.

Soroka, I., An Application of Statistical Procedure to Quality Control of Concrete, *Materials and Structures—Research and Testing,* RILEM, Vol. 1, No. 5, Sept.–Oct. 1968, pp. 437–441.

Soroka, I., and Sereda, P. J., The Interrelation of Hardness, Modulus of Elasticity and Porosity in Various Gypsum Systems, *Journal of the American Ceramic Society,* Vol. 51, 1968, pp. 337–340.

Soroka, I., On Compressive Strength Variation in Concrete, *Materials and Structures—Research and Testing,* RILEM, Vol. 4, No. 21, May–June 1971, pp. 155–161.

Soroka, I., *Portland Cement Paste and Concrete,* Macmillan Press, London, 1979.

Soroka, I., and Sereda, P. J., The Structure of Cement—Stone and the Use of Compacts as Structural Models, *Proceedings of the Fifth International Symposium on The Chemistry of Cement,* Part III, Properties of Cement Paste and Concrete, Tokyo, Dec. 1969, pp. 67–73.

Soroushian, P., Choi, Ki-Bong, and Alhamad, A., "Dynamic Constitituve Behavior of Concrete, *ACI Journal,* Proceedings, Vol. 83, No. 2, Mar.–Apr. 1986, pp. 251–259.

Soshiroda, T., Anisotropy of Concrete, *Proceedings of the International Conference on Mechanical Behavior of Materials,* Vol. IV, Mechanical Behavior of Materials, Society of Materials Science, Japan, 1972, pp. 300–307.

Spektor, B. V., Infra-Red Heating in the Production of Concrete (in Russian), *Beton i Zhelozobeton,* No. 7, 1956, pp. 264–266 (C & CA Library Translation, No. 79).

Spellman, D. L., and Stratfull, R. F., An Electrical Method for Evaluating Bridge Deck Coatings, *Quality Assurance, Concrete, Construction, Bridge Deck Coating, and Joint Seals,* Highway Research Record No. 357, Highway Research Board, Washington, DC, 1971, pp. 64–71.

Spetla, Z., and Kadlecek, V., Effect of the Slenderness on the Direct Tensile Strength of Concrete Cylinders and Prisms, *RILEM Bulletin,* No. 33, Dec. 1966, pp. 403–412.

Spinnger, S., and Tefft, W. E., A Method for Determining Mechanical Resonance Frequencies and for Calculating Elastic Moduli from These Frequencies, *Proc. ASTM,* Vol. 61, 1961, pp. 1221–1238.

Spinner, S., and Valore, R. C., Jr., Comparison of Theoretical and Empirical Relations Between the Shear Modulus and Torsional Resonance Frequencies of Bars of Rectangular Cross Section, *Journal of Research,* National Bureau of Standards, Vol. 60, No. 5, May 1958, pp. 459–464.

Spinner, S., Knudsen, F. P., and Stone, L., Elastic Constant—Porosity Relations for Polycrystalline Thoria, *Journal of Research of the National Bureau of Standards,* C. Engineering and Instrumentation, Vol. 67C, No. 1, Washington, DC, Jan.–Mar. 1963, pp. 39–46.

Spooner, D. C., *Measurement of the Tensile Strength of Concrete by an Indirect Method—The Cylinder Splitting Test,* Technical Report TRA 419, Cement and Concrete Association, London, May 1969.

Spriggs, R. M., Effect of Open and Closed Pores on Elastic Moduli of Polycrystalline Alumina, *Journal of the American Ceramic Society,* Vol. 45, No. 9, Sept. 1962, p. 454.

Spriggs, R. M., and Vasilos, T., Effect of Grain Size on Transverse Bend Strength of Alumina and Magnesia, *Journal of the American Ceramic Society,* Vol. 45, No. 5, May 1963, p. 225.

Stankowski, T., and Gerstle, K. H., Simple Formulation of Concrete Behavior Under Multiaxial Load Histories, *ACI Journal,* Proc. Vol. 82, No. 2, Mar.–Apr. 1985, pp. 213–221.

Stanton, T. E., Tests Comparing the Modulus of Elasticity of Portland Cement Concrete as Determined by the Dynamic (Sonic) and Compression (Secant at 1000 psi) Methods, *ASTM Bulletin,* No. 131. Dec. 1944, pp. 17–20.

Stassi D'Alia, F., Interdépendance des caractéristiques de résistance des materiaux (Interdependence of the Strength Characteristics of Materials), *RILEM Bulletin,* No. 21, Dec. 1963, pp. 47–57.

Steinbach, W., Uber die Einwirkung Von Mineralolen auf die Festigkeit von Zementmortel (The Effect of Mineral Oils on the Strength of Cement Mortar), *Betonstein Zeitung,* Vol. 33, No. 10, Oct. 1967, pp. 462–469.

Stock, A. F., Hannant, D. J., and Williams, R. I. T., The Effect of Aggregate Concentration upon the Strength and Modulus of Elasticity of Concrete, *Magazine of Concrete Research,* Vol. 31, No. 109, London, Dec. 1979, pp. 225–234.

Stone, D. E. W., and Clarke, B., Ultrasonic Attenuation as a Measure of Void Content in Carbon-Fibre Reinforced Plastics, *Nondestructive Testing,* Vol. 8, 1975, pp. 137–145.

Stone, W. C., and Carino, N. J., Deformation and Failure in Large-Scale Pullout Tests, *ACI Journal,* Proc. Vol. 80, No. 6, Nov.–Dec. 1983, pp. 501–513.

Stone, W. C., and Carino, N. J., Comparison of Analytical with Experimental Internal Strain Distribution for the Pullout Test, *ACI Journal,* Proc. Vol. 81, No. 1, Jan.–Feb. 1984, pp. 3–12.

Stone, W. C., and Giaz, B. J., The Effect of Geometry and Aggregate on the Reliability of the Pullout Test, *Concrete International: Design and Construction,* Vol. 7, No. 2, Feb. 1985, pp. 27–36.

Stork, J., Proportioning Concrete Mixes Using Digital Computers, *Computer Applications in Concrete Designing and Technology,* ACI Publication SP-16, Detroit, MI, 1967, pp. 41–76.

Storozhenko, V., and Meshkov, S., Infrared Thermography in Non-destructive Testing: Achievements and Development Perspectives, International Symposium, *Non-*

destructive Testing in Civil Engineering (NDT-CE), Vol. 1, DGZfP, Berlin, 1995, pp. 927–928.

Stowe, R. L., and Thornton, H. T., Jr., *Engineering Condition Survey of Concrete in Service,* Technical Report REMR-CS-1, USAE Waterways Experiment Station, Vicksburg, MS, Sept. 1984, p. 109.

Stroeven, P., Structural Investigations of Concrete by Means of Sterological Techniques, *RILEM Seminar on Fresh Concrete,* Paper 5.20, Leeds, Mar. 22–24, 1973.

Sturrup, V. R., Evaluation of Pulse Velocity Tests Made by Ontario Hydro, *Effects of Concrete Characteristics on the Pulse Velocity—A Symposium,* Bulletin No. 206, Highway Research Board, Washington, DC, 1959, pp. 1–13.

Sturrup, V. R., Vecchio, F. J., and Caratin, H., Pulse Velocity as a Measure of Concrete Compressive Strength, *In Situ/Nondestructive Testing of Concrete,* V. M. Malhotra (Ed.), ACI Publication SP-82, American Concrete Institute, Detroit, MI, 1984, pp. 201–228.

Sturman, G. M., Shah, S. P., and Winter, G., Effect of Flexural Strain Gradients on Microcracking and Stress–Strain Behavior of Concrete, *ACI Journal,* Proc. Vol. 62, No. 7, July 1965, pp. 805–822.

Suaris, W., and Fernando, V., Detection of Crack Growth in Concrete from Ultrasonic Intensity Methods, *Materials and Structures—Research and Testing,* RILEM, Vol. 20, No. 117, May 1987a, pp. 214–270.

Suaris, W., and Fernando, V., Ultrasonic Pulse Attenuation as a Measure of Damage Growth During Cyclic Loading of Concrete, *ACI Materials Journal,* Vol. 84, No. 3, May–June 1987b, pp. 185–193.

Swamy, R. N., *The Inelastic Deformation of Concrete,* Highway Research Record No. 324, Highway Research Board, Washington, DC, 1970, pp. 89–99.

Swamy, R. N., Aggregate–Matrix Interaction in Concrete Systems, *Structure, Solid Mechanics and Engineering Design,* Proceedings of the Southampton 1969 Civil Engineering Materials Conference, Part 1, Wiley-Interscience, London, 1971, pp. 301–316.

Swamy, R. N., Fracture Phenomena of Hardened Paste, Mortar and Concrete, Mechanical Behavior of Materials, *Proceedings of the International Conference on Mechanical Behavior of Materials,* Vol. IV, Society of Materials Science, Japan, 1972, pp. 132–142.

Swamy, R. N., Fracture Mechanics Applied to Concrete, *Developments in Concrete Technology*—1, F. D. Lydon (Ed.), Applied Science Publishers, London, 1979, pp. 221–228.

Swamy, R. N., The Nature of the Strength of Concrete, *Progress in Concrete Technology,* V. M. Malhotra (Ed.), Department of Energy Mines and Resources, Ottawa, Canada, MRP/MSL 80-89 (TR), 1980, pp. 189–222.

Swamy, R. N., and Al-Hamed, A. H., Evaluation of Small Diameter Core Tests to Determine In Situ Strength of Concrete, *In Situ/Nondestructive Testing of Concrete,* V. M. Malhotra (Ed.), ACI Publication SP-82, American Concrete Institute, Detroit, MI, 1984a, pp. 411–440.

Swamy, R. N., and Al-Hamed, A. H., The Use of Pulse Velocity Measurements to Estimate Strength of Air-Dried Cubes and Hence In Situ Strength of Concrete, *In*

Situ/Nondestructive Testing of Concrete, V. M. Malhotra (Ed.), ACI Publication SP-82, American Concrete Institute, Detroit, MI, 1984b, pp. 247–276.

Swamy, R. N., and Ali, A. M. A. H., Assessment of In Situ Concrete Strength by Various Non-destructive Tests, *NDT International,* Vol. 17, No. 3, June 1984, pp. 139–146.

Swartz, S. E., Lu, L. W., and Tang, L. D., Mixed-Mode Fracture Toughness Testing of Concrete Beams in Three-Point Bending, *Materials and Structures—Research and Testing,* RILEM, Vol. 21, No. 121, Jan. 1988, pp. 33–40.

Szabo, G., Die Grundlagen einer neuen Festigkeitstheorie (Foundation of a New Theory of Strength), Bauverlag, Wiesbaden, 1966.

Szabo, G., Uber die Gestaltabhangigkeit der Betondruckfestigkeit (How the Compressive Strength of Concrete Depends on the Shape of the Specimen), *Betonstein-Zeitung,* Vol. 33, No. 4, Apr. 1967, pp. 164–166.

Szoke, D., Report on the International Symposium on Testing In Situ of Concrete Structures, *Materials and Structures—Research and Testing,* RILEM, Vol. 9, No. 53, Sept.–Oct. 1976, pp. 361–368.

Szoke, D., RILEM Tentative Recommendation, General Recommendation for Loading Test of Load-Bearing Structures In Situ, *Materials and Structures—Research and Testing,* RILEM, Vol. 12, No. 70, July–Aug. 1979, pp. 307–319.

Taerwe, L. R., Compliance Criteria for Concrete Strength, *Reliability Theory and Its Application in Structural and Soil Mechanics,* P. Thagt-Christensen (Ed.), NATO Advanced Study Institute, Ser. E, No. 70, 1983, pp. 365–375.

Tait, H. B., Concrete Quality Assurance Based on Strength Tests, *Concrete International: Design and Construction,* Vol. 2, No. 9, Sept. 1981, pp. 79–87.

Takabayashi, T., Comparison of Dynamic Young's Modulus and Static Young's Modulus for Concrete, *RILEM International Symposium on Nondestructive Testing of Materials and Structures,* Vol. I, Paris, 1954, pp. 34–44.

Takabayashi, T., and Ishida, H., Measurement of Thickness of Concrete Pavement and Structural Members, *In Situ/Nondestructive Testing of Concrete,* V. M. Malhotra (Ed.), ACI Publication SP-82, American Concrete Institute, Detroit, MI, 1984, pp. 659–672.

Takeda, J., and Tachikawa, H., Deformation and Fracture of Concrete Subjected to Dynamic Load, *Mechanical Behavior of Materials,* Proceedings of the International Conference on Mechanical Behavior of Materials, Vol. IV, The Society of Materials Science, Japan, pp. 267–277.

Talaber, J., Borjan, J., and Jozsa, Z., Betontechnologiai parameterek hatasa a roncsolasmentes szilardsagbecslo osszefuggesekre (Effects of Parameters of Concrete Technology on the Strength Estimation from Nondestructive Testing), *Tudomanyos Kozlemeny,* 29, Budapest Muszaki Egyetem Epitomernoki Kar, Budapest, 1979.

Talbot, A. N., *Tests of Reinforced Concrete Beams,* Bulletin, University of Illinois Engineering Experiment Station, Sept. 1904, pp. 1–20.

Talbot, A. N., and Richart, F. E., *The Strength of Concrete—Its Relation to the Cement, Aggregate and Water,* Bulletin No. 137, Engineering Experiment Station, University of Illinois, Urbana, IL, Oct. 1923.

Tamas, F. D., Acceleration and Retardation of Portland Cement Hydration by Additives, *Symposium on Structure of Portland Cement Paste and Concrete,* Highway Research Board, Special Report 90, Washington, DC, 1966, pp. 392–397.

Tamas, F. D., and Fabry, M., The Change in Reactivity of Silicate Anions During the Hydration of Calcium Silicates and Cement, *Cement and Concrete Research,* Vol. 3, No. 6, Nov. 1973, pp. 767–776.

Tamas, F. D., and Varady, T., Role of Poly-Reactions in the Hydration of Cement, *Hungarian Journal of Industrial Chemistry,* Vol. 3, Veszprem, 1975, pp. 347–354.

Tamas, F. D., Sarkar, A. K., and Roy, D. M., Effect of Variables upon the Silylation Products of Hydrated Cements, *Hydraulic Cement Pastes: Their Structure and Properties,* Proceedings of a Conference held at University of Sheffield, Apr. 8–9, 1976, Cement and Concrete Association, Wexham Springs, Slough, England, 1976, pp. 55–72.

Tamura, H., and Yoshida, M., Nondestructive Method of Detecting Steel Corrosion in Concrete, *In Situ/Nondestructive Testing of Concrete,* V. M. Malhotra (Ed.), ACI Publication SP-82, American Concrete Institute, Detroit, MI, 1984, pp. 689–702.

Tanigawa, Y., Yamada, K., Kumagai, S., and Kosaka, Y., Effect of Age on Strength of Concrete Estimated by Combined Non-destructive Testing Method, *Transactions of the Japan Concrete Institute,* Vol. 2, 1980, pp. 163–170.

Tanigawa, Y., Baba, K., and Mori, H., Estimation of Concrete Strength by Combined Nondestructive Testing Method, *In situ/Nondestructive Testing of Concrete,* V. M. Malhotra (Ed.), ACI Publication SP-82, American Concrete Institute, Detroit, MI, 1984, pp. 57–76.

Taplin, J. H., A Method for Following the Hydration Reaction in Portland Cement Paste, *Australian Journal of Applied Science,* Vol. 10, No. 3, Sept. 1959, pp. 329–345.

Tassios, T. H. P., and Demiris, C. A., *Standard Nails Extraction—A New Nondestructive Method for Concrete Strength Determination* (Hellenic Method), Publication 21, National Technical University, Athens, 1968.

Tassios, T. P., and Koroneous, E. G., Local Bond-Slip Relationships by Means of the Moire Method, *ACI Journal,* Proc. Vol. 81, No. 1, Jan.–Feb. 1984, pp. 27–34.

Tasuji, M. E., Slate, F. O., and Nilson, A. H., Stress-Strain Response and Fracture of Concrete in Biaxial Loading, *ACI Journal,* Proc. Vol. 75, No. 7, July 1978, pp. 306–312.

Taylor, H. F. W., Nanostructure of C–S–H: Current Status, *Advanced Cement Based Materials,* No. 1, 1993, pp. 38–46.

Taylor, M. A., General Behavior Theory for Cement Pastes, Mortars, and Concretes, *ACI Journal,* Proc. Vol. 86, No. 10, Oct. 1971, pp. 756–762.

Taylor, M. A., and Broms, B. B., Shear Bond Strength Between Coarse Aggregate and Cement Paste or Mortar, *ACI Journal,* Proc. Vol. 61, No. 8, Aug. 1964, pp. 939–958.

Taylor, M. A., Jain, A. K., and Ramey, M. R., Path Dependent Biaxial Compressive Testing of an All-Lightweight Aggregate Concrete, *ACI Journal,* Proc. Vol. 69, No. 12, Dec. 1972, pp. 758–764.

Tegart, W. J., *Elements of Mechanical Metallurgy,* Macmillan Company, New York, 1967.

Tennessee Valley Authority, *Concrete Production and Control,* Technical Report No. 21, Washington, DC, 1947.

Teodoru, G. V., Mechanical Strength Property of Concrete at Early Ages as Reflected by Schmidt Rebound Number, Ultrasonic Pulse Velocity, and Ultrasonic Attenuation, *Properties of Concrete at Early Ages,* J. F. Young (Ed.), ACI SP-95, American Concrete Institute, Detroit, MI, 1986, pp. 139–153.

Teodoru, G. V., The Use of Simultaneous Nondestructive Tests to Predict the Compressive Strength of Concrete, *Nondestructive Testing,* H. S. Lew (Ed.), ACI Publication SP-112, American Concrete Institute, Detroit, MI, 1988, pp. 137–152.

Teodoru, G. V., *Zerstorungsfreie Betonprufungen* (Nondestructive Tests for Concrete), Beton-Verlag, Dusseldorf, 1989, p. 158.

Terrier, P., and Hornain, H., Sur l'application des méthodes mineralogiques a l'industrie des liants hydrauliques (Applications of Mineralogical Methods in the Cement Industry), *Revue des Materiaux de Construction,* Ciments et Betons, Nos. 618–621, Mar.–June 1967.

Testa, R. B., and Stubbs, N., Bond Failure and Inelastic Response of Concrete, *Journal of the Engineering Mechanics Division, ASCE,* Vol. 103, No. EM2, Paper 12866, April 1977, pp. 295–310.

Teychenne, D. C., Recommendations for the Treatment of the Variations of Concrete Strength in Codes of Practice, *Concrete,* BRE Building Research Series, Vol. 1, Construction Press, Lancaster, 1973, pp. 137–145.

Teychenne, D. C., Franklin, R. E., and Erntroy, H. C., *Design of Normal Concrete Mixes,* Department of the Environment, Her Majesty's Stationary Office, London, 1975.

Tharmabala, T., Reel, R., Chung, T., and Carter, C. R., Bridge Deck Condition Survey by Radar: Ontario Experience, *Structural Materials Technology—An NDT Conference,* R. J. Scancella and M. E. Callahan (Eds.), Technomic, Atlantic City, NJ, 1994, pp. 141–145.

Thaulow, N., Estimation of the Compressive Strength of Concrete Samples by Means of Fluorescence Microscopy, *Nordisk Betong,* Vol. 26, No. 2–4, Stockholm, 1982, pp. 51–52.

Thaulow, S., *Field Testing of Concrete,* Norsk Cementforening, Oslo, 1952.

Thaulow, S., Tensile Splitting Test and High Strength Concrete Test Cylinders, *ACI Journal,* Proc. Vol. 53, No. 7, Jan. 1957, pp. 699–706.

Theocaris, P. S., and Prassianakis, J. N., The Mohr Envelope of Failure for Concrete: A Study of Its Tension–Compression Part, *Magazine of Concrete Research,* Vol. 26, No. 87, London, June 1974, pp. 73–82.

Timoshenko, S., *Strength of Materials,* Vol. 1, Van Nostrand Co., Princeton, NJ, 1957.

Timoshenko, S., and Goodier, J. N., *Theory of Elasticity,* 2nd ed., McGraw-Hill, New York, 1951.

Todd, J. D., The Determination of Tensile Stress–Strain Curves for Concrete, *Proceedings, Institution of Civil Engineers,* Vol. 4, No. 2, Part 1, London, Mar. 1955, pp. 210–211.

Tom, J. G., and Magoun, A. D., *Evaluation of Procedures Used to Measure Cement and Water Content in Fresh Concrete,* National Cooperative Highway Research Program Report 284, Transportation Research Board, Washington, DC, June 1986.

Tomsett, H. N., The Practical Use of Ultrasonic Pulse Velocity Measurements in the Assessment of Concrete Quality, *Magazine of Concrete Research,* Vol. 32, No. 110, Mar. 1980, pp. 7–16.

Tomsett, H. N., Non-destructive Testing of In Situ Concrete Structures, *NDT International,* Vol. 14, No. 6, Dec. 1981, pp. 315–320.

Torrens, R. J., The Log-Normal Distribution: A Better Fitness for the Results of Mechanical Testing of Materials, *Materials and Structures—Research and Testing,* RILEM, Vol. 11, No. 64, July–Aug. 1978, pp. 235–245.

Towne, D. H., *Wave Phenomena,* Addison-Wesley, Reading, MA, 1967.

Transportation Research Board, *Recent Developments in Accelerated Testing and Maturity of Concrete,* Transportation Research Record No. 558, Transportation Research Board, Washington, DC, 1975.

Transportation Research Board, *Durability of Concrete Bridge Decks,* National Cooperative Highway Research Program Synthesis of Highway Practice No. 57, Transportation Research Board, Washington, DC, 1979.

Troxell, G. E., Davis, H. E., and Kelly, J. W., *Composition and Properties of Concrete,* 2nd ed., McGraw-Hill, New York, 1968.

Tucker, J., Jr., The Maximum Stresses Present at Failure of Brittle Materials, *Proceedings, ASTM,* Vol. 45, Philadelphia, 1945, pp. 961–973.

Turrizziani, R., and Rio, A., High Chemical Resistance Pozzolanic Cements, *Chemistry of Cement,* Proceedings of the Fourth International Symposium, NBS Monograph 43, Vol. II, Washington, DC, pp. 1067–1073, 1960.

Tynes, W. O., and Mather, B., *Air Entrainment in Mass Concrete,* Technical Report C-69-1, USAE Waterways Experiment Station, Vicksburg, MS, Feb. 1969.

Uchikawa, H., Uchida, S., and Hanehara, S., Flocculation Structure of Fresh Cement Paste Determined by Sample Freezing—Back-Scattered Electron Image Method, *Il Cemento,* New Series, Vol. 84, Jan.–Mar. 1987, pp. 3–21.

Ujhelyi, J., A beton levegotartalmanak hatasa (The Effect of Air Content of Concrete), *Magyar Epitoipar,* No. 8, 1980, Budapest, pp. 469–481.

Ujhelyi, J., A beton strukturajanak es nyomoszilardsaganak a tervezese (Design of the Structure and Compressive Strength of Concrete), Dissertation for the degree of Doctor of Technical Sciences, Budapest, Aug. 1989.

Umoto, T., and Kobayashi, K., In Situ Test to Determine Fiber Content of Steel Fiber Reinforced Concrete by an Electro-Magnetic Method, *In Situ/Nondestructive Testing of Concrete,* V. M. Malhotra (Ed.), ACI Publication SP-82, American Concrete Institute, Detroit, MI, 1984, pp. 673–688.

U.S. Bureau of Reclamation, *Concrete Manual,* 7th ed., U.S. Government Printing Office, Washington, DC, 1966.

U.S. Department of the Interior, Water and Power Resources Service, *Concrete Manual,* 8th ed., rev., John Wiley, New York, 1981.

Uzhpolevitchius, B. *Nondestructive Methods of Control and Estimation of Concrete Strength in Reinforced Concrete Construction* (in Russian), Vilnius "Mokslas," 1982.

Van Brakel, J., Modry, S., and Svata, M., Mercury Porosimetry: State of the Art, *Powder Technology,* 29, 1981, pp. 1–12.

Van Cauwelaert, F. V., and Eckmann, B., Indirect Tensile Test Applied to Anisotropic Materials, *Materials and Structures—Research and Testing,* Vol. 27, No. 165, Jan.–Feb. 1994, pp. 54–60.

Van de Geer, J. P., *Introduction to Multivariate Analysis for the Social Sciences,* W. H. Freeman, 1971.

Vassitch, P., Outline of the Various Possibilities of Nondestructive Mechanical Tests on Concretes. New Proposals of the Author, *RILEM International Symposium on Nondestructive Testing of Materials and Structures,* Vol. II, Paris, 1954, pp. 301–306.

Venuat, M., Effect of Elevated Temperatures and Pressures on the Hydration and Hardening of Cement, *VIth International Congress on the Chemistry of Cement,* Moscow, Sept. 1974.

Verbeck, G. J., and Foster, W., Long-Time Study of Cement Performance in Concrete With Special Reference to Heats of Hydration, *Proceedings ASTM,* Vol. 50, 1950, pp. 1235–1257.

Verbeck, G., Cement Hydration Reactions at Early Ages, *Journal of the PCA Research and Development Laboratories,* Vol. 7, No. 3, Sept. 1965, pp. 57–63.

Verbeck, G. J., and Helmuth, R. H., Structures and Physical Properties of Cement Pastes, *Proceedings of the Fifth International Symposium on the Chemistry of Cement,* Part III, Properties of Cement Paste and Concrete, Tokyo, Dec. 1969, pp. 1–32.

Viest, I. M., Discussion of "Shear Strength of Lightweight Reinforced Concrete Beams" by J. A. Hanson, *ACI Journal,* Proc. Vol. 55, No. 9, Mar. 1959, pp. 1062–1065.

Vile, G. W. D., The Strength of Concrete Under Short-Term Static Biaxial Stress, *The Structure of Concrete,* Proceedings, International Conference, London, Sept. 1965, Cement and Concrete Association, London, 1968, pp. 275–288.

Villareal Rivera, R., Accelerated Splitting-Tension Test for Determining Potential 28-Day Splitting-Tensile Strength and Modulus of Rupture of Concrete, *Recent Developments in Accelerated Testing and Maturity,* Transportation Research Record No. 558, Transportation Research Board, Washington, DC, 1975, pp. 114–120.

Voellmy, A., Examination of Concrete by Measurements of Superficial Hardness, *RILEM International Symposium on Nondestructive Testing of Materials and Structures,* Vol. II, Paris, 1954, pp. 323–336.

Voellmy, A., High Concrete Quality in Cold Weather (General Report), RILEM Symposium, *Winter Concreting Theory and Practice,* Session D, The Danish National Institute of Building Research, Special Report, Copenhagen, 1956.

Voellmy, A., Festigkeitskontrolle von Betonbelagen (Strength Control of Concrete Pavements), *Betonstrasse-Jahrbuch,* Bd. 3, Fachverband Zement, Koln S., 1957–1958, pp. 179–204.

Von Euw, M., and Gourdin, P., Le calcul prévisionnel des résistances des ciments portland (Estimation of the Strength of Portland Cements), *Materials and Structures—Research and Testing,* RILEM, Vol. 3, No. 17, Paris, Sept.–Oct. 1970, pp. 299–312.

Vuorinen, J., Some Notes on the Use of Accelerated Curing of Test Specimens for Concrete Quality Control, *RILEM Bulletin,* No. 31, June 1966, Paris, pp. 205–208.

Waddell, J. J., *Practical Quality Control for Concrete,* McGraw-Hill, New York, 1962.

Wagner, W. K., Effect of Sampling and Job Curing Procedures on Compressive Strength of Concrete, *Materials Research and Standards,* ASTM, Vol. 3, No. 8, Aug. 1963, pp. 629–634.

Walker, H. N., *Void Parameters of Concrete Cores from a Section of I-64,* Transportation Research Record No. 504, Transportation Research Board, Washington, DC, 1974, pp. 27–36.

Walker, H. N., *Evaluation and Adaption of the Dobrolubov and Romer Method of Microscopic Examination of Hardened Concrete,* Report No. VHTRC 79-R42, Virginia Highway and Transportation Research Council, Apr. 1979.

Walker, H. N., and Marshall, B. F., Methods and Equipment Used in Preparing and Examining Fluorescent Ultrathin Sections of Portland Cement Concrete, *Cement, Concrete, and Aggregates,* CCAGDP, Vol. 1, No. 1, 1979, pp. 3–9.

Walker, S., Modulus of Elasticity of Concrete, *Proceedings, ASTM,* Vol. 19, Part II, 1919, pp. 510–585.

Walker, S., *Application of Theory of Probability to Design of Concrete for Strength Specifications,* NRMCA Publication No. 57, Nov. 1955.

Walker, S., and Bloem, D. L., *Studies of Flexural Strength of Concrete,* Part 1: Effects of Different Gravels and Cements, Joint Research Laboratory Publication No. 3, National Sand and Gravel Association and National Ready Mixed Concrete Association, College Park, MD, July 1956.

Walker, S., and Bloem, D. L., Studies of Flexural Strength of Concrete—Part 2: Effects of Curing and Moisture Distribution on Measured Strength of Concrete, *Proceedings, HRB,* Vol. 36, Highway Research Board, Washington, DC, 1957a, pp. 334–346.

Walker, S., and Bloem, D. L., Studies of Flexural Strength of Concrete—Part 3: Effects of Variations in Testing Procedures, *Proceedings, ASTM,* Vol. 57, Philadelphia, 1957b, pp. 1122–1139.

Walker, S., and Bloem, D. L., Variations in Portland Cement, *Proceedings, ASTM,* Vol. 58, 1958, pp. 1009–1032.

Walker, S., Bloem, D. L., and Gaynor, R. D., Relationships of Concrete Strength for Maximum Size of Aggregate, *Proceedings HRB,* Vol. 38, Highway Research Board, Washington, DC, 1959, pp. 367–385.

Walker, S., and Bloem, D. L., Effects of Aggregate Size on Properties of Concrete, *ACI Journal,* Proc. Vol. 57, No. 9, Sept. 1960, pp. 283–298.

Walker, S., and Bloem, D. L., Author's closure to the discussions of the paper "Effects of Aggregate Size on Properties of Concrete," *ACI Journal,* Proc. Vol. 57, Mar. 1961, pp. 1248–1258.

Walz, K., Die Prufung von Kies und Splitt fur Strassenbeton (Test of Gravel and Crushed Stone for Paving Concrete), *Die Betonstrasse,* Vol. 14, Nos. 11 and 12, 1939, pp. 215–220, 229–235.

Walz, K., Beton- und Zementdruckfestigkeiten in den USA und ihre Umrechnung auf deutsche Prufwerte (Strengths of Cement and Concrete in the USA and Their Conversion to German Testing Values), *Betontechnische Berichte 1962,* Beton-Verlag, Dusseldorf, 1963, pp. 123–140.

Walz, K., and Wischers, G., Konstruktions-Leichtbeton hocher Festigkeit (Structural Lightweight Concrete of High Compressive Strength), *Betontechnische Berichte 1964,* Beton-Verlag, Dusseldorf, 1965, pp. 127–186.

Ward, M. A., The Testing of Concrete Materials by Precisely Controlled Uniaxial Tension, Ph.D. Thesis, University of London, 1964.

Ward, M. A., and Cook, D. J., The Development of a Uniaxial Tension Test for Concrete and Similar Brittle Materials, *Materials Research and Standards,* MTRSA, Vol. 9, No. 5, May 1969, pp. 16–20.

Ward, M. A., and Newman, K., The Behavior of Plain Concrete Subjected to Uniaxial Tension and Compression, *An Inter-American Approach for the Seventies,* Materials Technology—I, Second Inter-American Conference on Materials Technology, Mexico City, Mexico, Aug. 24–27 1970, pp. 9–23.

Washa, G. W., and Fluck, P. G., Effect of Sustained Loading on Compressive Strength and Modulus of Elasticity of Concrete, *ACI Journal,* Proc. Vol. 46, May 1950, pp. 693–700.

Watanabe, F., Complete Stress–Strain Curve for Concrete in Concentrical Compression, *Proceedings of the International Conference on Mechanical Behavior of Materials,* Vol. IV, Mechanical Behavior of Materials, Society of Materials Science, Japan, 1972, pp. 153–161.

Watstein, D., Effect of Straining Rate on the Compressive Strength and Elastic Properties of Concrete, *ACI Journal,* Proc. Vol. 49, No. 8, Apr. 1953, pp. 729–744.

Webster, F., Optimum Sample Sizes for Concrete Cylinder Tests Using Information Theory, *ACI Journal,* Proc. Vol. 68, No. 5, May 1971, pp. 373–379.

Weed, R. M., Optimum Performance Under a Statistical Specification, *Quality Assurance: Performance, Sealers, and Materials Variation,* Transportation Research Record No. 697, Transportation Research Board, Washington, DC, 1979, pp. 1–6.

Weibull, W., *Investigations into Strength Properties of Brittle Materials,* Handlingar, Nr. 149, Ingeniors Vetenkaps Akademien, Stockholm, 1938.

Weibull, W., *A Statistical Theory of the Strength of Materials,* Handlingar, Nr. 151, Ingeniors Vetenkaps Akademien, Stockholm, 1939.

Weil, G. J., Infrared Thermographic Techniques, *CRC Handbook on Nondestructive Testing of Concrete,* V. M. Malhotra and N. J. Carino (Eds.), CRC Press, Boca Raton, FL, 1991, pp. 305–316.

Weisz, G., Egy olajkorrozios eset tanulsagai, *Melyepitestudomanyi Szemle,* Vol. X, No. 12, Budapest, Dec. 1960, pp. 563–566.

Welch, G. B., Discussion of the paper "Water–Cement Ratio Versus Strength—Another Look" by H. J. Gilkey, *ACI Journal,* Proc. Vol. 58, Part II, Dec. 1961, pp. 1866–1868.

Welch, G. B., Tensile Strains in Unreinforced Concrete Beams, *Magazine of Concrete Research,* Vol. 18, No. 54, Mar. 1966, pp. 9–18.

Wells, P. N. T., *Biomedical Ultrasonics,* Academic Press, London, 1967.

Werner, D., Giertz-Hedstrom, S., Die Abhangigkeit der technischwichtigen Eigenschaften des Betons von den physikalisch-chemischen Eigenschaften des Zements I (Relationship Between the Technically Important Properties of Concrete and the Physico-chemical Properties of Cement I), *Zement,* Vol. 20, 1931, pp. 984–987, 1000–1005.

Whitehurst, E. A., Soniscope Tests Concrete Structures, *ACI Journal,* Proc. Vol. 47, Feb. 1951, pp. 433–444.

Whitehurst, E. A., *Evaluation of Concrete Properties from Sonic Tests,* ACI Monograph No. 2, American Concrete Institute, Detroit, MI, 1966.

Whitney, C. S., Plastic Theory of Reinforced Concrete Design, *Transactions, ASCE,* Vol. 107, 1942, pp. 251–326.

Williams, R. I. T., *The Effect of Cement Content on the Strength and Elastic Properties of Dry Lean Concrete,* Technical Report, TRA/323, Cement and Concrete Association, London, Nov. 1962.

Williamson, R. B., and Tewari, R. P., Effects of Microstructure on Deformation and Fracture of Portland Cement Paste, *Electron Microscopy and Structure of Materials,* Gareth Thomas (Ed.), University of California Press, Berkeley, CA, 1972, pp. 1223–1233.

Wills, M. J., Accelerated Strength Tests, *Significance of Tests and Properties of Concrete and Concrete Making Materials,* ASTM STP 169B, Chap. 13, Philadelphia, 1978, pp. 162–179.

Winslow, D. N., Diamond, S., A Mercury Porosimetry Study of the Evolution of Porosity in Portland Cement, *Journal of Materials,* JMLSA, Vol. 5, No. 3, Sept. 1970, pp. 564–585.

Winslow, D. N., and Diamond, S., The Specific Surface of Hydrated Portland Cement Paste as Measured by Low-Angle X-Ray Scattering, *Journal of Colloid and Interface Science,* Vol. 45, No. 2, Nov. 1973, pp. 425–426.

Winslow, D. N., and Diamond, S., Specific Surface of Hardened Portland Cement Paste as Determined By Small-Angle X-Ray Scattering, *Journal of the American Ceramic Society,* Vol. 57, No. 5, May 1974, pp. 193–197.

Wischers, G., Einfluss der Zusammensetzung des Betons auf seine Fruhfestigkeit (Influence of Concrete Composition on Early Concrete Strength), *Betontechnische Berichte 1963,* Beton-Verlag, Dusseldorf, 1964, pp. 137–151.

Wischers, G., Herstellung und Verwendung von Transportbeton (Production and Application of Ready-Mixed Concrete), *Zement Taschenbuch 1964/65,* Verein Deutscher Zementwerke, pp. 347–369.

Witte, L. P., and Price, W. H., "Discussion of Reference Stanton (1944)," ASTM Bulletin 131, Dec. 1944, pp. 20–22.

Wittmann, F. H., Surface Tension, Shrinkage and Strength of Hardened Cement Paste, *Materials and Structures—Research and Testing,* Vol. 1, No. 6, Nov.–Dec. 1968, pp. 547–552.

Wittmann, F. H. (Ed.), *Fracture Mechanics of Concrete,* Elsevier, Amsterdam, 1983.

Woods, H., Starke, H. R., and Steinour, H. H., Effect of Cement Composition on Mortar Strength, *Engineering News-Record,* Vol. 109, No. 15, 1932, pp. 435–437.

Woods, K. B., and McLaughlin, J. F., *Application of Pulse Velocity Tests to Several Laboratory Studies of Materials,* Bulletin 206, Highway Research Board, Washington, DC, 1959, pp. 14–27.

Wright, P. J. F., The Effect of the Method of Test on the Flexural Strength of Concrete, *Magazine of Concrete Research,* Vol. 4, No. 11, Oct. 1952, pp. 67–74.

Wright, P. J. F., Entrained Air in Concrete, *Proceedings, Institution of Civil Engineers,* Part I, Vol. 2, No. 3, London, May 1953, pp. 337–358.

Wright, P. J. F., Comments on an Indirect Tensile Test on Concrete Cylinders, *Magazine of Concrete Research,* Vol. 7, No. 20, July 1955, pp. 87–96.

Wright, P. J. F., Variations in the Strength of Portland Cement, *Magazine of Concrete Research,* Vol. 10, No. 30. Nov. 1958, pp. 123–132.

Wyllie, M. R. J., Gregory, A. R., and Gardner, L. W., Elastic Wave Velocities in Heterogeneous and Porous Media, *Geophysics,* Vol. 21, pp. 41–71.

Xi, J. P., and Jennings, H. M., Relationship Between Microstructure and Creep and Shrinkage of Portland Cement Paste, *Materials Science of Concrete,* J. Skalny (Ed.), Vol. 3, American Ceramic Society, Columbus, OH, 1993, pp. 37–69.

Yener, M., and Chen, W. F., Evaluation of In-Place Flexural Strength of Concrete, *ACI Journal,* Proc. Vol. 82, No. 6, Nov.–Dec. 1985, pp. 788–796.

Yokobory, T., The Strength, Fracture and Fatigue of Materials, *Brittle Fracture,* Chap. VI, P. Noordhof, Gronigen, 1965, pp. 117–150.

Yoshida, H., *Uber des elastische Verhalten von Beton* (Elastic Properties of Concrete), Verlag Julius Springer, Berlin, 1930.

Yoshimoto, A., Ogino, S., and Kawakami, M., Microcracking Effect on Flexural Strength of Concrete After Repeated Loading, *ACI Journal,* Vol. 69, No. 4, Apr. 1972, pp. 233–240.

Zaitsev, J. W., and Wittman, F. W., Crack Propagation in a Two-Phase Material Such as Concrete, *Advances in Research on the Strength and Fracture of Materials,* D. M. R. Taplin (Ed.), Vol. 3B, Pergamon Press, New York, 1978, pp. 1197–1203.

Zia, P., Torsional Strength of Prestressed Concrete Members, *ACI Journal,* Proc. Vol. 57, No. 10, Apr. 1961, pp. 1227–1359.

Zielinski, A. J., Reinhardt, H. W., and Kormeling, H. A., Experiments on Concrete Under Uniaxial Impact Tensile Loading, *Materials and Structures, Research and Testing,* Vol. 14, No. 80, Mar.–Apr. 1981a, pp. 103–112.

Zielinski, A. J., Reinhardt, H. W., and Kormeling, H. A., Experiments on Concrete Under Repeated Uniaxial Impact Tensile Loading, *Materials and Structures, Research and Testing,* Vol. 14, No. 81, May–June 1981b, pp. 163–169.

Zielinszki, S., The Development of the Setting of Roman and Portland Cements in Pastes, in Mortars, and in Concrete, *Proceedings International Association for Testing Materials,* Vol. I, Copenhagen, 1909, pp. 1–55.

Zielinszki, S., and Zhuk, J., Roman cementek osszehasonlito vizsgalata (Comparative Investigation of Roman Cements), *Third International Congress of the Association for Testing Materials,* Budapest, 1901.

Zivica, V., The Properties of Cement Paste with Admixtutre of Polyvinyl Acetate Emulsion, *RILEM Bulletin,* No. 28, Sept. 1965, Paris, pp. 121–128.

Zoldners, N. G., Calibration and Use of Impact Test Hammer, *ACI Journal,* Proc. Vol. 54, Aug. 1957, pp. 161–165.

Zoldners, N. G., and Soles, J. A., An Annotated Bibliography on Nondestructive Testing of Concrete, 1975–1983, *In Situ/Nondestructive Testing of Concrete,* V. M. Malhotra (Ed.), ACI Publication SP-82, American Concrete Institute, Detroit, MI, 1984, pp. 745–826.

ABOUT THE DISK

The enclosed disk contains the program *Proportioning for Concrete Strength in the 21st Century* (*Prop 21*). Detailed information about how to use the program can be found in User's Manual that is installed with the program.

MINIMUM SYSTEM REQUIREMENTS

- IBM PC or compatible computer with 386 or higher processor
- 8 MB RAM
- 9 MB hard disk space
- 3.5″ floppy disk drive
- Windows 3.1 or higher

INSTALLATION

The setup program will install the PROP21 program onto your hard drive in the default directory C:\PROP21. To run the setup program, do the following:

1. Insert the enclosed disk into the floppy disk drive of your computer.
2. Windows 3.1 and NT 3.51: From the Program Manager, choose File, Run.
   Windows 95 and NT 4.0: From the Start Menu, choose Run.
3. Type **A:\SETUP** (where A is the letter of your floppy disk drive). Press Enter.

4. The opening screen of the setup program will appear. Follow the instructions given on the screen.

USER SUPPORT

If you have comments or questions about the software, please contact the author at

popovics@duvm.ocs.drexel.edu.

If you need basic assistance or have a damaged disk, please contact Wiley Technical Support at

Phone: (212) 850-6753
Fax: (212) 850-6800 (Attention: Wiley Technical Support)
Email: techhelp@wiley.com

To place additional orders or to request information about Wiley products, please call (800) 225-5945.

INDEX

Visual inspection, 54, *see also*
    Nondestructive tests of strength
Voids, *see* Pore size distribution;
    Porosity
Voigt element, 436, 437

w/c, *see* Water-cement ratio
$w/(c + p)$, *see* Water-cement ratio
Warm water curing, 20–27
Water
    effect on strength, *see* Curing,
        wetness
    free, 175, 176, 217, 236, 253
    sensitivity, 327, 329, 335, 339,
        340, 344, 366, 368–372
Water-cement ratio, 88, 91, 109, 111,
    156, 176
    in absolute volume, 305, 306
    critical, 320, 324, 327, 328
    effect on strength, *see also*
        Strength vs. water-cement
        ratio
      compressive, 206

    flexural, 208
    other, 137, 138
    effective, 304
    at interface, 299
    test, 56
    vapor, 258, 259
(Water + air)-cement ratio, 374–376
Water-cementitious materials ratio,
    see Water-cement ratio
Water content, 316, 353, 356, 357,
    363–366, 383–384
Weakest link, 115, 136, 142, 156,
    176, 177, 249–251, 299
Wetting, effect on strength, 239–245
    compressive, 239, 240
    flexural, 241
    gradient, 241–244
Windsor probe, 52

X-ray, 55

Young's modulus, *see* Modulus of
    elasticity